Cities in the Developing World

Cities in the Developing World

Issues, Theory, and Policy

Edited by

Josef Gugler

OXFORD UNIVERSITY PRESS

1997

Oxford University Press, Great Clarendon Street, Oxford OX2 6DP

Oxford New York

Athens Auckland Bangkok Bogota Bombay
Buenos Aires Calcutta Cape Town Dar es Salaam
Delhi Florence Hong Kong Istanbul Karachi
Kuala Lumpur Madras Madrid Melbourne
Mexico City Nairobi Paris Singapore
Taipei Tokyo Toronto

and associated companies in
Berlin Ibadan

Oxford is a trademark of Oxford University Press

Published in the United States
by Oxford University Press Inc., New York

British Library Cataloguing in Publication Data
Data available

Library of Congress Cataloging in Publication Data
Data available
Cities in the developing world: issues, theory, and policy / edited
by Josef Gugler
1. Cities and towns—Developing countries. 2. Urbanization—
Developing countries. I. Gugler, Josef.
HT 169.5.C55 1997 307.76'09172'4—dc21 96–38159
ISBN 0–19–874215–0 (pbk.)
ISBN 0–19-874216–9

10 9 8 7 6 5 4 3 2 1

Typeset by Hope Services (Abingdon) Ltd.
Printed in Great Britain
on acid-free paper by
Bookcraft (Bath) Ltd.
Midsomer Norton
Somerset

To the citizens of the South
who have humoured our enquiries

Contents

The Contributors

Heribert Adam is Professor of Sociology at Simon Fraser University and past president of the International Sociological Association's Research Committee on Minority, Ethnic and Race Relations. He has held fellowships and visiting appointments at Cairo, Yale, Marburg, Siegen, and several South African universities. His interests focus on the history of apartheid and its transformation together with the theoretical analysis of nationalism, comparative ethnicity, and immigration policies.

Maria Helena Moreira Alves is Professor of Latin American Politics and Political Economy at the Universidade do Estado do Rio de Janeiro. She has devoted much of her time to popular education with labour unions and community-based organizations. At this time she co-ordinates a project, funded by the MacArthur Foundation, on the ecological and socio-economic impact of fruit monoculture in the Aconcagua Valley, Chile, in comparative perspective.

Charles M. Becker, President of the Economics Institute, Boulder, simultaneously holds positions at the University of Colorado as Professor Adjunct in the Department of Economics and as Research Associate at the Institute of Behavioral Science. He specializes in the urban and regional economics, economic development, and economic demography. Much of his research has been funded by the World Bank, the Carnegie Foundation, and the Investor Responsibility Research Center.

Hale Cihan Bolak directs the Graduate Psychology Program at New College of California, San Francisco, and continues to teach in Women's Studies at the University of California, Santa Cruz. In her teaching and research, she privileges a cross-cultural and interdisciplinary perspective in feminist studies. Her work focuses on feminist psychology and on gender in the Middle East. Her research has been funded by the Ford Foundation and the Social Science Research Council.

York W. Bradshaw is Director of African Studies at Indiana University, Bloomington. His research interests include urbanization and development, with a particular focus on education and health in developing countries. He is currently examining the importance of non-government organizations in African development.

Ray Bromley is Professor of Planning, Geography, and Latin American Studies at the State University of New York at Albany. He specializes in urban studies and planning, and has worked extensively on urban poverty, micro-enterprise, and regional planning issues in Latin America. He is currently engaged in a study of the history and dynamics of neighbourhood change in the South Bronx, New York.

Hemalata C. Dandekar is Professor of Urban Planning, College of Architecture and Urban Planning, and Director of the Center for South and Southeast Asian Studies at the University of Michigan. A licensed architect with a doctorate in urban and regional planning, she works on issues of housing, women, and urban–rural linkages, with particular reference to India. Her current research focuses on housing and infrastructure concerns in rapidly growing metropolitan areas.

Deborah Davis is Professor and Chair, Department of Sociology, Yale University. Since 1979 when she first did field-work in China, her primary interest has been the transformation of urban society after 1949. Currently she is working on emerging patterns of occupational mobility and on the impact of the consumer revolution on the scope and vitality of an urban public sphere.

Susan Eckstein is Professor of Sociology at Boston University. She has written extensively on urban poor in Mexico and on outcomes of revolutions in Latin America. Currently she works on protest and resistance movements in Latin America, and on the political economy of the Cuban revolution. She is President of the Latin American Studies Association.

Manuel Antonio Garretón Merino is Professor and Chair of the Department of Sociology at the University of Chile. He has held visiting appointments at a number of universities in Latin America, Europe, and the United States. He has written extensively on political processes, social actors, and culture in Chile and Latin America. Currently he works on the changing relations between state and society and the consequences for social action.

Christiaan Grootaert is Senior Economist in the Environment Department of the World Bank. He has worked extensively on poverty and human-resource issues in Africa, Asia, and Eastern Europe. Currently he is undertaking empirical research on the determinants of child labour in four countries.

Josef Gugler is Professor of Sociology at the University of Connecticut. Previously he served as Director of Sociological Research at the Makerere Institute of Social Research, Uganda. His research has taken him to Cuba, India, Kenya, Nigeria, Tanzania, and Zaïre. At this time he works on the role of gender in migration and on the relationship between literature and politics in Africa.

Jorge E. Hardoy, until his death in 1993, was President of the Instituto Internacional de Medio Ambiente y Desarrollo, an international non-profit NGO based in Buenos Aires that he had founded. He edited the journal *Medio Ambiente y Urbanizacion*. He was also President of the National Commission for Historic Monuments in Argentina, former President of the Inter-American Planning Society, twice a Guggenheim Fellow, and an adviser to the The Brundtland Commission for its report *Our Common Future*.

John Humphrey is a sociologist working at the Institute of Development Studies at the University of Sussex. He has studied work and employment issues in Brazil for many years, focusing in particular on industrial relations in the motor industry and unemployment in São Paulo. His most recent research was concerned with the use of Japanese management techniques in India and the role of clusters and networks in the promotion of small enterprise growth.

Lea Jellinek has spent the past 25 years observing and documenting how poor people in urban and rural Indonesia are affected by economic development. She has analysed social and environmental impacts as a consultant for the World Bank, the United Nations, ADB, and German (GTZ) and Australian Aid (Ausaid). Between consultancies she has lectured at the Royal Melbourne Institute of Technology, Melbourne University, and other Australian universities.

Michael Johnson lectures on the politics of the Middle East in the School of African and Asian Studies at the University of Sussex. In the early 1970s he was a member of the University of Manchester's research team in Lebanon. Later he taught at the University of Khartoum. He is currently conducting comparative research on ethnic and communal conflict in the Middle East, India, and South Africa.

Ravi Kanbur is the Chief Economist of the Africa Region at the World Bank. Previously he was Professor of Economics and Director of the Development Economics Research Centre at the University of Warwick. He has also taught at the University of Oxford, Cambridge, Essex, Princeton, and Columbia.

Fernando Kusnetzoff is an architect and planner. Dean of the College of Architecture and Urban Planning at the University of Chile until the overthrow of the Allende Government, he subsequently taught at the College of Environment Design of the University of California at Berkeley. Currently he works as a consultant for the UN Centre for Human Settlements (Habitat), and for the Interamerican Development Bank.

Michael F. Lofchie is Professor of Political Science at the University of California, Los Angeles, where he served as Director of the James S. Coleman African Studies Center from 1978 to 1989. He has been a member of the Board of Directors of the African Studies Association and currently serves as the Vice-Chair of the US–China African Studies Exchange Committee.

Larissa Lomnitz is Professor of Social Anthropology at the National University of Mexico. She has done research on the urban poor, the industrial bourgeoisie, and the political culture of Mexico. Kinship and network analysis run through all these studies.

Kosta Mathéy is Professor Adjunto at the School of Tropical Engineering and Architecture of the Technical University in Havana. He is the Chairperson of TRIA-LOG, the Association for Scientific Research into Planning and Building in the Developing World. He worked as a consultant to UNCHS, ILO, EEC, GTZ, IHS, KFW, and others and taught in Latin America, Africa, and Asia.

Kogila Moodley holds the David Lam Chair of Multicultural Studies at the University of British Columbia. She taught Sociology at the newly founded Indian University in Durban before emigrating to Canada in 1968. Her academic interests have focused on comparative race relations and the socialization of students and teachers in multi-ethnic contexts. She has carried out research in South Africa, Germany, and the United States.

Andrew R. Morrison is Associate Professor of Economics at Tulane University. He has worked extensively on labour-market issues in developing countries. In 1988 he received the Dorothy Thomas Award from the Population Association of America for quantifying the effects of interregional migration on Gross National Product.

Harald Moßbrucker is an anthropologist who studied the social and economic organization of Amerindians and their migration in Peru (1983–4, 1987–9, 1994) and Mexico (1991–3). At this time he works for the European Union on a potable water and sanitation project in Honduras.

Nici Nelson teaches in the Department of Anthropology at Goldsmiths' College, University of London. She lived for four years in Nairobi, Kenya, and travels there frequently. She works in the areas of African urbanization and women and development.

Rita Noonan is a Ph.D. candidate in Sociology at Indiana University, Bloomington. Her research interests include gender, development, and social movements, with a particular focus on Latin America. She is completing her dissertation which examines the gendered nature of state-sponsored medicine in Costa Rica and women's grass-roots efforts to broaden definitions of health needs.

Bryan Roberts is Professor of Sociology at the University of Texas at Austin where he holds the C. B. Smith, Sr. Chair in US–Mexico Relations. He obtained his doctoral degree from the University of Chicago, then returned to the United Kingdom to teach at the University of Manchester from 1964 to 1986. He has carried out research in Central America, Peru, and Mexico, and is currently working on migration, poverty, and enterprise in the US–Mexico transborder region.

Janet W. Salaff is Professor of Sociology, University of Toronto, and cross-appointed to the Centre for Urban and Community Studies. She mainly does research on family formation among Chinese peoples on the Pacific Rim. Currently she is engaged in a study of how Hong Kong families approach the reversion to China in 1997 and focuses with colleagues at Hong Kong University on the effect of networks on political views and the decision to emigrate.

David Satterthwaite is Director of the Human Settlements Programme of the International Institute for Environment and Development and edits its journal *Environment and Urbanization*. He was an adviser to the Brundtland Commission for its report *Our Common Future* and more recently has worked with the World Health Organization, UNICEF, and the UN Centre for Human Settlements (Habitat) on the links between environment, health, and development in cities.

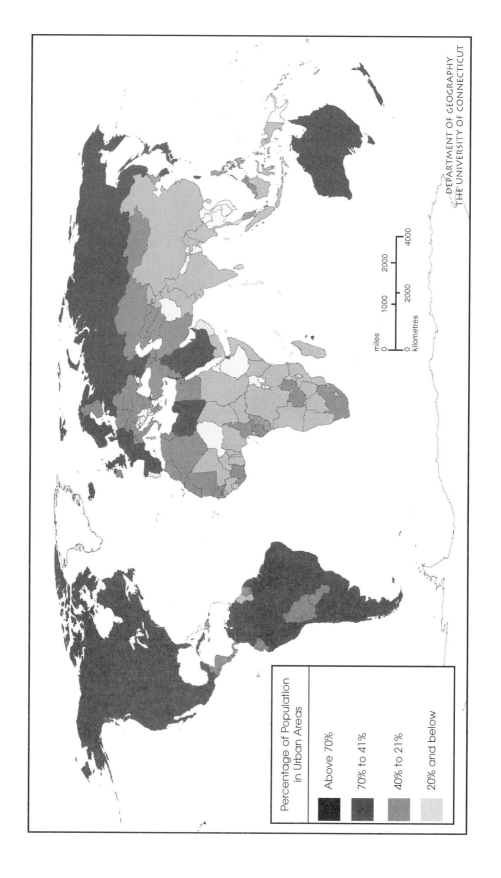

Percentage of Population
in Urban Areas

Above 70%

70% to 41%

40% to 21%

20% and below

miles
0   1000   2000

kilometres
0   2000   4000

Figure 1 Level of Urbanization by Country, 1993

Introduction

MORE than one and a half billion people live in the cities and towns of the developing countries of Asia, Oceania, Africa, Latin America, and the Caribbean. Their numbers are growing rapidly in most of these countries—they will be more than four billion within little more than a generation. Urban dwellers are spread rather unevenly across the South: in much of Asia population densities are high, but levels of urbanization are still relatively low; Latin America, in contrast, has a rather small but highly urbanized population; Africa, finally, has low levels of population density as well as urbanization (Figure 1, Table1).

The urban transition constitutes a profound human transformation, comparable to the domestication of plants and animals ten thousand years ago that made a sedentary life possible. This second transformation began five thousand years ago, when the first urban settlements were established in the valleys of the Tigris and the Euphrates. However, as recently as the beginning of this century only one in eight people lived in an urban area. The twentieth century is the century of the urban transition: by the end of the century about half the world's population will live in urban settlements.

The last phase of the urban transformation is taking place in the developing countries. Already, two-thirds of the world's urban population live in the developing world. By some estimates, the population of São Paulo has outstripped that of Tokyo; and Mexico City, Calcutta, and Bombay each are home to more people than inhabit New York. The magnitude of this transformation of the 'South', the sheer number of people involved, is without precedent in human history. And the poverty of these regions makes it a difficult transition. Large numbers of rural–urban migrants have to adapt to the urban environment. Rapidly growing urban populations have to find employment in urban labour-markets characterized by widespread unemployment and underemployment. They increase demands for urban housing and services already considered inadequate, and, for better or worse, exacerbate popular pressures on political systems.

Urbanization constitutes a multi-faceted phenomenon, and the developing countries make up a large and diverse universe. The task of understanding the urbanization of the 'South' may be approached by composing a picture from studies more limited in scope. This is the approach taken in *The Urban Transformation of the Developing World*, the companion volume that presents accounts of the urbanization process in major countries and regions and embarks on the task of comparative study.

The present collection takes a different approach. It brings together leading experts on various facets of the urban experience. The authors come from the far corners of the globe and offer diverse perspectives. They range across the social sciences: sociologists and anthropologists predominate, but architects, economists, geographers, planners, political scientists, and a psychologist have contributed as well. Their studies have been grouped to explore five aspects of urbanization: the strategies of rural–urban migrants and migration policies; the characteristics and problems of the urban labour-market; the survival strategies of the urban poor and attempts at social engineering so as to transform urban society; housing needs, environmental problems, and the policies designed to address them; and the various ways in which political actors in the urban arena co-opt, confront, and impose themselves on each other to

TABLE 1 Demographic, Economic, and Human Development Indicators for Large Developing Countries[a]

Country/region	Total population	Annual population growth-rate (%)		Urban population[b] (% of total population)		Urban population annual growth-rate (%)		Real GDP[c] per capita (US$)	Infant mortality (per 1,000 life births)
	1993	1960–93	1993–2000	1960	1993	1960–93	1993–2000	1993	1993
East Asia									
China[d]	1,196.4	1.8	1.0	19	29	3.1	3.8	2,330	44
North Korea	23.0	2.3	1.7	40	61	3.6	2.3	3,000	24
South Korea	44.1	1.7	0.9	28	79	5.0	2.3	9,710	11
Taiwan[e]	20.9	1.8[f]	—	—	94	3.0[f]	—	—	5
South-East Asia									
Burma	44.6	2.2	2.1	19	26	3.1	3.7	650	82
Indonesia	191.7	2.1	1.5	15	34	4.7	4.2	3,270	56
Malaysia	19.2	2.6	2.1	27	52	4.8	3.6	8,360	13
Philippines	64.8	2.6	2.0	30	52	4.3	3.9	2,590	43
Singapore	2.8	1.6	0.9	100	100	1.6	0.9	19,350	6
Thailand	57.6	2.4	1.0	13	20	3.8	2.8	6,350	36
Vietnam	71.3	2.2	2.1	15	20	3.2	3.5	1,040	42
South Asia									
Bangladesh	115.2	2.5	2.2	5	17	6.3	5.4	1,290	106
India	901.5	2.2	1.8	18	26	3.4	3.0	1,240	81
Nepal	20.8	2.4	2.6	3	13	6.8	6.9	1,000	98
Pakistan	132.9	3.0	2.8	22	34	4.3	4.6	2,160	89
Sri Lanka	17.9	1.8	1.2	18	22	2.4	2.7	3,030	17
West Asia									
Afghanistan	17.7	1.5	6.0	8	19	4.3	8.2	800	163
Iran	64.2	3.4	2.2	34	58	5.0	3.1	5,380	34
Iraq	19.5	3.2	2.9	43	74	4.9	3.6	3,413	58
Kazakhstan	17.0	1.6	0.6	45	59	2.5	1.4	3,710	27[g]
Lebanon	2.8	1.3	2.3	40	86	3.7	2.9	2,500	34
Saudi Arabia	17.1	4.4	3.1	30	79	7.6	3.6	12,600	28
Syria	13.7	3.4	3.4	37	51	4.4	4.4	4,196	39
Turkey	59.6	2.4	1.8	30	66	4.9	3.7	4,210	64
Uzbekistan	21.9	2.9	2.2	34	41	3.4	2.8	2,510	28[g]
Yemen	13.2	2.8	3.7	9	32	6.8	6.6	1,600	119

Africa								
Algeria	26.7	2.8	2.2	30	54	4.6	3.6	5,570	54
Angola	10.3	2.3	3.5	10	31	5.7	6.0	674	123
Cameroon	12.5	2.6	2.9	14	43	6.2	4.9	2,220	62
Côte d'Ivoire	13.3	3.9	3.3	19	42	6.4	4.9	1,620	91
Egypt	60.3	2.4	2.0	38	44	2.9	2.6	3,800	66
Ethiopia	51.9	2.5	3.0	6	13	4.7	5.2	420	118
Ghana	16.4	2.7	3.0	23	35	4.0	4.5	2,000	80
Kenya	26.4	3.6	3.1	7	26	7.6	6.0	1,400	69
Madagascar	13.9	2.9	3.2	11	26	5.7	5.8	700	91
Malawi	10.5	3.4	2.1	4	13	6.8	5.0	710	142
Mali	10.1	2.6	3.1	11	26	5.2	5.7	530	158
Morocco	25.9	2.5	1.9	29	47	4.0	3.0	3,270	67
Mozambique	15.1	2.2	3.3	4	31	9.0	7.4	640	147
Nigeria	105.3	2.8	2.9	14	38	5.8	5.0	1,540	84
South Africa	39.7	2.5	2.2	47	50	2.8	3.1	3,127	52
Sudan	26.6	2.7	2.7	10	24	5.3	4.7	1,350	77
Tanzania	28.0	3.1	2.8	5	23	8.2	5.9	630	85
Uganda	19.9	3.4	3.1	5	12	6.2	5.6	910	115
Zaïre	41.2	3.0	3.1	22	29	3.8	4.4	300	92
Zimbabwe	10.7	3.2	2.2	13	31	6.0	4.6	2,100	67
Latin America								
Argentina	33.8	1.5	1.2	74	88	2.0	1.5	8,350	24
Brazil	156.5	2.4	1.6	45	77	4.0	2.4	5,500	57
Chile	13.8	1.8	1.5	68	84	2.5	1.6	8,900	15
Colombia	34.0	2.3	1.5	48	72	3.6	2.2	5,790	37
Cuba	10.9	1.4	0.7	55	75	2.3	1.2	3,000	12
Ecuador	11.0	2.8	2.0	34	57	4.4	3.3	4,400	49
Guatemala	10.0	2.9	2.9	32	41	3.6	4.1	3,400	48
Mexico	90.0	2.7	1.9	51	74	3.9	2.5	7,010	35
Peru	22.9	2.6	1.9	46	71	3.9	2.5	3,320	64
Venezuela	20.9	3.1	2.1	67	92	4.1	2.5	8,360	23
Developing countries[h]	4,299	2.2	1.8	22	36	3.8	3.5	2,696	70
Industrial countries	1,209	0.8	0.4	61	73	1.4	0.7	15,136	13
WORLD	5,508	1.9	1.5	34	44	2.7	2.5	5,428	63

[a] This table includes all countries with a population of at least 10 million in mid-1993 (and Lebanon and Singapore each of which are the focus of a chapter).

[b] Percentages of the population living in urban areas are based on different national definitions of what is 'urban', and cross-country comparisons should be made with caution. For national definitions of urban area, see United Nations (1995), *1993 Demographic Yearbook* (New York: United Nations), notes to table 6.

[c] Real Gross Domestic Product is based on conversion of GDP in terms of purchasing power parity.

[d] The data for China includes Taiwan.

[e] The data for Taiwan is from Republic of China (1995), *Statistical Yearbook of the Republic of China 1995* (Taipei: Directorate-General of Budget, Accounting and Statistics).

[f] The data is for 1965–93.

[g] The data is for 1994, from World Bank (1996), *World Development Report 1996: From Plan to Market* (New York: Oxford University Press).

[h] The summary measures for developing countries and the world are appropriately weighted values, except for total population.

Sources: United Nations Development Programme (1996), *Human Development Report 1996* (New York: Oxford University Press), except as noted.

shape local and national policies. These studies are preceded by a section exploring theoretical perspectives on urbanization—'modernization theory', 'dependency theory', 'world system theory', the 'urban bias' approach—and related policies. Introductions to each of these six sections put the contributions into context. Together they provide a multi-faceted picture of the cities of the 'South'.

A third of the studies are general reviews of the literature and data on selected issues. The others are based on research in particular settings that serves to illuminate a specific topic. Much of this research is comparative in nature: a third of the authors have returned to their research sites to report on developments stretching over several decades, a quarter draw on research across different societies. The research settings range across the 'South', but they are by no means randomly distributed. Regional biases arise from the predilection of researchers and the priorities of their sponsors, e.g. the interest of US political scientists in Latin America after the triumph of the Cuban revolution. Not only the extent to which research is pursued in a region, but also the topics it addresses, the theoretical perspective it employs, and the methods it uses, vary across regions. Relatedly, certain issues are more topical in some settings than in others, e.g. government control in China, ethnicity in Africa, squatter movements in Latin America. If Latin America research preponderates in this collection, this reflects the amount of important research accomplished in the region in recent years. This regional bias may be argued to be desirable to the extent that Latin American patterns prove to presage developments elsewhere: as states mature that gained their independence much more recently, as income levels rise in even poorer economies, and as the urban transition proceeds in as yet predominantly rural societies.

Half the contributions appear here for the first time. Most of the others were first published within the last ten years, and all but two have been updated and/or revised. They thus complement as well as update the 1992 edition of the volume *Cities, Poverty, and Development: Urbanization in the Third World* I co-authored with Alan Gilbert.

Over the years I have accumulated many intellectual debts: debts to colleagues across the social sciences who have made me sensitive to the multiple approaches needed to understand the urbanization process, whose conflicting perspectives have convinced me that we need to assimilate what is valuable in new research, rather than succumb to the grandiose claims of passing intellectual fashions; and debts to students on three continents who have challenged me to integrate multiple disciplinary approaches and conflicting theoretical perspectives. The selections I have made for this volume, my suggestions to the contributors for revisions, and the perspective the chapter introductions provide on the contributions, are the outcome of this intellectual journey. I am grateful to Loretta Bass, for assistance most efficient and invariably cheerful, and to staff at the Homer Babbidge Library, University of Connecticut, who time and again went well beyond the call of duty.

This volume foremost represents the work of the scholars who have agreed to make original contributions or to revise and update earlier work. I wish to thank them for taking on a task rather undervalued in academia: contributing to an edited collection. Appreciation is due to some for bearing with me through the successive drafts I urged on them and for the hard work they invested in revising, and revising yet again, to others for turning in dazzling first drafts.

All of us are profoundly indebted to the men and women in the 'South' who were prepared to let us enter their lives, who gave us of their time to answer our queries, some of whom put themselves at risk by responding to us. This collection is dedicated to them.

Josef Gugler

Storrs
September 1996

I Development Theory and Policy

Introduction

URBANIZATION and industrialization are frequently assumed to be intimately connected, and urbanization is seen as a prerequisite for development.[1] Such assumptions are problematic. Not only do cities pre-date industrial manufacture by several millennia, but even today only a small proportion of urban dwellers work in industry. Most cities are foremost centres of public administration and commerce. And while the more urbanized countries tend to be more developed, whether measured in terms of economic output or human welfare (see Table 1), this correlation does not establish the direction of causality. It is easily assumed that a larger, more productive urban population creates higher incomes. However, the reverse causal relationship is quite plausible: the richer countries have greater resources to concentrate in urban areas in terms of public bureaucracies, of élite and middle-class standards of living that support a large service sector, and of public works.

The two views of urbanization just sketched may be identified with 'modernization theory' and the 'urban bias' approach, respectively. From a modernization perspective, the transition from a rural to an urban society is conducive to the transformation from 'traditional' to 'modern' in terms of both the technology of production and the orientation of individuals and social institutions. Proponents of the urban bias thesis counter that policy-makers tend to favour urban areas over rural, that they disproportionately allocate resources to the cities, especially capital cities, with results that are both inequitable and inefficient.

'Dependency theory' and the 'world-system' approach have had a major impact on discussions of development or rather, it was frequently argued, underdevelopment over the last two decades .Concern with the impact of the powerful countries of the 'First World' on the dependent countries of the 'Third World' eventually prompted enquiries into its consequences in terms of the level of urbanization, the degree of urban primacy, and the high level of non-industrial employment characteristic of cities in developing countries. These concerns were commonly cast in terms of 'overurbanization', expressly denying a functional relationship between urban growth and development. We will return to the notion of overurbanization in Chapter 7.[2]

Nearly all research on these issues adopts or rejects one particular perspective and marshalls evidence to support or reject it. York Bradshaw (1987) broke this pattern when he put the explanatory power of all three approaches to the test in a cross-national study: he tested the simultaneous effects of variables derived from modernization, urban bias, and economic dependency perspectives on a country's 'overurbanization'—defined in terms of the level of urbanization relative to GNP per capita, and the effect of a similar set of variables including overurbanization on economic growth.

York Bradshaw and Rita Noonan replicate his pioneering work in Chapter 1, using more recent data and expanding the number of variables. They find that two distinct dependency variables have an impact on overurbanization; that economic growth is affected by variables derived from modernization theory, the urban bias approach, and dependency theory; and that all three perspectives also serve to explain female labour-force participation. Bradshaw and Noonan thus demonstrate the need to simultaneously consider a range of explanatory variables derived from different theoretical positions.

The urban bias approach specifically relates to the rural–urban divide, rural–urban migration, and the rate of urbanization. The underlying argument was enunciated in 1967 in a document discussed at a meeting of the leaders of the Tanganyika African National Union (TANU), the country's only political party. It came to be known as the Arusha Declaration for the town in which they met (Nyerere, 1968). It warned presciently that the real exploitation in a poor, largely rural country such as Tanzania might become that of the town dwellers exploiting the peasants.

In 1977 Michael Lipton coined the expression 'urban bias' in his *Why Poor People Stay Poor: Urban Bias in World Development*. He argued that the most important class conflict in poor countries is between the rural classes and the urban classes, that urban bias provides the main explanation 'why poor people stay poor'. Lipton's concern with urban bias grew out of his work in India. However, as he noted, his case is stronger for many other countries, especially in Africa South of the Sahara.

Michael Lofchie, in Chapter 2, describes the various policies that contributed to urban bias in Africa South of the Sahara during the two decades after independence came to most of the region around 1960. He traces the origins of these policies to the Development Economics of the 1950s and 1960s that advocated a government-induced transformation from the primacy of agriculture towards the primacy of industrial production via an import-substituting industrialization strategy.

Lofchie emphasizes that such an explanation is not complete: why were these urban-biased policies continued long after their adverse economic effects had become apparent? To answer this question, he points out that the very pursuit of urban-biased policies created a whole array of interest groups prepared to defend them. While Robert Bates, in his *Markets and States in Tropical Africa*, argued that governments were far more dependent for their survival upon the support of urban constituents than upon that of politically weak rural dwellers, Lofchie argues that urban-biased policies were responsible for establishing these powerful urban interests in the first place.

By the early to mid-1980s, African governments began to abandon urban-biased policies in favour of liberal, market-oriented approaches to growth that do not discriminate against the rural sector. Lofchie traces this profound policy change to the severe economic crisis that had befallen virtually every African country as well as an intellectual revolution within Economics: Development Economics as a separate set of principles applicable to largely agricultural countries had been discredited.

Confronted with neglect, or even outright exploitation, peasants are left with three options.

Some rebelled, like the farmers in the south-west of Nigeria who organized the *agbekoya* ('farmers shall not suffer') rebellion in 1968–9 (Beer, 1976). Such rebellions have difficulty reaching across largely rural areas, at best obtain limited concessions, and cannot alter the fact that power is urban-based. Other peasants withdrew from the market and reverted to subsistence farming (Hyden, 1980, 1983). In some cases policy-makers had to seek an accommodation so as to stimulate food supplies for the cities and the production of export crops—i.e. to reduce urban bias—but eventually urban-based power reasserted itself. Finally, many peasants opted to partake of urban privilege by moving to the city. To this option we will turn in Part II.

Notes

1. For a comprehensive review of the burgeoning literature on what is still commonly referred to as Third World urbanization, see Kasarda and Crenshaw (1991). The findings of a large international project surveying past urban research and proposing an agenda for future research in Asia, Africa, and Latin America are presented by Stren (1994–5). On the 'gendering' of various aspects of urban development in the research of the past three decades, see Moser (1995) who provides an exhaustive bibliography.

2. The companion volume provides ready examples of the application of various theoretical approaches: Murphey's (1996) account of the history of the city in Monsoon Asia draws on modernization theory; Abu-Lughod (1996) focuses on the political economy of petroleum in the urbanization of Arab states; and Gugler (1996) emphasizes the impact of urban bias on post-independence urbanization in Africa South of the Sahara.

References

Abu-Lughod, Janet (1996), 'Urbanization in the Arab World and the International System', in Josef Gugler (ed.), *The Urban Transformation of the Developing World* (Oxford: Oxford University Press), 184–208.

Bates, Robert (1981), *Markets and States in Tropical Africa* (Berkeley: University of California Press).

Beer, Christopher (1976), *The Politics of Peasant Groups in Western Nigeria* (Ibadan: Ibadan University Press).

Bradshaw, York W. (1987), 'Urbanization and

Underdevelopment: A Global Study of Modernization, Urban Bias, and Economic Dependency', *American Sociological Review*, 52: 224–39.

Gugler, Josef (1996), 'Urbanization in Africa South of the Sahara: New Identities in Conflict', in Josef Gugler (ed.), *The Urban Transformation of the Developing World* (Oxford: Oxford University Press), 210–51.

Hyden, Goran (1980), *Beyond Ujamaa in Tanzania: Underdevelopment and an Uncaptured Peasantry* (London: Heinemann; Berkeley: University of California Press).

—— (1983), *No Shortcuts to Progress: African Development Management in Perspective* (London: Heinemann; Berkeley: University of California Press).

Kasarda, John D., and Crenshaw, Edward M. (1991), 'Third World Urbanization: Dimensions, Theories, and Determinants', *Annual Review of Sociology*, 17: 467–501.

Lipton, Michael (1977), *Why Poor People Stay Poor: Urban Bias in World Development* (London: Maurice Temple Smith; Cambridge, Mass.: Harvard University Press).

Moser, Caroline O. N., with Linda Peake (1995), 'Seeing the Invisible: Women, Gender and Urban Development', in Richard Stren, with Judith Kjellberg Bell (eds.), *Urban Research in the Developing World*, iv. *Perspectives on the City* (Toronto: Centre for Urban and Community Studies, University of Toronto), 279–349.

Murphey, Rhoads (1996), 'A History of the City in Monsoon Asia', in Josef Gugler (ed.), *The Urban Transformation of the Developing World* (Oxford: Oxford University Press), 17–58.

Nyerere, Julius K. (1968), *Ujamaa—Essays on Socialism* (Dar es Salaam: Oxford University Press).

Stren, Richard (1994–5) (ed.), *Urban Research in the Developing World*, 4 vols. (Toronto: Centre for Urban and Community Studies, University of Toronto).

1

Urbanization, Economic Growth, and Women's Labour-Force Participation: A Theoretical and Empirical Reassessment

York W. Bradshaw and Rita Noonan

'THE urban explosion in today's Third World is nothing less than the evolution of a society during its structural transformation from an agrarian to an industrial-service economy' (Rogers and Williamson, 1982: 468). This provocative statement could still serve as a centre-piece for the ongoing debate over the meaning of Third World urbanization. While few would dispute that underdeveloped nations are undergoing rapid urbanization, debate over its causes and effects is intense and often bitter. Much of the debate revolves around theories of modernization, urban bias, and economic dependency.

In 1987 Bradshaw published a quantitative cross-national study that simultaneously tested each of these perspectives between 1960 and 1980. Modernization scholars assert that urban expansion is part of the natural transition from a traditional (agrarian) society to a modern (industrial) nation (Berliner, 1977: 448–9). Accordingly, rapid urbanization is a positive feature that should be encouraged. Urban bias theorists argue that government policies biased in favour of metropolitan areas have prompted migration from country to city and thereby increased the size of urban regions. This stress on urban devel-

opment may promote temporary economic growth in poor countries, but it will not produce the type of long-term, equitable development that is possible only through aiding agriculture (Lipton, 1977, 1984). Dependency and world-system arguments postulate that foreign investment reduces the amount of land available to farmers, forcing migration to urban areas; moreover, investment in urban areas creates industry that may attract rural migrants. Rapid urbanization distorts urban labour-markets and impedes comprehensive economic development (Ledogar, 1975; Kentor, 1981; Timberlake and Kentor, 1983; Bradshaw, 1987).

Bradshaw found support for all three perspectives, but support for the dependency and urban bias views was especially strong. The issue needs to be reassessed due to recent changes in the structure of the world economy. Since the 1970s and particularly in the 1980s, the global debt crisis may have significantly influenced patterns of Third World urbanization and overall development. Austerity measures imposed by the International Monetary Fund (IMF), combined with pressure for debt repayment, have seriously eroded the capacity of urban areas to feed,

This is a substantially revised, extended, and updated version of an article written by the senior author in the *American Sociological Review*, 52 (1987), 224–39. We wish to thank Josef Gugler for very helpful comments on an earlier draft of this essay.

house, and employ growing populations. This has increased 'overurbanization' and created economic imbalances that retard economic expansion. Global debt and austerity have also forced more women into the labour-force to earn money during difficult economic times. Women are increasingly moving to urban areas in order to work in various formal and informal activities.

These developments compel us to elaborate the theoretical arguments first presented by Bradshaw (1987) to include changes over the last two decades. After the theoretical discussion, we present a new quantitative analysis of urbanization, economic growth, and women's participation in the labour-force that examines several important arguments, especially those associated with the global debt crisis. The data analysis contains several other features that contribute to the research design, including a rigorous examination of influential cases and a regional focus on Africa and Latin America. This analysis uses data from 1970–1992, a period that covers the beginning of the debt crisis through contemporary times. The findings produced in our study underscore the importance of the growing debt crisis for patterns of urbanization and development, findings often ignored in traditional quantitative analyses.

Theories of Urbanization and Economic Development

Modernization and Urbanization

According to classical economists, rural inhabitants are pulled to urban areas by high industrial wages (Berliner, 1977; Spengler and Myers, 1977). Two prominent economists state that 'industrialization (and manufacturing employment growth) has been the "engine of urbanization" in the past and will continue to be so in the future' (Kelley and Williamson, 1984: 179). Migrants respond not simply to the actual wage differential between country and city, but also to the *expected* differential (Todaro, 1969, 1977; see also Rogers and Williamson, 1982: 472–3; Stark, 1982: 66). People will continue to migrate townward as long as their expected urban wages exceed their current rural wages.

Despite the increasing costs associated with urban expansion (Rogers and Williamson, 1982;

Kelley and Williamson, 1982, 1984), modernization theory generally views urbanization as a positive phenomenon for several reasons. First, urbanization supposedly facilitates economic growth by increasing 'modern sector' output in developing countries. Berliner's comment from nearly 20 years ago is still a good description of this argument:

The traditional view [of economics is] that labor mobility contributes significantly to the efficiency of resource allocation. By migrating from regions of low marginal productivity and low wages to regions with higher marginal productivity and wages, mobile labor increases the total output of the society. Of two societies alike in all other respects, the one with the higher degree of mobility would enjoy the higher income. (Berliner, 1977: 448)

Accordingly, large-scale migration to urban areas is a prerequisite for modern (industrial) society.

Secondly, some economists and sociologists also have argued that city life is conducive to the formation of modern ideas necessary for economic growth and overall development (Inkeles and Smith, 1974; see also Hoselitz, 1962). Third World cities contain modernizing institutions such as the school, the factory, and the mass media, all of which inculcate modern values that facilitate economic development. This view also asserts a positive relationship between increased levels of urbanization and economic growth.

Thirdly, recent scholarship on women and development suggests that increasing industrialization, education, and urbanization may provide more opportunities for women to advance economically and socially (see Sershen, 1992 for a recent review of this literature; see also Ward, 1984). More women will be able to participate in the wage-earning labour-force when they have greater access to education and industry. This does not mean, however, that women's participation in the labour-force is always a positive feature of development. Poor women often are exploited in various types of work, especially in factories, agricultural plantations, and related activities. None the less, education and industrial activities are viewed as two of the only ways for women to advance in economically underdeveloped countries. These activities are also normally associated with lower fertility rates, giving women additional opportunities outside the household.

To conclude, it is clear that the developing world continues to urbanize at a rapid rate. Despite such growth, it is also clear that slums, unemployment, and crime in Third World cities call into question the positive features of the modernization perspective.

Urban Bias and Urbanization

Urban bias theorists (e.g. Lipton, 1977, 1984) argue that many underdeveloped nations implement investment, tax, pricing, and other policies that disproportionately favour urban areas. The state enacts these policies because of pressure from various urban-based groups such as industrialists, small-scale capitalists, and urban workers. This bias in favour of city areas has created a disparity between country and city with respect to consumption, wage, and productivity levels (Lipton, 1977: 145–59). Such disparities translate into a higher standard of living for urban citizens and draw migrants from poorer (rural) areas. Studies of urban bias have produced conflicting results; some provide modest support for this perspective (Lipton, 1973, 1977, 1984; Nolan and White, 1984), while others challenge its basic tenets (Seers, 1977; Byres, 1979; Herring and Edwards, 1983).

Perhaps the most important indicator of urban bias entails the investment of substantially more domestic capital in non-agriculture than in agriculture, relative to output per sector (Lipton, 1977: 189–215; 1984: 140–1). This practice is not only inequitable, it is inefficient. Specifically, Lipton (1977: 445) shows that the investment-to-output ratio in non-agriculture is about twice as high as in agriculture, prompting Griffin (1977: 108) to note that 'a reallocation of investable resources in favour of agriculture would raise total national output, increase the productivity of labour in agriculture relative to that in the rest of the economy and reduce rural–urban income inequality'. Such reallocation would provide farmers with additional machinery, roads, dams, barns, and sheds, all of which would improve the productivity and welfare of rural dwellers.

National tax policies may also contribute to urban bias in underdeveloped countries. State policies allegedly overtax rural citizens relative to (1) urban citizens with a similar income and (2) the proportion of government benefits allocated to the countryside (Lipton, 1977: 270–86; 1984: 147). Unfortunately, reliable data on taxation in many poor nations is scarce, precluding any firm conclusions about this strand of the urban bias argument. Moreover, studies focusing on the few countries with data on taxation have produced ambiguous and often conflicting results (see Ghandi, 1966; Lipton, 1973; Byres, 1979; Herring and Edwards, 1983).

It is possible, however, that the real source of taxation against agriculture may come not through direct taxes but through the practice of 'price twists' (Lipton, 1977: 287–327). Price twists occur when states formulate pricing and related policies that transfer resources from the agricultural to the urban sector. State-controlled marketing boards, for example, have consistently twisted prices against peasant farmers in many poor countries. Such boards purchase agricultural products like coffee, tea, and cocoa from local farmers at an artificially low price and then resell these crops on the export market at the prevailing world price (Bates, 1981: 11–29; see also Ellis, 1982, 1983, 1984). The surplus accumulated by the state (through its marketing board) is used to finance industrial and other programmes that benefit urban élites.

Lipton argues that urban bias perpetuates economic inefficiency and therefore inhibits long-term economic development in poor regions. Because the majority of Third World citizens live in rural areas and the investment-to-capital ratio is greater in non-agriculture than in agriculture, it would seem logical to reallocate domestic resources in favour of farmers. Thus, in contrast to modernization arguments, the urban bias thesis asserts that rural dwellers should remain in agricultural activities instead of migrating to urban areas, where they will join the informal labour-market or turn to crime, begging, or prostitution (Lipton, 1977: 216, 233–4). This is an unproductive way to utilize human capital and will contribute to economic stagnation. Wellisz (1971: 44) outlined this very argument more than 25 years ago by noting that excessive urban growth 'stands for a "perverse" stream of migration, sapping the economic strength of the hinterland, without correspondingly large benefits to urban production. Instead of being a sign of development, overurbanization is a sign of economic illness.'

Economic Dependence and Urbanization

Dependency-oriented research on urbanization can be divided into three parts. First, traditional dependency arguments argue that foreign investment in Third World agriculture displaces peasants and 'pushes' them to the city (Ledogar, 1975; Walton, 1977; Evans and Timberlake, 1980; Kentor, 1981; Timberlake and Kentor, 1983). The severity of this situation is compounded by the fact that many Third World farmers produce a variety of primary commodities (e.g. coffee, tea, and cotton) that are 'purchased' by the government and exported. These farmers are damaged by low international prices or by other unfavourable circumstances such as drought or famine which affect agricultural production, thereby enhancing townward migration and inhibiting economic growth. Moreover, these farmers suffer especially deleterious conditions if they do not produce different types of agricultural crops for export. Specialization in one or two crops leaves farmers vulnerable to adverse conditions that affect those crops. Of course, the national economy is also vulnerable to this overspecialization.

Secondly, the dependent development perspective is concerned with factors that lead to 'modern-sector' development and economic expansion in developing countries (Evans, 1979). Foreign investment in manufacturing and related activities supposedly facilitates overall economic expansion, although it does so at the expense of equality. Cities throughout the developing world have increased their level of urbanization largely because of external capital. At the same time, foreign investment in manufacturing may also increase urban population growth. Rural inhabitants may *perceive* that increased industrial activity will establish a superior quality of life in cities relative to that found in rural areas. This argument is related to the so-called 'bright lights' theory of urbanization, which asserts that people in the countryside are attracted to the excitement and supposed opportunity offered by the city.

Thirdly, recent developments in the world economy may have profoundly altered arguments associated with Third World urbanization and development. Quantitative studies completed in the last few years show that the global debt crisis and IMF pressure increase overurbanization (Bradshaw, Noonan, Gash, and Sershen,

1993), inhibit economic growth and physical quality of life (Bradshaw and Huang, 1991), and increase political protest and domestic riots (Walton and Ragin, 1990). Since the 1970s and especially during the 1980s, the growing debt crisis has led to the implementation of severe structural adjustment policies mandated by the IMF and other transnational financial institutions. These structural adjustments compel poor countries to impose strict austerity programmes that result in less government spending for health, education, food subsidies, and other services. Ideally, the savings from such austerity programmes are supposed to service the growing international debt. In reality, however, debt service payments and structural adjustments have created great hardship for Third World residents, especially women, children, the poor, and the aged.

The primary problem is that Third World cities continue to grow at a time that city services are being curtailed. As a result of IMF pressure, food subsidies have been reduced or eliminated in many countries; employment and wages have failed to expand; government spending on health, education, and housing has been slashed; and currency devaluation increases the cost of living and makes it very difficult to pay for vital imports (e.g. medicine). UNICEF and other groups have severely criticized the IMF for these policies because they harm Third World children and overall quality of life. In fact, UNICEF (1989: 30) asserts that 'hundreds of thousands of the developing world's children have given their lives to pay their countries' debts, and many millions more are still paying the interest with their malnourished minds and bodies'. Urban areas are especially vulnerable due to growing numbers of poor and desperate people who require various services. Bradshaw, Noonan, Gash, and Sershen (1993: 637–8) assert directly that 'a country is considered overurbanized when it lacks the capacity to support its urban population, and *structural adjustment policies directly inhibit urban areas from adequately supporting their populations*' (emphasis in the original).

Moreover, and in sharp contrast to IMF intentions, there is strong cross-national evidence showing that structural adjustment retards economic growth (see Bradshaw and Huang, 1991). Western-oriented 'shock treatments' may well impose inappropriate economic conditionality on developing economies that already face significant problems. A negative association between IMF-imposed austerity measures and economic growth may call into question conventional arguments advanced by dependency and world-system theories. Some studies assert that foreign investment generally reduces economic development (e.g. Bornschier and Chase-Dunn, 1985), while others postulate that such investment actually facilitates development (Firebaugh, 1992). However, debate over the effects of foreign investment could be misplaced in the current era. The dependency of IMF pressure and debt service payments has superseded the traditional arguments over direct foreign investment.

In addition to increasing overurbanization and slowing economic growth, global indebtedness and IMF pressure may have a growing impact on the propensity of women to be economically active (see Gladwin, 1991; Inter-American Development Bank, 1990). Fewer government services, economic contraction, higher prices, and more overurbanization may well force a greater number of women into the labour-force, especially into the informal sector (Hirata and Humphrey, 1991). Because they bear primary responsibility for family care, women will have to engage in various survival strategies to help support their families. While a few women will be able to obtain wage employment in the 'modern' sector, the vast majority of women will have to settle for low-paying jobs in informal activities. The problem is compounded by the fact that men usually dominate higher paying jobs in the export-crop sector of poor countries. By contrast, women often are 'relegated' to subsistence agricultural production (Lele, 1991; United Nations, 1991: 81–99).

To conclude, we argue that traditional dependency and world-system arguments related to overurbanization, underdevelopment, and women's labour should be reassessed after taking into consideration recent developments in the global economy. The data analysis below examines the various dependency-oriented arguments as well as the modernization and urban bias perspectives.

Hypotheses, Operationalization, and Model

Before testing the relative merits of each theory, it is important to review and summarize the central hypotheses of each theoretical perspective. *Modernization theory* asserts that industrial employment attracts people to urban areas, where they work in modern-sector occupations that facilitate national economic expansion. In addition, urbanization, education, and industrial employment should increase labour opportunities for women.

Urban bias arguments state that the disparity in welfare between country and city increases rural-to-urban migration and thereby expands urbanization. The disparity ultimately reduces economic growth and efficiency in the developing world.

Dependency theories advance various types of arguments. Traditional dependency and world-system arguments claim that foreign investment in agriculture and agricultural specialization increase urban growth and decrease economic expansion. Dependent development arguments assert that foreign investment in manufacturing increases both urbanization and economic growth. Finally, recent dependency arguments revolving around the global debt crisis postulate that foreign debt and IMF pressure increase overurbanization and retard economic growth. Moreover, global indebtedness may create conditions that compel more women to participate in the labour force.

The different theoretical perspectives should be tested simultaneously to prevent spurious results. For example, an investigator should not say that foreign investment in agriculture prompts townward migration without controlling for the pull of industrial employment and for the disparity in welfare between country and city. Likewise, an investigator should not say that the welfare disparity causes townward migration without controlling for the impact of global indebtedness and IMF pressure. With few exceptions (e.g. London, 1987), quantitative cross-national studies have not simultaneously tested for a variety of theoretical perspectives.

A simultaneous test is possible using a panel regression model that has been employed in many other cross-national studies (see

Bornschier and Chase-Dunn, 1985 for a review). This model regresses each dependent variable at a later point in time on its value at an earlier point in time and on the independent variables. A panel design (1) tests the stability of the dependent variable over a specified time period and (2) controls for the prior effect of the dependent variable and thus prevents false inferences due to reciprocal causation. The original version of this paper (Bradshaw, 1987) utilized data at basically two points in time, 1960 and 1980. This time period permitted examination of a 20-year span of time.

As argued earlier, however, there have been profound changes in the global economy starting in the 1970s and continuing through the present time. Thus, our essay utilizes data between 1970 (just before the debt crisis began) and 1992 (the latest period for complete data on most variables of interest). We attempt to replicate the results found in Bradshaw (1987) for this new time period, and then we add some additional variables that are important. Our sample is based on 57 poor and middle-income countries that have complete data on all variables of interest.[1] To improve the theoretical, substantive, and methodological focus of the analysis, we also utilize a rigorous process of (1) identifying and controlling for influential cases and (2) examining the different effects for Latin America and Africa. The data sources for each variable are included in the Appendix.

Measurement

Dependent Variables

Overurbanization. This concept evaluates whether countries are too urbanized relative to

TABLE 1.1   Relative Urbanization Residuals, Urban Population, and Level of Economic Development, 1970–1992

Country	1992			1970			Change, 1970–1992
	residual[a]	urban[b]	gnpc[c]	residual	urban	gnpc	residual
Large positive residuals							
Nicaragua	29.806	61	340	5.847	47.2	581.1	23.959
Venezuela	27.358	91	2,910	13.982	76.2	1,836.5	13.377
Peru	24.277	71	950	15.863	59.5	659.4	8.414
Uruguay	23.276	89	3,340	27.237	82.1	1,224.4	−3.960
Chile	22.324	85	2,730	27.809	75.2	811.0	−5.486
Colombia	19.192	71	1,330	23.844	59.8	431.8	−4.652
Sierra Leone	14.198	34	160	−3.107	18.1	191.5	17.305
Brazil	14.104	77	2,770	11.695	55.9	680.4	2.409
Dominican Republic	13.765	62	1,050	2.117	40.3	488.2	11.648
Argentina	12.297	87	6,050	21.807	78.4	1346.9	−9.510
Large negative residuals							
Portugal	−42.849	35	7,450	−29.039	26.2	1,250.0	−13.810
Thailand	−33.714	23	1,840	−14.165	13.2	268.9	−19.549
Papua New Guinea	−30.923	16	950	−24.435	9.8	392.7	−6.488
Rwanda	−20.547	6	250	−5.574	3.2	965.0	−14.973
Burundi	−17.916	6	210	−5.269	2.2	89.8	−12.643
Sri Lanka	−16.185	22	540	−7.880	21.9	128.8	−24.074
Greece	−13.521	64	7,290	−8.964	52.5	1,761.7	−4.557
Burkina Faso	−12.302	17	300	−2.765	6.8	100.8	−9.537
Malawi	−11.916	12	210	−3.038	6.4	100.1	−8.873
Costa Rica	−9.660	48	1,960	−5.349	39.7	712.8	−4.319

[a] residual = relative urbanization residual.

[b] urban = percentage of the total population living in urban areas.

[c] gnpc = gross national product per capita in US dollars.

their level of development. Since Timberlake and Kentor's (1983) seminal study, numerous scholars have utilized their measure of overurbanization (e.g. Bradshaw, 1987; Walton and Ragin, 1990). Specifically, the variable is a residual calculated by regressing level of urbanization (per cent of the population urbanized) on the natural logarithm of GNP per capita in 1970 and in 1992. The following equation illustrates the computation of the residual in 1970 (the same procedure is also used for 1992):

$$\text{Level of Urbanization 1970} = [a + B(\log \text{GNP per Capital 1970})].$$

This residual evaluates overurbanization while remaining orthogonal to GNP per capita. Most important, the residual is sound from a theoretical and substantive standpoint because it captures the notion of overurbanization. The indicator measures whether a nation's level of urbanization is higher or lower than expected, given its level of economic development. Countries with positive residuals would have a higher level of urbanization than expected, while countries with negative residuals would have a lower level of urbanization than expected.

The overurbanization residual can be illustrated by examining some important descriptive data. The countries with the ten largest positive and negative residuals in 1992 are ranked in Table 1.1 according to their 1992 values. The countries' 1970 values are also listed so that comparisons can be made over the 22-year period. In addition to the residuals, the table shows the percentage of the population residing in urban areas, the GNP per capita, and the overurbanization change scores between 1970 and 1992. The residuals underscore that absolute level of urbanization is not the key issue in assessing overurbanization. Instead, the central concern is the level of urbanization *relative to* the level of economic development.

Economic growth. Following some other quantitative cross-national studies, this variable is measured as the *percentage change* in GNP per capita between 1970 and 1992.

Women's labour-force participation. Cross-national data on gender over time for a broad sample of countries is difficult to obtain. For example, data on women in the industrial and

agricultural sectors for a broad sample is unavailable after 1980. Fortunately, however, there is data measuring the percentage of the total labour-force comprised of women. We use this variable for 1970 and 1985. It is well-known that this variable is not perfect, primarily because women's labour is undercounted. Specifically, women's work in rural areas (e.g. gathering fuelwood) is seldom counted as 'economic activity', and unpaid housework is never counted as such. The United Nations (1991: 85) comments: 'Many ambiguities arise in applying the concept of economic activity, especially where it relates to activities on the borderline between subsistence production and housework. Where does non-economic housework end and economic activity begin?' This underscores the difficulties associated with how women's work is perceived and defined by society. If women are not working in the formal (wage) labour-force, then they often are defined as not working at all. Despite such shortcomings, this data on women is the best available and widely used by numerous scholars (e.g. Ward, 1984) and organizations (e.g. United Nations, 1991).

Independent Variables

Foreign investment dependence. This analysis uses three different measures of external investment: (1) the *total* stock of direct, private foreign investment in 1967 divided by total GNP in the same year; (2) the stock of direct, private foreign investment in *manufacturing* in 1967 divided by total GNP in the same year;[2] and (3) the stock of direct, private foreign investment in *agriculture* in 1967 divided by total GNP in the same year. The investment data is standardized on GNP so that foreign investment is measured relative to the size of a nation's economy (see Delacroix and Ragin, 1981). The 1967 date is the closest date to 1970 that has each type of investment variable.

Concentration of agricultural exports. This variable is measured as the percentage contribution of the top three agricultural exports to the total agricultural exports in 1970. Agricultural exports include all export commodities in Standard International Trade Classification (SITC) sections 0, 1, 2, and 4. These categories include all primary commodities comprised of food and beverages and of non-food agricultural products.

Typically, Third World farmers produce a limited number of products from one or more of these categories. Items from SITC section 3 are omitted because they include fuels, minerals, and metals, none of which is directly germane to the theories under investigation.

IMF pressure index. The IMF loans money to poor countries and is also involved in almost all international debt renegotiations and restructurings. To capture the structural adjustments, pressure, and conditionality required by the IMF, Walton and Ragin (1989, 1990) recently introduced an excellent 'IMF pressure index'. It is based on: (1) the number of debt renegotiations between a country and an international financial body (private bank or multilateral lender); (2) the number of debt restructurings experienced by an indebted country; (3) the number of times a country utilized the IMF Extended Fund Facility; and (4) the total IMF loans received by a country as a percentage of its allotted quota. The first three measures are for the period 1975–85 and the fourth is for 1985. Following Walton and Ragin, we created an IMF pressure index by transforming the four variables to Z-scores and then summing them.

Debt dependence change. In addition to pressure imposed by the IMF, countries are also under constant pressure to service their foreign debts. Just as IMF pressure has increased since the mid-1970s, pressure to service the debt has also expanded dramatically during this time period. Thus, we include a variable that measures the percentage increase in total debt service payments/exports, 1975–85 (see Walton and Ragin, 1990).

Industrial employment. This variable measures the percentage of the total work-force employed in industrial jobs in 1970.

Urban bias and the rural–urban disparity. Lipton (1984: 141) notes that there is too little reliable data to test many tenets of the urban bias paradigm. Complete data on the sectoral allocation of domestic investment, on agricultural taxes, and on prices is not available for a broad sample. Fortunately, however, there is sufficient data to measure Lipton's primary indicator of the disparity in welfare between rural and urban areas.

Lipton (1977: 145) describes this disparity and suggests how to measure it: 'The most significant single indicator of the gap [between rural and urban areas] is here termed the *disparity*: the ratio between productivity—output per worker—outside agriculture and productivity inside agriculture.' Lipton shows that this ratio is typically between three and seven in underdeveloped countries, indicating that non-agricultural occupations have a higher level of output per worker (and greater capacity for consumption) than agricultural activities. Following Lipton, we create a variable that measures the ratio of output per worker in non-agriculture (i.e. non-agriculture GDP/per cent of labour-force in non-agriculture) to output per worker in agriculture (agriculture GDP/per cent of labour-force in agriculture) for 1970.

Female secondary school enrolment. When assessing female labour-force participation, it is important to control for the educational attainment of girls and women. Thus, we include a variable measuring the percentage of *eligible* female secondary school students who are actually enrolled in secondary school in 1970.

Women's industrial employment. Another important control variable in some equations is the percentage of the female labour-force employed in industrial occupations, measured here in 1970. Industrial activities represent one way to bring more women into the wage labour-force.

Fertility-rate. A final control variable is the national fertility-rate in each country in 1970.

Regional effects. Throughout the analysis, we are interested in different regional effects especially for Latin America and Africa. Previous studies have demonstrated that these regions offer a good source of comparison and contrast (see Bradshaw and Wahl, 1991). Our urbanization residuals also underscore the fact that several African and Latin American countries are among the most and least overurbanized in the world. To further investigate these issues, we utilize (1) a dummy variable for sub-Saharan African countries and (2) a dummy variable for Latin American countries (including Caribbean nations). In footnotes 3 and 4, we also report the results of several dummy slope variables that investigate whether particular variables exert a different effect in Africa and Latin America than in the rest of the sample.

Analysis of the Causes of Overurbanization

We follow a deliberate strategy throughout the data analysis by including the original variables from Bradshaw (1987) along with the IMF and debt dependence measures. This enables us to compare the importance of the original variables to the debt-related measures. We also identify influential cases for every equation in the analysis by utilizing a Cook's D test. After identifying the influential cases in every equation, we control for them by including each as a separate dummy variable (see Bradshaw and Huang, 1991). Of course, the influential cases often vary from equation to equation. Failure to control for these cases can severely distort the analysis, especially with such a small sample size.

The analysis begins by examining modernization, urban bias, and dependency/world-system theories of urbanization. Equation one of Table 1.2 lends support only to dependency arguments, as the lagged dependent variable, total foreign investment, and the IMF pressure index have a positive effect on overurbanization. Panama is the only 'influential case' that demonstrates statistical significance in the equation. Equation two drops the non-significant variables from the first equation and also adds a dummy variable for both Africa and Latin America, both of which are positively significant. This indicates that these two continents are more overurbanized relative to other regions.[3] On balance, results from the first two equations suggest that dependency arguments have become more important than in Bradshaw's initial study (1987), which showed support for all three theoretical perspectives. A major reason for this change is the growing pressure applied by the IMF and other institutions.

Equation three substitutes manufacturing investment for total direct foreign investment. It fails to achieve statistical significance, but the effect of IMF pressure is positive and strong. This result also obtains in equation four, where IMF pressure, Africa, Latin America, and two

influential cases (Trinidad and Tobago and Portugal) are significantly positive. Equation five continues the same pattern as earlier, with agricultural investment failing to achieve significance while all other variables remain the same from the standpoint of statistical significance. There is no reason to run a sixth equation, as it would be redundant given the results of the fourth equation, which tests IMF pressure separately.

To summarize, results reported in Table 1.2 reveal that IMF pressure is the primary variable of interest. This variable was not even examined in Bradshaw's (1987) initial study, which utilized data prior to the international debt crisis. Inclusion of IMF pressure eliminates the effect of many other indicators in the analysis. Theoretically, such findings are profound. They suggest that modernization and urban bias arguments have become less relevant as the 'new dependency' of international debt becomes increasingly important.

Analysis of the Effects of Economic Growth

Results reported in Table 1.3 focus on the impact of some of our variables on economic growth between 1970 and 1992. Equation one includes variables for modernization theory, the urban bias perspective, and economic dependency. It also includes a variable for *change* in overurbanization because previous studies have demonstrated convincingly that this variable severely retards economic expansion. (Note that the overurbanization change variable stops in 1987 so that it does not overlap totally with the dependent variable.)

The findings shown in equation one support several theoretical perspectives. The significant positive effects of industrial labour and urban–rural disparity confirm modernization and urban bias arguments, respectively. At the same time, the significant negative coefficients of IMF pressure and overurbanization change substantiate dependency theories. Conditions placed on underdeveloped countries by international financial institutions impede economic growth, as does the increasing rate of overurbanization throughout the developing world. These same findings obtain in equation two, where Latin

TABLE 1.2   Unstandardized Regression Coefficients (top numbers) and Standardized Regression Coefficients (in parentheses) for Determinants of Overurbanization, 1992 (N = 57)

Independent variables	Dependent variable				
	Overurb. 1992	Overurb. 1992	Overurb. 1992	Overurb. 1992	Overurb. 1992
Overurbanization, 1970	.792[a] (.154)	.851[a] (.125)	.697[a] (.142)	.771[a] (.116)	.645[a] (.154)
Industrial labour-force, 1970	.090 (.265)	—	.141 (.260)	—	.247 (.257)
Rural–urban disparity, 1970	.166 (.394)	—	.209 (.395)	—	.412 (.381)
Total investment dependence, 1967	.219[c] (.126)	.197[c] (.104)	—	—	—
Mfg. investment— dependence, 1967	—	.808	(.567)	—	—
Agric. investment dependence, 1967	—	—	—	—	-.370 (.412)
Concentration of agric. exports, 1970	.002 (.055)	—	-.052 (.053)	—	-.019 (.054)
IMF pressure, 1975–1985	.578[c] (.367)	.567[c] (.338)	.744[b] (.354)	.814[b] (.323)	.792[b] (.360)
Debt dependence change, 1975–1985	-.577 (.469)	—	-.481 (.470)	—	-.597 (.480)
Influential cases					
Panama	-27.311[c] (15.811)	-25.932[c] (13.186)	—	—	—
Trinidad and Tobago	12.457 (13.836)	15.811 (11.857)	8.476 (15.576)	27.230[a] (10.682)	20.101[c] (13.343)
Portugal	-16.107 (12.782)	—	-25.291[b] (12.368)	-24.184[b] (12.511)	-24.184[c] (12.511)
Regional effects					
Africa	—	7.786[a] (2.972)	—	8.447[a] (3.019)	—
Latin America	—	7.875[b] (3.482)	—	9.686[b] (3.255)	—
Constant	-4.414 (7.369)	-8.840[a] (2.404)	-1.563 (7.305)	-7.010[a] (2.261)	-2.693 (7.396)
R-squared	.640[a]	.663[a]	.630[a]	.633[a]	.621[a]
Adjusted R-squared	.562[a]	.614[a]	.559[a]	.597[a]	.548[a]

[a] Coefficient is at least 2.5 times its standard error.
[b] Coefficient is at least 2.0 times its standard error.
[c] Coefficient is at least 1.5 times its standard error.

America and Africa also demonstrate a negative impact on economic growth.[4]

This pattern of results has both theoretical and substantive significance. They suggest that failure to consider several perspectives will provide an incomplete explanation of economic growth in today's rapidly changing world. They also suggest, following other recent studies (e.g. Bradshaw and Huang, 1991), that IMF pressure directly inhibits economic expansion. It should also be remembered that IMF pressure indirectly slows economic growth by contributing to

TABLE 1.3   Unstandardized Regression Coefficients (top numbers) and Standardized Regression Coefficients (in parentheses) for Determinants of Economic Growth, 1970–1992 (N = 57)

Independent variables	Dependent variable	
	Economic growth	Economic growth
Economic development, 1970	.003	−.191
	(.304)	(.297)
Industrial labour-force, 1970	.078[b]	.062[c]
	(.034)	(.033)
Rural–urban disparity	.096[b]	.085[c]
	(.047)	(.046)
Overurbanization change, 1970–1987	−.055[a]	−.040[b]
	(.017)	(.017)
Total investment dependence, 1967	−.009	
	(.008)	
IMF pressure, 1975–1985	−.063[c]	−.059[c]
	(.039)	(.037)
Debt dependence change, 1975–1985	−.022	
	(.051)	
Influential cases		
Korea	16.692[a]	16.333[a]
	(1.063)	(1.025)
Regional effects		
Africa	—	−1.141[b]
		(.466)
Latin America	—	−.640[c]
		(.377)
Constant	.004	1.942
	(1.366)	(1.468)
R-squared	.861[a]	.874[a]
Adjusted R-squared	.837[a]	.853[a]

[a] Coefficient is at least 2.5 times its standard error.
[b] Coefficient is at least 2.0 times its standard error.
[c] Coefficient is at least 1.5 times its standard error.
Note: Economic development is logarithmically transformed throughout the analysis.

overurbanization (Table 1.2). Overall, therefore, the unrelenting pressure applied by the IMF and other organizations harms the prospects for balanced development in the Third World.

Analysis of the Causes of Women's Labour-Force Participation

With few exceptions (e.g. Ward, 1984), quantitative cross-national studies have not examined the determinants of women's labour-force participation rates. This is unfortunate because modernization arguments claim that women will become more active in the labour-force when countries have more female education, higher levels of economic development, rapid urbanization, and more industrial opportunities for women. Recent dependency arguments assert that the growing debt crisis will force more and more women into both the formal and especially the informal labour-market. In addition, overurbanization should have a negative impact on women's devel-

opment by placing more pressure on health facilities and other services that disproportionately
impact women (Massolo, 1991).

We test these perspectives in Table 1.4, where
equation one examines modernization-oriented
variables.[5] In addition to the positive effect of the
lagged dependent variable, female industrial
employment and economic development also
exhibit a propitious impact on women's economic activity rates. These findings support
modernization arguments. By contrast, the negative effect of overurbanization change, 1970–85,
contradicts the generally positive effect of modernization theories. Equation two drops female
education (which failed to achieve significance in
equation one) and adds the fertility-rate. The
new variable was not included initially because it
has a high correlation with female school enrolment (r = .791). Fertility-rate fails to achieve statistical significance and the other variables retain
their original direction and are statistically significant.

Equation three includes the significant variables from the previous equation and adds the
dependency indicators: debt dependence change,
IMF pressure, investment dependence, and concentration of agricultural exports. Debt dependence change has a significant positive effect and
agricultural exports exhibit a significant negative
impact, while all the other variables retain their
original direction and level of significance. The
debt dependence indicator suggests that pressure
to service national debts has increased women's
economic activity rates. Not surprisingly, however, concentration of agricultural exports
inhibits women's involvement in the labour-
force. This is consistent with our theoretical discussion asserting that men have greater access to
jobs in the export-crop sector, an especially
important issue for highly indebted countries
forced to export more products to service their
debts. Equation four retains the significant variables from equation three and adds dummy variables for Africa and Latin America. Neither
variable attains statistical significance. The previously significant effect of overurbanization
change also narrowly misses statistical significance in this equation (the regression coefficient
is 1.45 times the standard error).

To conclude, the four equations in Table 1.4
again underscore the importance of examining
different theories of urbanization and development. Both modernization and dependency theories are relevant when studying the determinants
of female labour-force participation. Again, if
the debt crisis had not been examined, the equation would have been misspecified.

Discussion

This essay opened with the proposition that
urban growth characterizes societies undergoing
a transition from an agrarian to an industrial/ser-
vice economy. Our analysis shows that rural
dwellers in the developing world are leaving agricultural activities and migrating to urban areas,
increasing the urban population relative to
national economic development. The determinants of such overurbanization have been the
subject of several articles examining different theoretical perspectives (Kentor, 1981; Timberlake
and Kentor, 1983; London, 1987; Bradshaw,
1987). Although these studies have found support for theories of modernization, economic
dependency, and urban bias, none has examined
the global debt crisis and IMF pressure. Our analysis established that the global debt crisis, specifically pressure applied by the IMF, has a strong
positive impact on overurbanization. Structural
adjustments mandated by the IMF have eroded
the capacity of Third World cities to support
their growing populations.

Pressure from the IMF and growing overurbanization also have a strong negative effect on
economic growth. Structural adjustments and
rapid urban expansion inhibit Third World economic production. Moreover, debt dependence
change facilitates women's labour-force participation and growing overurbanization slows it. In
addition, a number of modernization-oriented
variables also contribute to participation of
women in economic activities, including level of
economic development and women's industrial
employment.

Our analysis offers a number of theoretically
important findings. First, the causes of overurbanization are explained by different strands of
the dependency argument. The positive effect of
total foreign investment supports traditional
dependency arguments, and the positive effect of
IMF pressure supports recent dependency arguments that underscore the growing importance
of the IMF and other transnational financial

TABLE 1.4   Unstandardized Regression Coefficients (top numbers) and Standardized Regression Coefficients (in parentheses) for Determinants of Female Labour-Force Participation, 1985 (N = 55)

Independent variables	Dependent variable			
	Female labour-force partic.	Female labour-force partic.	Female labour-force partic.	Female labour-force partic.
Female labour-force, 1970	.932[a]	.909[a]	.928[a]	.933[a]
	(.029)	(.031)	(.032)	(.033)
Female industrial labour-force, 1970	.180[a]	.164[a]	.206[a]	.187[a]
	(.045)	(.046)	(.049)	(.048)
Female secondary education, 1970	.021	—	—	—
	(.028)			
Fertility, 1970	—	−.089	—	—
		(.308)		
Economic development, 1970	1.269[b]	1.132[b]	1.376[a]	1.044[c]
	(.617)	(.565)	(.502)	(.582)
Overurbanization change, 1970–1985	−.058[c]	−.076[b]	−.102[b]	−.071
	(.039)	(.037)	(.043)	(.048)
Total investment dependence, 1967	—	—	−.005	—
			(.014)	
Concentration of agric. exports, 1970	—	—	−.027[b]	−.025[b]
			(.012)	(.012)
IMF pressure, 1975–1985	—	—	.066	—
			(.102)	
Debt dependence change, 1975–1985	—	—	.213[c]	.221[c]
			(.128)	(.129)
Influential cases				
Côte D'Ivoire	−6.313[a]	—	—	—
	(2.422)			
Jamaica	9.812[a]	11.357[a]	9.443[a]	9.188[a]
	(2.504)	(2.445)	(2.704)	(2.598)
Portugal	—	7.053[a]	—	—
		(2.508)		
Guatemala	—	—	−5.370[b]	−5.476[b]
			(2.676)	(2.683)
Regional effects				
Africa	—	—	—	−1.107
				(1.317)
Latin America	—	—	—	.068
				(.921)
Constant	−8.493[b]	−6.264	−7.622[b]	−5.432
	(3.620)	(4.945)	(3.272)	(3.835)
R-squared	.970[a]	.970[a]	.971[a]	.971[a]
Adjusted R-squared	.965[a]	.966[a]	.965[a]	.965[a]

[a] Coefficient is at least 2.5 times its standard error.
[b] Coefficient is at least 2.0 times its standard error.
[c] Coefficient is at least 1.5 times its standard error.
Note: Economic development is logarithmically transformed throughout the analysis.

institutions. Secondly, the determinants of economic growth are explained by dependency and urban bias arguments. The negative effect of IMF pressure supports recent dependency arguments revolving around the debt crisis; the negative effect of overurbanization change supports dependency and urban bias arguments, which assert that rapid Third World urban expansion will inhibit economic growth. And the positive effect of industrial labour substantiates modernization arguments, which assert that modern-sector employment should facilitate economic growth. Thirdly, the determinants of women's labour-force participation are also explained by different theoretical perspectives. The positive impact of women's industrial employment supports modernization arguments; the positive effect of debt dependence change supports recent dependency arguments; and the generally negative effect of overurbanization change supports dependency and urban theories, which again stress the harmful impact of rapid urbanization.

What do these findings tell us about the study of urbanization and underdevelopment in the world system? More than anything else, they underscore the importance of two points. First, various theories must be utilized to provide a complete explanation of the causes and effects of Third World urbanization. Modernization, urban bias, and dependency theories all receive support at some point in the analysis, although the last two perspectives receive more support than the first theory. Secondly, the findings point to the growing importance of the global debt crisis. If we had not focused on this topic, then our analysis would have distorted the global picture with respect to urbanization and underdevelopment. We encourage other scholars to focus more attention on international indebtedness and related issues.

Appendix

Variables	Sources
Relative urbanization, 1992 Bank,	Level of urbanization and GNP per capita (World Bank 1994)
Relative urbanization, 1987	Level of urbanization (World Bank, 1989a); GNP per capita (World Bank, 1988a)
Relative urbanization, 1985	Level of urbanization (World Bank, 1987); GNP per capita (World Bank, 1987)
Relative urbanization, 1970	Level of urbanization (World Bank, 1984); GNP per capita (World Bank, 1989b)
Industrial labour-force, 1970	World Bank, 1984
Rural–urban disparity, 1970	GDP by industrial origin (World Bank, 1976, 1980, 1984); Composition of the labour-force (World Bank, 1984)
Concentration of agricultural exports, 1970	World Bank, 1984
GNP per capita, 1967	World Bank, 1969
GNP per capita, 1970	World Bank, 1989b
GNP per capita, 1987	World Bank, 1988a
GNP per capita, 1992	World Bank, 1994
All investment variables, 1967	Investment capital (Ballmer-Cao and Scheidegger, 1979); Total GNP (World Bank, 1969)
Debt renegotiations, 1975–1985	World Bank, 1986
Debt restructurings, 1975–1985	IMF, 1985a
Use of extended fund, 1975–1985	IMF, 1985b
Loans/IMF quota, 1985	IMF, 1985b
Debt servicing, 1975, 1985	World Bank, 1988b
Women's total labour-force, 1970	ILO, 1977
Women's total labour-force, 1985	ILO, 1986, 1988
Women's industrial labour-force, 1970	ILO, 1977
Female secondary education, 1970	World Bank 1984
Fertility-rate, 1970	World Bank, 1984

Notes

1. The sample of 57 countries includes: Benin, Burkina Faso, Burundi, Cameroon, Côte D'Ivoire, Ethiopia, Ghana, Kenya, Liberia, Madagascar, Mali, Rwanda, Senegal, Sierra Leone, Somalia,

20   York W. Bradshaw and Rita Noonan

Sudan, Togo, Uganda, Zaire, Zambia, India, Pakistan, Sri Lanka, Indonesia, Republic of Korea, Papua New Guinea, Philippines, Thailand, Argentina, Bolivia, Brazil, Chile, Colombia, Costa Rica, Dominican Republic, Ecuador, El Salvador, Guatemala, Haiti, Honduras, Jamaica, Mexico, Nicaragua, Panama, Paraguay, Peru, Trinidad and Tobago, Uruguay, Venezuela, Algeria, Egypt, Morocco, Syria, Tunisia, Greece, Portugal, and Turkey. Two countries, Papua New Guinea and Haiti, drop out of the analysis of female labour participation in Table 1.4 because of lack of data.
2. Although a very small amount of foreign investment in manufacturing may go to rural areas, it is safe to assume that the overwhelming portion of such investment is allocated to urban regions. It should also be noted that other studies of development have used a measure of foreign investment in manufacturing, but these studies have not focused on urbanization (see Bornschier, 1981; Bornschier and Chase-Dunn, 1985).
3. In an earlier analysis, we created dummy slope variables to test whether IMF pressure has a different effect for Latin America (IMF pressure * Latin America dummy variable) and Africa (IMF pressure * Africa dummy variable). Neither variable demonstrated statistical significance.
4. An earlier analysis again included slope dummy variables for IMF pressure in Latin America and in Africa. They failed to demonstrate statistical significance. Earlier equations also included three other variables when examining the determinants of economic growth: agricultural concentration, manufacturing investment, and agricultural investment. They all had a very weak effect that failed to achieve statistical significance. We therefore excluded these variables from the equation to preserve degrees of freedom.
5. As noted earlier, data limitations require us to use 1985 as the date for the dependent variable.

References

Ballmer-Cao, Thanh-Huyen, and Scheidegger, Jurg (1979), *A Compendium of Data for World-System Analyses*, Volker Bornschier and Peter Heintz (eds.) (Zurich: Sociological Institute of the University of Zurich).

Bates, Robert (1981), *Markets and States in Tropical Africa: The Political Basis of Agricultural Policies* (Berkeley: University of California Press).

Berliner, Joseph (1977), 'Internal Migration: A Comparative Disciplinary View', in Alan Brown and Egon Neuberger (eds.), *Internal Migration: A Comparative Perspective* (New York: Academic Press), 443–61.

Bohrnstedt, George W., and Knoke, David (1988), *Statistics for Social Data Analysis*, 2nd edn. (Itasca, Ill.: Peacock).

Bornschier, Volker (1981), 'Dependent Industrialization in the World Economy: Some Comments and Results Concerning a Recent Debate', *Journal of Conflict Resolution*, 25: 371–400.

—— and Chase-Dunn, Christopher (1985), *Transnational Corporations and Underdevelopment* (New York: Praeger).

Bradshaw, York W. (1987), 'Urbanization and Underdevelopment: A Global Study of Modernization, Urban Bias, and Economic Dependency', *American Sociological Review*, 52: 224–39.

—— and Huang, Jie (1991), 'Intensifying Global Dependency: Foreign Debt, Structural Adjustment, and Third World Development', *Sociological Quarterly*, 32: 321–42.

—— and Wahl, Ana-Maria (1991), 'Foreign Debt Expansion, the International Monetary Fund, and Regional Variation in Third World Poverty', *International Studies Quarterly*, 35: 251–72.

—— Noonan, Rita, Gash, Laura, and Sershen, Claudia Buchmann (1993), 'Borrowing Against the Future: Children and Third World Indebtedness', *Social Forces*, 71: 629–56.

Byres, T. J. (1979), 'Of Neo-Populist Pipe-Dreams: Daedalus in the Third World and the Myth of Urban Bias', *Journal of Peasant Studies*, 6: 210–44.

Delacroix, Jacques, and Ragin, Charles (1981), 'Structural Blockage: A Cross-National Study of Economic Dependency, State Efficacy, and Underdevelopment', *American Journal of Sociology*, 86: 1311–47.

Ellis, Frank (1982), 'Agricultural Price Policy in Tanzania', *World Development*, 10: 263–83.

—— (1983), 'Agricultural Marketing and Peasant-State Transfers in Tanzania', *Journal of Peasant Studies*, 10: 214–42.

—— (1984), 'Relative Agricultural Prices and the Urban Bias Model: A Comparative Analysis of Tanzania and Fiji', *Journal of Development Studies*, 20 (3): 28–51.

Evans, Peter (1979), *Dependent Development: The Alliance of Multinational, State, and Local Capital in Brazil* (Princeton: Princeton University Press).

—— and Timberlake, Michael (1980), 'Dependence, Inequality and the Growth of the Tertiary: A Comparative Analysis of Less Developed Countries', *American Sociological Review*, 45: 531–52.

Firebaugh, Glenn (1992), 'Growth Effects of Foreign and Domestic Investment', *American Journal of Sociology*, 98: 105–30.

Ghandi, Ved (1966), *Tax Burden on Indian Agriculture* (Cambridge, Mass.: Harvard Law School).

Gladwin, Christina H. (1991) (ed.), *Structural*

Adjustment and African Women Farmers (Gainesville: University of Florida Press).

Griffin, Keith (1977), 'Book Review—*Why Poor People Stay Poor: A Study of Urban Bias in World Development* by Michael Lipton', *Journal of Development Studies*, 14 (1): 108–11.

Herring, Ronald, and Edwards, Rex (1983), 'Guaranteeing Employment to the Rural Poor: Social Functions and Class Interests in the Employment Guarantee Scheme in Western India', *World Development*, 11: 575–92.

Hirata, Helena, and Humphrey, John (1991), 'Workers' Response to Job Loss: Female and Male Industrial Workers in Brazil', *World Development*, 19: 671–82.

Hoselitz, Bert (1962), *Sociological Aspects of Economic Growth* (Glencoe, Ill.: Free Press).

ILO (International Labour Office) (1977), *Labour Force Estimates and Projections, 1950–1977* (Geneva: ILO).

—— (1986), *Economically Active Population—Estimates and Projections 1950–2025* (Geneva: ILO).

—— (1988), *Year Book of Labour Statistics* (Geneva: ILO).

IMF (International Monetary Fund) (1985*a*), *Recent Developments in External Debt Restructuring*, Occasional Paper No. 40 (Washington, DC: IMF).

—— (1985*b*), *Annual Report, 1985* (Washington, DC: IMF).

Inkeles, Alex, and Smith, David (1974), *Becoming Modern: Individual Change in Six Developing Countries* (Cambridge, Mass.: Harvard University Press).

Inter-American Development Bank (1990), *Poor, Female and Working in Latin America* (Washington, DC: Inter-American Development Bank).

Kelley, Allen, and Williamson, Jeffrey (1982), 'The Limits to Urban Growth: Suggestions for Macromodeling Third World Economics', *Economic Development and Cultural Change*, 30: 595–623.

—— —— (1984), *What Drives Third World City Growth? A Dynamic General Equilibrium Approach* (Princeton: Princeton University Press).

Kentor, Jeffrey (1981), 'Structural Determinants of Peripheral Urbanization: The Effects of International Dependence', *American Sociological Review*, 46: 201–11.

Ledogar, Robert (1975), *Hungry for Profits: U.S. Food and Drug Multinationals in Latin America* (New York: IDOC/North America).

Lele, Uma (1991), 'Women, Structural Adjustment, and Transformation: Some Lessons and Questions from the African', in Christina H. Gladwin (ed.), *Structural Adjustment and African Women Farmers* (Gainesville: University of Florida Press), 46–80.

Lipton, Michael (1973), 'Transfer of Resources from Agriculture to Non-Agricultural Activities: The Case of India', Communication 109, Institute of Development Studies, University of Sussex.

—— (1977), *Why Poor People Stay Poor: A Study of Urban Bias in World Development* (Cambridge, Mass.: Harvard University Press).

—— (1984), 'Urban Bias Revisited', *Journal of Development Studies*, 20 (3): 139–66.

London, Bruce (1987), 'Structural Determinants of Third World Urban Change: An Ecological and Political Economic Analysis', *American Sociological Review*, 52: 28–43.

Massolo, Alejandra (1991), 'De La Tierra A Los Tortibonos: La Lucha Urbana De Las Mujeres En La Ciudad De México', in María del Carmen Feijoó and Hilda María Herzer (eds.), *Las Mujeres Y La Vida De Las Ciudades* (Buenos Aires: Instituto Internacional de Medio Ambiente Y Desarrollo), 63–90.

Nolan, Peter, and White, Gordon (1984), 'Urban Bias, Rural Bias or State Bias? Urban–Rural Relations in Post-Revolutionary China', *Journal of Development Studies*, 20 (3): 52–81.

Rogers, Andrei, and Williamson, Jeffrey (1982), 'Migration, Urbanization, and Third World Development: An Overview', *Economic Development and Cultural Change*, 30: 463–82.

Seers, Dudley (1977), 'Urban Bias: Seers versus Lipton', Discussion Paper 116, Institute of Development Studies, University of Sussex.

Sershen, Claudia Buchmann (1992), 'The Debt Crisis, Structural Adjustment, and Women's Status: A Cross-National Analysis', MA thesis (Department of Sociology, Indiana University).

Spengler, Joseph, and Myers, George (1977), 'Migration and Socioeconomic Development: Today and Yesterday', in Alan Brown and Egon Neuberger (eds.), *Internal Migration: A Comparative Perspective* (New York: Academic Press), 11–35.

Stark, Oded (1982), 'Research on Rural-to-Urban Migration in LDCs: The Confusion Frontier and Why We Should Pause to Rethink Afresh', *World Development*, 10: 63–70.

Timberlake, Michael, and Kentor, Jeffrey (1983), 'Economic Dependence, Overurbanization and Economic Growth: A Study of Less Developed Countries', *Sociological Quarterly*, 24: 489–507.

Todaro, Michael (1969), 'A Model for Labor Migration and Urban Development in Less Developed Countries', *American Economic Review*, 59: 138–48.

—— (1977), *Economic Development in the Third World* (New York: Longman Press).

UNICEF (United Nations Children's Fund) (1989), *The State of the World's Children, 1989* (Oxford: Oxford University Press).

United Nations (1991), *The World's Women: Trends and Statistics, 1970–1990* (New York: United Nations).

Walton, John (1977), 'Accumulation and Comparative Urban Systems: Theory and Some Tentative Contrasts of Latin America and Africa', *Comparative Urban Research*, 5 (1): 5–18.

Walton, John, and Ragin, Charles (1989), 'Austerity and Dissent: Social Bases of Popular Struggle in Latin America', in William Canak (ed.), *Lost Promises: Debt, Austerity, and Development in Latin America* (Boulder, Colo.: Westview Press), 216–32.

—— —— (1990), 'Global and National Sources of Political Protest: Third World Responses to the Debt Crisis', *American Sociological Review*, 55: 876–90.

Ward, Kathryn B. (1984), *Women in the World-System: Its Impact on Status and Fertility* (New York: Praeger).

Wellisz, Stanislaw (1971), 'Economic Development and Urbanization', in Leo Jakobson and Ved Prakash (eds.), *Urbanization and National Development* (Beverly Hills, Calif.: Sage), 39–55.

World Bank (1969), *World Bank Atlas 1969* (Washington, DC: World Bank).

—— (1976), *World Tables*, 1st edn. (Baltimore: Johns Hopkins University Press).

—— (1980), *World Tables*, 2nd edn. (Baltimore: Johns Hopkins University Press).

—— (1984), *World Tables*, 3rd edn. (Baltimore: Johns Hopkins University Press).

—— (1986), *World Debt Tables: External Debt of Developing Countries, 1985–86* (Washington, DC: World Bank).

—— (1987), *World Development Report, 1987* (Oxford: Oxford University Press).

—— (1988a), *World Bank Atlas, 1988* (Washington, DC: World Bank).

—— (1988b), *World Debt Tables: External Debt of Developing Countries, 1987–88* (Washington, DC: World Bank).

—— (1989a), *World Development Report, 1989* (Oxford: Oxford University Press).

—— (1989b), *World Tables, 1988–89* (Baltimore: Johns Hopkins University Press).

—— (1994), *World Development Report, 1994* (Oxford: Oxford University Press).

2

The Rise and Demise of Urban-Biased Development Policies in Africa

Michael F. Lofchie

HUMAN suffering supplies the imagery that continues to attract global attention to the problem of sub-Saharan Africa's dismal economic performance during the past thirty years. The cold statistics that provide a quantitative portrait of this problem are widely available in a host of official and academic studies. Agricultural stagnation has been the core problem. Nearly fifteen years ago, the Economic Research Service of the US Department of Agriculture published a report which showed that, during the roughly twenty-year period from the early 1960s to the end of the 1970s, food production per capita fell at the average rate of about 1 per cent per year.[1] The inevitable results were sharp increases in food imports that consumed scarce reserves of foreign exchange and, in countries unable to acquire food from international sources, intermittent famine.

During the same period, sub-Saharan Africa also experienced stagnation in its exports of primary agricultural commodities with the result that its share of world trade in these goods declined by about one-third.[2] This trend led the World Bank to conclude that for the majority of African countries poor export crop production accounted for a greater proportion of falling export earnings than more commonly cited external factors such as declining world prices for agricultural goods. Recent research indicates that the tendency towards poor agricultural performance continued unabated during the 1980s. Between 1980 and 1989, total agricultural production per capita fell approximately 8 per cent and food production per capita, about 9 per cent.[3]

Even these figures do not tell the entire story. Where agricultural growth had occurred, it was almost entirely concentrated in a small number of countries, such as Kenya and Côte d'Ivoire, that were the continent's few 'success stories'. In continent-wide aggregate figures, the agricultural growth of a few successful countries obscured the full magnitude of decline elsewhere.

Though less visually dramatic than faltering food production, Africa's loss of world market share in agricultural exports was more harmful. Since the majority of African countries depend upon agricultural exports to finance their imports, the effects of stagnant export revenues permeated every aspect of political and economic life. Scarcities of hard currency contributed to the deterioration of physical infrastructure because it became increasingly difficult to import construction equipment or rolling stock. Educational systems deteriorated for lack of instructional materials and classroom supplies; medical care deteriorated because of scarcities of equipment and medications; and urban industries operated at lower and lower levels of installed capacity as a result of persistent shortages of capital goods, energy, and raw materials.

The author wishes to thank Peter Kilby for his extremely helpful comments on an early draft of this essay.

Poor agricultural performance touched off a vicious cycle of decline. As export earnings stagnated, the resultant worsening of physical infrastructure became an ever greater constraint on the production and marketing of agricultural products. Scarcity of hard currency also lowered the ability to import fertilizers and herbicides, thereby adding input scarcities to the economics of declining production. Inadequate export earnings also had ripple effects on food production. Contrary to the widespread imagery that views Africa as capable of self-sufficiency in food production, the performance of this sector had become highly dependent upon imported technology such as transportation and processing equipment, chemical inputs, and modern facilities for storage and packaging.[4]

Agricultural decline contributed to Africa's political difficulties. Civil wars occurred where rival élites fought over control of diminishing economic resources, and political repression intensified as incumbent regimes struggled to maintain power. Corruption also assumed growing proportions as individual officeholders sought to extract personal wealth in the midst of general decline. In the most deeply afflicted societies such as Liberia, Ethiopia, Somalia, Sudan, Angola, and Mozambique, the fabric of civil society became so severely shredded that political recovery, which must precede and lay the basis for economic recovery, seemed all but impossible.

Urban Bias

The basic cause of Africa's poor agricultural performance has been a systematic tendency towards urban bias in development policy, the tendency to 'squeeze agriculture' for resources to finance urban industrial development.[5] This bias grew out of a conviction that agriculture could not contribute to sustained development and that the key to modernization lay in attaining rapid industrialization. Many of Africa's first-generation political leaders believed that the agricultural sector had to be subjected to an economic squeeze so that it would yield up the economic resources necessary to finance an industrial revolution.[6] For some of these leaders, the goal of rapid industrialization seemed to rule out the Western experience of industrialization as a relevant model because this had been such a slow process, sometimes requiring several centuries to be complete. The Soviet experience of the 1930s, with its seemingly successful imposition of the squeeze agriculture approach and its emphasis on the centrality of the state in the allocation of economic resources, appeared more appropriate.[7]

During the early 1980s, however, this approach came under attack and today the picture of Africa as a continent where inappropriate agricultural policy contributes to broader economic malaise is no longer an accurate one. Efforts at reform of development policy, loosely grouped under the rubric 'structural adjustment', are presently underway in a host of countries. The core of the reform effort has been an attempt to move away from the anti-agricultural bias of the early post-independence period, towards a policy framework that treats the agricultural sector in a less repressive manner.[8] Although it would be premature to suggest that development policy in Africa has now come full circle, returning to the colonial emphasis on agricultural exports, conceptions of development that subordinate agriculture to the needs of urban industrial growth are now considered discredited.

To understand the evolution of development policy in Africa, it is useful to pursue three questions. What was the urban bias in development policy? Why was the bias against agriculture first introduced and why was it continued for so long after its adverse economic effects had become apparent? Why did the region's agricultural policies begin to change in the early to mid-1980s?

Urban Bias in Development Policy

During the two decades following independence, the bias against agriculture was evident throughout a wide range of government policies. It was a glaring feature of systems of taxation designed to extract as much revenue as possible from agricultural producers. It was also apparent in the tendency to concentrate public services in the large cities, especially national capitals. The bias against agriculture also provided the underlying rationale for central planning of national economies which was justified on the premiss that the necessary transfer of resources from agriculture to non-agricultural activities would not take place without governmental intervention.

The policies employed to squeeze the agricultural sector were first systematically identified in the World Bank's 1981 report generally known as the 'Berg Report'.[9] These were monetary and tax policy, reliance upon statist systems of crop procurement and marketing, and industrial and trade policy. The cumulative effect of policies in these areas was a tendency to impose upon agricultural producers a burden of implicit and explicit taxation so great that it became a strong disincentive to production.

Monetary and Tax Policy

The greatest burden was borne by the producers of exports. The principal tax derived from a continent-wide tendency towards a monetary policy characterized by currency overvaluation.[10] This practice was so widespread and, in many cases, so extreme that it has become almost universally identified as the litmus test of urban bias in development policy.[11] There is ample basis for this view. Overvaluation is an implicit tax on the producers of agricultural exports, since it reduces the prices (in domestic currency) of the goods they seek to sell on world markets.

African governments also imposed various explicit forms of taxation on agriculture. The most common were export duties, but there were a variety of other taxes including port charges, processing fees, and document and licensing costs. As a result of these various taxes and charges, a permanent and sizeable wedge was driven between the world price of exports and the prices received by producers. Because of its disincentive effects, this wedge could be considered the fundamental policy error of the post-independence period.

Overvaluation also had serious effects on food production. By lowering the domestic prices of imported food items, especially imported cereals or other goods that could be easily substituted for locally produced food staples, it compelled local food producers to sell their products at prices that often failed to provide an adequate economic incentive. Overvaluation also suppressed food prices in most subtle ways. Sometimes the producers of export crops, discouraged by the low prices they were receiving, abandoned the production of exports and began to produce food crops that could be sold directly for cash.

Crop Procurement and Marketing

The disincentive effect of urban bias in monetary policy was exacerbated by the manner in which the majority of African governments chose to procure and market their most important agricultural commodities. The chosen institutional vehicle for these purposes was the government marketing board. During the late colonial period, governmental marketing agencies had been given monopoly status to engage in the purchasing and vending of the most important export crops and, following independence, this practice was typically extended to domestic food staples.

The official rationale for the marketing board system had to do with the opportunity such agencies would provide for price stabilization (since the international prices of primary agricultural commodities tend to fluctuate), and with the need to protect highly vulnerable small-scale agricultural producers from exploitation by predatory private traders. The marketing boards were further justified on the basis of a concept of market failure: markets, it was widely believed, would fail to provide an adequate supply of public goods such as agricultural research.

Whatever the official rationale of the marketing board system, its effects were generally the opposite. In practice, price stabilization meant price suppression as these boards were used to force prices down to levels far lower than might have prevailed under more open marketing arrangements. And the idea of protecting vulnerable agriculturists from economic exploitation became a mockery: peasant farmers everywhere in Africa were the victims of the ubiquitous tendency towards bureaucratic corruption on the part of marketing board employees. Farmers often had to pay bribes at each level of the crop processing system. In addition, the marketing boards often treated the essential services they were obligated to perform, such as research or input provision, either as additional sources of rent-seeking or with cavalier indifference. And their payments to farmers were frequently months if not years in arrears.

The agricultural marketing boards furnished one of Africa's most glaring examples of urban bias in operation. At the same time that they were lowering rural incomes by suppressing producer prices, they also became major sources of high paid urban employment. Since, as government monopolies, they had no incentive to keep their costs under control, they could engage in rampant overstaffing, pay high salaries, and offer expensive emoluments such as overseas 'fellowships' to their personnel. They could also distribute lucrative contracts to politically favoured sub-contractors who did the actual work of processing, packing, storing, and transporting of crops.

The greatest cost of the marketing board system, however, may have been hidden: a tendency to prevent the flexibility necessary to take advantage of new niches of opportunity in the agricultural market-place. A marketing board tends to have a vested interest in entrenching production of the crops it controls. For this is essential if these boards are to justify their scale of operation and establish a political rationale for increases in the size and benefit levels of their operational staff. By its very nature, then, a bureaucratic approach to crop marketing tends to introduce an element of inflexibility into a country's pattern of agricultural production, preventing short- or long-term innovations in such economically vital areas as crop mix. Africa's marketing boards tended to lock in place the production of crops that are overproduced for domestic and world markets.[12] The costs of this inflexibility are all but impossible to calculate for this would require an estimate of what Africa's export levels might have been under a more open and competitive system.

Industrial and Trade Policy

Africa's bias against agriculture is also reflected in the industrial and trade policies African governments tended to implement following independence. The most widely adopted approach to industrial development was import-substituting industrialization (ISI). The ISI approach, which had been attempted earlier in Latin America and some countries in Asia, was based upon a simple and rather compelling idea; namely, that it was possible to industrialize by launching local production of commodities that were imported in volumes large enough to sustain a domestic industry. The ISI strategy typically called for industrialization to begin with easily produced consumer goods such as textiles, soft drinks, cig-

arettes, and footwear, and then to move towards production goods such as agricultural or industrial machinery.

ISI policies added a further layer of discrimination against agriculture by requiring trade policies that provided high levels of protection for industries oriented towards production for the domestic market. Since Africa's import-substituting industries were generally of the 'easy' import-substitution variety, producing consumer items such as textiles, soap, soft drinks, beer, and cigarettes, the high levels of protection they enjoyed tended to raise the retail prices of these goods.[13] The burden of paying for higher priced but lower quality ISI goods tended to fall on the nation's consumers, a majority of whom were small-scale rural farmers.

Protected industries can be likened to agricultural marketing boards as a tangible manifestation of urban bias in operation. Not only were they capitalized and operated largely on the basis of the hard currency earnings from agricultural exports, but the real wage levels of both workers and management were usually far higher than the incomes of peasant families. Indeed, the wage differential tended to be so great that the ISI strategy of development explains why Africa had, at one and the same time, the world's highest rates of urbanization and labour shortages during periods of peak need in the countryside.[14]

The Causes of Urban Bias in African Development

There are two very different explanations of Africa's policy bias against agriculture. One identifies the origins of economic policy almost wholly in the political realm, asserting that Africa's post-independence economic framework can best be viewed as the outcome of powerful urban pressure groups such as civil servants and industrial workers. A second explanation looks primarily towards intellectual currents, tracing the bias against agriculture to the influence of a school of economic thought that enjoyed great intellectual stature in developing countries during the 1950s and 1960s; namely, development economics.[15]

The Political Explanation

Scholarly discussion of the political origins of urban bias in sub-Saharan Africa has been dominated for more than a decade by Robert Bates's classic work, *Markets and States in Tropical Africa*.[16] Since its publication in 1981, this book has provided our profession's definitive understanding of the relationship between political factors and the policies African governments adopted towards their agricultural sectors. Against the background of Africa's calamitous economic decline following independence, its provocative opening question—'why should reasonable men adopt public policies that have harmful consequences for the societies they govern?'[17]—remains as fresh as ever.

Bates argues that agricultural policy in Africa was decisively shaped by powerful urban economic interests. In his view, the policy framework just described can be best understood as a political means of transferring resources away from agricultural producers, who lacked the ability to pressure their governments effectively, to a variety of urban groups which, though heterogeneous, could organize more effectively and had a common political interest extracting resources from agriculture. The city was relatively well-to-do because it was powerful; the countryside, poor because it was powerless.

According to Bates, the reasons for Africa's approach to agricultural policy were overwhelmingly political: governmental élites were far more dependent for their day-to-day political survival upon the support of urban constituents than upon that of politically weaker rural dwellers. To build and maintain a viable coalition of urban interests required economic resources. And generating economic benefits for urban clienteles meant that economic costs had to be imposed upon the countryside. Industries benefited from overvalued exchange rates that artificially lowered the cost of capital goods and other inputs; urban workers and civil servants benefited from policies that boosted employment and lowered the cost of food staples and imported necessities; and legions of economically needy political supporters benefited from the seemingly infinite array of jobs created in ministries, parastatal corporations, and protected industries. The burden of these resource transfers was borne by rural populations who

seemed to lack the capacity to mobilize effective political opposition.

The analysis Bates developed in *Markets and States* is very much in the tradition of the urban bias approach first pioneered by Michael Lipton in the mid-1970s.[18] These two highly influential authors share a set of presuppositions. Both argue that the most potent interest groups in developing countries are located in the major cities, principally national capitals, and both feel that governments concerned with political survival must accommodate the economic interests of urban workers, managers, and civil servants.

The sources of political power for these urban pressure groups are numerous but the single most important is the factor of sheer geography. Since urban interests are in direct physical proximity to the power centres of governments, they cannot only convey their demands to government more easily, but with almost equal ease, threaten the survival of governments through strikes, demonstrations, riots, or other forms of political protest. For rural populations, on the other hand, geography is often a decisive handicap to political action. Rural dwellers are not only spread across vast regions, making organization more difficult and expensive, but have commensurately greater difficulty gaining access to the urban centres where government offices are located.

Bates's argument was so compelling that, throughout most of the 1980s, it proved highly immune to its critics.[19] By the end of the decade, however, it became clear that his paradigm would benefit from qualification and reformulation. Among the most telling critics of that paradigm, not surprisingly, was Bates himself who, in 1991, presented an inventory of theoretical issues he considered still unresolved.[20] One of the most important of these had to do with the need for further clarification of the relationship between urban pressures and post-independence public policy. Acknowledging the importance of empirical findings that contradicted portions of his original argument, 'countervailing facts', Bates's essay called for a 'refinement of interest group theories of [African] politics'.[21]

There were at least two significant weaknesses in Bates's original formulation. The first was its inability to deal with exceptional cases. Like any highly generalized paradigm, the Bates approach was most powerful in explaining the pattern of development policy in countries that conformed fairly closely to the continent-wide pattern, but it had great difficulty in explaining countries that were exceptional.

Exceptional Cases

The tendency towards a policy bias against agriculture was not universal and a few countries stand out as exceptions to the general trend. Of these, the most important are Côte d'Ivoire and Kenya, both of which came to be known as African 'success stories' during the 1960s and 1970s. Agricultural growth in Côte d'Ivoire was so rapid that this country came to be widely known as the Ivorien Miracle. Throughout the two decades from 1960–1982, its agricultural growth rate averaged almost 4.5 per cent.[22] By the early 1980s, Côte d'Ivoire had become the world's largest exporter of cocoa and one of the world's top five coffee exporters. Perhaps more importantly, Côte d'Ivoire's ability to generate high levels of foreign-exchange earnings through consistent increases in exports of high-value commodities enabled it to sustain very high rates of industrial growth which, during the period 1960–70 averaged over 11 per cent and, during the period 1970–82, about 8.5 per cent.[23]

Kenya's success in attaining high levels of agricultural and industrial growth during this period was no less dramatic. From 1965 to 1973, its agricultural growth rate was 6.2 per cent, and from 1973 to 1983, 3.4 per cent.[24] By the end of the 1970s, Kenya had become one of the world's largest exporters of high-quality tea and a major source of premium-grade coffee. As was the case with Côte d'Ivoire, the foreign-exchange earnings generated by rising levels of high-value exports permitted a rapid growth of the industrial sector, which, during these two periods, grew at rates of 12.4 and 5.3 per cent, respectively.

A common denominator linking these two countries was avoidance of the squeeze agriculture approach. Unlike the majority of African countries, both had maintained prudent exchange-rate policies that prevented an overvaluation 'tax' on agricultural exporters.[25] While both governments had introduced parastatal systems of crop procurement and marketing for their major agricultural exports, their parastatals seemed far less susceptible to the problems of inefficient management that were so ubiquitous elsewhere on the continent. And while both governments also favoured an industrial strategy of import-substitution, both had pursued this approach to development in a moderate manner that balanced industrial protection with a sensitivity to the need to provide economic incentives to agricultural producers.

The theoretical significance of the exceptional cases is considerable. Taken together, they suggest that the urban bias approach in its unqualified form ignores a potentially more important factor, the economic interests of the politically dominant class. In both Côte d'Ivoire and Kenya, the political leaders of the governing parties were large-scale landowners, heavily invested in plantation-scale production of exportable commodities. As a result, their class interests operated as a fundamental constraint on the pursuit of policies intended to shift economic wealth from the agricultural to the industrial sector.[26] The economic interests of the Kenyan and Ivorien governing élites lay, instead, in pursuing policies that facilitated the on-going profitability and growth of agricultural exports. This interest led to policies that produced a rural, not urban bias.

In the end, however, the rural sector bias of Côte d'Ivoire and Kenya only underscores the far more pervasive tendency towards urban bias in the majority of African nations. The fact that these two countries avoided an extreme urban bias approach only heightens the importance of refining our understanding of why the majority of other African countries behaved so differently.

Bates's political explanation, while providing a useful point of departure, is not entirely helpful. For his paradigm suffers from a second critical weakness, the fact that it makes no distinction between policy origination—the causal factors that first bring a policy into being—and policy persistence—the factors that explain why a policy remains in place over an extended time. Bates had probably intended his theory as an explanation of both. But a cursory examination of the pattern of interest-group pressures immediately following independence suggests that these were neither as uni-directional nor as politically decisive as Bates's original argument suggested.

Policy Origination versus Policy Persistence

An alternative viewpoint has been suggested by Merilee Grindle who rejects the premiss that urban pressure groups with an interest in industrial protectionism were politically decisive in the post-independence period. She believes that there were interest-group pressures in multiple directions, both for and against agricultural exports. In addition, she believes, the political leaders of Africa's newly independent countries were sufficiently popular that they enjoyed considerable latitude in their choice of development policy.[27] Describing these leaders as 'relatively autonomous at the outset', Grindle raises the classic question of direction of causality. Did the interests of urban pressure groups give rise to the policy? Or did the policy create the clienteles?

She argues that the direction of causality was contrary to that suggested by Bates. Having considerable freedom of choice, African leaders adopted economic policies that, in their judgement, offered the best prospect of sustained growth. These policies, over time, fostered the emergence of a range of urban interest groups which subsequently applied pressure for the perpetuation and enlargement of the policy framework.

Much historical evidence supports Grindle's interpretation. At independence, some of the most powerful interest groups in African societies were associated with export-oriented sectors and thus had economic interests in continuing the relatively open trade policies installed by the colonial powers. In a number of cases, for example, export-oriented farmers had powerful organizations to sponsor their interests.[28] There was also a formidable array of non-agricultural interests attached to the extractive sectors. The African working class, for example, included large numbers of workers dependent upon an export orientation: miners, plantation labourers, railroad and dock workers, and workers involved in the processing of agricultural exports. Such groups often provided the core membership of Africa's trade-union movements and their economic interests clearly lay in a set of policies that would promote the well-being of rurally-based exporters.

This profile of interest-group pressures at the time of independence requires a dramatic revision in our understanding of the origins of Africa's anti-agricultural bias. Far from being an outcome of urban pressures, the import-substituting development strategy with its cumbersome baggage of agricultural taxation and trade restrictions sometimes had to be forcibly imposed against the interests of groups whose livelihood depended upon an export orientation.[29] This helps explain why newly independent African governments so often clashed with their trade-union movements and why, in two of the best documented cases of government–worker relationships, Ghana and Tanzania, the ISI economic framework required considerable repression not only of farmer interests but of important trade-union organizations as well.[30]

Bates's argument requires reformulation. His approach does not provide an explanation of the origins of anti-agricultural policy in Africa but it does help provide an explanation for why those policies persisted for so long. Once African governments had implemented a set of policies designed to stimulate urban industrial development, they created a whole array of interest groups that depended upon an urban-biased policy framework.

Over and above these two weaknesses, there is one additional difficulty with a purely political analysis of Africa's anti-agricultural policy bias: its tendency to portray political leaders in somewhat cynical terms as motivated solely by political power. Pressure-group analysis conveys the impression that the perversity of economic policy made little difference to governmental leaders. It would be unfair if that impression were allowed to stand uncorrected. Africa's urban-industrial development strategy may have had some basis in urban political pressures, but it was also solidly grounded in the dominant economic theory of the period.

The Role of Development Economics

Bates's approach, then, does not provide a satisfactory explanation of the origins of urban bias in contemporary Africa. To find such an explanation, it is essential to venture beyond the political realm, to the realm of economic thought. Ideas matter, sometimes greatly. For they inspire and empower political choice. The adoption of an urban-industrial approach to development can be better explained as a product of a set of economic ideas that thoroughly dominated discussions of development during the 1950s and 1960s. These ideas, generally termed 'development economics', tended to portray the 'squeeze agriculture' approach as offering the best prospect of overall economic development.

At the risk of vast oversimplification, it could be said that the sub-discipline of Development Economics rested on one essential conviction; namely, that economic relationships between industrially developed countries and poor, primarily agricultural societies operated to the detriment of the latter group. The intellectual core of development economics was the concept of unequal exchange, sometimes referred to as 'declining terms of trade'. This concept held that the real prices of the manufactured goods exported by industrial nations would increase over time relative to those of the primary agricultural commodities exported by underdeveloped nations. Underdeveloped countries would have to pay more and more for their imports while receiving less and less in purchasing power for their exports. As a result of this inherent inequality, wealth would flow from the poorer nations to the wealthier ones, resulting in growing prosperity for the industrial nations and enduring poverty for the rest.

Development economists believed emphatically that exports of primary agricultural commodities would not generate sustained economic growth. They held that the best hope of attaining growth lay in an economic transformation, away from the primacy of agriculture and towards the primacy of industrial production. Since market forces could not be depended upon to induce this transformation, the development economists concluded that government would have to play a major role, principally in providing a nurturing environment for new industries. As a result, they became advocates of the ISI development strategy.

The development economists' policy prescriptions had a powerful influence on the governments of newly independent African countries. This influence was transmitted in a variety of ways. Development economists were often active as expatriate economic advisers to African nations, sent there by such prestigious organizations as the Ford and Rockefeller Foundations or by international aid agencies. They were also

extremely influential within international lending institutions such as the World Bank that enjoyed financial leverage over the trajectory of post-independence development. In addition, development economists often taught in the economics departments of African universities where they had considerable influence over the academic curriculum. In Western universities, African students in the social sciences were often directed towards classes in development economics which, it was uncritically assumed, were more relevant to their countries than orthodox Ricardian ideas. Through these varying lines of influence, African heads of state and senior economic personnel became intensely exposed to the ideas of the development economists and to the conviction that agriculture should be sacrificed to the needs of industrial development. The influence of these ideas, then, provides an explanation for the policy choices of African governments.

Trade Pessimism

Development economics has sometimes been referred to euphemistically as 'trade pessimism' in order to highlight its intellectual departure from the classical Ricardian doctrine that trade conferred mutual benefits on all trading partners and to underscore the distinctive belief that trade in primary commodities would only contribute to the persistent poverty of agricultural countries.[31] Although the case for this conviction consisted of several building-blocks, its foundation stone was the work of Hollis Chenery, who was, for many years, a Vice-President for Development Policy at the World Bank. His most important argument against agricultural exports was that these would not contribute to diversified economic growth and, in particular, to industrial development.[32] For this to occur, economic resources must be deliberately withdrawn from presently dynamic sectors (i.e. the production of agricultural exports), for investment in sectors that had greater future potential.

Declining Terms of Trade

The development economists' pessimism about trade was most explicit in its insistence upon the declining terms of trade. This was possibly the most commonly cited building-block in the intellectual case against those who favoured policies to increase agricultural exports. During the 1950s and 1960s, there was a vast outpouring of economic research to show that the real prices of primary agricultural commodities had consistently declined relative to those of manufactured goods. Many of the development economists sought, in their empirical research, to document a decline in the terms of international trade and to show that this represented a major transfer of economic resources away from the world's poorer nations to the richer ones.[33]

Among the foremost proponents of this notion was the Jamaican economist W. Arthur Lewis who, during the late 1950s and early 1960s, was an influential economic adviser to the government of Ghana. Lewis put the idea in the following terms.

On the demand side, the world just has all the tea, coffee, cocoa and sugar it wants for the time being: more precisely, the demand for these products is growing slowly. The industrial countries have also developed synthetic substitutes for rubber and cotton, which have eaten into the markets for these two. So the countries which specialize in producing tea, cocoa, coffee, cotton and rubber are trying to increase their sales faster than world demand increases.[34]

Lewis was among the most influential advocates of ISI in Africa.

Agricultural Pessimism

The development economists' pessimism about agricultural trade was based upon doubts about the possibility of productivity increases in the agricultural sector. Ironically, the source of this concern was a psychological interpretation of peasant behaviour. The core of peasant culture was considered to be a tendency towards risk aversion. Development economists who accepted this notion sought assiduously to show that peasant agricultural practices were strictly guided by the principle of avoiding risk and that decisions about such economically vital matters as crop mix and productive technology were based upon a determination to avoid crops that had a high probability of failure rather than on the possibility of increasing profit and income.[35] By impli-

cation, the possibilities of growth through agricultural innovation were poor.

Low Marginal Productivity of Labour

A number of the development economists, most prominently Lewis himself, believed that the shift of economic resources, especially labour, from agriculture to industry could be pursued without major detrimental effects on the agricultural sector. The reason was that Africa's rural areas had abundant supplies of unused or under-utilized workers. Since the marginal productivity of labour in agriculture was very low, the effect of moving workers from agriculture to industry would be minimal.[36] Lewis believed that the most pressing challenge of development was to find ways to shift labour away from the agricultural sector to manufacturing activities where their productivity would be greater.

The conclusion was unmistakable: to break out of the cycle of underdevelopment meant, somehow, to break free of a dependence on primary agricultural production. To do so required that industries be launched. The obvious means of doing so was to establish protected markets that would provide insulation from global competition. The only questions that remained were institutional: how best to structure the new industries, and how to effect the transfer of resources, especially capital, from agriculture to industry.

The first of these questions required an assessment of the developmental potential of the private sector. For many of the development economists, competitive private enterprise did not seem to be a viable approach in Africa. Not only were internal markets too small to sustain economic competition between firms, but there was a scarcity of entrepreneurial experience. State enterprise seemed to be the most promising alternative. An influential voice for this idea was Barbara Ward who, like Lewis, functioned for a time as an economic adviser to the government of Ghana.[37]

The second of the institutional questions was more challenging: how to tax the rural sector in a way that would minimize the likelihood of rural political protest. The difficulty lay in the fact that Africa's peasant farmers seemed to present a segment of society that would be extremely difficult to tax. Not only did individual peasant farmers have low levels of cash income that could be easily hidden from the tax system but, since they are widely scattered over large areas, conventional methods of taxation such as income taxes would require a massive bureaucracy that would inevitably prove highly expensive and vulnerable to corruption.

Taxation by overvaluation of the exchange rate seemed to solve these problems. It involved only minimal bureaucratic difficulties, since it did not require the creation of a large taxation bureaucracy. And, because it captured the foreign-exchange earnings from agriculture directly in the Central Bank, it was difficult for export farmers to evade. Overvaluation also seemed to offer an important political advantage. Since it is relatively invisible, hidden from public view in the opaque accounting mechanism central banks employ to handle foreign-exchange transactions, it seemed less likely to generate political protest. Perhaps most importantly, overvaluation offered an efficient way to transfer the foreign-exchange earnings from agricultural exports directly to the needs of the new industrial sector, since it would lower the cost of capital goods, raw materials, and other imported inputs.

In sum, the ideas of the development economists presented a formidable case for imposing a heavy burden of taxes on agricultural exports in order to implement an import-substituting strategy. Against the background of doubt as to the explanatory value of the urban bias/interest group approach to the origins of post-independence economic policy, the intellectual influence of this economic school provides a powerful explanation for why so many African governments opted for a development strategy based upon a policy bias against the agricultural sector.

Why Did the Policy Framework Change?

The great difficulty with having identified persuasive explanations of why a policy originates and why it then tends to persist over time is the risk of being theoretically empty-handed when it comes to explaining rapid policy change. Yet, the dominant political reality of sub-Saharan Africa

in the 1980s has been the breathtaking speed with which the urban-biased policy framework is being dismantled. African governments are actively disassociating themselves from this strategy and replacing it with liberal, market-based approaches to growth that are economically neutral as between urban and rural populations.[38]

One by one, the policy building-blocks of the old urban-biased development system are being dislodged. Among the first to be removed is the practice of currency overvaluation. Virtually all African countries, including those in the CFA zone, are devaluing their currencies. In some cases, governments are institutionalizing freely floating exchange rates as a precaution against a recurrence of overvaluation.[39] The tendency to offer protection to loss-producing urban enterprises that sap agricultural wealth is also disappearing quickly. Indeed, among other benefits, trade liberalization has meant that the import requirements of the agricultural sector are no longer assigned a lower priority than those for urban producers and consumers.

It would be premature to suggest that urban bias is altogether a thing of the past. But the tendency to treat agricultural wealth as a source of finance capital for urban industries, as a source of subsidies for urban consumers or as a source of revenue to pay high salaries to urban civil servants is rapidly losing its salience. Numerous governments are taking rapid steps towards liberalization of agricultural markets. Producer prices are being rapidly decontrolled and producers are increasingly free to market their goods through either state or private marketing systems. Most African governments have also reduced the size and role of state marketing agencies, sometimes eliminating once powerful parastatal marketing bodies, while others have sought to supplement economic changes with greater opportunities for rural political participation.

The process of economic liberalization is so widespread that it raises questions of profound importance for students of policy reform. The most difficult of these is the very basic issue of why governments change their policy framework. For certain countries, a superficial answer is readily available. In Ghana, for example, a puritanical military regime replaced a corrupt civilian government and also swept aside an entire older generation of military officers which had governed the country throughout the 1970s. Here, reform took place because a reform-minded group of leaders forcibly replaced one that was disinclined towards change.

But in many other countries, no such abrupt succession occurred and important reforms began without sudden changes in political leadership. A persuasive explanation of policy change, therefore, must rest on more robust explanatory principles than idiosyncratic events such as a change of regime. The best answer to the question of 'why change?' is that there is no single answer. A variety of factors, both external and internal, have come into play and, at some moment, converged in their effects.

Among the sources of policy change are the changing realities of the real world. The most salient of these within Africa is the depth of the continent's economic crisis. In country after country, economic hardship has eroded popular faith in the previous policy framework and in the politicians who implemented it. Economic difficulties provoke demands for change and for leaders who will bring it about especially among those who are experiencing real declines in their material well-being. Thus, an economic crisis tends to bring about the dissolution of the urban political coalition that froze the previous policy in place for so long.

Real-world events outside Africa have also played a prominent role. Among these, the collapse of the Soviet political and economic system has been monumentally consequential. In the late 1950s and early 1960s, the Soviet model may have represented a credible alternative for countries seeking rapid industrialization and may have provided empirical reinforcement for those who favoured the squeeze agriculture model. By the late 1980s, it no longer performed either function. In addition, the rise of the Asian economies provided compelling evidence that countries which oriented their economic policies towards production for international trade were vastly outperforming those which sought growth through domestic protectionism.

Of the many forces for policy change, however, one of the most powerful is exactly the same as that which explained the choice of initial policies: the influence of economic ideas in first inspiring and then empowering political choices. The essential point here is utterly simple: during the mid-1970s, the economics profession underwent a profound paradigmatic change. The trade and agricultural pessimism of the development economists came under severe criticism from orthodox economists who believed in the universality of a positive correlation between trade and economic growth.[40] By the end of the decade, an intellectual revolution had taken place. Development economics as a theory advocating industrial growth through heavy taxes on agriculture and state sponsorship of protected urban industries had largely been discredited.[41]

The continent-wide trend towards liberalization, then, has been prompted in large measure by the impact of a new set of intellectual convictions about the economic sources of development. As a result of this intellectual revolution, approaches to development based on statist institutions and industrial protectionism have begun to come to an end, not only in Africa but throughout the developing world, because they have been intellectually discredited.

During the 1980s, countless Africans including innumerable students, scholars, journalists, bureaucrats, and politicians had been personally exposed to the contrast in economic performance between statist and market-based systems. They could draw only one possible conclusion about the policy changes necessary to rekindle a process of economic growth in their own societies. Many would also have been exposed to the intellectual revolution in economic ideas that, at one and the same time, launched, accompanied, and resulted from the superior performance of economies open to trade.

The list of those to whom this theoretical revolution is intellectually indebted is extensive but a few scholars stand out, most particularly Jagdish Bhagwati, Anne O. Krueger, and Deepak Lal. These scholars can be credited with having successfully re-established the validity of the Ricardian view of trade, and especially its core proposition that trade confers benefits on all participants.

Within this group, Anne O. Krueger is uniquely important. In the mid-1970s, she and Bhagwati, working under the sponsorship of the National Bureau of Economic Research (NBER) of Cambridge, Massachusetts, organized and co-edited a series of studies of the relationship between trade and economic growth in developing countries. This series, which included both case studies and more general theoretical volumes, was completed in 1978 with the publication of Krueger's synthesizing monograph, *Liberalization Attempts and Consequences*. The Krueger–Bhagwati series constituted an intellectual landmark. It not only presented a formidable critique of import substitution but set forth a commensurately powerful case for the classic doctrine of comparative advantage and openness to trade, even as this applied to countries dependent upon the export of primary agricultural commodities.

The Case Against Industrial Protectionism

To understand Krueger's ideas, it is best to begin with her response to Chenery, contained principally in an article entitled 'Comparative Advantage and Development Policy Twenty Years Later'.[42] Acknowledging the intellectual force and real-world policy impact of Chenery's argument—'his essay stood for well over a decade as the definitive statement of the profession's understanding of the trade-policy and growth relationship'—Krueger suggests that the rebuttal of economic arguments that call for industrial protectionism must begin with the economic data.[43]

Three economic findings show decisively that openness to trade produces benefits for poor agricultural countries as well as rich industrial nations. Countries that practice trade openness and emphasize exports enjoy higher rates of economic growth than countries that do not. Countries that shift from a protectionist position to greater openness to trade enjoy an acceleration in their rates of economic growth when they do so. Finally, the new higher rates of economic growth associated with trade openness are sustained over longer periods of time than can be explained by a simple one-time shift in the allocation of economic resources. This strongly suggests that deeper, systemic considerations have come into play.

The empirical evidence that Krueger and other free-trade scholars have generated to support these propositions is overwhelming.[44] It demonstrates convincingly that, while the development economists were elaborating a theoretical approach to growth based on the need to protect

domestically oriented industries, the economic realities of both the developed and the developing worlds were providing the empirical basis for a contrary truth.

Krueger believes that the challenge to political economy, now that the strong relationship between openness to trade and economic growth has been established, is to ascertain the theoretical reasons why the protectionist position was so badly mistaken. Since the bulk of her extraordinarily prolific career has been devoted to this challenge, it would be impossible here to do more than provide a threadbare sketch of a few of her key arguments.[45]

Krueger's case against industrial protectionism considers both strictly economic matters and considerations of political economy. Some examples of purely economic considerations are the constraints that derive from the small size of markets available to import-substituting firms. Since such firms produce for domestic consumption, their potential markets are inherently smaller than those of firms that have an international orientation. As a result, their irreducible base costs per unit of production, termed 'indivisibilities', are higher than those of firms that can sustain longer production runs. The larger size of the potential markets available to export-oriented firms thus permits a more sustained growth of value added for the same amount of increase of fixed and human capital than would be the case for firms limited to domestic markets.

The key economic presupposition underlying the work of Krueger has to do with the element of competition. Production for the intensely competitive international market-place makes it essential for business firms to use their resources as efficiently as possible; to employ the least expensive mix of inputs to attain a given value of output. In African countries, capital tends to be scarce and, therefore, expensive relative to the abundant supply of unskilled and, therefore, low-cost labour. If international competition had been allowed to influence the continent's factor proportions, it would have generally promoted the development of industries that are intensive in their use of unskilled labour rather than those more capital intensive.

Import substitution led Africa in exactly the opposite direction. The tendency towards currency overvaluation which was inherent in the ISI approach as a means of subsidizing capital costs, encouraged capital intensiveness. Even the most casual observers of the continent have noted, often with utter bewilderment, the socio-economic anomaly of capital-intensive urban industries coexisting with vast reserves of urban and rural unemployed and underemployed workers.

Krueger is rare among orthodox economists in her remarkable sensitivity to the impact of political considerations on economic outcomes. That sensitivity is basic to the concept that many would regard as her most enduring contribution to the political economy of development; namely, the idea of 'the rent-seeking society'.[46] The basic idea is disarmingly simple. When a government imposes a limitation on trade, as in the form of quantitative restrictions on imports, it is creating an artificial scarcity. The opportunity to supply the goods that have been made scarce, as in the form of an import licence, then becomes a valuable commodity. Those who control this commodity—i.e. the bureaucrats with authority to dispense trade licences and foreign-exchange allocations—are no less likely to market them at a favourable price than those who possess other scarce commodities.

Political scientists have commonly used the term 'rent-seeking' as if it were merely an economic euphemism for corruption such as the bribery involved in obtaining an import licence from the official in charge of dispensing it. This usage fails to capture the full economic ramifications of the process, the tendency to retard economic growth by distorting the way in which a society's resources are allocated. In controlled trade regimes, a huge economic loss derives from the fact that vast resources are spent on obtaining the lucrative governmental positions that dispense goods, such as licences, that can provide rents. The resources that are used for those purposes are expended wastefully. This occurs because the prevailing pattern of incentives encourages investments in non-productive pursuits at the expense of productive enterprise, as when a family allocates its savings to educate a child for a government position simply because that position has a rent-seeking potential. Those finances would then be unavailable to finance improvements in its farm or manufacturing firm.

Bureaucratic rent-seeking is by no means confined to developing countries. But controlled trade regimes with quantitative import restrictions have been more common in these societies

than in the more industrial countries of the world. And such trade regimes provide exceptionally fertile ground for rent-seeking practices. This is especially so in Africa where the import-substitution approach to development was pursued more vigorously and has remained in place longer than in other of the world's developing regions. The tendency of controlled trade regimes to lend themselves to economically corrosive rent-seeking practices provides the intellectual epicentre of the political economy explanation of Africa's underdevelopment.

Krueger herself believes that, absent rent-seeking, purely economic considerations such as inefficient use of inputs would provide an adequate explanation for the dismal performance of controlled trade regimes. But from the standpoint of African political economy, the order of intellectual priority must be reversed. Remove strictly economic factors and the ubiquitous tendency towards rent-seeking, by itself, explains why import substitution failed the continent so badly.

Krueger's analysis provides our deepest explanation of why African agriculture performed so poorly during the post-independence period. It was not simply that agriculture was called upon for the economic resources required to create new industries or to provide them with imported inputs. Rather, it was the fact that a controlled trade regime gave rise to an economic environment which led to the misallocation of economic resources everywhere in the economy, both in agriculture and in industry. Agriculture and industry both declined. But the agricultural loss was more important because this had been the most productive sector of the economy.

One of the most valuable aspects of Krueger's rent-seeking model is that it provides a way of understanding the process whereby the urban coalition that dominated African politics for nearly a generation lost its political pre-eminence. The starting-point of the political decline of the urban coalition was the national economic decline. Economic crisis reduced the level of resources available to hold the urban coalition together. Growing levels of unemployment forfeited the support of industrial workers hitherto cushioned from economic pain by rigid protectionism and the security of guaranteed employment in state-owned industries. The owners, managers, suppliers, and sub-contractors of parastatal industries became similarly disaffected when these industries exhausted their ability to confer economic benefits. Virtually all urban wage-earners, ranging from middle-class civil servants to low-paid service-sector workers became politically disenchanted when they experienced a decline in the purchasing power of their salaries and were forced to deal with the day-to-day unavailability of even the most basic goods. Even university graduates, long Africa's presumptive élite, began to question economic systems that could no longer guarantee secure employment upon graduation. Once these powerful groups began to face the prospect of economic hardship, the stage was set for policy reform.

A government that is perched on the edge of political transformation because its supportive coalition has come unglued is also in a position to be influenced by other factors that make further continuation of the existing system impossibly difficult. International pressures are among these. With the end of the Cold War, the ability of African governments to exploit geo-political leverage to extract financial assistance has lessened and diminishing aid flows have exposed underlying economic weaknesses and stimulated domestic pressures for reform. The end of the Cold War also opened the door to more effective pressure from reform-minded international lending institutions such as the World Bank and International Monetary Fund.

Conclusion

The one unmistakable conclusion that can be drawn from the preceding discussion is that urban industrial prosperity cannot be obtained through policies that suppress the rural sector. The fact of the matter is that, in modern Africa as in 1930s Russia and developing countries more generally, agricultural growth seems to be the precondition for rapid growth in manufacturing and trade.[47] This is the now famous agricultural paradox: even as agricultural growth occurs, agriculture's share of GDP declines because agricultural production accounts for a smaller and smaller proportion of a society's total economic output.

The importance of agricultural growth in economic development has led to a widespread mis-

conception about the nature of the reform process. This misconception holds that governments which, in the past, have overtaxed agriculture in favour of industry should now be encouraged to correct this problem by pursuing policies that redirect economic resources back towards the production of exportable agricultural commodities, thereby substituting a new 'export bias' for the old import-substituting bias.

This would involve a gross misunderstanding of the process of economic liberalization. Structural adjustment is not about substituting export promotion (rural bias) for domestic protection (urban bias). Liberalization is about attaining neutrality in a country's policy framework, a system that does not discriminate in favour of either rural or urban dwellers.[48] Trade liberalization is a major step in this direction because it contributes to the emergence of an economic environment in which governments allow economic resources to flow freely towards their most productive uses.

The idea that most fully captures this conception is Joseph A. Schumpeter's classic notion of 'creative destruction'. Schumpeter believed that in dynamic and growing economies there would be a constant process of withdrawing economic resources from less productive uses and reinvesting them in more productive activities.[49] The experience of the ISI countries demonstrates that once a government makes protection available, the groups that benefit will lobby vigorously against the removal of that protection. Even present advocates of import substitution accept the validity of this argument.[50]

All of this bears directly on the question of what can be expected from an agricultural recovery. There is a widespread belief that as societies which have experienced economic decline began to recover, their future will recapitulate the past; that is, their new economic profile will closely resemble that of the pre-crisis era: once great cocoa or coffee or tea exporters will enjoy a rapid recovery of the cocoa or coffee or tea sector.

While economic reform may indeed stimulate some recovery of traditional agricultural exports, this should by no means be the first criterion of success. It would be utterly unrealistic in the global economic environment of the 1990s to expect that a recovered economy would resemble its historic profile. Too much has changed in the international market-place during the thirty years since African governments first began to implement the 'squeeze agriculture' approach. Markets that African countries once dominated have been forfeited permanently to dynamic producers in other regions. And technological evolution has shifted demand away from primary agricultural exports that were once vitally important as industrial inputs. The ladders of opportunity for economic growth in the 1990s are fundamentally different than they were in the 1960s.

These changes underscore the necessity for economic resources to be highly nimble and, in this respect, they give fresh importance to Schumpeter's notion as a caution against policies that repeat the past. Developmental approaches which lock productive resources into inefficient and unproductive uses would only doom Africa to another generation of economic failure.[51]

Notes

1. United States Department of Agriculture, Economic Research Service, *Food Problems and Prospects in Sub-Saharan Africa: The Decade of the 1980s*, Foreign Agricultural Research Report No. 166 (Washington, DC: US Dept. of Agriculture, 1981), see fig. 1, p. 2.
2. World Bank, *Accelerated Development in Sub-Saharan Africa: An Agenda for Action* (Washington, DC: World Bank, 1981), table 5.1, p. 46.
3. United States Department of Agriculture, Agriculture and Trade Analysis Division, *World Agriculture Trends and Indicators, 1970–89* (Washington, DC: Economic Research Service, 1990), 50–1.
4. Indeed, some observers have held that because of the relatively low value to weight ratio of food grains, these crops have a higher ratio of imported inputs to market price than exportable commodities. See Jennifer Sharpley, *Kenya: Macro Economic Policies and Agricultural Performance* (Paris: Organization for Economic Cooperation and Development, Development Centre, 1984).
5. For an excellent recent discussion of the literature on urban bias, see Ashutosh Varshney, 'Urban Bias in Perspective', *Journal of Development Studies*, 29: 4 (July 1993). This entire issue of the journal is devoted to this topic.
6. For a discussion of the economic literature on this approach, see Wyn F. Owen, 'The Double Developmental Squeeze on Agriculture', *American Economic Review*, 61: 1 (Mar. 1966), 43–70.
7. This point is discussed in John W. Mellor, 'Agriculture on the Road to Industrialization', in John P. Lewis and Valeriana Kalab (eds.), *Development Strategies Reconsidered* (New Brunswick [US] and Oxford [UK], 1986), see esp. 71–2.
8. For a discussion, see World Bank, *Adjustment in Africa: Reforms, Results and the Road Ahead* (Oxford and New York: Oxford University Press, 1994).
9. Cited above.
10. Inasmuch as the terminology is sometimes ambiguous, it may be useful to state that a currency is overvalued when the exchange rate is too low; that is, when too few units of local currency are exchanged for a unit of hard currency such as the US dollar.
11. The francophone African countries which have joined the West African Currency Union were considered an exception to this generalization because of the existence of a fixed exchange rate between their currency and the French franc. But most observers felt that these countries did devise ways to overvalue the CFA and in early 1992 there was a devaluation of about 50%.
12. One of the most widely and uncritically accepted axioms among scholars of African development has to do with the problem of economic dependency characteristic of Africa's 'mono-crop' economies. Scholars who propound this view generally do not recognize or acknowledge that this tendency was an economic outcome of the policies adopted by African governments.
13. For a classic case study of this phenomenon, J. Clark Leith, *Ghana* (New York: National Bureau of Economic Research, 1974).
14. For an excellent discussion, see Charles M. Becker, Andrew M. Hamer, and Andrew R. Morrison, *Beyond Urban Bias in Africa: Urbanization in an Era of Structural Adjustment* (Portsmouth, NH, and London: Heinemann and James Currey, 1994), esp. ch. 1, 'African Economic Development and Urbanization'.
15. The list of economists whose work would fit this general description includes, at the very least, those discussed in the chapter on development economics in W. W. Rostow's *Theorists of Economic Growth from David Hume to the Present* (New York and Oxford: Oxford University Press, 1990). Rostow's list includes: Colin Clark, Albert Hirschman, W. Arthur Lewis, Raul Prebisch, Paul N. Rodenstein-Rodan, and Hans Singer. Hollis Chenery, sometimes considered the most influential of this group, is discussed elsewhere in the Rostow volume.
16. Robert H. Bates, *Markets and States in Tropical Africa* (Berkeley and Los Angeles: University of California Press, 1981).
17. Ibid. 3.

18. Michael Lipton, *Why Poor People Stay Poor: Urban Bias in World Development* (Cambridge, Mass.: Harvard University Press, 1976).

19. For a comprehensive discussion of Bates's work, see *World Development*, 21: 6 (June 1993), 1033–81, 'Special Section: Robert Bates, Rational Choice and the Political Economy of Development in Africa'.

20. Robert H. Bates, 'Agricultural Policy & the Study of Politics in Post-Independence Africa', in Douglas Rimmer (ed.), *Africa: 30 Years On* (Portsmouth, NH: Heinemann, 1991), 115–29.

21. Ibid. 122.

22. World Bank, *World Development Report 1984* (New York and Oxford: Oxford University Press, 1984), table 2, p. 220.

23. Ibid.

24. World Bank, *World Development Report 1985* (New York and Oxford: Oxford University Press, 1985), table 2, p. 176.

25. In the Ivory Coast case, this was of course an integral feature of its participation in the French CFA zone.

26. For Côte d'Ivoire, see Aristide Zolberg, *One Party Government in the Ivory Coast* (Princeton: Princeton University Press, 1964), 26–7; and Jennifer A. Widner, 'The Origins of Agricultural Policy in Ivory Coast', *Journal of Development Studies*, 29: 4 (July 1993), 25–59. For Kenya, Michael F. Lofchie, *The Policy Factor: Agricultural Performance in Kenya and Tanzania* (Boulder, Colo., and London: Lynne Rienner Publishers, 1989), ch. 7.

27. Merilee Grindle, 'The New Political Economy: Positive Economics and Negative Politics', in Gerald M. Meier (ed.), *Politics and Policy Making in Developing Countries: Perspectives on the New Political Economy* (San Francisco: ICS Press, 1991).

28. Among the many examples are the following: the Kilimanjaro Native Cooperative Union representing coffee-growers in Tanzania; the Kenya Farmers Association representing coffee-growers in Kenya and the United Ghana Farmers Council representing cocoa-producers in Ghana.

29. For a study of governmental repression of export farmers, see Gwendolyn Mikell, *Cocoa and Chaos in Ghana* (New York: Paragon House, 1989).

30. See e.g. William H. Friedland, *Cooperation, Conflict and Conscription: Tanu–TFL Relations, 1955–1964* (New York State School of Industrial and Labor Relations; Cornell University Reprint Series, No. 222; 1967) and St. Claire Drake and Leslie Alexander Lacy, 'Government versus the Unions: The Sekondi–Takoradi Strike, 1961', in Gwendolen M. Carter (ed.), *Politics in Africa: 7 Cases* (New York: Harcourt, Brace & World, Inc., 1966). In both Ghana and Tanzania, transportation unions involved in the movement of exportable goods through the ports were among the earliest targets of governmental repression.

31. For one important example of this view, see Hans W. Singer and Javed Ansari, *Rich and Poor Countries* (London: George Allen & Unwin Ltd, 1982).

32. 'Comparative Advantage and Development Policy', *American Economics Review* (Mar. 1961). Cited here as reprinted as ch. 7 of Hollis Chenery, *Structural Change and Development Policy* (New York: Oxford University Press, 1979).

33. Today, most economists consider that the economic evidence for a structural decline in the terms of trade between industrial and agricultural nations is ambiguous.

34. W. Arthur Lewis, *Some Aspects of Economic Development* (Accra: Ghana Publishing Corp. for the University of Ghana, 1969), 8.

35. See e.g. Subrata Ghatak and Ken Ingersent, *Agriculture and Economic Development* (Baltimore: Johns Hopkins University Press, 1984), esp. ch. 6, 'Resource Use Efficiency and Technical Change in Peasant Agriculture'.

36. Lewis's classic statement on this topic was 'Economic Development with Unlimited Supplies of Labour' (Manchester, Manchester School, V. 22, No. 2), 131–91.

37. Barbara Ward, *The Rich Nations and the Poor Nations* (New York: W. W. Norton & Co. Inc., 1962), 99.

38. World Bank and UNDP, *Africa's Adjustment and Growth in the 1980s* (Washington, DC: World Bank, 1989), annex, p. 32.

39. The preferred means of doing so is through the creation of licensed currency bureaus that can legally trade in foreign currencies. The owners, employees, and clients of these bureaus quickly become a lobby for a free currency system.

40. An excellent discussion can be found in P. F. Leeson, 'Development Economics and the Study of Development', in P. F. Leeson and M. M. Minogue, *Perspectives on Development* (Manchester and New York: Manchester University Press, 1988).

41. See e.g. Dudley Seers, 'The Birth, Life and Death of Development Economics', *Development and Change*, 10: 4 (Oct. 1979), 707–19; Albert Hirschman, *Essays in Trespassing* (New Haven: Yale University Press, 1981), ch. 1, 'The Rise and Decline of Development Economics'; and Polly Hill, *Development Economics on Trial* (New York: Cambridge University Press, 1986).

42. See Anne O. Krueger, *Perspectives on Trade and Development* (Chicago: University of Chicago Press, 1990), ch. 3, 'Comparative Advantage and Development Policy'. This article first appeared in a Festschrift in honour of Hollis Chenery.

43. Ibid. 50.
44. See Deepak Lal and Sarath Rajapatirana, 'Foreign Trade Regimes and Economic Growth in Developing Countries', *World Bank Research Observer*, 2: 2 (July 1987), 189–217.
45. Like Chenery, Krueger was for many years associated with the World Bank, as Vice-President for Research. She is presently Professor of Economics at Duke University.
46. Anne O. Krueger, 'The Political Economy of the Rent-Seeking Society', *American Economic Review*, 64: 3 (June 1974), 291–303.
47. The discussion that follows draws heavily upon C. Peter Timmer, 'The Agricultural Transformation', in H. Chenery and T. N. Srinivasan (eds.), *Handbook of Development Economics*, i (North Holland: Elsevier Science Publishers, 1989).
48. For an excellent discussion of this topic, see Lal and Rajapatirana, 'Foreign Trade Regimes and Economic Growth in Developing Countries', see esp. 208 ff.
49. Joseph A. Schumpeter, *Capitalism, Socialism and Democracy* (New York: Harper and Brothers Publishers, 1950), see ch. 7, 'The Process of Creative Destruction'.
50. See Sanjaya Lall, 'Structural Problems of African Industry', in Frances Stewart, Sanjaya Lall, and Samuel Wangwe (eds.), *Alternative Development Strategies in Sub-Saharan Africa* (New York: St Martin's Press, 1992), 125.
51. For an excellent treatment of urban bias in development politics, see Ashutosh Varshney, *Democracy, Development and the Countryside: Urban–Rural Struggles in India* (New York: Cambridge University Press, 1995). For additional documentation of the impact of urban bias on the export crop sector, see Azita Amjadi, Ulrich Reinke, and Alexander Yeats, *Did External Barriers Cause the Marginalization of Sub-Saharan Africa in World Trade?* (Washington, DC: World Bank, International Economics Department, Policy Research Working paper 1586, 1996).

II Rural–Urban Migration

Introduction

THE pace of urban growth in the South is without precedent. Between 1960 and 1993, the urban population in developing countries grew by 3.8 per cent a year, i.e. it tripled in 30 years, just about a generation (Table 1). This growth was the result, in the main, of the conjunction of natural population growth and rural–urban migration. Between 1975 and 1990, rural–urban migration accounted, on average, for 50 per cent of urban growth in nine Asian countries, for 75 per cent in four African countries, and for 49 per cent in eleven Latin American countries (Findley, 1993). The impact of rural–urban migration on urban labour-markets is even more pronounced, since the majority of rural–urban migrants arrive ready to enter the labour-force. And it is rural–urban migration that accounts, by and large, for the increase in the level of urbanization in developing countries from 22 to 36 per cent between 1960 and 1993 (Table 1).

Until the last century, many rural populations had little connection with urban centres; they lived in quite self-centred societies. They operated subsistence economies and maintained only limited external contacts. However, the expansion of the capitalist system, under way for half a millennium and accelerated by the Industrial Revolution and Western imperialism, incorporated ever more outlying regions into the emerging world economy. By now the process is virtually complete. All over the world, rural populations have been drawn into the urban nexus. Throughout the world today, few are the rural dwellers who have not sold and bought in markets or shops, who have not seen what a school certificate can do for the future of a child, who have not listened to a first-hand account of work in the city. Some improve their condition while staying where they were born, or moving to other rural areas as farmers, traders, or artisans. But rural prospects appear bleak to many, the urban scene more promising.

Most people move to cities for economic reasons. A substantial body of research on rural–urban migration provides conclusive evidence on this point in several ways. When migrants are asked why they moved, they usually cite the better prospects in the urban economy as their chief reason. Also, migration streams between regions have been shown to correspond to income differentials between those regions. And over time, as economic conditions at alternative destinations change, migration streams alternate accordingly.[1]

The sight of severe and widespread urban poverty in developing countries easily leads to the assumption that migrants do not know what to expect, that illusions about the prospects lying ahead bring them to an urban environment in which they find themselves trapped. However, numerous studies report that most migrants consider that they have improved their condition, and that they are satisfied with their move. In fact, most migrants are quite well informed before they move. Many rural communities have developed migration strategies which are informed by the experience of migrants who have kept in touch, who return to the village on visits or to stay, and by villagers who have visited kin and friends in the city. These strategies are modified over time as experience dictates. Potential migrants are thus presented with quite well-defined options. In turn, their decision to migrate is rarely an individual one, rather it is usually a family decision. Indeed, much rural–urban migration of individuals is part of a family strategy to ensure the viability of the rural household. And migrants typically receive considerable assistance in the move, in adapting to the urban environment, in securing a foothold in the urban economy.

The decision to migrate is based on a comparison of rural and urban opportunities. And these differ for different categories of people. Young adults predominate among migrants in search of employment: they tend to have less at stake in the rural economy than their elders and are favoured in the urban labour-market. The movement of women is often inhibited by patriarchal control in the rural areas and discrimination in the urban economy (Gugler and Ludwar-Ene, 1995). Well-educated migrants find better opportunities in the urban job market than in the rural economy. Others have the right connections and come with reasonable assurance that the assistance of their kinsman, fellow-villager, or patron will prove effective. For most rural dwellers, however, prospects in the city are dim. Some venture forth nevertheless, to succeed or to fail, or to return to the village. But most stay on the land—as long as it supports them. We will return to these issues when we discuss the urban labour-market in Part III.

In many cases migration is not just a once and for all move. Rather there are a series of moves over a lifetime. The preponderance of men over women in the cities of South Asia, the Middle East, and much of Africa reflects the tendency of male migrants to move on their own. Some of these men are young and single, some stay in the city only for a short while. However, given widespread urban unemployment, men who have secured a job tend to hang on to it: they become long-term workers, even while leaving wives and children in their rural area of origin. As Thomas Weisner (1972) put it, they have 'one family, two households'. If the Industrial Revolution established the distinction of work-place and home, the separation of men from their wives and children has been drastically magnified for many families in poor countries.

Hemalata Dandekar, in Chapter 3, provides an account of migration from Sugao to Bombay. For more than a century, men have moved between the village and the city, more than 150 miles away. Most of the early migration was seasonal. By the 1940s, the villagers had come to appreciate reliable wage-work in the city, but temporary jobs had become difficult to find. Men came to spend their entire working life in the city, returning to their family only once or twice a year on vacation. The proportion of men working outside the village expanded over the years: by 1990 almost one-third of the men were living outside Sugao while maintaining their families in the village.

Many migrants would prefer to live and work in Sugao: they complain about the miserable living conditions in Bombay, they and their families endure long separations, but the village economy offers them little opportunity. Their urban opportunities, however, remain narrowly circumscribed: few men have left Sugao permanently. And the few women who left permanently because their marriages had failed and they had little claim to assets at their in-laws, returned to their parents' or a brother's home in their parental village. Those women who do move to the city are almost exclusively wives accompanying their husbands. But few men can afford to support a wife in the city, and the number of women who have obtained work in Bombay is negligible.

The position of these migrants in the urban economy is precarious, but their claim to a piece of village land assures them of the ultimate sanctuary. They use their savings to improve their village home, to install piped-water supplies and electricity connections, to acquire one or two milk animals, and to improve the cultivation of their land. Migrants from the more affluent families have the economic backing, education, and connections to bide their time and secure better paying, less arduous jobs in Bombay. Their remittances and savings in turn allow their families to purchase land and agricultural equipment and to consolidate their dominant position in the village.

Settling down in town with a family is usually for the long term, perhaps for a working life. However, it does not necessarily signify a permanent move away from the village. Many urban families maintain strong ties with a rural community which they continue to consider their home and where they anticipate retiring eventually. In Chapter 4 I describe such a strategy of temporary family migration. I found it well-established in south-eastern Nigeria in 1961—and enduring when I returned there in 1987. Urban dwellers were fully committed to their urban career, all the while maintaining strong ties with the husband's community of origin. They visited there, welcomed visitors more or less enthusiastically, let themselves be persuaded to contribute to development efforts, built a house, planned to retire, and wanted to be buried 'at home'. They partook of the city as well as the village. Their strategy may be characterized as 'life in a dual system'.

This dual commitment to an urban working life as well as to a rural community of origin may be explained in structural terms. The city offers an improved standard of living, but the position of most urban dwellers is precarious: they remain dependent on the village for their security. But most of those securely established in the urban arena—senior civil servants in the early days after independence, affluent businessmen in the late 1980s—remain committed to the village as well. Life in a dual system has become an established cultural norm.

Temporary migration is predicated on maintaining a rural base. The 'dual-household' migrant leaves wife and children on the farm to grow their own food, perhaps to raise cash crops as well. The strategy is a function of high fertility and limited educational and earning opportunities for women. An added factor is the lack of compensation for those who leave land that is communally controlled, as is the case in much of Africa South of the Sahara. The 'dual-system' strategy entails social and economic investments that allow the family to return to the village. Either strategy assures the migrant of a measure of security, meagre but more reliable than what the city offers most of its citizenry.

In every Latin American country, as well as in the Philippines and Thailand, women outnumber men in the urban population, often by a substantial margin. This reflects a pattern of permanent rural–urban migration in which women predominate. Recent data indicate that this pattern is appearing elsewhere. This suggests a historical transition from a preponderance of men to a preponderance of women in net rural–urban migration: a 'gender transition in rural–urban migration' (Gugler, 1996). It would appear that later marriage, reduced fertility, and greater independence are modifying cultural definitions of women's roles as wives and mothers and set them free to move. Never married, separated, divorced, and widowed women usually are faced with limited rural opportunities and are attracted by cities that offer them better opportunities in terms of work as well as (re)marriage.

In parts of Africa, in much of Asia, and in most of Latin America, migrants by and large have little prospect of maintaining access to agricultural land because of population pressure and/or institutional constraints. They have become entirely dependent on their urban earnings. Instead of planning for a return to the village, they press for the provision of social security. And they search for sources of earnings other than paid employment. Escaping the vagaries of employment is a major motive in establishing one's own business. Ownership of a home, however rudimentary, assures accommodations and the possibility of income from rent. The fact that most migrants in Latin America have severed their rural connection explains, in part, why they have established powerful trade unions—we will examine the case of Brazil in Chapter 24—and secured social security protection to various degrees. Squatter movements in Latin America—which we will encounter in Chapter 19—are similarly propelled by the need to establish an urban base.

Some of the communities that have maintained their pre-Colombian identity present notable exceptions to the Latin American pattern of permanent rural–urban migration. Thus Laite (1981) describes and analyses a widespread pattern of temporary migration of men from highland Peru that has been established since the Inca period. Harald Moßbrucker, in Chapter 5, presents a detailed analysis of migration from Amerindian villages in Peru and Mexico. Comparisons across the villages as well as across the migrants' destinations, Lima, Mérida, and Cancún, reveal major differences in migration processes, urban integration, and ties between migrants and villagers.

Moßbrucker shows how population pressure on limited resources, severe inequality, and deep conflicts induced early and heavy emigration from one of the Peruvian villages, Santiago de Quinches. The constellation in San Miguel de Vichaycocha was quite different. Virtually all villagers belong to the village commune which assures them of access to rich grazing lands as well as land for subsistence farming. Migration was short-term until the 1950s, and emigration since then has remained moderate.

Yucatán has been transformed by a huge expansion in tourism since the 1970s. Short-term migration of young men has given way to long-term migration by families. The village of Sudzal has seen massive emigration. There has been less emigration from Huhi: the village is better connected to Mérida, 40 miles away, and many men commute to work in the city. Both villages have come to manufacture for urban markets. In Huhi production is organized in workshops by local entrepreneurs, whereas in Sudzal it takes the form of domestic outwork by women. The dichotomy between city and countryside has thus been all but dissolved in Yucatán. For an account that emphasizes a similar interpenetration of Jakarta and its surrounding countryside, see Hugo (1996).

The strength of the urban–rural connection differs among Amerindians. In Peru, the kinship group usually coincides with the village so that migrants' relations with kin focus on their village of origin. Their associations in the city, the 'clubs', maintain and support home ties. In the case of Quinches, these ties are particularly strong: emigrants still own a great deal of land in the village and many are business partners as well as kin to cattle ranchers and traders there. Emigrants from Vichaycocha are less connected with the village, but ready admission to the village commune has encouraged some to return to the village. In Yucatán, in contrast, migrants' kinship relations do not encompass the village, they have not established their own associations, and they are not involved in business ventures with villagers.

Most governments in the South grope for policies to slow down rural–urban migration and to direct it away from the largest cities. Rapid urban growth and the appearance of mega-cities confront them with huge demands on public resources to maintain even a minimum level of infrastructure and public services, demands that they often meet only very partially. They are concerned about popular pressures for such resources, and about the discontent of the unemployed and underemployed that may erode the legitimacy of the regime, escalate into strikes and civil unrest, and threaten the very survival of the regime.

Charles Becker and Andrew Morrison, in Chapter 6, survey and evaluate a wide range of public policies affecting rural–urban migration. Direct controls on migration are the least significant, except under strong authoritarian regimes—as long as they last. Policies reducing the natural rate of population growth, on the other hand, can have a major impact on rural–urban migration as well as directly on the growth of the urban population. In terms of redirecting rural–urban migration, attempts to promote secondary cities and market towns have been constrained by the huge infrastructure investments they demand. Where capital cities have been relocated, the costs have been truly staggering. And the results of tax incentives to affect the location of private firms have been problematic. More promising are policies that decentralize government—they are, however, unlikely to be embraced enthusiastically by central government officials.

Policies specifically targeting the rural or the urban sector can have a major impact on rural–urban migration. However, the most important determinants of rural–urban migration, Becker and Morrison suggest, are macro-economic policies. Import-substituting industrialization and populist policies that emphasized income redistribution and rapid growth encouraged rural–urban migration, especially in Latin America—until they collapsed under the strain of huge budget and trade deficits. Whether the structural adjustment programmes that ensued will significantly slow rural–urban migration remains to be seen.

There is general agreement that rural development, in order to close the rural–urban gap, constitutes a policy of equity that will truly stem rural–urban migration. Admittedly, initial improvements in education and in communications may encourage migration as long as rural–urban discrepancies remain large. To reduce significantly the gap requires that considerable resources be made available to a huge rural population. Rural incomes have to be raised directly, e.g. through a reduction in taxes or through an increase in the prices agricultural products fetch, and/or indirectly through investments that raise the productivity of the rural labour-force whether in agriculture or in small-scale industry. Given the size of the rural population in most developing countries, this requires an enormous reallocation of resources, a reallocation that has to contend with the determined opposition of urban interests. A long-run rural development policy—focusing on labour-intensive investments to retain a growing rural population—can constitute a feasible compromise. Such a policy appears, however, to be beyond the horizon of most governments, preoccupied as they are with their very survival in a much more immediate future.

Note

1. For a comprehensive discussion of rural–urban migration in developing countries, see Gilbert and Gugler (1992 [1982]: 62–86).

References

Findley, Sally E. (1993), 'The Third World City: Development Policy and Issues', in John D. Kasarda and Allan M. Parnell (eds.), *Third World Cities: Problems, Policies, and Prospects* (Newbury Park, Calif.: Sage), 1–31.

Gilbert, Alan, and Gugler, Josef (1992 [1982]), *Cities, Poverty, and Development*: *Urbanization in the Third World*, 2nd edn. (Oxford: Oxford University Press).

Gugler, Josef (1996) 'Gender and Rural–Urban Migration: Regional Contrasts and the Gender Transition', typescript

Gugler, Josef, and Ludwar-Ene, Gudrun (1995), 'Gender and Migration in Africa South of the Sahara', in Jonathan Baker and Tade Akin Aina (eds.), *The Migration Experience in Africa* (Uppsala: Nordiska Afrikainstitutet), 257–68.

Hugo, Graeme (1996), 'Urbanization in Indonesia: City and Countryside Linked', in Josef Gugler (ed.), *The Urban Transformation of the Developing World* (Oxford: Oxford University Press), 132–83.

Laite, Julian (1981), *Industrial Development and Migrant Labour in Latin America* (Manchester: Manchester University Press; Austin: University of Texas Press).

Weisner, Thomas Steven (1972), 'One Family, Two Households: Rural–Urban Ties in Kenya', Ph.D. diss. (Harvard University).

3

Changing Migration Strategies in Deccan Maharashtra, India, 1885–1990

Hemalata C. Dandekar

MIGRATION out of the countryside, primarily to the megalopolis of Bombay, is an established routine for people from villages in the city's hinterland. This has been true from the turn of the century for people from the Deccan Plateau of Maharashtra, the state in which Bombay is located. The growth of the cotton textile industry in Bombay was a significant factor in attracting a stream of migrants from the coastal belt to the south of the city known as the Konkan and from the Deccan Plateau to the south-east. For those from the Deccan Plateau, migration has grown and changed as the fortunes of the Bombay textile industry have waxed and waned. Once that industry enjoyed a central position in the economic base of the city, but it has precipitously declined in importance, its fate sealed by a crippling industry-wide strike in the 1980s. Villagers from the Deccan region have adjusted and responded to the opportunities in, and needs of, the city's economy. They have responded differentially in two ways: (1) by their socio-economic status in the village, which affected their experience of the constraints of the village economy and their ability to link with the city economy; and (2) by their gender, which circumscribed their degree of spatial and social mobility. Both factors are reflected in their decisions about migration, providing a revealing view of the survival strategies rural people have employed in the face of the dual forces of rapid industrialization centred in urban areas and a slow, negligible level of diversification of the rural economy.

To reveal those strategies in the 'fine grain', some of the findings from a longitudinal case study, spanning almost fifty years of the life of a single village, are presented in this essay. The village is located more than 150 miles from Bombay. In the 1970s, this was an uncomfortable ten-hour bus ride from Bombay; now it is a seven-hour, but still uncomfortable, journey. This village, here referred to as Sugao, is in the Satara District of Maharashtra State. The vignette sketched here is that of a typical small village in transition, caught in the middle of strong currents of modernization and development.

The significance of this movement of people between village and city, over time, both for the individuals who migrate and their families, and as importantly for the village they leave behind, is apparent from quantifiable indicators and qualitative observations, as well as anecdotal information. The flow of village people to the city is thoughtful, responsive, optimizing, and

Adapted, updated, and elaborated by the author from her *Men to Bombay, Women at Home: Urban Influence on Sugao Village, Deccan Maharashtra, India, 1942–1982* (Ann Arbor: University of Michigan Center for South and Southeast Asian Studies, 1986).

purposive, and its analysis provides a glimpse into what these moves have meant to the general tone of the migrants' individual lives in the city and to the tenor and beat of the village entity.

It is clearly important to understand this phenomenon of migration and its effects on the people and the societies that are involved. It is now generally understood that much of the presently industrializing world is 'on the move'. Migration and urbanization have been one of the most significant forces of transformation during the decades since the end of the Second World War. In this context the choice of Sugao village in Maharashtra State is neither arbitrary nor accidental. With a 1991 population of close to 79 million people, 39 per cent of whom live in urban areas and represent over 14 per cent of the total urban population of the country, Maharashtra is the most urbanized and industrialized among India's major states.[1]

A pragmatic reason for the choice of Sugao for such a study is that base-line data going back to 1942 are available for this village. Full census surveys of the village were executed in 1942–3 and 1957–8, and were repeated by the author in 1977 and under the author's supervision in 1990. These, along with many shorter visits by the author between 1976 and 1994, provide a continuous profile of village transitions.[2] Sugao appears quite typical of villages in Satara District, and the Deccan region in general. Its location is not particularly good: it is not on any major road or rail line; nor particularly bad: it is not too far from major transport and road facilities; with a dirt road connecting it to the Pune–Satara state highway. The nearest railway station is at Wathar, more than 15 miles away. The most convenient way to reach Sugao is thus by road. Although the surface of the two-lane state highway has been improved over the years, its volume of truck traffic has also increased, making for poor driving conditions. The two-and-a-half miles of road going off the highway to Sugao are still unpaved dirt.

In terms of political, strategic locus, too, the village is fairly typical. Sugao does not have a leader who has achieved political prominence, an established way through which public funds and programmes may be channelled to specially targeted villages. And, as is typical of most villages, the people of Sugao as a whole are fairly apathetic about government programmes and their relationship with the government. Village leaders rarely approach development agencies on behalf of the village, and the officials in charge seldom visit Sugao. Thus what one observes, in terms of development and change in Sugao, is what is occurring as a result of more general forces affecting Maharashtra's countryside.

The Economy of Sugao

Sugao has been inhabited for several centuries. It was a site where nomadic traders would camp on their journey west to trade grain grown on the Deccan for rice grown on the coastal plain. Taxation papers indicate a stable and well-established settlement with wool and cotton weaving industries in 1690.[3] The old trading connection with the coast was still alive in 1977 when the three or four Sugao goat wool weavers who were still operating their hand looms in the village carried all their blankets over the mountains for sale to the rice farmers of the Konkan.

Sugao is nestled in the foothills of low, flat-topped spurs of the Mahadev Hills. It is in a zone of moderate rainfall (averaging 28 inches per annum) well distributed over the year, allowing rain-fed cultivation of diverse crops. If the rains are good, Sugao has a reliable base of agricultural production. Its picturesque setting and patches of cool, green, irrigated land make it appear inviting during most seasons of the year, in contrast to villages further east, which are in a drought-prone region. Consequently, migration attributable to drought is not part of the typical migration pattern of Sugao people.

But the limited amount of arable land and the prevailing mode of cultivation, combined with the growth of the population, have meant that the people could not continue to support themselves with agriculture. Land in the Deccan, historically vested in individual peasant proprietors, has been gradually subdivided into ever-smaller holdings and sub-groups of original families. Sugao is no exception to this. The land-to-man ratio has decreased from 1.35 acres/person in 1942 to 0.84 acres/person in 1977 and 0.75 acres/person in 1990. Most Sugao villagers working in agriculture are owner-cultivators. Over the years land has been divided and subdivided among family members so that 399 of the 493 (81 per cent) Sugao families owned land in 1977. By 1990 the number of households in the village had increased to 601 of whom 492 (82 per cent) owned land. Many of these individual land holdings are minuscule, and the owners are virtually landless. However, even those with tiny parcels hold on to them and cultivate them diligently, both because the land is ancestral and sacred, and because they know that it is usually the only asset that will provide some security in old age. Selling land is a last resort.

Most farms in Sugao are not large or fertile enough to support a family fully. Faced with this reality, Sugao farmers, like those in other villages in Satara District, have two major options, to remain in the village and augment the income from their subsistence farm with locally available work, or to migrate. In 1977 almost half the men were living outside Sugao while maintaining their family and household in the village. Out of a total of 399 land-owning farm families, 75 had no men living in the village. Another 22 farms belonged to women who were alone, having separated from or been abandoned by husbands or having been widowed. Thus about a quarter of Sugao farms were cultivated primarily by women. In 1990 the proportion had decreased to about a fifth because men were back in the village as a result of the textile industry strike. But although the incidence of women managing family farms by themselves had dropped, the overall outflow of people from the village had not. Forty-five per cent of villagers were living outside Sugao, primarily in Bombay, in 1990.

Data on Sugao's last fifty years provide a typical profile of the migration that people from villages in the Deccan have made for economic opportunities. Given the historical linkages with the west coast, Deccan people have been attuned to emerging possibilities for earning a livelihood first from trade and then from work emerging in the developing parts of the coastal strip. Migration from Sugao has predominantly been by way of a single leap to the growing megalopolis of Bombay. The postulated theories of step-wise migration through an established urban hierarchy of towns such as nearby Wai and the nearest metropolis of Pune did not hold true in this region. The reasons for this are clear from interviews with Bombay migrants. Industrial development in Maharashtra remained concentrated in the megalopolis of Bombay until planned decentralization in the 1960s forced a relocation of old and a diversification of new industrial growth in second-order cities such as Pune and Nasik. The economies of smaller towns nearer to Sugao, such as Wai or even the more industrializing Satara to the south, did not expand to create the kind of work and pay scales that could be obtained in Bombay. In addition, villagers perceived fewer opportunities and more constraints to entrepreneurship. In interviews they claimed that some of the characteristics inherent in village interactions, such as the need to give credit to certain families or status groups, carry over into the nearby small towns, making it difficult to run a successful business. When launching a new venture the impersonal anonymity of Bombay is valuable. Most important, however, is that a preponderance of Sugao men's jobs have been obtained through the contact network of village relatives and friends already established in the city. Given Sugao's contacts with Bombay since at least the turn of the century, networks exist in the city for locating jobs for newcomers. In addition, a new migrant could live in communal places such as the Sugao *talim* (exercise place) while he looked for work. This enabled those who did not have a relative to stay with while they looked for jobs to move to the city anyway. Out-migration has meant that Sugao's resident population has increased only from 1,621 in 1942 to 2,583 in 1977 (1.7 per cent per year) and from 2,583 in 1977 to 2,888 in 1990 (0.9 per cent) per year (Table 3.1). This is a slower growth rate than for Maharashtra State overall, which has grown by more than 2.4 per cent per year since 1951 and whose rural population grew at the rate of 1.8 per cent from 1981–91. It appears that Sugao's capacity to sustain people has been reached and the resident population has levelled off as more people choose, or are forced, to migrate out.

Men to Bombay

Migration of men from Satara District to Bombay for work is relatively long standing. The first Satara District *Gazetteer* notes, in 1885, the movement of the most able men from the district for work in the docks and railways of Bombay.[4] At that time the migration pattern was seasonal.

Jobs in Bombay were relatively easy to obtain, and workers migrated to the city for the six months after the main growing season in winter,[5] when agricultural tasks were few. As for Sugao village specifically, interviews in the 1930s of Sugao men whose family members had come to the city indicate that a few had migrated to Bombay at the turn of the century, drawn to newly available jobs as dock workers in the port and as porters in the railways. Our case studies indicate that some of the families that are now well-to-do gained an economic edge as a result of their early entry into this migration cycle to Bombay. The expansion in the 1930s by Indian entrepreneurs of the textile industry, first established in Bombay in the 1850s, resulted in many more Sugao men finding jobs in the mills of Bombay.[6]

Most of the early Sugao migrants were men from higher caste Maratha families who migrated to the city seasonally while their women remained in Sugao, working within the joint or extended family, primarily in agriculture.[7] By 1942 the main occupation of the Sugao male migrant to Bombay was that of textile worker. By this time, temporary jobs had become much more difficult to obtain, and migration of men to work in the city was no longer seasonal. To keep their positions, men remained at work in Bombay year round, returning to their family once or twice a year for vacations. Jobs in the textile mills had become not only more difficult to obtain but also more desirable. The villagers

had begun to appreciate the value of reliable wage-work in Bombay and the cash income it provided. The migration pattern that emerged encompassed the entire life cycle of an increasingly larger number of Sugao men, who moved out of the village at age 18 or 20, spent all their working life in the city, and returned to Sugao upon retirement, bringing their savings to invest in the village.

The connection to the coast has thus deepened in modern times. One of the most significant characteristics of Sugao today, apparent even from the most casual stay in the village, is the extent to which migration from the village in general, and linkages with the city of Bombay in particular, affect the very fabric of village society. Migration out of Sugao has been increasing. The percentage of migrants from the resident population of Sugao has continually increased from 17 per cent in 1942 to 24, 29, and 45 per cent in 1958, 1977, and 1990, respectively. Few families remain untouched. As early as 1942, almost half (48%) of Sugao families had at least one person who migrated out of the village for work, and 14 per cent had two or more people located outside. By 1977 almost as many adult men were earning a living outside the village as there were men living in the village. More than half (57%) of the families had at least one migrant, and 27 per cent had two or more. This trend has continued into 1990. The primary movement of Sugao men continues to be to Bombay.

Table 3.1 provides some details about who in

TABLE 3.1   Caste Distribution of Sugao Population in Village and At Large, 1942–1990 (percentage)

Caste	1942		1958		1977		1990	
	village	at large	village	at large	village	at large	village	at large
Marathas	67	57	68	59	70	64	72	62
Dhangars	10	8	8	10	7	9	7	10
Artisans	11	11	10	9	11	9	9	10
Neo-Buddhists	9	20	9	18	6	14	5	14
Others	3	4	5	4	6	4	7	4
TOTAL	100	100	100	100	100	100	100	100
No. of Persons	1,621	271	2,040	488	2,583	737	2,888	1,304
Migrants[a]		14		19		22		31

[a] Proportion of migrants as a percentage of the total village and at large population.

Sources: Full-census surveys of households in Sugao in 1942–3 and 1958–9 by the Gokhale Institute of Politics and Economics (Pune), similar surveys of the village by the author in 1976–7, and supervised by the author in 1990.

the village community actually migrates. The table illustrates that the proportional caste representation within the village is more or less reflected in the proportion of migrants from that caste. The exception is that of the Neo-Buddhas or former Mahar community (which used to be considered to be an untouchable group). More obviously in the 1940s and 1950s, but continuing to today, migration from this group is higher than in other castes, the asset base of the group, whose status remains one of the lowest in the village, is very poor. Government efforts to reserve jobs have made the move to the city an attractive alternative to the low-paying day-labour work available in Sugao. A similar but less pronounced increase in relative proportions migrating out of village can be seen on the part of the Dhangar (sheep-herding and blanket-weaving) community. This group of Dhangars has higher status in Sugao than in other parts of the state; it claims to be descended from the original settlers of the village.[8] The Sugao Dhangars have lost one of their traditional sources of earning income in the village, that of weaving goat-wool blankets on hand looms in their houses and selling them to the farmers in the Konkan. Blanket-weaving activity had become non-competitive and died out by 1990. As their land holdings have become subdivided and diminished in size, the Dhangar community has relied more and more on migration out of the village for survival. Other artisan activity and small village industry that offered a means of survival at a subsistence level to various artisan caste groups in the village, through a system of reciprocity and pay-

ment-in-kind for services rendered (called *baluta* in Maharashtra), had also died out completely by 1990 in Sugao.[9] Therefore one finds that for this group too, life in the village, as it becomes more and more monetized with few options for exchanging services for necessities, becomes difficult. The proportion of villagers from these communities has decreased relative to 1942 and their proportion of migrants has increased.

However, a significant fact revealed by the four point linked surveys, is that, overall, very few men leave Sugao permanently (Table 3.2). The men who left permanently between 1942 and 1958 constituted 1.2 per cent of the 1942 population and female out-migration was 1.7 per cent of the same. Overall out-migration was thus 0.18 per cent per annum. The men who left between 1958 and 1977 constituted some 4.6 per cent of the 1958 population of 2,040 persons, and female out-migration was 5.5 per cent of the same. Overall out-migration was thus 0.53 per cent per annum. Between 1977 and 1990 it was 6.2 per cent of total Sugao population in 1977 of 2,583 persons, and female out-migration was 4.5 per cent of the same. Overall out-migration was thus 0.83 per cent per annum.[10] Although this constitutes an increase from .07 per cent per annum in the first period to .245 per cent per annum in the second to 0.47 per cent per annum in the third, the increase is relatively small and the village population of men has thus remained quite stable. This fact and the persistence of a circular migration to, and back from, the city is important, for it highlights the continuity of village relationships and connections despite what appears

TABLE 3.2   Permanent Out-Migration of Sugao Residents between 1942 and 1990

Place migrated to	1958			1977			1990		
	male	female	total	male	female	total	male	female	total
Bombay	2	1	3	26	21	47	77	33	110
Wai Taluka	6	7	13	12	10	22	5	6	11
Satara District	2	—	2	6	2	8	7	4	11
Pune	—	—	—	4	2	6	7	4	11
Other	—	1	1	1	3	4	27	22	49
Father's House	—	11	11	—	15	15	—	7	7
Don't Know	10	7	17	44	59	103	38	41	79
TOTAL	20	27	47	93	112	205	161	117	278

Sources: Same as Table 3.1.

at first glance to be a tremendous outflow and turnover in the village community. At least at present, increasingly larger numbers of village men are moving to the city for a majority of their working lives, as they see few opportunities for themselves in the current rural economy. However, their base in the city economy is quite fragile and they are not remaining there in any numbers when their productive work life is over.

The cost of living in the cities continues to rise and available city amenities and services are less attractive than they used to be, even apart from the price. Also, it is now possible to obtain some of these amenities, at a much lesser cost, in the village itself as it becomes physically more upgraded and comfortable. Remittances from city workers have been instrumental in improvements of a significant proportion of housing in the village. Almost half the households now have water piped into their units; more than two-thirds of them have electricity; and one or two of the 19 bio-gas digesters constructed by 1990 have attached enclosed toilets. Many of the men who migrate say that they would prefer to stay and work from or in the village if they could. But development and growth in India to date have not significantly enhanced job opportunities in rural areas outside of agriculture. For those who do not own adequate amounts of cultivable land, migration to the city is often the most attractive option. The effects of an urban-oriented development strategy are thus apparent in the flow of men to Bombay out of the microcosm of Sugao village.

Women at Home

The quite small incidence of permanent male migration out of Sugao is paralleled by an incidence of women who leave the village permanently. Between 1942 and 1958, 1.7 per cent of the total 1942 population migrated (as opposed to 1.2% for men). Between 1958 and 1977 this figure rose to 5.5 per cent (as opposed to 4.6% for men), but between 1977 and 1990 the figure for women was 4.5 per cent (as opposed to 6.2% for men) (Table 3.2). Interviews revealed that many of these women had come to the village because they had married Sugao men. The two largest family groups in Sugao (the Jadhavs and the

Yadavs), which constitute a significant proportion of the major Maratha community in the village, do not intermarry. So there is a substantial marrying into Sugao of Maratha women from other villages and a marrying out of Sugao daughters. But if, for one reason or another, the marriage fails (for instance, a woman can be abandoned for not bearing a son), these women leave Sugao, often returning to their maternal village to their parents' or a brother's home. The increases in the percentage of women leaving Sugao in the latest two periods may be mirroring the fact that overall women's worth in the marriage transaction has suffered as a result of the monetization of the rural economy and the demise of extended-family cultivation of family farms. This phenomenon as manifest in Sugao has been discussed elsewhere.[11] However, as has been noted in these earlier publications, since women's claims to village assets, polity, and identity are quite weak, this departure of vulnerable women with lower status does not cause a sense of instability in the village psyche.

The extent of the migration of men out of villages in the Deccan is apparent in the sex ratios. Comparing the number of females per 1,000 males in the village and in the surrounding rural areas indicates the extent of male out-migration from a village relative to that of the region—the greater the out-migration, the more females there are per male. In 1971 rural areas in Wai Taluka (1,129 females per 1,000 males) were sending out more males than was average for rural areas of Satara District (1,062 females per 1,000 males) in which Wai Taluka is located. Sugao, located in Wai Taluka, had 1,226 females per 1,000 males in 1977, above the average for rural areas of Wai Taluka in 1971. The village was thus sending more migrants to Bombay than its neighbours in the Taluka, probably because of its long-established network of contacts with the city. However, it was located in the middle of the spectrum of sending villages. In the same vicinity, in hamlets and smaller villages with less fertile land, practically all the men had gone to the city. By 1991 the picture had changed to some extent. The 1991 census indicates a ratio of 1,060 females per 1,000 males in rural areas of Satara District, close to what it had been in 1971. But the difference between the rural areas of the district as a whole and Wai Taluka had narrowed. In 1991 the rural areas of Wai Taluka had 1,107 females per 1,000 males, down by 22 perhaps reflecting the effect of the return of men to the area who had been in the textile industry. By 1990 Sugao had 1,192 females per 1,000 males as compared to 1,107 in rural areas of Wai Taluka.

Historically, single women have not moved to the city from Sugao to find jobs. It was higher caste Maratha men who first went to the city to find work. These men, who were from landed peasant families, adhered to the principle, in keeping with their higher caste status, that their women were to remain within the home involved in domestic chores. The work outside the house on a joint-family farm was acceptable, but the family would lose face and status if their women had to work as labourers on other's fields or if they were required to leave the village for city work. The results of this attitude and perception is apparent today. The numbers of women migrants working outside Sugao remains quite small. The women who do migrate are almost exclusively wives accompanying their husbands to the city. For men the migration is substantially different and predominantly for work. Over two-thirds of the migrant men are married, although the proportion of unmarried males is increasing slightly. In 1942 and 1958, 29 per cent of the migrant males were unmarried; by 1977, 31 per cent were unmarried, probably due to the rising age at marriage for Sugao as a whole.

The reason families are left behind in the village is largely economic. Very few of the migrants can afford to take their wives with them. Housing in the city is scarce and when available, its rents are high. The day-to-day necessities cost more, and men in lower wage jobs cannot support their wives in the city. Some of the men who did earn larger incomes felt that they would not be able to continue to help support their extended families if they brought their wives and children to the city. Thus personal happiness and comfort were sacrificed for the good of the larger joint family. Having 'one foot in the city and one in the village' is a necessity for many Sugao families that do not have enough assets to survive solely in one or the other place. The resulting prolonged separations cause emotional stress and hardship in the personal lives of individuals, particularly in the relationship between husband and wife. When a young wife is left in the village with a joint family that does not have much land, cultivation of which requires all

the family hands, the joint family tends to split into nuclear families. The reason for staying together, to maximize the output from the land using family labour, no longer exists. The productive asset, land and labour, and the products are no longer a joint undertaking. Conflicts erupt over how to share the proceeds from salaried work where there is direct accountability for who earned what. Joint families tend to split up into nuclear ones and the land is subdivided so that it is less and less viable as a farm. This can be seen quite clearly over the years in both the land-poor Dhangar community and the Neo-Buddha community.

Often when a joint family is cultivating land in Sugao and has two or three men in Bombay, one of the men's wives will be sent to the city to cook and keep house for the men. Wives of brothers in Bombay sometimes rotate trips to the city so that each brother's wife lives with him for some part of the year. In one or two of these families, the women have started a *khanawal* business in which they cook meals in return for cash payments from a number of village men living alone in Bombay. However, this requires considerable efforts on the part of the women and has never been considered a very high status activity. Often *khanawals* are established by women who have no one to look after them and their families in the village and who migrate to the city in distress to try to earn a living. Helping someone who is catering meals with the hope of building up enough assets to establish oneself similarly one day has been one way to survive for these women, most of whom are not well educated. It is an occupation just a step up from domestic work. Interviews with some of these Sugao women who were operating *khanawals* by themselves in Bombay revealed that a woman starting this business needs some protection and patronage from an established and more powerful man if she is going to succeed in keeping the business profitable.[12]

Over the last fifty years few women from Sugao have managed to obtain white-collar jobs in the city. There are exceptions, most of them quite recent. In the last decade a few of the younger and more educated women from Sugao have migrated to cities to obtain advanced training or employment commensurate with their education. A notable case is that of a women who has become a doctor and obtained both her own flat and a commercial space to set up her own practice in Bombay. She is really a product of the city. Her father was in the police service and had no sons, and emphasized education for his daughters. Two sisters from another family moved to the city. The elder sister obtained a job in the Bombay police force; the younger, with the help of her sister, became trained as a nurse and secured a steady, salaried job in a private hospital in the city. Yet another exceptional woman has put her post-graduate degree in commerce to use in managing her doctor-husband's thriving hospital in a small town near Sugao. But cases like these are very much the exception. The incidence of single young Sugao women leaving the village for better work opportunities is still so low as to be statistically negligible. What these exceptions do reveal, however, is that the more educated Sugao woman, like the man, finds more lucrative work in the city or small town than in the village.

City Jobs and City Life

Although the major impetus in the last fifty years for migration out of Sugao village has been the opportunities for work in the textile industry in Bombay, the type of jobs that Sugao men have been able to obtain outside the village has been changing gradually. In 1942 most of the Sugao male migrants worked in the textile mills (67%) and other factories (4%). These jobs were generally manual labour, which had little status and relatively low pay. By 1958 a smaller proportion of men were in the textile mills (58%) and by 1977 this had dropped to 41 per cent. At this time another 10 per cent were in other manufacturing industries. Thus, there was an appreciable drop in the number of Sugao workers employed in textile mills and other factories from 70 per cent in 1942 to 51 per cent in 1977. This move to work in other activities and sectors has been important in their survival. Over the years, men from the village have become better educated, have developed more contacts in the city, and have been able to find better, more diversified work in the city. Thus, the initial jobs in the textile industry have been an avenue for upward mobility for village men. Nevertheless, textile mill work has remained a mainstay occupation

for Sugao migrants, although it dropped from 67 per cent of all migrants to Bombay in 1942 to 41 per cent in 1977. The fact remained that in absolute numbers more men from Sugao were in the textile industry in 1977 (195) than in 1942 (135). Much has been written about the specific reasons for the decline of the textile industry in Bombay.[13] The details continue to be debated but it seems clear that the long-term prognosis for the viability of textile industry in Bombay is not good and more and more Sugao men will need to look for work elsewhere.

This diversification had already started by 1977, when an increasing number of men, almost 11 per cent of the migrants, were able to find employment in government services such as the military and the police. They have security of tenure and offer highly valued benefits like a life-long pension and government housing for the duration of the work. Those who had found jobs in the armed forces, mainly in the army, have been posted to various regions in India. In 1942 a negligible number of Sugao people had found jobs in such occupations. The government's policy of reserved quotas of government jobs for depressed and backward castes had enabled the more educated men from the Neo-Buddha community and from artisan castes such as cobblers and carpenters to obtain such work. The policy of reserving jobs for particular groups has been reinforced in recent years. This has increased the percentage of reserved slots for members of groups classified as depressed. This trend of diversification is expected to continue. In 1977 as many as 12 per cent of the Sugao migrants were in white-collar and teaching positions, whereas almost none had held such white-collar jobs in 1942. This diversification has continued as the textile industry has shrunk in Bombay. Sugao men's long-term connections in the city have enabled them to find alternative job opportunities. It appears that overall the male migrants from Sugao have enjoyed a fair degree of upward mobility in their employment in the last five decades.

Some of this upward mobility of Sugao men can be attributed to better training and education. In 1977, 10 per cent of migrant males had been educated beyond high school, either with some technical training (4%) or some college education (3%). There were in fact some migrants who held college degrees (3%). Only 8 per cent of the men were illiterate, a significant shift from 1942, when 52 per cent of the migrants were illiterate and not one had been educated beyond the seventh grade. A few of the better educated individuals from the more affluent families in the village have remained in the village to manage their family farms. But among the poorer families almost all the educated men leave the village, as there are few opportunities for them in the village itself. It is clearly a push migration out of the limited village reality where there are precious few options for work if you do not own substantial amounts of land. Thus one observes a diversification of flow of men by caste from predominantly higher caste Maratha men in the earlier years, at the turn of the century, to include other groups at present. There is, however, a distinction by caste and status, and differences in objectives in this stream of village migrants. When migrants from the wealthier families in Sugao choose to leave the village, it is often for better opportunities in the city. Once there, they have the economic backing, education, and influential network of relatives and family friends needed to bide their time and obtain better paying, less arduous jobs. Those from the less well-to-do, lower caste communities are more often forced to leave the village to survive. They take what work they can. The fortunate find places in government services, benefiting from job quotas and reservations. Others may remain in menial and insecure petty trade and services for all their lives and pass that legacy on to their children.

The diversification of work from the textile industry enabled Sugao men not only to weather the textile industry strike but also to have some confidence in getting a more secure foothold in the city. A salaried job that offers life-long benefits is not the only kind of success. Some entrepreneurs have made good in business enterprises in Bombay. One Sugao migrant invested his savings from contract hamali or coolie work in a trucking business. He now owns several trucks. Another established a flower-vending stall in the middle-class shopping area of Dadar early in the 1940s. This has been a very successful venture and the family has consolidated its housing in the village as well as acquired assets and a home in the city. This entrepreneur now not only operates the flower-vending stall in Dadar, but owns the shop and a comfortable apartment in the city.

His house in Sugao has been upgraded to a two-storey, cement concrete structure, boasting some of the most modern amenities in the village, including an indoor toilet. Another entrepreneur from the minority Muslim community in Sugao, a group that has traditionally performed the function of butchers for the village, now owns a few trucks and has parlayed a business of transporting goods into a lucrative activity from which he has earned enough to buy an apartment in Bombay. He has little inclination to return to Sugao to his one-room, decrepit, ancestral home there, preferring to work in Bombay and live with his mistress, a woman who was married into a higher status Maratha family in the village. His elderly widowed mother and his wife and children live a poverty-stricken life in Sugao and he contributes rather modest remittances to their maintenance. Establishing a business in Bombay is risky, but it has been the road to substantial success for a handful of Sugao men. Some, like this Muslim man, have been able to flaunt village conventions and societal traditions as a result of this success.

On the other hand, some of those migrants who had anticipated in 1977 that they would make a break with the village have not done so as their foothold in Bombay was jolted loose by the effects of the textile industry strike on their textile mill jobs. In the early 1980s, I asked one of the Sugao men who was a foreman and a leader of those in the textile industry, someone who had said he would remain in Bombay after he retired from mill work, what would happen to Sugao men if there were a textile industry strike? He answered, 'People like us, from the Deccan— we can go home, live off the land. Konkan people have no land, so what can they do? They have to stay, but in Bombay they have to give money to the khanawal bai just to eat and survive. They have a real problem. It will be difficult for the people of the Deccan, but they will go home. The land will support them.' And in fact this is exactly what this particular individual did when the textile strike took place. The demise of the industry in Bombay that followed in the years following our conversation fundamentally affected his position and stature in Bombay. He was no longer the leader with powers to provide access to mill jobs nor one whose advice or support was important should one get into some trouble while in mill work. He returned home to the village and built himself a small but comfortable two-room house, separate from his brothers, on the outskirts of the village, on ancestral land. There he is living out his retirement years. The city job was the means by which he could do this and live quite comfortably. However, in Sugao, he is not a member of the leadership that counts. There is not the same group of men clustering to talk to him, seeking his advice, that one encountered when one visited him in Bombay. In Sugao, he is one more Bombay man come home to spend the rest of his life. His opinions, his experience, and his vision do not count for much in the context of village realities. But his claim to a piece of village land, to village ancestry, has allowed him to operationalize a right to be there. Without the land as a tangible connection, he would not be as welcome.

Land is in fact the ultimate reliable sanctuary in the eyes of the village people. The major asset the city worker on wage-work takes back to Sugao is the provident fund (a lump-sum payment made at retirement) and gratuity. Even if there is not much land to return to in Sugao, most migrants find that, once their jobs in Bombay end, they have no means with which to go on living there. Few are on salaried jobs that pay a life-long pension. Thus, most workers do not continue to earn enough to enable them to live in the city when they leave their jobs. This is particularly true as few migrants are able to own their living accommodation in Bombay. They generally plan to return to Sugao when they retire, and many of them have said that, if they could earn even close to what they do in the city in the village, they would prefer to live and work there. Even if all they have in Sugao is an acre of indifferent land and a tiny dilapidated house, they can invest their savings in home improvements, in acquiring one or two milk animals, and in better cultivation of their land. The village offers a healthier environment, better food, and a more congenial social life. The climate is good and the lack of amenities such as piped water and sanitary facilities is not a problem as the returnee was reared in those conditions. Also, some of these conditions are being ameliorated with private investments in amenities such as piped water connections to a house and electricity.

The villagers' preference for rural life is easy to understand when one sees the lack of amenities, the pollution and congestion of the rapidly urbanizing parts of India today. Urban life in India has become more difficult for people from most social strata. The ability of the city governments to provide even the most basic of services and amenities has been stripped by demands of a burgeoning urban population. It is a daily battle for most of the poor and much of the middle class to take care of the daily necessities, to feed and clothe themselves, and get to work and back on congested roads and in crowded public transport. Certainly, survival and perhaps economic gains have resulted from Sugao migrants' forays into the city. However, the costs borne by the migrant in quality of life in the city, and by him and his family in psychological and emotional terms, cannot be ignored. In interviews of Sugao men in Bombay a recurring refrain was the misery of daily living conditions in the city and a stark articulation of their understanding of the lack of economic options open to them in Sugao.

The Economic Impact of Migration on Sugao

In Maharashtra today, increased urbanization and migration to cities appears to be a fact of life in the immediate future of much of the rural population. This is to be observed in the lives of Sugao families. In the last fifty or so years that we have obtained a fine-grain understanding of Sugao life and Sugao people, the village has become more and more closely tied to the world around it. With planned decentralization of industries, corridors of industrialization have been made to spread from Bombay out to the rapidly growing cities of Pune and Nasik to the south and north-east. But the shifting of industrialization from a megalopolis, to two emerging metropolises has done little to change the basic economy of village entities such as Sugao. Fingers of urban-industrial development spread out along corridors such as the Pune–Satara highway. They edge by Sugao. The nearest village on the Pune–Satara highway has become a very large truck-stop experiencing all the monetary benefits and social problems that result from participation in the 'highway culture'. But few new jobs have become available for Sugao people in the immediate vicinity of the village allowing them to enjoy the relative tranquility of the village environment as they earn a living in off-farm occupations. Most Sugao men would seize such opportunities were they to become available. They are acutely aware of the personal, physical costs they incur as they travel to Bombay to make a living there. They are financially better off with this movement, but they know in intimate detail the price they pay for their survival and success.

The fact that earning a surplus in the city is getting more and more difficult to do is reflected in how much migrants are sending home to the village. Although the number of families with migrants has increased, the proportion of migrants who send remittances home has declined. In 1942, 85 per cent of the migrants sent money home; in 1958, it was 81 per cent; and in 1977, only 74 per cent. This decrease may be linked to the decline in the real earnings of migrants working in Bombay. Usually the individual who leaves the villages manages to support himself in the city even if he cannot send money home. So, in effect, migration for some families simply takes a person off the land, leaving one less mouth to feed on village assets. The average remittance to Sugao per year for the 209 families that did receive money from outside the village was Rs. 1,088 per family in 1977. This accounted for 25 per cent of the total income for the village for the year. Some of the families in the land-poor Dhangar and Mahar communities were completely dependent on monthly money orders from outside for daily consumption over and above what they produced for themselves.

As the population of rural areas such as Satara District has increased, the dependence of rural economies on the city has increased. Currently Sugao lands support only 25 per cent of its population in comfort. This, combined with the availability of jobs in the city, particularly in Bombay, and the resulting mass migration of men for work, has resulted in the increasing integration of the Sugao economy with that of the region. Sugao men living in Bombay send packages home containing consumer goods such as clothing, tea, sugar, cosmetics, and toys. In 1977, 7 per cent of families in the village claimed that they received all their consumer goods from Bombay: an additional 13 per cent said they received about half; and another 7 per cent received from one-tenth to one-quarter. Thus

more than 25 per cent of the village families were getting at least one-tenth of their consumer goods from the city. By 1990 a negligible number (2) of the families in the village received all their consumer goods from the village; 9 per cent said they received more than half; and more than a quarter (27%) claimed they received from one-tenth to one-quarter. Thus, almost 37 per cent of the 1990 Sugao families were getting at least 10 per cent of their consumer goods from the city (an increase over 25% in 1972) but a lower percentage (10% versus 20%) were getting more than 50 per cent of their consumer needs satisfied by the city. The connections to the city have become tangible for more families but the capacity, or need, to send goods from outside to the village has declined. At the village's present level of development, and despite migration to Bombay, Sugao farms can no longer fully support its residents. In-kind remittances from the city are a norm for a substantial proportion of village families.

The more affluent families with members working in the city do not depend on the monthly remittances to meet their daily consumption needs. Migrant remittances and savings enable these families to make lump-sum investments in major purchases—land, agricultural equipment, and housing—enabling them to increase their asset base and consolidate their dominant position in the village. A major share of the money spent in Sugao on building new, or upgrading existing, housing, investing in new industry, and other fixed assets such as land comes from the accumulated savings, gratuity, and pension payments of migrants' earnings in the city economy. Those families that have remained on the land in Sugao and have consolidated their asset base and begun to diversify are also investing in their homes. Next to land and its improvements, upgrading the quality of housing and improving the amenities provided is a major focus of investing wealth in Sugao. This is reflected in the overall improvement of housing stock in the village. The 1990 survey of Sugao revealed the increase in amenities and comforts that had become available in the village. Piped water to individual households not available in 1977 now reaches 45 per cent of the village households, and 69 per cent of the households now have household electrical connections, as opposed to 35 per cent in 1977. There are now 19 new bio-gas digesters installed and functioning in the village, with one or two attached to indoor toilets and plumbing, and providing gas for cooking. In 1977 there were none. In 1980 an investigation by the author had revealed an interest in the concept of bio-gas but no investment on the part of the more affluent families who had the resources to make such an investment economical.[14] The rising price of energy, coupled with an improvement in the design of small, household-size bio-gas digesters to fit village resources and spatial conditions, have facilitated this investment. In this decision to invest in bio-gas digesters, as in other decisions like it to invest or not in a technology endorsed by the development wings of state and national government, Sugao people have displayed acute judgement about 'fit'. They display sound economic understanding of technology and wait until the fit is right with their resources before they make the investment. One thus finds that the rising prosperity of the well-to-do families in the village has resulted in purchases of consumer goods that were earlier confined to urban areas. For example, by 1990, 10 households owned a burshane (bottled gas) stove for cooking, and there were 14 television sets in individual homes, as opposed to 1977, when there were none.

Thus the migrants from Sugao make a substantial contribution to the basic economic survival of a number of families in the village. Remittances supplement earnings eked out of meager land holdings. They also improve the asset base of some families by allowing lump-sum investments in agricultural machinery, livestock, agricultural inputs, and housing. Whether for survival or 'development', the migrants and their families pay the price—thirty years or more of separation during the most productive years of a man's life. During this period the village loses its best-educated, most able men to the city. Over the years the migrant's lot in the city has become more and more difficult. Increased congestion, a deteriorating environment, tighter job markets, and an income that is decreasing in real terms are the realities of city life today. This decline is reflected in the falling off of remittances per migrant in real terms, although they still constitute a large share of the total village income. The city continues to absorb the surplus manpower resulting from rural population increases. However, it is at an increasing cost—physical, psychological, and economic—to the migrant and his family as well as to the village.

Conclusion

The size of the village resident population appears to have stabilized over the last thirteen years (Table 3.1). Informal conversations with Sugao farmers, both men and women, in 1992 revealed a new sense of vigour among the younger, more educated farmers who own some irrigated land. They are experimenting with crops targeted to evolving urban tastes and preferences. For example, they are concentrating on producing new crops such as fresh fruit or expanding production of traditionally minor crops such as salad greens or tomatoes. Those who remain on the land and are doing well in agriculture are responding to market demands. They articulate optimism about the potential to make a good profit from the land. But the experience of a great number of poorer people, those who do not have a significant asset base of land, is different. They choose to move away from the village or are pushed from it. Not to move means being marginalized in agricultural labour in Sugao as there are rather few new options in the village.

Life in the city for most migrants is quite difficult, as it is getting to be for those in most strata of the urban Indian population. In the modern Indian metropolis one must be very wealthy to be able to insulate oneself from the tribulations of commuting to work on congested urban transport, feeding, clothing, and sheltering oneself in contexts where the supply of housing and services have been unable to keep up with the demand, and in protecting oneself from the insidious effect of pollution of the urban environment, from the poor quality of the air, the contamination and unreliability of the water supply, from ambient noise, and a host of other unpredictable changes in the physical contexts of one's daily life. As one observes the miserable living conditions of the urban Indian poor, one realizes that the life which one sees them lead is a far cry from what one had wished for from development.

The choices Sugao people have made about where to live at what stage of their lives, and how to earn a living and survive, reflect the reality of changes in the village that have resulted through planned and unplanned changes in modern life. They are the result of the industrialization and urbanization processes put into place during the past five decades. What is clear from this case is that the effect of these forces on the rural peasantry has been significant in the disruption of old ways of subsistence through small-scale agriculture. The push out of the village of those with little land is apparent. The choice to persist in returning to the village reveals the limits of the benefits to be derived from the urban sojourn.

Notes

1. Census of India 1991, Final Population Totals, Paper 1 of 1992, Vol. 1, Series 1: India. Present-day Maharashtra is still predominantly rural, as some 61% of its population continues to live in the countryside. But, given its urban growth rate of 3.8% between 1981 and 1991, demographic projections indicate that by the year 2000 more than half of the state's population will be living in cities. The *Times of India*, 16 July 1994, Daily Edition shows in a table providing data on employment in major, medium, and small industrial units that Maharashtra tops the list of states for numbers of units of major and medium industries and for total employment in all industries.

2. The Gokhale Institute of Politics and Economics, Pune, provided the author with raw data, in the form of completed questionnaires from comprehensive socio-economic, full-census surveys collected in Sugao during 1942–3 and 1957–8. The survey was replicated by the author in 1977 during an eighteen-month stay in the village. This was followed by three months in Bombay to complete a purposive sample of Sugao migrants there. Subsequently, the author made numerous visits to the village and to the migrants in Bombay. In 1990, under the author's supervision, a full census survey of Sugao was completed by the same senior field-worker who had executed the 1942 and 1957 surveys for the Gokhale Institute and who had helped the author with coding the 1977 survey. Thus a great deal of consistency and continuity has been achieved in the collection, processing, and tabulation of data on Sugao. In addition to the author's own field observations, use was made of anecdotal information collected in the early 1930s by the Gokhale Institute about families who had some members living in Bombay during the early 1900s.

3. Records in the Poona Archives (Innam Daftar, Satara District, Section 44) show that in 1690 there were 43 houses in the village which were taxed and 5 wool-weaving and 6 cotton-weaving looms.

4. Maharashtra State *Gazetteers*, Satara District, Directorate of Government Printing, Maharashtra State, 1963, which is a revised edition of Vol. XIX of the original *Gazetteer* of the Bombay Presidency relating to Satara, notes on p. 187 that 'except for the tendency on the part of well-built, able-bodied males who go to Bombay in search of work chiefly in the docks or railways as porters and others who join the armed forces, no drift towards industrial cities as such is noticeable among the Kunbis, Marathas or Mahars in the district'.

5. Kharif season is the main growing season in winter, following the monsoon rains.

6. Chandavarkar in *The Origins of Industrial Capitalism in India: Business Strategies and the Working Classes in Bombay, 1900–1940* (Cambridge: South Asia Studies, 1994) notes (p. 24) that, when the first cotton mills were built in Bombay in the 1950s, the industry was, significantly, pioneered and developed largely by Indian enterprise. Although its location became more dispersed as the industry developed, Bombay remained, until the end of 1940, the largest centre of India's most important industry, employing almost a quarter of the labour-force working in perennial factories.

7. This caste selection seems to have been typical for textile labour recruitment in Bombay. Kunj Patel (*Rural Labor in Industrial Bombay* (Bombay: Popular Prakashan, 1963), notes (p. 43) in her survey of 500 workers from the Konkan that almost 63% were Maratha. Caste remains an important factor in village life, its influence permeating many interactions and affecting people's opportunities and decisions. Sugao is dominated by the Maratha caste, which made up 72% of the population in 1990. The caste composition of the village has remained fairly constant over the years, but the Marathas have consolidated their strength relative to the other groups. Table 3.1 shows the locally accepted status hierarchy, with the major caste groups besides Marathas being Dhangars (weavers), a mixture of Artisan castes of varying status, and Neo-Buddhists (formerly Mahars and considered untouchable).

8. See ch. 2, 'The Settlement and Power', in H. C. Dandekar, *Men to Bombay, Women at Home: Urban Influence on Sugao Village, Deccan Maharashtra, India*, 1924–1982 (Ann Arbor: University of Michigan Center for South and Southeast Asian Studies, 1986), 53–73, for a description of resource distribution, settlement

location, and status of various groups in the village over time. Although the Dhangars are typically classified with other artisan groups in Sugao, their higher status (by virtue of early occupation of the village site) is substantiated by the fact that their houses are large, solidly constructed, and located in the middle of the Maratha enclave. It is unlikely that the group could have obtained such strategic home sites and valuable irrigated land in a village first settled by the higher caste Marathas.

9. For a detailed description of rural industries and non-farm activity that traditionally existed in Sugao, see H. C. Dandekar, with S. Brahme, 'Role of Rural Industries in Rural Development', in R. P. Misra and K. V. Sundaram (eds.), *Rural Area Development* (New Delhi: Sterling Publishers, 1979), ch. 8, 122–43.

10. In this tally, individuals who had left Sugao, having sold their house and land, and who village residents reported as having settled elsewhere, were counted as permanent out-migrants.

11. See, for instance, H. C. Dandekar, 'The Impact of Bombay's Textile Industry on Work of Women from Sugao Village', *Third World Planning Review*, 5: 4 (Nov. 1983), 371–82; 'Indian Women's Development: Four Lenses', *South Asia Bulletin*, 6: 1 (Spring 1986), 25–9.

12. For details of women's work in the city and in Sugao, see Dandekar, 'The Impact of Bombay's Textile Industry on Work of Women from Sugao Village'.

13. See e.g. Prafull Anubhai, 'Sickness in Indian Textile Industry: Causes and Remedies', *Economic and Political Weekly*, 26 Nov. 1988, M–147 to M–157 and discussion by Bagaram Tulpule, 18 Feb. and 22 July, and Prafull Anubhai's responses, 8 July 1989.

14. For a description of bio-gas and its reception by Sugao people, see H. C. Dandekar, 'Gobar Plants: How Appropriate Are They?', *Economic and Political Weekly*, 15: 20 (17 May 1980), 887–93.

4

Life in a Dual System Revisited: Urban–Rural Ties in Enugu, Nigeria, 1961–1987

Josef Gugler

We are strangers in this land.
If good comes to it may we have our share.
But if bad comes let it go
to the owners of the land
who know what gods should be appeased.

(Achebe, *No Longer at Ease*)

IN 1961–2, I carried out a study of the ties urban residents in what was then Eastern Nigeria maintained with their rural areas of origin. At the time, these urban residents invariably stressed that they were strangers to the city. Regardless of his birthplace, every man could point without hesitation to a community he considered his 'home place', i.e. the community in which his forefathers lived, even if he himself was not born there. Women, upon marriage, would adopt their husband's 'home place'. People commonly referred to their 'home town', but in fact it would be a rural community except for the few who descended from families long established in the towns that arose as trading centres in the eighteenth and nineteenth century: Onitsha, the commercial metropolis on the Niger legendary throughout West Africa, the towns on the coast that had their heyday in the nineteenth century before the British managed to bypass them and trade directly with suppliers inland. Most urban residents identified with their rural home, felt that they belonged there, and affirmed their allegiance. The home community conversely referred to the men in the city as 'our sons abroad'. They were expected to maintain contact and to return eventually. Only a few had broken contact altogether, and the hope that they would return one day was not abandoned until word of their death was received. As the Igbo saying has it: 'Nwa Egbe anaghi ato na uzo ije' or 'The son of a hawk does not remain abroad'.

These same urban residents were fully committed to urban life. Most of the married men had their wife and children living with them in town. Nearly all expected to spend their entire

This is a revised version of an article that appeared in *World Development*, 19: 5 (1991), 399–409. I wish to thank, without implicating, William G. Flanagan, Ulf Himmelstrand, Katharine P. Moseley, Edward Mere Ntomchukwu, Monibo Abiriba Sam, and two *World Development* reviewers who commented on various drafts, and Christine S. Gugler who edited a draft. By permission of Pergamon Press.

working life away from their home place. Losing their urban employment or trade was the worst calamity that could befall them, and employers reported that labour turnover was indeed minimal. I concluded that Eastern Nigerians in the city lived in a dual system: they belonged not only to the city in which they lived, but also to the village from which they or their progenitors had come (Gugler, 1971). In terms of social network analysis, the social networks of Eastern Nigerians resident in urban areas were focused on two poles, and these poles were well connected.

In 1987 I returned to Enugu, the focus of much of my earlier research.[1] I had carried out a survey of urban–rural ties among several occupational groups there in 1961, when Enugu was the capital of Eastern Nigeria. The 1987 study is a replication of that survey. While Eastern Nigeria had been divided into four states in the mean time, Enugu, now the capital of Anambra State, remained by far the largest city in the region.[2]

We quota-sampled men from the same five occupational groups:

- *market traders*
  1961: 49 traders at the main open-air market
  1987: 61 traders at several open-air markets
- *blue-collar workers*
  1961: 49 workers at the railroad workshops
  1987: 50 workers at the railroad workshops
- *retired junior staff*
  1961: 38 junior staff retired from the Enugu colliery
  1987: 59 junior staff retired from the Enugu colliery
- *senior civil servants*
  1961: 58 senior civil servants of the government of Eastern Nigeria
  1987: 42 senior civil servants of the government of Anambra State
- *self-employed*
  1961: 26 professional and contractors
  1987: 31 entrepreneurs and contractors

The circumstances of these occupational groups had changed substantially in the meantime. In 1987 the market traders appeared to have more substantial stock; the railway workers were paid irregularly; the pensions of the retired colliery workers were 13 months in arrears; the

buying power of the senior civil servants' salaries had been sharply eroded; while some of the entrepreneurs and contractors had attained levels of affluence beyond the wildest dreams of the preceding generation.

The first three occupational groups were interviewed by college students. We used an English-language questionnaire, but in many cases the interviewers had to translate the questions into Igbo, the mother tongue of most respondents. The same questionnaire was self-administered by the senior civil servants, professionals, entrepreneurs, and contractors. Refusals were rare. Incomplete or missing answers did not present a systematic problem; the number of valid answers for each question are given in the tables which follow. Some answers to the questions which sought an opinion regarding visits between city and home place were ambiguous; these answers also did not form a pattern, and they are excluded from further analysis. In general, the quality of the information obtained through the interviews was better than that from the self-administered questionnaires, as might be expected.

I was prepared to record a loosening of the ties that had been so remarkably strong a generation earlier. In 1961 Enugu had been very much a 'new town'—not yet two generations old. Established in 1915 by the British following the discovery of coal deposits, it had become a colonial headquarters. In 1961, one year after Nigeria had gained its independence, Enugu was the capital of the Eastern Region, a quiet town with about 100,000 inhabitants. In 1987, a generation later, Enugu had been transformed by dramatic growth. Nobody knew the population of the capital of Anambra State, so half a million had to suffice as a round estimate. In 1961 there were children who protested against the family visits to the home village: 'I don't want to go bush!' What decisions were they making now that they had become adults? Surely, the citizens of this city had become truly 'urban'?

Informal interviews suggested otherwise, and the survey confirms that the next generation of Enugu residents continues to maintain strong ties with their community of origin as well. The 1961 and 1987 surveys document four dimensions of these ties in terms of

- residential history and aspirations,
- the locus of economic support and investment,

- the social field maintained and drawn upon, and
- the attitudes Enugu residents express about their visits to their home place and the visits they receive from relatives.

First Generation Urban Residents Committed to Rural Retirement and Burial

Most of our respondents were born in the community of their forefathers (see Table 4.1), nearly always a rural community. The high rate of rural–urban migration, as evidenced by the rapid growth of Enugu over the last generation, provides part of the explanation. In addition, many women, especially in the low-income groups, used to return to the village to deliver their children. Thus in 1987, just as in 1961, more than nine out of ten market-traders, blue-collar workers, and retired junior staff were born 'at home'. Even among the senior civil servants, men who have virtually all completed a college education, three-quarters were born at their ancestral home. The entrepreneurs and contractors of 1987 present something of a contrast to those of 1961, with about half born away from their ancestral home—they thus appear to be well established in town: their mothers were already urban-based and, unlike poorer women, had access to superior delivery services in town.

While our respondents are overwhelmingly born in rural areas, most have never returned to live in their home place since they left school or

reached age 15. Again, there is some gradation by occupation. Strikingly, the proportion who had returned to live at home had increased between 1961 and 1987. Some had sought refuge and subsistence during the civil war that ravaged the region between 1967 and 1970, and others had fallen back on the village economy in the hard times that followed the end of the petroleum boom of the 1970s.

Our respondents are committed to an urban career. Since on retirement many would make attempts to find other sources of income in the city, the question on retirement plans had to be phrased: 'Where do you intend to settle after you have completely retired from active life?' The overwhelming response was the desire to settle at the home place on retirement. It is particularly striking that nine out of ten entrepreneurs and contractors wanted to retire to a home place where many of them were not born. This in spite of the fact that many of them have the resources to prepare for a secure retirement in the urban setting.

Is the intent to retire at home realized in actual return migration? A 1973–4 survey of three Igbo villages found that one out of every three sample households included a pensioner (Lagemann, 1977: 230). But this evidence, the only available, is too circumscribed to warrant generalization. In any case, many of the implications of temporary, as distinct from permanent migration hold, whether the intention to return is eventually realized or not; they hold as long as migrants act on the assumption that, one day, they will settle down 'back home'.

Among the senior civil servants, approxi-

TABLE 4.1   Residential History and Aspirations by Occupational Group, 1961 and 1987 (percentage and number of valid responses)

		Market traders	Blue-collar workers	Retired junior staff	Senior civil servants	Self-employed
Born in home place	1961	98% (49)	96% (49)	97% (38)	79% (58)	81% (26)
	1987	92% (60)	94% (50)	97% (59)	76% (42)	52% (29)
Never resided at home since leaving school/age 15	1961	80% (49)	67% (49)	95% (38)	95% (55)	96% (20)
	1987	67% (61)	40% (50)	75% (59)	83% (42)	97% (31)
Wants to settle at home on retirement	1961	98% (49)	92% (49)	97% (38)	66% (50)	63% (24)
	1987	97% (60)	98% (50)	91% (58)	68% (41)	90% (29)
Wants to be buried at home	1961	96% (49)	98% (49)	95% (37)	84% (56)	88% (26)
	1987	100% (59)	100% (50)	100% (59)	98% (41)	97% (30)

Sources: Surveys in Enugu.

TABLE 4.2   Locus of Economic Support and Investment by Occupational Group, 1961 and 1987 (percentage and number of valid responses)

		Market traders	Blue-collar workers	Retired junior staff	Senior civil servants	Self-employed
Wife (wives) living at home most of the time	1961	9% (32)	39% (36)	33% (36)	2% (50)	13% (23)
	1987	24% (34)	13% (45)	9% (58)	5% (39)	3% (30)
Builder: has built a house at home	1961	100% (17)	100% (8)	94% (32)	68% (19)	63% (19)
	1987	100% (27)	100% (22)	100% (52)	93% (27)	77% (26)
Non-builder: would build first house at home	1961	72% (29)	95% (39)	100% (4)	68% (25)	100% (2)
	1987	96% (27)	96% (26)	100% (6)	100% (7)	50% (4)

Sources: Same as in Table 4.1.

mately one-third do not want to retire to their home community. They nearly all, however, want to be buried at home. The desire to be buried at home is just about universal across the occupational groups. As a rule, kin can be counted on to take the dead back to the village. Under the influence of the missionaries, burials had taken place at the local church cemetery, but burials have been moved back to the ancestral compound in recent years, a pattern that indicates the profound spiritual significance of being buried 'at home'. The prospect of being buried at home is a powerful motive to maintain and enhance one's standing in the eyes of home people, and is frequently mentioned as a rationale for building a house at home. The first-born son will inherit the very place where his father is buried and establish his own home there.

Urban Income and Rural Property

Enugu residents derive their economic support largely from urban income. In 1961 one-third or more of the blue-collar workers and the retired junior staff had a wife living at home most of the time (see Table 4.2). She would run the farm, produce subsistence for herself and the children staying with her, probably sell some crops, and perhaps send food to her husband in the city. By 1987 such a separation of husband and wife had become very much the exception. The market-traders were most likely to report a wife living at home, but even among them this was the case for only one out of four. Others, however, managed to run a farm from the city with hired help, both

husband and wife visiting frequently to supervise. The pattern may have become more widespread since such visits are now easier than they were a generation ago, as I shall explain below.

In 1961–2, whenever I enquired about land rights, the response was unanimous: every man could claim land in his home area. Among the Igbo—who are the great majority of Enugu residents and of our respondents—farmland used to be controlled by the community, i.e. by the lineage, usually the maximal lineage, but sometimes the major or minor lineage (Jones, 1949: 314). In 1961 informants both in the city and in villages were emphatic that a man did not lose the land rights he enjoyed in his home community if he left, however long he might stay away. Chubb (1961: 115) shares this view. In fact, it would seem that only the lineage member's claim to some of the lineage's farmland was maintained, but that land that was not occupied and farmed reverted to the community and could be assigned to other households in the lineage (Jones, 1949: 314). Thus the migrant had his claim to a share in the communal farmland upheld, but might lose his rights to a specific piece of land.

Dike (1983: 858) agrees that lineage land tenure was the most accepted form of land holding. He indicates, however, that population increases have led to the partitioning of lineage land among the extended families. Dike (1983: 866) further states that the traditional assertion that land does not belong to an individual no longer holds, and that family members now prefer that their land be partitioned among them. A sample survey of three Igbo villages (Lagemann, 1977: 188) accords with this general picture: between 70 and 78 per cent of the fields were inherited individual property. In some areas, land tenure has been individualized to such an extent that many adult males are said to be landless (Okafor, 1993: 339). Elsewhere, transfers of land-use rights on a short-term basis are reported to be common, farmers to feel that it is appropriate to supply land to those who need it, especially if any relationship of kinship or friendship is involved (Goldman, 1993: 279–81).

I have been unable to locate research that would document the extent to which people in the region have lost access to farm land. My impression is that landlessness is not the issue here that it has become in some parts of Africa. Certainly, as far as land for residential purposes is concerned, the survey shows that Enugu residents were able to secure village land to build a house. Such residential land, formerly held by the lineage, is now, according to Dike (1983: 869), mostly individually owned. Among the low-income groups in the survey, the respondents most likely to be dispossessed, the proportion who had built a house increased substantially between 1961 and 1987; and all these builders had built in their home community. Those who had not yet built a house in 1987 were near-unanimous that they would build their first house at their home place. While such a commitment was common in 1961, it was far from universal among the market-traders. The proportion of senior civil servants who had built a house had increased dramatically between 1961 and 1987, and nearly all these builders had built in their home community.

Building a house in the village is a poor invest- . ment compared with the income from building in Enugu where rents are high. Of course, land has become expensive in élite neighbourhoods. Throughout the history of Enugu, however, land for building could be acquired cheaply from villages on the margin of the agglomeration. In addition, the colonial government at various times acquired land from neighbouring villages and sold it on freehold to prospective builders (Gbanite, 1981: 124–68). In 1961 such land was easily accessible. The high returns available on investment in urban housing may explain why the minority who had built in the city rather than in the home community was made up largely of entrepreneurs and contractors. But even among them, a large majority had discarded such economic considerations to build a house that estab-

lishes their standing in the home community, provides comfort during visits and retirement, and will serve its most important function the day its owner is buried there. Dike (1982: 91) speaks of the feeling of shame and disgrace if the migrant's village house or that of a close kinsman collapses out of neglect or if the graves of the ancestors are exposed to rain.

The Web of Kin and Home People

The evidence is overwhelming that virtually all respondents maintain significant relationships with their rural home community, even though the minority who had a wife (and children) living at their home place had become very small by 1987. Nearly all respondents had visited their home places within the last 12 months both in 1961 and in 1987 (see Table 4.3). In the interim, visits had become more frequent and shorter. For some, visits serve to maintain farm operations, in particular to hire and to supervise labour. Continued farming generates income: produce is brought to town for consumption or sale. In addition, farming upholds the claim to specific fields where these are still communally controlled. Also, many senior civil servants, entrepreneurs, and contractors regularly drive home for weekends at their country home—the weekend is extended for civil servants since the establishment of the work-free Saturday in 1976. In effect, distances have diminished for Enugu residents in two respects. Roads were greatly improved during the petroleum boom of the 1970s. And the region from which Enugu draws its population has shrunk: the capital of Anambra State recruits from a smaller hinterland than the capital of Eastern Nigeria. Enugu is in the centre of the state, and any destination within the state can be reached by car in less than two hours. For the large majority of Enugu residents who hail from within Anambra State, the average travelling time to their home place by car is less than one hour.

Close ties with the home place easily lead to marrying a woman from there. Such affinal connections in turn reinforce ties with the home community. In 1961 the majority of married respondents reported that their wife has come from the same home place. By 1987 marriages

TABLE 4.3 Social Field by Occupational Group, 1961 and 1987 (percentage—except for average number of relatives living with respondent—and number of valid responses)

		Market traders	Blue-collar workers	Retired junior staff	Senior civil servants	Self-employed
Visited home within last 30 days	1961	57% (49)	31% (48)	82% (38)	46% (56)	69% (26)
	1987	56% (61)	70% (50)	73% (59)	90% (41)	97% (31)
Visited home within last 12 months	1961	92% (49)	88% (48)	97% (38)	88% (56)	96% (26)
	1987	98% (61)	100% (50)	100% (59)	100% (41)	100% (31)
Spent more than one week at home during last visit	1961	10% (49)	35% (48)	18% (38)	26% (54)	4% (23)
	1987	18% (61)	8% (49)	22% (58)	3% (38)	3% (31)
Wife (wives) originate(s) from same place	1961	75% (32)	78% (36)	81% (36)	54% (50)	52% (23)
	1987	41% (34)	84% (45)	74% (58)	49% (37)	43% (28)
Wife visited home within last 30 days	1961	34% (29)	55% (22)	42% (31)	40% (42)	33% (18)
	1987	38% (29)	43% (40)	64% (53)	71% (31)	65% (26)
Wife visited home within last 12 months	1961	79% (29)	95% (22)	90% (31)	81% (42)	89% (18)
	1987	100% (29)	98% (40)	100% (53)	100% (31)	96% (26)
Wife spent more than one week during last visit	1961	60% (25)	65% (17)	60% (30)	44% (34)	40% (20)
	1987	46% (28)	54% (39)	26% (53)	19% (21)	32% (22)
Number of relatives living with respondent	1961	1.1 (32)	0.8 (49)	0.6 (38)	0.8 (57)	1.4 (24)
	1987	1.0 (61)	1.0 (50)	0.4 (59)	1.0 (42)	1.2 (29)
Active member of an ethnic association	1961	78% (49)	86% (41)	84% (38)	57% (58)	85% (26)
	1987	77% (61)	84% (50)	81% (59)	81% (42)	81% (31)
Member of another club or organization	1961	12% (49)	20% (49)	3% (38)	45% (56)	68% (25)
	1987	23% (60)	4% (50)	3% (59)	55% (42)	84% (31)

Sources: Same as in Table 4.1.

with a woman from home were reported somewhat less frequently, presumably because of increased resistance to parental matchmaking. Still, five out of six blue-collar workers, three out of four retired junior staff, and close to half the others had found a wife at home or among home people in town.

Women visit their husband's home—considered their home irrespective of their own origin—less frequently than their husbands. Are the expenditures for the cost of travel and the gifts visitors are expected to bring an obstacle? Are men more involved in the affairs of their home place? Do women visit what used to be their home place before they married, the home place of their fathers, instead? On average, the reported duration of wives' last visit was longer than that of men. It would be tempting to conclude that wives visit less frequently but stay longer, that work schedules impose a different pattern on men. The data may simply indicate, however, that wives tend to visit on those major occasions, Christmas foremost among them, when men also extend their stay at home.

Certainly, in 1987 virtually all wives were reported to have visited their husband's home place within the last 12 months.

Frequent visits from relatives, whether they live at the home place or elsewhere, are the counterpart to such visits home. Furthermore, Enugu residents commonly have relatives living with them. On average, one relative lived with each respondent in 1961. He or she was usually related to the husband rather than the wife. The majority of these relatives were at school, and the host typically paid their fees: such support for relatives is made conspicuous by having the beneficiary stay with his or her sponsor and help around the house. A second group of living-in relatives were apprentices of the respondent; they were particularly common with traders. Others were unemployed or self-employed. The 1987 data show little change and repeat the previous pattern: the entrepreneurs and contractors have the most relatives living with them, and the retired junior staff the fewest. In 1961, 63 per cent of the retirees had no relations living with them, and this proportion had increased to 81 per cent in 1987. As one put it in 1987: 'Since I retired I do not get visitors. I have no money.'

The relationship with the home community as well as with co-villagers resident in town provides the basis for ethnic associations, which in turn reinforce these connections. The great majority of our respondents were active members of an ethnic association, i.e. they attended and contributed regularly. The senior civil servants who lagged behind the other groups in 1961 had caught up in 1987. In contrast, membership in a social club, or a cultural, education, or sports organization has remained less common, except among the more affluent.

Ethnic associations are more accurately characterized by the terms current in francophone countries: 'associations d'originaires' or 'associations de ressortissants'. Based on common origin, they bring together people from a particular area. In Enugu they typically represent a village group.[3] Ethnic associations provide support for the urban dweller: they give assistance in litigation, illness, and death (return of the corpse to the home place, repatriation of the family); they are channels of information on urban conditions, in particular employment opportunities; and they act as arbitrators in conflicts between members. The main concern of these associations is, however, with the affairs of the home community. If many an association calls itself an 'Improvement Union', this refers to the improvement of the home area, not of urban living conditions. Ethnic associations, along with those who return from the city to visit or retire, transmit new ideas and aspirations, constitute an urban lobby for village interests, counsel on village developments, and finance a major part of such developments. Throughout the rural areas, roads and bridges, schools, and maternity clinics have been built with funds collected through such associations, in a few instances even post offices, high schools, hospitals, water or electricity supply systems.[4] Of course, such investments serve not only those who stay in the village, but their urban cousins as well: better access roads, pipe-borne water, electricity, and a post office make visits more comfortable, and the prospect of retirement in the village more attractive.

To find employment is the crucial task confronting most urban residents. In the survey, open-ended questions enquired about the identity of employers and how jobs were obtained (see Table 4.4). In the low-income groups, kin play a major role in securing employment in two ways: kin assist in finding employment and even employ their 'brothers'. A substantial minority reported having been employed by a relative in 1961 as well as 1987. A few had been employed by an unrelated co-villager in 1961, virtually none in 1987. Friends who were neither relatives nor co-villagers virtually never appeared as employers. In terms of help received to secure employment, a similar gradation from kin to co-villager to friend had become more clear-cut in 1987. Among the market traders and the blue-collar workers, those who reported finding work

TABLE 4.4 Employment Assistance by Occupational Group, 1961 and 1987 (percentage and number of valid responses)

		Market traders	Blue-collar workers	Retired junior staff	Senior civil servants	Self-employed
At one time employed by relative	1961	41% (49)	33% (49)	13% (38)	0% (58)	4% (26)
	1987	36% (61)	38% (50)	17% (59)	2% (42)	7% (30)
At one time employed by co-villager	1961	18% (49)	14% (49)	16% (38)	0% (58)	4% (26)
	1987	8% (61)	0% (50)	0% (59)	0% (42)	0% (30)
At one time employed by friend	1961	0% (49)	2% (49)	0% (38)	0% (58)	0% (26)
	1987	0% (61)	0% (50)	0% (59)	0% (42)	0% (30)
Found employment through relative	1961	20% (49)	27% (49)	32% (38)	0% (58)	8% (26)
	1987	63% (57)	56% (50)	34% (59)	2% (42)	14% (29)
Found employment through co-villager	1961	6% (49)	31% (49)	32% (38)	0% (58)	0% (26)
	1987	4% (57)	8% (50)	5% (59)	0% (42)	3% (29)
Found employment through friend	1961	2% (49)	10% (49)	24% (38)	0% (58)	0% (26)
	1987	4% (51)	8% (50)	5% (59)	0% (42)	0% (29)

Sources: Same as in Table 4.1.

through a relative had increased dramatically, whereas the number of blue-collar workers and retired junior staff who reported having found work through co-villagers had shrunk drastically in 1987. Thus co-villagers have become rather insignificant in terms of offering either employment or assistance in finding employment. The assistance of relatives, on the other hand, has become more important. The greater reliance on relatives may be due to a weakening of ties with co-villagers, the greater number of kin to be drawn on in the larger city, and/or the greater difficulty of finding employment during the economic crisis that followed the petroleum boom of the 1970s.

Senior civil servants, in contrast, virtually never report the assistance of kin, co-villagers, or friends in securing employment. In part this is because most have a simple employment history: they moved from college into the civil service. Also, of course, many are reluctant to acknowledge the assistance they did receive, and find it easy to evade the question in a self-administered questionnaire. The fact that the entrepreneurs and contractors should virtually all present themselves as 'self-made men' is more puzzling.

Structural Constraints and Cultural Norms

Life in a dual system might be conceptualized as a phase in the urban transition. Modernization theory leads us to expect that this stage of change will pass as a better-educated population increasingly adapts to urban life. We have found instead an enduring pattern of a dual commitment to an urban working life as well as to a rural community of origin. The structural reasons for this persistent pattern are not hard to find.

Enugu residents articulate their home ties invariably in terms of loyalty. However, I eventually came to see in 1961 that the home community provided economic security for urban workers, and that this security was crucial for most. At a time when a social-security system covering illness, disability, and old age was in its infancy, and when there was no unemployment compensation, the low-income earners remained dependent for their security on the village.[5] However uncertain their urban prospects, they

could return to the village at any time, find an abode, and claim land to farm. Since 1961, life in Enugu has become less secure. In 1987 conditions were precarious for each of the five groups we interviewed. Large-scale retrenchments had taken place in the Nigerian Coal Corporation in 1984, and further retrenchments had been announced both there and at the Nigerian Railways Corporation, whose workers were paid irregularly. Large numbers of senior civil servants had been dismissed over-night by a previous military government in 1975—many without a pension. In 1987 all federal employees who had been in service 35 years or more were retired. Those retiring with a pension found that it stretched further in the village—when they received it: the pension payments of the Nigerian Coal Corporation were 13 months in arrears. As for the self-employed, whether small market traders or wealthy entrepreneurs and contractors, they were subject to major economic crises and sudden edicts of the military government. Many of the entrepreneurs and contractors we tried to contact using the telephone directory had gone out of business. Indeed, the rigours of the structural adjustment programme initiated under the auspices of the International Monetary Fund in 1986 is said to have induced substantial numbers of able-bodied men to leave Enugu and return to what livelihood they could find in their home village.

The 1961 survey revealed—contrary to the stereotype invariably presented by Enugu residents—a substantial minority among the professionals, the contractors, and especially the senior civil servants who had loosened their home ties, in particular with respect to the two interrelated issues of where they built their houses and where they planned to retire. In 1987, in contrast, the responses of the high-income groups were much more in line with those of the low-income groups. The most affluent could no longer take for granted the security they enjoyed in 1961, and this may constitute sufficient explanation. The proposition was commonly advanced, however, that the attachment of Igbo to their home community was reinforced by the civil war. A number of well-off individuals told us how the experience of the civil war made them feel that their urban existence was insecure. In those difficult days, their home village had proved the one place that provided succor in calamity. Some told of their embarrassment in returning to their home place where they had not built a house—and of the house they had there now. Isichei (1976: 202–3) takes the position that the erosion of the élite's ties was halted for a generation by the war. I am sceptical of such an interpretation which fails to consider the changed circumstances of the urban élite. Indeed, some of the entrepreneurs and contractors spoke of investments in production facilities in their home place. Several were concerned that common crime as well as political reversals threatened their urban property, whether it be a house or a factory.[6]

Still, it is hard to see what security the village provides the millionaire contractor I interviewed. I met him one evening as he returned from his home village. He was accompanied by his youngest wife and three elderly women who burst into ululations, excited that their successful 'son' (nephew, relative) should have taken them along to the city, admiring his magnificent villa, and appreciative of his hospitality. This millionaire is firmly based in the city; his three other wives have their own villas in Enugu. At the same time he remains committed to the village, practices loyalty to his home place, continues to set store in what his home people think of him. Life in a dual system has become an established cultural norm.

The norm of loyalty to the home community is anchored in necessity—the insecurity threatening most urban workers—but it is acknowledged by most of the affluent as well. They incur considerable expenses in maintaining the village con-

nection: their gifts to relatives and their contributions to village developments have to be commensurate to their wealth. These expenses, not to speak of the cost of their rural houses, dwarf the value of the limited land rights they maintain. A few invest in production facilities, taking advantage of easy access to land and cheap labour. Some find the village a ready-made clientele for their business. Others may use their standing in the community to establish major land holdings and venture into commercial farming. A few look forward to using the village as a political base when civilian rule returns. But for most of the more affluent urban residents, the village connection presents a financial drain with little economic return. They do, however, enjoy social status in the village as 'sons' who have made good and become benefactors. Indeed, chieftaincy—an institution established in much of the region only in colonial days—is bestowed on some and enhances their status not only in the village but also in the urban arena.

The strength of the cultural norm is demonstrated by the responses to a series of questions about attitudes towards visits (see Table 4.5). Most report that they like visiting home and, when expressly queried about anything they might dislike about such visits, do not express any reservations. When specifically asked, many acknowledge that such visits involve too much spending of money. Their numbers, however, dropped substantially between 1961 and 1987 in every occupational group. When it comes to visits by relatives, most low-income earners again

TABLE 4.5   Attitudes toward Visits by Occupational Group, 1961 and 1987 (percentage and number of valid responses)

		Market traders	Blue-collar workers	Retired junior staff	Senior civil servants	Self-employed
Likes visiting home without reservations	1961	90% (49)	79% (48)	87% (38)	61% (54)	81% (26)
	1987	88% (59)	84% (50)	82% (57)	88% (40)	70% (27)
Visiting home involves spending too much money	1961	73% (49)	96% (48)	74% (38)	77% (56)	58% (26)
	1987	38% (60)	62% (47)	41% (58)	65% (40)	38% (29)
Likes relatives' visits without reservations	1961	88% (48)	69% (49)	100% (38)	45% (51)	52% (21)
	1987	78% (55)	86% (49)	90% (49)	69% (35)	50% (26)
Relatives do not visit too often	1961	90% (49)	58% (48)	84% (38)	70% (56)	60% (25)
	1987	91% (54)	73% (49)	88% (59)	88% (41)	80% (30)
Relatives' visits not too long	1961	94% (49)	94% (49)	95% (38)	78% (55)	84% (25)
	1987	90% (51)	93% (45)	98% (57)	90% (40)	90% (29)

Sources: Same as in Table 4.1.

do not express any reservations. High-income earners are less enthusiastic, but even among them only a small minority complain when asked whether relatives visit too often or stay too long.

Identities of Origin and the Political Economy

Life in a dual system is the response to specific economic and political conditions but it in turn has major consequences for the political economy. First of all, effective urban–rural ties have a major effect on the migration process. Villagers are well informed about urban opportunities. Migration based on unrealistic expectations is unlikely, and therefore urban unemployment remains circumscribed. At the same time, considerable resources are mobilized in support of migrants. More or less distant relatives, in the village as well as in town, sponsor the education and training of potential migrants, facilitate their move to the city, integrate them into a network of kin and people from home, and support them in their search for earning opportunities. The urban–rural connection, while it cannot guarantee the success of the rural–urban migrant, makes a successful outcome more likely and reduces the stress involved in the venture.

The frequent crises urban low-income earners are exposed to are cushioned by support from relatives and other people from 'home'. When the need arises, they can fall back on the rural community. The prospect of retiring in one's 'home place' mutes pressures for comprehensive social-security legislation. These urban workers who maintain a rural base thus do not constitute a proletariat, but rather to use Wallerstein's (1967: 501) phrase, a 'quasi-proletariat'.

'Sons abroad' are ready intermediaries for the rural population. They are considered members of the village, and their counsel is trusted. Their influence was powerfully demonstrated in many parts of Colonial Africa in the 1950s when they rallied the rural masses to the nationalist cause. The vagaries encountered by every Nigerian census since 1962 were caused in part by urban residents concerned about the welfare of their home areas. Urban residents persuaded villagers to let themselves be counted (as recently as 1953 there had still been considerable evasion of cen-

sus takers), and they returned *en masse* to their villages on census day to swell the population enumerated and thus to strengthen the village's political representation and its claim on public services.

Resources are transferred by urban residents to the rural economy. They are expected to bring gifts, cash, and goods when they visit; they sponsor the education and training of relatives; they assist villagers who come to town for medical treatment or dealings with the government; they provide employment to the builders of their house 'at home'. Collectively urban residents, through their ethnic associations, foster rural developments projects: they initiate and organize such projects, lobby the government, and provide substantial financial support. I have little doubt that ethnic associations play a key role in rural development in Anambra State and beyond. Lest the picture appear too rosy, I hasten to add that conflicts within these associations and complaints about corruption are common.

The transfer of resources to the rural community has contradictory effects on rural emigration. Increased education and training are commonly assumed to enhance access to urban employment opportunities and thus to foster rural–urban migration. While this effect certainly holds for the individual village, assessing the aggregate effect is more difficult. To the extent that urban contributions raise levels of education throughout a region, the likely effect is an increase in the educational and training requirements for specific jobs and only a limited expansion in urban employment. In any case, most transfers will tend to have a dampening effect on emigration in as much as they raise rural standards either directly through income transfers or indirectly by improving rural earning opportunities. Gifts for relatives and the provision of public amenities have the direct effect. More significant in the long run are improved roads that facilitate the sale of rural products; the generation of employment—this was limited to the employment of farm labour and the building of houses in the past, but isolated cases of urban entrepreneurs investing in production facilities in their home place hold out the promise of a more productive village economy; and for such production, the provision of electricity.

Life in a dual system fosters ethnic identity and ethnic alignment in several ways. The urban social network in which the migrant operates is largely demarcated in terms of origin and affects occupational distribution patterns in turn. Conflicts of interest in the urban arena are therefore readily perceived in ethnic terms. At the same time, the efforts of urban residents to secure public resources for their home areas foster political alignments on the basis of origin. Finally, ethnic associations offer privileged means for political mobilization—along ethnic lines (Gugler, 1975a). What level of ethnic identity is salient and at what level people align will depend on the political arena. They may support their village against a neighbouring village, Enugu State against neighbouring Anambra State, Igbo against another ethnic group, or Nigeria against Cameroon (Gugler, 1975b). To refer to such conflicts as 'tribal' is doubly misleading. The major ethnic groups recognized in Nigeria today have been forged quite recently in the competition of national electoral politics of the 1940s—in the urban arena. And key interest groups on closer inspection frequently turn out to represent much smaller identities and alignments—of origin.

Conclusion

In 1962 the children of the men I interviewed were proclaiming: 'I don't want to go bush!' But a generation later I found that adults continued to be committed to both the city in which they live and work and their village of origin.

This dual commitment is rooted in rural collectivities that continue to control resources, in particular access to the ancestral lands. The village assures a refuge in a political economy that fails to provide economic security to most of the urban population and that threatens all with an uncertain political future. The continuing involvement of urban residents in the village is reinforced by an ideology of loyalty to home that tends to hold sway even over those few who succeed in establishing themselves securely in the urban environment.

The continuing rural commitment of urban residents, in turn, has consequences for the political economy. Urban residents articulate political conflicts in terms of more or less narrowly defined identities of origin. They are committed

to the improvement of their home communities rather than the city. Their position is well summed up in the prayer said at the meeting of an Igbo improvement union in the city—as rendered by Achebe (1960: 5–6), Nigeria's distinguished writer, an Igbo himself, in a novel that gives play to urban–rural ties: 'We are strangers in this land. If good comes to it may we have our share. But if bad comes let it go to the owners of the land who know what gods should be appeased.'

Notes

1. I wish to express my appreciation to Gudrun Ludwar-Ene who collaborated in the direction of the 1987 survey and contributed to the analysis, to John W. Curtis who directed the processing of the survey data and contributed to the analysis, and to Eucharia Akpa, Samuel Chukwu, Chukwucmeka Kalu, Julian Meniru, Mackyngs Nezianya, and Theophilus Nwalozie who carried out the interviews. The research was funded by the German Research Council through the Special Research Programme 'Identity in Africa' at Bayreuth University.

2. The administrative role of Enugu was diminished once more in 1991 when Anambra State was divided in turn and Enugu became the capital of Enugu State.

3. Whether an ethnic association represents a village, a village group, or a larger administrative region is a function of the number of people from a given region resident in the city: associations need a critical mass of members to function, say a minimum of 15, yet beyond a certain size, say 100 active members, they tend to split—along lines of origin. Quite frequently Enugu residents are more or less active members of a second association. During the First Republic, i.e. until the military took over in 1966, ethnic associations representing villages, village groups, and administrative regions were organized in pyramidal structures culminating in pan-ethnic organizations such as the Ibo State Union. For a more detailed discussion focusing on the political role of ethnic associations, see Gugler (1975b).

4. For two detailed case studies of ethnic associations, see Smock (1971).

5. Most federal and state workers do have pension rights since colonial days. And since 1961 workers in establishments with more than nine employees and their employers contribute to the National Provident Fund which provides a lump-sum payment of retirement. However, in both cases, the effect of rampant inflation had become devastating

by the time of the second survey (Williamson and Pampel, 1993: 171–5).

6. Ludwar-Ene's (1993) parallel study in Calabar, also in south-eastern Nigeria but little affected by the civil war, found similarly strong ties with the rural community of origin.

References

Achebe, Chinua (1960), *No Longer at Ease* (London: William Heinemann), quoted from 1975 reset edn.

Chubb, L. T. (1961), *Ibo Land Tenure*, 2nd edn. (Ibadan: Ibadan University Press).

Dike, Azuka A. (1982), 'Urban Migrants and Rural Development', *African Studies Review*, 25 (4): 85–94.

—— (1983), 'Land Tenure System in Igboland', *Anthropos*, 78: 853–71.

Gbanite, Ekene Michael (1981), 'Third World Urbanisation: Enugu, Nigeria', Ph.D. diss. (New York: New School of Social Research).

Goldman, Abe (1993), 'Population Growth and Agricultural Change in Imo State, Southeastern Nigeria', in B. L. Turner II, Goran Hyden, and Robert W. Kates (eds.), *Population Growth and Agricultural Change in Africa* (Gainesville: University Press of Florida, 250–301.

Gugler, Josef (1971), 'Life in a Dual System: Eastern Nigerians in Town, 1961', *Cahiers d'études africaines*, 11: 400–21.

—— (1975a), 'Migration and Ethnicity in Sub-Saharan Africa: Affinity, Rural Interests and Urban Alignments', in Helen I. Safa and Brian M. duToit (eds.), *Migration and Development: Implications for Ethnic Identity and Political Conflict* (The Hague/Paris: Mouton), 295–309.

—— (1975b), 'Particularism in Sub-Saharan Africa: "Tribalism" in Town', *Canadian Review of Sociology and Anthropology*, 12: 303–15.

Isichei, Elizabeth (1976), *A History of the Igbo People* (London/Basingstoke: Macmillan Press).

Jones, G. I. (1949), 'Ibo Land Tenure', *Africa*, 19: 309–23.

Lagemann, Johannes (1977), *Traditional African Farming Systems in Eastern Nigeria: An Analysis of Reaction to Increasing Population Pressure* (Munich: Weltforum Verlag).

Ludwar-Ene, Gudrun (1993), 'The Social Relationships of Female and Male Migrants in Calabar, Nigeria: Rural versus Urban Connections', in Gudrun Ludwar-Ene and Mechthild Reh (eds.), *Gros-plan sur les femmes en Afrique, afrikanische Frauen im Blick, Focus on Women in Africa*, Bayreuth African Studies 26 (Bayreuth: Eckhard Breitinger), 31–47.

Okafor, Francis C. (1993), 'Agricultural Stagnation and Economic Diversification: Awka-Nnewi Region, Nigeria, 1930–1980', in B. L. Turner II, Goran Hyden, and Robert W. Kates (eds.), *Population Growth and Agricultural Change in Africa* (Gainesville: University Press of Florida), 324–57.

Smock, Audrey C. (1971), *Ibo Politics: The Role of Ethnic Unions in Eastern Nigeria* (Cambridge, Mass.: Harvard University Press).

Wallerstein, Immanuel (1967), 'Class, Tribe and Party in West African Politics', in Seymour M. Lipset and Stein Rokkan (eds.), *Party Systems and Voter Alignments: Cross-National Perspectives* (New York: Free Press), 497–518.

Williamson, John B., and Pampel, Fred C. (1993), *Old Age Security in Comparative Perspective* (New York/Oxford: Oxford University Press).

5

Amerindian Migration in Peru and Mexico

Harald Moßbrucker

SINCE the 1970s, social scientists have paid increasing attention to the massive rural exodus in Latin America. Anthropologists have been in the forefront; they were among the first to ask whether migration to the cities might not contribute to a process of acculturating the Amerindians.

Even today the question remains open. But in the 1980s, the prevailing view was that Amerindian migrants differed from other people fleeing the land by the high degree to which they had organized themselves. This opinion is based on observations in cities like Lima, La Paz, and Mexico City, where Amerindian immigrants solve problems on their own, from housing construction and infrastructure to employment. Intensive field research has also shown that Amerindian migrants generally keep in close contact with their home regions. Associations of migrants from the same village are thought to play a central role in these people's efforts to fit into urban life and to weave the web of relationships that persist between the city and the home village (Altamirano and Hirabayashi, 1991).

For the cities and regions mentioned, the research results are largely uncontroversial. But as this essay will show, the pattern does not always hold in predominantly Amerindian areas. Rather, migration strategies and solutions to problems vary with a number of factors that require individual examination. Among these factors are the changing world economy, government policies, environmental and infrastructural conditions, and the migrants' cultural heritage.

This study focuses on two regions where the populace is predominantly Amerindian, and many people have left the countryside for the cities: Lima, the capital of Peru, and its hinterland, peopled by descendants of the Quechua culture, and the northern part of the Yucatán Peninsula in Mexico, with its two urban centres Mérida and Cancún, inhabited mainly by members of the Mayan culture. A concluding section will compare the two areas and analyse the reasons for their similarities and differences.

Lima and its Hinterland

Lima is located in a narrow desert strip along the coast. To the city's east, the Cordillera Occidental of the Andes rises to heights of more than 20,000 feet. Extending across the steep slopes in many places are terraced fields built even before the arrival of the Spaniards and still in use today, while on the *puna* (plateau), large herds of wool-producing livestock graze: sheep, llamas, and alpacas.

The Free University Berlin awarded me a junior scholar grant for 1987–9, which enabled me to conduct field-work in Peru from October 1987 to December 1988. A research grant from the German Research Council for 1991–4 made it possible for me to work in Yucatán from December 1991 to May 1993. I wish to express my appreciation to both institutions. Translation by Steven Gilbert and Josef Gugler.

Aside from a few isolated mines on the *puna*, the mountain region is agrarian, sending what it produces to the city—which explains why the city and the countryside have been closely linked ever since the Spanish conquest. With its seven million inhabitants, the metropolis provides an enormous market for agricultural products and absorbs the bulk of the people who emigrate from the villages in the department. We will focus on two such villages.

Santiago de Quinches

The village of Santiago de Quinches is about 110 miles south of Lima at an altitude of 10,000 feet. Its inhabitants support themselves by farming, cultivating alfalfa, and raising cattle to sell meat and cheese. At first just a hamlet, the village grew rapidly to become a regional centre with 1,810 inhabitants by 1940. In 1987 only 850 people remained there, while the number of emigrants was steadily rising; in that same year, more than 2,000 of them were in Lima.

The village's resources are distributed very unequally. The cattle ranchers, just half of the inhabitants, own more land than the farmers, who often work as day-labourers for the ranchers. More than one-quarter of the irrigable, cultivable land, the sole agricultural resource, is worked by peasant share-croppers while it remains the property of emigrants living in Lima. Income distribution in the village is correspondingly unequal. At the one extreme are the big cattle ranchers, who have plenty of money and satisfy their own needs primarily through subsistence production on the fields they own. At the other extreme are the day-labourers who, without any fields of their own, depend entirely on the cattle ranchers' patronage.

At the turn of the century, Quinches, unlike many other villages, did not have a *comunidad*. Only in 1936 did a group of cattle ranchers organize it to seize the communal grazing land for themselves and keep others out. Previously, owners of large cattle herds running the *munipicio* had controlled the grazing land. But after fighting bitterly with them for several decades, the new *comuneros* eventually enjoyed complete victory (Moßbrucker, 1990: 14–16). Today, the *comuneros*, organized in the *comunidad*, are the most powerful group in the village. At first they were very interested in admitting new members so as to strengthen their group, but they soon erected barriers in order to keep non-*comuneros* from gaining the right to use the communal grazing land.

Because of the particular way the socioeconomic relations within the village have evolved, a complex network of relationships has emerged that range from open hostility to close co-operation. Each family is woven into this net at several levels, so that an individual's or a family's loyalty often depends on the specific situation in which they find themselves. As a consequence of emigration, this network has now extended beyond the village to Lima and other places on the coast.

It is unclear just when emigration began. A few families from Quinches were living in Lima even at the turn of the century. Back then, the few who did leave were often young and well-to-do, going off to study at the university. But since about 1940 emigration has surged. The migrants themselves cite their desire for education as the primary reason—and migrants from Quinches are in fact often well educated. The second most frequently given reason is their desire for work, and the third, familial considerations. Many mention education and work as equally important rationales.

Although the people pouring into Lima from Quinches found many different ways of earning their living, they have gravitated towards certain kinds of work. More than half are in trade, and of these, many either have booths in Lima's major markets, where they sell various kinds of food—cheese from Quinches, among other things—or they run small stores, mostly out of their own homes. Most of them deal with meat: either in a stand in one of the major markets or as cattle traders. Many migrants work as butchers in the large slaughterhouse in Yerbateros, and they often have close business and familial relationships with other people in the meat trade: slaughterhouse butchers and butcher-shopkeepers are often members of one and the same family. Roughly 20 per cent of the migrants have found employment in academic professions, principally as teachers, while less than 10 per cent work as labourers. Less than 20 per cent of the female migrants refer to themselves as housewives.

Migrants from Quinches so often land jobs in or connected with the meat business because their network of social relations built on an initial foothold. Until the late 1950s, some of the peasants from the village made their way to the *lomas*—coastal hills where a bit of grass grows during the foggy months—to graze their herds. They established contact with city people who bought milk and cheese. Sometimes herd owners brought their own livestock, or herds increased by special purchases, to Callao, where Lima then had its sole abattoir, to sell them to the butchers. Through such contacts, several Quinchinos managed early on to find jobs there. Once enough *paisanos* (people from the same region or village—since endogamy is common, a *paisano* is usually also a relative) had landed jobs in the slaughterhouse, new arrivals had no difficulty finding work in the same business. In time, the Quinchinos evolved into a strong group within the slaughterhouse. They played an active role in constructing a second one, in Yerbateros, where they wielded considerable influence right from the beginning.

From the 'chance' discovery of specific opportunities for work, the Quinchinos had managed to establish an economic niche. Over time, they greatly enlarged it and expanded into related areas. Today, their network ranges from buying the livestock through the slaughtering to the selling of the meat.

Many families in Lima remain in close contact with their relatives in the village. *Regalos* (gifts) are exchanged regularly, some of considerable value, such as a slaughtered sheep, or several sacks of potatoes or corn. For many villagers, the gifts received—clothing, rice, sugar—constitute an important economic contribution, but the relationship has great social significance as well.

The most important aspect of the connection between Quinches and Lima is the fact that migrants still own a great deal of land in the village. The landowners put their property in the custody of people whom they trust, primarily relatives of the first and second degree. The share-cropping system does not, in this instance, create a relationship of exploiter to exploited; rather, the system is used to strengthen existing social relations with a certain person or household. Entrusting the land to others for their use is thus usually an important component of more comprehensive reciprocal relations (Moßbrucker, 1992*b*).

To understand why so many people left Quinches, it is necessary to examine more closely the situation in the village. Because the ways of ensuring social reproduction vary within the village, the interests of individual households differ. Households are rich or poor, they are headed by farmers or cattle ranchers. From about 1920 until 1960, Quinches experienced a 'population explosion' which increased pressures on resources. Villagers became more prone to resort to violence in their conflicts over land. An ambiance of profound mutual mistrust resulted. To this day, it severely impedes co-operation at the village level. The villagers failed to create an institution that could reconcile conflicting objectives and intentions and then carry out communal projects without quarrels between opposed factions. Indeed, the majority of peasants never grasped the advantages of co-operation.

This overall situation—population pressure, scarcity of resources, and tense social relations—induced people to emigrate early on. Emigration was all the more attractive because the Quinchinos were already familiar with the coast and could in some cases even count on friends there. In Lima they were not confronted with poverty and despair, rather they often could—through their contacts—find work and shelter even before making a permanent move.[1]

Early emigration also had its effects on the village. Because the community was divided into factions, the migrants managed to keep their fields and have them worked by relatives. They were also able to occupy 'economic niches' in Lima, benefiting not only themselves but also the *paisanos* who later followed. The new arrivals found accommodation with those already established. Thus, once a group of villagers had gained a foothold in the city, they attracted others ready to leave the village.

As more and more migrants found work in the processing, trading, and retailing of animal products, livestock breeding intensified in Quinches. The cattle ranchers and traders in the village were now doing business with their *paisanos* on the coast. This development probably explains why the migrants supported the ranchers in founding the *comunidad*.

If emigration relieved the pressure on resources, it did nothing for harmony. On the contrary: the conflicts now spread beyond the narrow confines of the village and were pursued with the support of migrants. Thus the migrants had as many as twelve 'clubs'—which were furthermore divided into two hostile camps when mass emigration first began.

As emigration progressed, a further pattern emerged. Because some of the younger people studied in Lima and returned to the village as teachers, where they enjoy high prestige, anyone who does not spend several years outside the village after secondary school is now considered a failure. That the ones who stay in the village are 'dumbbells' is the opinion not just of many migrants but also of many of the 'dumbbells' themselves. Parents therefore move heaven and earth to send their children for further education to the coast. This involves substantial expenditures—and provides part of the explanation for the steady expansion of cattle ranching, even while the population declined.

Although many migrants keep their fields and thus the possibility of reproduction, few return to the *sierra*. Village opportunities to realize what is considered a decent life are limited, while in the city most migrants succeed in improving their situation as subjectively defined.

San Miguel de Vichaycocha

San Miguel de Vichaycocha is about 100 miles north of Lima and roughly 12,000 feet above sea level. Irrigable terraced fields cover a small area within the village, two-thirds of them in alfalfa. Fields belonging to the community, *moyas*, cover much more territory. They cannot be irrigated and are cultivated in a seven-year fallow cycle. About 60 square miles of communal grazing land on the *puna* are the most extensive resource, essential to the populace's principal livelihood, livestock breeding. Cattle, sheep, and llamas are most commonly raised, but also donkeys, goats, and hogs. The cattle are sold entirely, while the wool-producing animals are raised to provide meat and fibre for the villagers themselves as well as for the market. These products are marketed mostly by traders who come from outside the village, but local dealers or the breeders themselves sell wool-producing animals directly to the miners working in the area.

The village's demographic evolution has been smooth. The population grew from 290 in 1876 to 576 in 1940, then shrank only slightly, so that Vichaycocha still counted 496 inhabitants in 1988. Two factors were important: the village's considerable resources, and the *comunidad campesina*, which organizes the inhabitants in an all-encompassing manner and assures reasonably equitable participation in the lucrative livestock-breeding business.

Of course, economic inequality does exist here as well. Just 11 per cent of the ranchers own 35 per cent of the livestock, while 51 per cent of the owners have only 17 per cent of the animals. A very few families for various reasons do not belong to the *comunidad* and therefore have no livestock at all. They depend on working as day-labourers for the owners of large herds. Nevertheless, economic inequality here is not nearly so dramatic as in Quinches.

Besides the relative economic equality, the social cohesion among the villagers is remarkable. The *comunidad* serves as the axis around which virtually all socio-economic relations in the village are organized. This institution provides Vichaycocha with an effective mechanism for defusing conflicts, gaining consensus among groups large and small, carrying out community projects expeditiously, and defending the village against demands from the outside.

Several factors made it possible to establish and maintain the *comunidad* and to absorb population growth until well into the 1950s. First of all is the distribution of resources. The peasants, who have large grazing areas of high quality at their disposal, earn their living primarily by raising livestock. And *comunidad* control over the *moyas* assures general access to subsistence production. The *comunidad* even exerts some control over the irrigated fields in semi-private property.

Secondly, we need to consider the village's history. As recently as the 1920s, when the village's population was much smaller, the livestock breeders had no interest in using the grazing lands on the *puna*: they could earn enough money from cattle breeding and seasonal migration to the coast. When population pressure increased, the *comunidad* was an old and well-established institution, a ready mechanism to deal with the problem. Vichaycocha regained its rights to the grazing lands, previously leased to other villages. Then the *comuneros*, to secure more grazing land for themselves, restricted the number of outside shepherds and the number of animals they were allowed. The advantages of the *comunidad* were thus repeatedly demonstrated to the villagers.

Thirdly, location and climate make Vichay-cocha a perfect place for Andean livestock breeding. With so little farmland available, the tendency towards ranching is even more pronounced. The peasants' interests thus generally converge. The necessity of co-operation, inherent in livestock breeding, is further strengthened by the existence of the communally owned *moyas*.

Finally, seasonal and temporary migration provided a ready means to reduce pressures on local resources. When cotton farming expanded in the delta of the Rio Chancay at the turn of the century, villagers gained the opportunity to make some extra money as seasonal workers in the fields there. And since the 1920s, young men also had the option of leaving temporarily to work in the mines. With these opportunities, part of the economically active population withdrew from the village until their parents' passing made resources available to them. Thus, while the burden on the village economy eased, the community did not lose the migrants permanently, and its population remained relatively stable.

The distribution of resources, the specific history of the village, and the particular environment account thus for the *comunidad's* cohesion. The availability of resources and the partial absorption of people in seasonal and temporary migration go some way to explain why villagers started emigrating rather late, the first going off to Lima around 1930. In addition, the *comunidad* itself presents a major obstacle to emigration: the status of *comunero* requires Cochanos to assume responsibilities for the village early on, and leaving the village entails losing all the accumulated rights to use grazing land, keep animals, and work irrigated terraced fields.

The first sizeable wave of Cochanos—as inhabitants of Vichaycocha are called—to emigrate to Lima did not occur until the early 1950s. Because they started leaving late, they were also slow to find an economic niche for themselves. Initially therefore, migrants did not congregate in one business. Cochanos went to Lima, worked as waiters, factory workers, peddlers, and so forth—and then left the city again as soon as they learned about better jobs elsewhere.

This first phase came to an end in the late 1960s, when several Cochanos managed to gain a foothold in the clothing business around the central market in downtown Lima. In the 1960s and early 1970s, trade in the city enjoyed a boom. It was carried along by the sheer numbers of fresh immigrants and an economic dynamism marked by not just growth but also a redistribution of national income towards the middle class.

At first, just two Cochanos managed to break into the clothing business. Because they were able to expand quickly and thus needed workers, they employed in their stores more and more young *paisanos*. These employees, benefiting from the boom, often opened their own clothing stores within a very short time. They, in turn, employed more people from the village. It was not long before young people left Vichaycocha with the intention of finding jobs with relatives in the clothing business so that they could later open stores of their own. Today, some of those who came to Lima to start off as assistants and salesgirls have made fortunes—and not just by local standards. Although some did fail, potential migrants from Vichaycocha see the successful as their models.

By 1988, 35 per cent of the 238 adult migrants were working in the clothing trade; of those between the ages of 40 and 50, as many as 45 per cent were in this business. Just 15 per cent of the migrants earned their livelihoods in other trades such as street-vending and grocery stores, while 11 per cent were in academic professions, and students and labourers accounted for only 8 per cent each.

The predominance of clothing businesses as a source of earnings raises the question of where their start-up capital came from. After all, the trade requires a locale and equipment as well as an initial stock of goods. Contrary to my initial assumption, cattle wealth back home had hardly any bearing on opening a clothing business in Lima. Rather, the initial capital came partly from savings out of sales commissions and partly from credits: manufacturers provided merchandise on credit, *paisanos* merchandise as well as cash. Thus a migrant's socio-economic position in the village is not the prime determinant whether he opens a store: more important are good relations with his *paisanos* and whether he can 'make it' in the city.

The loans *paisanos* make to each other illustrate the close co-operation among Cochanos in the clothing business; it frequently begins with one *paisano* working for another. If the employee then goes out on his own, his former employer helps him with contacts and loans. They often go in together on larger purchases so as to qualify for rebates. Not only family ties make co-operation easier, but also physical proximity: almost all the Cochanos' clothing stores surround the central market-place and are thus within easy walking distance of each other.

Once Cochanos found their economic niche, they were able to benefit from their experience of co-operation in the village. Those already established helped their *paisanos* by giving them a place to stay and often a job as well. Thus the latecomers made their way into the clothing business much more easily, and the store owners found hard-working, reliable employees.

Because of Vichaycocha's particular social structure, the migrants kept hardly any property in the village: the consensus of interests within the *comunidad* enabled the peasants to ward off claims from migrants. With so little property in the village owned by those who have left, relations between migrants and village remain relatively weak. Migrants come from all directions to Vichaycocha to organize and celebrate the Feast of Señor de los Milagros—the Lord of Miracles, Lima's most important saint—which is organized and carried out mainly by Lima *residentes*. But hardly any migrants return for the Feast of San Miguel—the village saint—and they visit only seldom during the rest of the year.

There is thus close co-operation within the village which allows the peasants to resolve conflicts in such a way as to protect the majority's interests. The Cochano immigrants in Lima who are in the clothing business also work closely with one another. But relations between the villagers and the migrants are weak. None the less, it is Vichaycocha, and not Quinches, that has seen the return of several migrants. Decisive was the fact that Vichaycocha grants the right to become a *comunero* to anyone who grew up in the village, provided that he accepts the responsibilities of *cargos* (offices).

How the Migrants Changed Lima

A huge influx of migrants has radically changed Lima over the past half century. Not only did the metropolis expand further than ever thought possible and profoundly alter its ethnic composition, but its very economic and social role within the country has been transformed by the migrants.

In the past, the city was dominated by *criollos*. The administrative and commercial centre of the country, Lima took and redistributed the wealth created in other regions—keeping the lion's share for itself. Today the city still attracts the country's resources, but its people even more. And they, without any other means of making their way, created their own livelihoods, at first mostly in trade, but soon they began producing for customers mostly like themselves. Because of this development, traditional Western-style industry and the new, small-scale industries of the *provincianos* became so intimately interwoven that, for this reason alone, the distinction between 'formal' and 'informal' sectors is meaningless.

The city's metamorphosis was not only economic but also socio-cultural, with the *provincianos* taking over more and more of the city at the expense of the *criollos*. In this struggle, the migrants drew on the traditions of their native regions to create their own, novel, urban culture. It found its most salient, tangible expression in the associations of migrants. These associations, founded to preserve tradition and maintain relations with the home region, are often also very successful agencies for inventing and developing the new, urban culture—and its re-export back home. In addition to the economic interests which often link migrants with their villages, the social network serves as a conduit through which cultural models flow, greatly contributing to the urbanization of the rural areas.

I now turn to the northern region of Yucatán. We will see that while structures are similar on the meta-level to those in Peru—an agrarian crisis and the emergence of an urban sector absorbing large parts of the population—they are very different at the level of villages and cities.

Northern Yucatán

The Yucatán peninsula is mostly flat tableland with a subtropical climate. Until about twenty years ago, the entire north-eastern region, the most densely populated part, was dominated by the production of two varieties of agave, *fourcroydes* and *henequenana*, from whose fibres hawsers, ropes, and bags are manufactured. At the turn of the century, *henequen* farming made Yucatán's planters rich and turned the Mayan peasants, who had once had the area to them-selves, into quasi-serfs on the haciendas. When the government took over the plantations in 1937, a sad history of mismanagement and corruption began, which accelerated the inevitable decline of the *henequen*-based economy.

During *henequen*'s golden age, Mérida was the dominant city on the peninsula, its commercial, industrial, and political centre, and the focal-point of its socio-economic life. The *henequen* workers living in the hinterland earned a little on the side by farming the *milpa* (fields cleared by the slash-and-burn method) and the *solar* (a traditional garden with small livestock). The sparsely populated remainder of what are now the states of Yucatán and Quintana Roo were inhabited by corn farmers largely isolated from the outside world.

Today the overall situation is quite different. In the early 1970s, a gigantic tourist centre, generously supported by the Mexican Government, shot up on the previously sparsely-populated eastern cost of Quintana Roo. This project proved extremely successful, so that the entire northern part of the peninsula, which had been falling ever more deeply into poverty, began to prosper. Meanwhile, Mérida has been strengthening its position as the peninsula's centre of commerce and industry and is now also profiting from the increasing number of tourists, many of whom spend a few days in the city.

These two changes—the disappearance of the once-dominant *henequen* industry and the massive influx of tourists—have transformed the region. These developments are the subject of the next section, first in so far as they have affected two villages, then in the way they have touched Mérida, and finally as they are manifested in the tourist region whose centre is Cancún.

The Village of Huhi

Huhi is 40 miles south-east of Mérida, on the edge of what used to be the *henequen* belt. The village's population rose from 1,129 in 1940 to 2,169 in 1980, but dropped to 1,940 inhabitants in the following decade. Until recently, the people of Huhi had two major sources of income: they did some farming on the *milpa*, but mostly they worked the *henequen*—either on plantations belonging to the *ejido* (a form of collective ownership of land) or on the 'small estates', the remnants of former haciendas.

Even decades ago, a few people found temporary employment outside the village, but more recently, the few became many. At the beginning, just a small number of young men had worked as day-labourers for a few months at a time in Mérida or the United States. In the mid-1970s, more and more families left the village. And in the last few years, many men—25 per cent of heads of households in 1992—have begun working in Mérida while still living in the village, commuting daily.

Around 1980, several families began using sewing machines to stitch together synthetic-fibre bags copied from models from Mérida, Cancún, or the United States. Today there are fifteen such small businesses, each employing up to six people. They sell their product primarily in Mérida. Altogether, more than a third of all heads of households earn their income from Mérida, while just 38 per cent of them still work in agriculture: in the *milpa* and as day-labourers—*henequen* is no longer important. Among the young men, only a few are willing to farm the *milpa*: they would rather emigrate after graduating from secondary school, commute to a job in Mérida, or work for the bag manufacturers.

The women's situation is quite different. Some of them used to work manufacturing traditional bags out of *henequen* fibre (*sabucan*) or decorating traditional women's clothing (*huipiles*). But the manufacture of old-fashioned bags disappeared with the fibre itself, and only a few women still work on the clothes. Today, women do much more work in the *solar*, because men are often employed elsewhere. In addition, the women now have more work and responsibility in raising the children: in the past less attention was given to their education, and the boys would usually work with the men during the day. A few young women work also for the bag manufacturers or sell bags on a commission basis in the peninsula's cities.

The impact of the economic changes on the village's traditional sector thus contrasts by gender. The male domain—farming the *milpa* and *henequen*—is deeply depressed. It no longer attracts young men who now seek their livelihood elsewhere. The traditional female domain, by contrast, has emerged from the crisis stronger than ever. Women's economic-reproductive tasks, rather than diminishing in importance, have become more important in terms of both their contribution to the family's subsistence production in the *solar* and their responsibility rearing the next generation.

Between the village economy's traditional and modern sectors, there is a group of non-agricultural occupations supported by the local market: baker, butcher, retailer, bar owner, and tailor. These people benefit from the economy's increasing monetization, but their opportunities are circumscribed by limited local buying power.

The bag manufacturers constitute the dynamic part of the village economy. This sector is still expanding, because it has potential customers not only on the peninsula but also on the mainland. The bag manufacturers actually benefit from the traditional economy's decline, because they are able to draw on cheap labour in the village.

At the fringe of the village economy are the men who earn their income from Mérida. They do have some effect on Huhi: they spend money there for consumption, and they quite often hire labourers to plant *milpa*.

The dramatic macro-economic changes in Yucatán affected Huhi in two important ways. Some of its people emigrated, and the village became even more closely linked with Mérida, both geographically and socially. Many commuters work there, and 63 per cent of all the emigrants went to Mérida—barely a quarter went to Cancún or Cozumel, 6 per cent to the United States. Emigrants prefer Mérida because the village and the city have been closely linked since the turn of the century, when the people of the region took the new railroad to Mérida (Wells, 1992). Today, every family in the village has relatives there who can help in the search for a job or provide temporary shelter.

The Village of Sudzal

Sudzal is about 50 miles east of Mérida, near the small town of Izamal. Since no public transportation connects the village directly with Mérida, travelling to the city can take as long as three hours.

The population of the *municipio* grew in the forty years prior to 1980 from 1,007 inhabitants to 1,829. But in the following decade it dropped to just 1,329. But many families registered as residing in the village have actually been living in Cancún or Cozumel for quite some time. According to the official census, the village population was 1,037 in 1990. From a sample census of half the households it would appear that only 750 people actually lived in Sudzal in 1993.

Until recently, a small part of the *municipio* of Sudzal was planted with *henequen*; the rest is part of the corn belt and was used partly for the *milpa* and partly for livestock breeding. About fifteen years ago, these activities still provided the principal sources of income for the peasants of Sudzal. They were supplemented by tending and harvesting *henequen* on the *ejido* or working in one of the many fibre-stripping factories that used to operate in the region—Izamal and the surrounding area were once the heart of the *henequen* region.

Today, barely two-thirds of heads of households still farm the *milpa*. Because this work is primarily for subsistence and little agricultural product can be sold, the *milperos* have to find cash income from other sources. With the collapse of the *henequen* industry, the peasants now work on average three or four days a week as day-labourers on the cattle ranches that have sprung up in the *municipio* during the last twenty years. The ranchers need a great many seasonal workers to clear the ground for meadows; to build and repair fences; and to plant, tend, and harvest large corn fields—the corn to be fed to the animals.

About 10 per cent of the households in the village itself raise cattle. Two of these households have four-hundred head each, while the other herds number between ten and thirty. In addition, several influential people from Mérida have started cattle farms in the area around Sudzal. The big ranchers in the village also own its supermarket, two bars, and tortilla factories.

Very few men, just fifteen in 1993, work in a city, Mérida or Cancún, and return on weekends to their families in the village. Whoever leaves Sudzal and finds work in the city usually takes his family to the city shortly thereafter to live near his work-place, because commuting is too demanding in terms of expense and time.

For about twenty years, women's earnings have supplemented men's. In Yucatán, the demand for hammocks, made in the traditional craft industry, has soared as tourists have flooded in. Previously only a few hammocks were produced to satisfy local needs, but today most women in Sudzal, besides taking care of the household, keep busy weaving hammocks. They get the raw material from middlemen who come from Izamal and pick up the finished product at regular intervals, paying the producers about one-fifth of the retail price.

Economic changes thus have differentially affected men and women in the traditional sector of Sudzal as well. As in Huhi, the male sphere, i.e. *milpa* farming, has fallen on very hard times. But in Sudzal there are more opportunities for cash earnings through wage-labour. Some young men have consciously opted for the life of a *milpero*. The *milpa* has a future in a symbiotic relationship with cattle ranching—a relationship that is threatened by a contradiction: the ranchers offer wages for work, but they are taking more and more land away from the *milpa*.

As in Huhi, the traditional female sphere of economically reproductive work in the house has weathered the storm. However, because the economy is so polarized in Sudzal, women must contribute cash earnings to help support the family.

Since about 1980, Sudzal has witnessed a great wave of emigration, principally to Yucatán's east coast, where new tourist centres in the state of Quintana Roo have generated a huge number of jobs. When the enormous hotel complexes were going up in Cancún and Cozumel at the beginning of the 1970s, some young men from Sudzal went to work in construction. Back then, a young man typically left the village for a couple of months or maybe a few years to work somewhere and put money aside. Then he returned with his small savings and started his own family.

In the following years more and more men—and later women as well—went to Yucatán's east coast, pushed out by the deepening crisis in the *henequen* industry and pulled by the relatively high wages in the booming construction industry. Slowly a network of personal relations was built up—at first with foremen at the construction sites, later also with owners of restaurants and hotels—which provided jobs for new arrivals. Soon migrants preferred, rather than to return to the village, to stay in Cancún or Cozumel and bring the family there as well. Even today, most Sudzaleños in Cancún and Cozumel work in construction (the men) or in restaurants and hotels (men and women).

A survey of registered *ejido* members in Sudzal showed that only two-thirds of them still lived in the village in 1993. Sixty-seven per cent of the emigrants have settled on the east coast, in

Cancún and Cozumel, while only 14 per cent have moved to Mérida. The *ejidatarios* are generally older men who are firmly rooted in the life of the village. Among young people, the number of migrants is much higher, and their favourite destination is the east coast, attractive not only because of its relatively high living standard, but also because of its 'modern way of life'. Emigration poses no problems for young people: few responsibilities keep them in the village, and they can count on the support of relatives at their destination.

Modern Yucatán

The enormous economic change in the peninsula's north, along with the flood of migration it has caused, has transformed the region. People's economic and social lives used to revolve around the *milpa* and the work in the *henequen* industry, but *henequen* is gone, and the *milpa* is in deep trouble, so the very character of the villages has changed irreversibly.

Huhi illustrates what has happened to most villages within a 40-mile radius of Mérida. In the last quarter century, the city has grown from 200,000 inhabitants to about 650,000, almost half the state's total population. Most of Mérida's immigrants come from what used to be the *henequen* region and some from the corn belt (Fuentes Gómez, 1990; Moßbrucker, 1994). Immigrants from outside the peninsula are relatively few, just 10 per cent (INEGI, 1991).

But Mérida does not just absorb a considerable part of its hinterland's population outright; it also provides jobs for a great many commuters. Already at the beginning of the 1980s, men in the *henequen* region spent one-quarter of their workdays in Mérida (Baños Ramírez, 1990). Today, people living in the surrounding countryside like to joke that their villages have become dormitories. At present, up to half the region's income is earned in Mérida, either directly through employment or indirectly through the sale of products there (Moßbrucker, 1994). Urbanization of what used to be the *henequen* region has advanced to the point that Mérida and its hinterland are now in intimate symbiosis.

Unlike in Lima, small-scale industry and street-vending in Mérida are not organized primarily by Amerindian immigrants. A study of two districts in the east and south of the city showed that newly arrived immigrants often have poorly paid and physically demanding jobs.[2] Thus in Amalia Solorzano, a district founded in the eastern part of the city just fifteen years ago, almost 40 per cent of the men have work in factories or in construction; 18 per cent hold low-level office jobs. It is hard to find even the rudiments of an economy organized by the immigrants themselves, except various repair shops established by newcomers.

In Mercedes Barrera, founded about forty years ago on Mérida's south side, appreciable differences between migrants' and non-migrants' occupations have now disappeared. Among the immigrants, about 15 per cent are labourers and another 15 per cent office workers, while 10 per cent are general merchandise retailers or have repair shops.

It is striking how few women in Mérida contribute to the families' cash income. More than 80 per cent of the women in the two districts studied refer to themselves as housewives. Among the other 20 per cent, most work in a small general store. Because such a shop is usually located in an addition to the front of the house proper, the work in the store easily combines with household chores. One explanation for this work pattern can be found in what the women do in the sphere of reproduction and subsistence. They, like women in the countryside, are responsible not only for managing the household and raising the children but also for taking care of the *solar*. They thus do contribute to the family's income. This arrangement works because even within the city of Mérida, people kept to their custom of living in a one-storey house right next to a large *solar* where they would plant fruit trees and raise chickens, turkeys, and pigs. They could do so because land ownership was quite unregulated: almost every family could acquire some land by paying a small sum to a *ejido* near the city. Then the residents built their own houses on it, structures they gradually enlarged and strengthened. The easy access to land thus reduced the cost of migration in terms of both urban subsistence production and shelter. But the reorganization of the *ejidos* since 1992 has made it much more difficult to acquire land in this way.

Sudzal, on the other hand, illustrates what has happened in many villages in Yucatán's eastern corn belt and in Quintana Roo. They look to the peninsula's east coast, where there are enough jobs in the tourist industry and related activities, such as construction. As in Mérida, Amerindian immigrants have not started any businesses of their own on the east coast.

Because the home villages are so far from the work-places in Cancún and Cozumel, the immigrants settle permanently on the east coast with their families. The irregular work hours common in the service sector may also contribute to this pattern, as suggested by the fact that it also prevails in places fairly near Cancún, like Kantunil Kin (G. Moßbrucker, 1995). Cancún thus grew in just twenty years from a hamlet to a city with 167,730 inhabitants in 1990, according to official data. Its current population is much higher: many migrants remain registered as residents of their native villages, and immigration to Cancún has continued apace.

The northern part of the peninsula is thus dominated by two large economic and demographic centres today: the Mérida region in the west with Progreso and the *municipios* Kanasin and Uman, all in effect part of one conurbation, and the Cancún/Cozumel region in the east. They, more than anything else, drive the demographic and economic urbanization of the peninsula, which has shifted from farming to activities in the industrial and service sectors. This economy is dominated by national and international investors and is utterly dependent on events that occur outside of Mexico and beyond the control of Mexicans.

Migration in Peru and Yucatán: A Comparison

Mass emigration from rural areas has a common cause: the relative, if often not absolute, impoverishment of the countryside. In Peru, government policies favoured urban over rural residents. In Yucatán, developments in the world economy brought boom and bust to the rural economy, and then, in the last two decades, an urban boom. The scope for shaping events at the regional level is limited in both countries.

But the similarities do not extend beyond the macro level. The processes of migration itself already show major differences. Other dissimilarities can be observed in terms of the migrants' socio-cultural organization, their integration into the urban economy, and the ways they link the countryside with the city.

The success of migrants to Lima in establishing economic niches for themselves had an enduring impact on migration to the city. A promising niche continues to attract new migrants. Precisely how such a niche grows as more people from the same village join it reflects the organization of both the migrants and the village, as well as how each relates to the other. In many instances, the links are close, often re-enforced by economic relations. With the 'clubs', in which the immigrants in Lima organize themselves according to their village or region of origin, they have created a powerful institution for maintaining and supporting the relations between city and village.

The dichotomy between city and countryside has become less pronounced over the past few decades, especially between Lima and its hinterland: information, gifts, merchandise, relatives, and friends all travel continuously back and forth. But the intensity of these contacts depends to a large degree on just what the particular village's farmers, ranchers, or craftsmen produce and the economic niche opened up by the migrants. Villages like Vichaycocha maintain only social relations with the city: while meat is the main product of the village, its emigrants have not gained any foothold in the meat business. Quinches, on the other hand, with an animal population far below that of Vichaycocha, finds itself involved in intense relations with emigrants through their position in the meat business. In the apple-growing valleys on the coast near Lima families frequently maintain two households: some members live in the city to market the fruit, while others stay in the village to tend the orchards (Fuenzalida et al., 1982; Alber, 1990).

The migrants in Yucatán, by contrast, keep in touch with home almost exclusively for reasons of kinship that have no economic rationale. They have no 'clubs' or similar associations, and they are not concentrated disproportionately in particular occupations. Mérida and Cancún do not offer them economic niches such as those so important in Lima. Most of the construction workers, factory labourers, and porters in Yucatán are migrants not because of a complex network of social and economic relations like that found in Lima among paisanos, but because these unattractive jobs are readily available. In a relatively open labour-market, manual labour is the bottom rung of a ladder that the immigrant may in time be able to climb.

In Yucatán, especially in what used to be henequen country, the dichotomy between city and country has all but dissolved. For one thing, good infrastructure and cheap transportation have encouraged commuting to Mérida. Also, while racism is omnipresent in Lima, it is insignificant in Mérida and Cancún and is not experienced as an obstacle by immigrants.

The integration of the migrants in the city also proceeds differently. In Lima, the 'clubs' do not just provide access to economic opportunities, they also cater to people's socio-cultural needs. They often are the focus for part of migrants' social life and give them a sense of security in the urban context. Kinship ties are also very important for integration. They are re-enforced by the concentration of kin in specific occupations and particular neighbourhoods: migrants from the same village or region typically settle in the same neighbourhood until no more housing is available, at which point new arrivals establish a new neighbourhood, eventually also dominated by people of common origin. This housing pattern is yet another consequence of the specific form migration takes.

In Yucatán, migrants can rely only on the social network of close relatives or, under certain circumstances, the patronage system established by the PRI, Mexico's ruling political party. Here, migrants from Huhi or Sudzal are not concentrated in one part of the city, nor are there any 'clubs' providing social or cultural support—but then that is unnecessary, because urban society in Yucatán is different from that in Lima, more open.

In Lima, most provincianos work within economic structures organized by themselves. Drawing on family members as a source of cheap labour, they are now well established in branches of manufacturing and commerce connected with domestic consumption. In manufacturing, the provincianos have made such great strides in the past few decades that they have almost completely displaced the Western-style competition in branches such as clothing and shoes. They have enjoyed even greater success in street-vending, unsurpassed in their flexibility in offering a wide range of goods at convenient locations. The *provincianos* have the advantage that large parts of the national consumer goods market have come to be dominated by demand from people like themselves. They are familiar with their customers' tastes and employ appropriate sales strategies. The success of this 'informal sector', celebrated by De Soto (1986), results principally from the failure of traditional, Western-style businesses to absorb the migrants pouring into the city from the villages. Cultural likes and dislikes probably also played a role (Parodi, 1986).

In Yucatán, by contrast, the tourist industry boom since the 1970s has enabled the 'formal' sector to absorb wave after wave of migrants as well as commuters. Thus migrants did not need to organize their own businesses. Cultural forces have probably been a factor in Yucatán as well, especially for the migrants from the *henequen* region. These people had any sense of personal initiative physically beaten out of them by the hacienda owners, and when the government nationalized the business, its patronage, corruption, and fraud only made things worse. Personal resourcefulness does not emerge readily in a socio-cultural milieu that has been hostile to creativity for so long.

The migrants in the two regions are thus organized in very different ways. The *provincianos* in Lima managed to fashion their own urban culture that helps them to create and maintain a positive identity distinct from the *criollos*. This identity finds its ready expression in voting patterns. The *provincianos* carried APRA and the United Left to victory in the 1980s, and they were crucial to Fujimori's victory in 1990 (Degregori and Grompone, 1991).

The situation in Yucatán is altogether different. The migrants cling to features of Mayan culture, and they are also adapting them to an urban way of life, but they have not created any formal organizations that could actively champion their culture and identity. Kin groups take care of their immediate socio-cultural needs, while they define their overarching identity in terms of the Mexican nation and the state of Yucatán (Moßbrucker, 1992*a*). PRI need not worry that Mayan migrants establishing a socio-cultural identity opposed to official doctrine will pose an electoral threat.

Three connected factors may serve to explain the differences between the two processes of migration. First, a region's historic evolution sets it on a particular course that favours or even determines certain options, while excluding others. In Peru, the corporate community, as Wolf (1955) called it, arose at the village level out of the collapse of the Kuraqa political system. Even if the corporate community was not always as solid as Wolf suggested, it did contain democratic elements and tied people closely to their particular village or territory.

In Yucatán, on the other hand, people moved continuously from one village to the next even during the colonial era (Farriss, 1984: 199–223), so that social relations centred on the kin group rather than the village early on. The endogenous social structure of the villages in the densely populated Mérida region was further disturbed by the haciendas of the eighteenth century and later destroyed altogether by the Mayan uprising, *La Guerra de Castas* of 1847–53, and the subsequent rapid expansion of the *henequen* plantations. The *henequen* boom delivered the *coup de grâce* to the surviving Mayan villages, transforming their inhabitants first into dependent workers and later, with the institution of the collective *ejidos*, into the subjects of government patronage. But the Mexican Government, because it had to bear the huge costs of subsidizing the Yucatecan economy, had a well-defined interest in developing a new economic strategy: it succeeded with the establishment of the tourist economy in the northern part of the peninsula.

In Peru, haciendas had also been expanding continuously since the nineteenth century, but a significant part of the Indian villages managed to survive, communities that even today are largely endogamous. The kinship group thus usually coincides with the village. When Amerindians left the land for a city that confronted them with racism rather than offering them ways to earn a living, needed to organize themselves, they and their socio-cultural background provided the ready means to do so.

Secondly, the world economy played a vital role in this historical process. The technical revolution in agriculture, in particular the invention of the mechanical sheaf binder, and the fast-growing shipping industry's requirement for hawsers, gave Yucatecan planters the opportunity to control a lucrative niche in the world market. The extremely labour-intensive cultivation and processing of *agave* had enduring consequences for all the people on the peninsula. Subsequently, changes in the world market plunged them into a deep crisis that transformed northern Yucatán anew. Finally, another profitable business opened up: rising incomes in the highly industrialized countries offered the opportunity to develop mass tourism which came to provide a great number of jobs in Cancún and Mérida.

Peru has had only very few opportunities in this century to participate in the world market with profitable, labour-intensive products. Mining in the highlands and oil drilling in the Amazon region have enriched various, mostly foreign companies, but provided little positive feedback for Peru's economy. The fishing industry expanded rapidly in the 1960s, and spawned the big city of Chimbote, but the boom did not last beyond the 1970s. Furthermore, the Peruvian economy has been subject to a series of crises, making it virtually impossible for government or private money to provide opportunities for peasants or urban immigrants. The migrants in Lima thus had no choice but to create a market serving local consumer needs.

Thirdly, Yucatán demonstrates the impact government intervention can have on the migration process. Mismanagement on the collective, actually government, *ejidos* accelerated the decline of the *henequen* economy, but government subsidies kept many peasants on the land—until the 1980s, when the government cut the subsidies and erected a tourist complex in Quintana Roo, a gigantic project whose effects rippled across the peninsula, creating good jobs.

In Peru, all the governments since the 1950s have tried in varying degrees to prevent unrest in the rapidly growing cities by pursuing a policy of keeping prices for agricultural products low. While farmers earned relatively little for their crops, various programmes to expand the domestic market created opportunities in the growing consumer goods market. Both factors encouraged the *provincianos* to leave the countryside and forge their own socio-economic structures in Lima.

In each case, migration, the migrants' social integration into the city, and the repercussions from this process on the home villages follow a specific pattern that has far-reaching consequences for the region and the nation. The *provincianos*' economic success in Peru, and their increased self-confidence in dealing with the *criollos*, have made them into a force to be reckoned with, a force that is yet to demonstrate its explosive potential.

In Yucatán, on the other hand, the economic-political élite managed to integrate into their businesses the people who had once worked in the *henequen* industry or farmed corn. The Mayans were not marginalized because racism on the peninsula is minimal in comparison with Peru—or other regions of Mexico, for that matter. They modified their culture, adapted it to urban life, and then brought the results back to their native villages, but they had no reason to build their own economic structures or to set themselves apart. They present no threat to the economic or political élite of Yucatán and Mexico. Quite the contrary, because they have experienced urban life on the peninsula as overwhelmingly positive, they are now what they were when the *henequen ejidos* enjoyed subsidies: close allies and grateful citizens of the state.

Notes

1. In this essay, I distinguish seasonal, temporary, and permanent emigration. By 'seasonal', I mean leaving the village from a few weeks to several months a year in order to earn some extra money; by 'temporary', mean going away for several years with the intent of returning to the home village when conditions change, for instance, when the parents die and the returnee can take over the herd and land; and I mean by 'permanent', leaving without any intent to return.
2. This study was carried out in 1992, the data analysed and published by Moßbrucker (1992*a*, 1993, 1994). See also the study by Fuentes Gómez (1990) on the southern part of the city.

References

Alber, Erdmute (1990), *Und wer zieht nach Huayopampa? Mobilität und Strukturwandel in einem peruanischen Andendorf* (Saarbrücken/Fort Lauderdale: Breitenbach Publishers).

Altamirano, Teófilo, and Hirabayashi, Lane (1991), 'Culturas regionales en ciudades de América Latina: Un marco conceptual', *América Indígena*, 51 (4): 17–48.

Baños Ramírez, Othón (1990), 'Los nuevos campesinos de México, el caso de Yucatán', in

Othón Baños Ramírez (ed.), *Sociedad, estructura agraria y Estado en Yucatán* (Mérida: Universidad Autónoma de Yucatán), 401–30.

De Soto, Hernando (1986), *El Otro Sendero* (Lima: Instituto Libertad y Democracia).

Degregori, Carlos Iván, and Grompone, Romeo (1991), *Demonios y redentores en el nuevo Perú: Una tragedia en dos vueltas* (Lima: Instituto de Estudios Peruanos).

Farriss, Nancy M. (1984), *Maya Society under Colonial Rule: The Collective Enterprise of Survival* (Princeton: Princeton University Press).

Fuentes Gómez, José Humberto (1990), *Estrategias de supervivencia y reproducción social de los pobladores de la colonia Emiliano Zapata Sur de Mérida* (Yucatán, México: Tésis de Maestría, Centro de Investigación y Estudios Superiores en Antropología Social).

Fuenzalida, Fernando, Valiente, Teresa, Villaran, José Luis, Golte, Jürgen, Degregori, Carlos Iván, and Casaverde, Juvenal (1982), *El desafio de Huayopampa: Comuneros y empresarios* (Lima: Instituto de Estudios Peruanos).

INEGI (1991), *Yucatán: Resultados definitivos. Datos por localidad (integración territorial)*, XI Censo General de Población y Vivienda, 1990 (México: Instituto Nacional de Estadistica, Geografía e Informática).

Moßbrucker, Gudrun (1995), 'Los Mayas frente a la modernidad: Los exrebeldes del norte de Quintana Roo', in Ruth Gubler and Ueli Hostettler (eds.), *The Fragmented Present: Mesoamerican Societies Facing Modernization* (Möckmühl: Verlag Anton Saurwein), 151–9.

Moßbrucker, Harald (1990), *La economia campesina y el concepto 'comunidad': Un enfoque crítico* (Lima: Instituto de Estudios Peruanos).

—— (1992a), ' "Etnia", "cultura" y migración entre los mayas de Yucatán', *América Indígena*, 52 (4): 187–214.

—— (1992b), 'Sharecropping: Traditional Economy, Class Relation, or Social System? Towards a Reevaluation', *Anthropos*, 87: 49–61.

—— (1993), 'Situación socioeconómica, organización política y desarrollo de la infraestructura en dos colonias de Mérida', *Revista de la Universidad Autónoma de Yucatán*, 185: 34–43.

—— (1994), *Agrarkrise, Urbanisierung und Tourismus-Boom in Yukatan/Mexiko* (Münster: LIT Verlag).

Parodi, Jorge (1986), *'Ser obrero es algo relativo . . .': Obreros, clasismo y política* (Lima: Instituto de Estudios Peruanos).

Wells, Allen (1992), 'All in the Family: Railroads and Henequen Monoculture in Porfirian Yucatán', *Hispanic American Historical Review*, 72 (2): 159–209.

Wolf, Eric (1955), 'Types of Latin American Peasantry: A Preliminary Discussion', *American Anthropologist*, 57: 452–71.

6

Public Policy and Rural–Urban Migration

Charles M. Becker and Andrew R. Morrison

RECENT years have witnessed a dramatic transformation in economic policies throughout the world. Economists of nearly all political orientations have come to emphasize the manipulation of markets to achieve social and economic goals while planning has waned in importance. This emphasis on markets and economic liberalization has meant profound changes in public policy, especially in developing and (formerly) socialist nations. It has affected public policy in a vast range of areas, most obviously in macro-economic, government budget and trade policies, but also stretching to urbanization, regional development, and population policies.

The following pages argue that the impacts of direct public policies on urbanization and rural–urban migration are important, but are overshadowed by the consequences of general policy shifts taking place across the world. We argue further that policies aimed at urban growth control are likely at best to have minimal effects; the consequences they do have are likely to reduce welfare, especially of the poor and middle classes. Finally, we argue that many policy-makers are profoundly unaware of the impact of economic policies on migration and urban development—and for that matter, that they ill understand the economic consequences of the dramatic demographic shifts presently occurring in many countries.

Before proceeding, we briefly describe the shift in broad economic policy now under way in many (though hardly all) countries throughout the world. We also comment on government policies towards urban growth and migration. A brief exposition of why people migrate is followed by a sketch of policy-makers' concerns. We then address explicit regional, migration, and settlement policies. Most of these policies were designed with an implicit ideology that pervasive government intervention can be used to improve people's welfare: we term this the 'planning approach'. Finally, we analyse policies which indirectly influence migration flows and urban growth through their impact on market forces.

From the Keynesian era until the 1980s, most economic policy-makers regarded direct government intervention as the best way to attack specific social problems. Markets were often distrusted, and market failures were seen as frequent causes of social problems. While economists of virtually all ideologies urged policy-makers to work on removing market failures, economists' incantations about 'getting prices right' were far less popular in political circles than their efforts to improve planning and direct expenditures to specific target groups.

But, during the 1970s, increasing evidence showed that incentives mattered enormously in determining economic outcomes, and that government policies aimed at one area could have great impacts on other variables. At the same time, uncontrolled government expansion in many nations led to fiscal crises, followed by severe constraints on what governments could achieve through direct policies. Across Latin America, Africa, South Asia, the socialist world, and much of the Mediterranean and Middle East, government overextension resulted in a loss of control, and in the paradox of big but weak government. The responses have been liberalization and shrinking government, or paralysis and stagnation. Elsewhere, notably in South-east Asia and China, fiscal stability and gradual liberalization have permitted governments to remain far more involved in planning.

A 'new orthodoxy' has emerged. It stresses limited direct government involvement in production, and the liberalization of international trade, domestic trade, and labour and capital markets. It emphasizes balanced budgets; removal of subsidies and tariffs; privatization of government enterprises; and the development of legal institutions and property rights which enable free, competitive markets to function more efficiently. While hardly all countries have shifted away from deep government involvement, few have not made some moves towards liberalization, and do not officially acknowledge the need for further reform. This policy transition has enormous implications for rural–urban migration, and the consequences of these policy shifts easily dwarf those aimed at direct control of population movements.

Despite this fact, most governments do have policies aimed directly at urbanization and rural–urban migration. Especially in regions experiencing rapid urban growth, governments aim to curtail this growth, and redirect such growth which does occur into smaller cities. The primary rationale for such policies is that governments feel obligated to provide more services to urban dwellers than to rural dwellers. Consequently, controlling urbanization (the proportion of national population living in cities) limits the strain on government budgets.

One might reasonably ask why urbanites should receive more services than rural inhabitants in nearly every country. Part of the answer is that habitation in close proximity to others requires more of some services—for example, stricter sanitation and paved streets are needed to prevent epidemics, while looser standards are adequate in rural areas. There also are efficiency

gains to providing many public services aimed at the entire nation (such as hospitals and universities) in urban areas, and it is also cheaper on a per capita basis to provide other public services (such as electricity and clean water) to densely settled areas, hence, governments naturally do so. But there are also political reasons: many governments have been toppled by urban riots; few have fallen to rural discontent. Political leaders also regard major cities as showcases of their achievements: better prosperous cities and invisible rural poverty than shabby capitals and rural investment with little political payoff.

The determination with which governments seek to slow urban growth depends on several factors. Curbing urban growth is low priority when urban jobs are growing rapidly; when foreign investment is high, so that public investment in infrastructure does not mean an end to industrial capital accumulation; when economic growth is sufficiently rapid to provide government with the resources it needs to make key infrastructural investments; and when agricultural development results in the rapid growth of smaller cities and towns, which serve as marketing depots and commercial centres for an increasingly prosperous countryside. It is also a lower priority in democratic societies, which cannot repress voting peasants with the ruthlessness exhibited by autocratic regimes; and it is less critical in socially peaceful societies—where urbanization and crime are not perceived as synonymous. Finally, restricting migration is a minor issue in countries where rural education is advanced, so that an influx of newcomers does not mean a tidal wave of unskilled indigents with few economic prospects. Of course, in many countries, few or none of these conditions exist, and that is where resistance to one of mankind's most profound and peaceful social events—migration in search of a better life—is greatest.

Why Do People Migrate?

People migrate to seek better lives, with higher and more stable incomes, better educational opportunities for themselves and their children, and less exposure to violence. Migrants therefore leave low income for more prosperous locations. Since real urban per capita incomes are frequently (though not always) considerably higher than rural per capita incomes, much migration is from rural to urban areas, and net flows generally are in this direction. Asymmetric patterns of economic development, with higher rates of investment and job creation in cities than in the countryside, also bring migrants flocking in from the countryside. Thus, there have been vast internal migration flows: again and again, people vote with their feet for better economic opportunities.

The earliest modern economic analyses of migration in developing countries were surplus labour models (Lewis, 1954; Ranis and Fei, 1961). In these models, agricultural labour has very low productivity; workers can leave agriculture with little effect on rural output. Given the low productivity of agricultural labour, rural–urban migration raises national output and growth rates, at least initially. Nor is it surprising—at least while these models held sway—that policy-makers were optimistic about migration's contribution to economic development.

By the late 1960s, this optimism had disappeared. Many policy-makers found themselves overwhelmed by mounting urban problems, including rising underemployment, and insistent demands for public-service provision. Concurrently, an important paradigm shift occurred with the probabilistic migration model (Todaro, 1969; Harris and Todaro, 1970). In the Harris–Todaro (HT) model, individuals focus on expected incomes, moving if they expect to be able to earn higher wages elsewhere. In the initial HT model, expected earnings are related to the prevailing wage multiplied by the probability of obtaining employment; if unemployment exists in destination areas, this probability is less than one (finding a job is not a certainty), and expected earnings are less than the wage earned by employed individuals. Since individuals make their migration decision based on an expected rather than a certain wage, individuals may 'lose the lottery' and not find work in destination areas. If the urban wage is sufficiently high, migration is consistent with large and growing unemployment in destination areas. Hence, what is rational for an individual seeking to maximize expected income is not necessarily optimal from a social perspective.

Since Harris and Todaro's pioneering work, three further changes have occurred in the economic modelling of migration. First, the number of factors individuals are seen to consider in the migration decision has grown. These factors include amenities (such as education, housing, and public services), levels of critical poverty in origin areas, and the availability of easily extracted rents in destination areas.

Secondly, economists have come to question the focus on individual decision-making. Clearly, migration is an extended household decision in much of the world. As it is crucial to know how household decisions are made, there have been several important attempts to 'open up' the household to analysis. Central to questions of household decision-making is the issue of power differentials within households, and whether decisions are made co-operatively or conflictively. An important insight from this household focus is that risk reduction may be a crucial motivation for migration in developing countries (Stark, 1991). Migration may work as a form of insurance against wide swings in household income. Suppose, for example, that an agricultural region is characterized by a high probability of drought; consequently, there is extreme variability in crop yield from year to year. If the entire household remains in this area, there is no way (in the absence of insurance markets) for the household to reduce its exposure to risk. Alternatively, a member of the household could migrate to work in another region. As long as the risks in this other region do not occur at the same time of the year, the household has succeeded in lowering its risk at any point in time. Households might even be willing to accept lower but less variable income.

The third change concerns the way economists view urban labour-markets and wage determination in developing countries. Harris–Todaro argue that wages are fixed by government policy-makers. Evidence, however, suggests that both 'modern-sector' wages and informal-sector incomes are strongly influenced by market forces. On the other hand, unemployment remains a problem, and it is apparent that simple models of perfect competition do not easily explain all labour-market behaviour. In response, economists have focused on the employment search process, which induces some employers to pay above the minimally achievable level, and on factors such as increased productivity and reduced labour turnover that result from higher wages.

The Concerns of Policy-Makers[1]

Before turning to various policies and their impacts, we need to sketch five major characteristics or tenets common to policy-makers throughout the developing and socialist worlds.

'Excessive' Urbanization

A common view is that city growth rates are excessive. This belief in part stems from concern over and fear of the large proportion of visibly unemployed or underemployed young individuals in many African, Asian, and Latin American cities. A common conclusion drawn is that rural–urban migration is excessive and should be curtailed.

Most economists, however, believe that urbanization is an inevitable consequence of successful economic development and of rural stagnation, not an undesirable force that must be suppressed. Migration instead should be recognized as an equilibrating response to disequilibrium elsewhere in the economy. People migrate to cities because they believe the opportunities to be better there. Nor is there any evidence that Third World migrants are mistaken in their belief. In Africa, for example, many of the highest urbanization levels are associated with countries that had experienced periods of rapid economic growth (43% urban in Cameroon; 42% in Côte d'Ivoire; 57% in the Congo; 48% in Gabon), had undergone dramatic rural declines (Liberia, 44%; Mauritania, 51%; Somalia, 25%; Central African Republic, 39%), or had experienced rural stagnation (Zambia, 43%).[2]

But even successful agricultural growth may give rise to urban growth, and hence eventually to urbanization. In each of the major success stories of African agriculture over the past 30 years, the urban population grew more rapidly than the continental average (Cameroon, 8.6% per annum between 1973 and 1980; Côte d'Ivoire, 7.3%; Kenya, 8.5%; Malawi, 7.9%; vs. an African average of 5.7%). Agricultural growth generates needs for agricultural implements and services; it also creates opportunities for agricultural processing industries, and for the production of consumer durables for farmers. In some instances, labour-saving agricultural technological improvements release labour to urban areas. Perhaps most importantly, agricultural growth means growing exports, and hence increasing capacity to import the goods needed to run urban industries and create urban jobs.

Cities in the developing world also grow rapidly because the entire population is growing rapidly, and because conditions continually foster far more rapid growth of economic and educational opportunities in most urban than in most rural areas. As capital cities tend to be the most favoured, they grow more rapidly still.

The economist's perspective on 'excessive urbanization' is straightforward. Urbanization and migration are human behavioural responses to incentives. Incentives may be distorted, thereby creating excessive urbanization—in which case, the appropriate policy response is to remove the underlying distortions, rather than attempting to restrict their consequences. Furthermore, a great deal of rural–urban growth is an inevitable consequence of economic development, and no single pattern linking economic growth to urban population growth can account for the unique development features of specific countries. Therefore, urban population growth rates in excess of industrial output growth, for example, do not 'prove' anything.

Primacy and Optimal City Sizes

Another strongly held view is that current city-size distributions are too primate—that is, that their urban populations are too concentrated in a few large cities. Policy-makers in developing countries are particularly interested in restricting the growth rates of their largest cities. Furthermore, many developing countries, especially those which are small and least developed, do have city-size distributions that are extraordinarily primate by world standards. In Africa, Sierra Leone, Malawi, Côte d'Ivoire, Zimbabwe, and Cameroon, cities are all exceptionally primate. In Latin America, Argentina, Peru, Uruguay, and Venezuela have very primate city systems. But primacy is not universal. In Africa, Botswana, Niger, Nigeria, and Lesotho all have unusually unconcentrated urban populations. In Latin America, Colombia and Brazil have quite decentralized city systems.

The problem is that there is no clear 'optimum' city-size distribution. In their analysis of India's decentralization schemes, Mills and Becker (1986) find little reason to assume that such policies enhance any reasonable measure of social welfare. City-size distributions evolve in response to market and government-generated incentives. While there are many externalities and income-distribution considerations that warrant public-sector intervention to restore socially efficient production patterns, none of these considerations involve direct restrictions on urban growth as a first-best policy. It is, for example, better to deal with the problems of urban congestion in mega-cities by better transport management rather than by limiting the size of large cities. Migration restrictions and land-use controls are second-best policies that may impose significant costs in other areas in order to achieve the primary objective.

Another common concern is that the largest cities exceed their optimal sizes. This concern is misguided for the simple reason that optimal size calculations are extraordinarily imprecise. If, for example, the optimal size is found to be (as is often the case) between 100,000 and 1,000,000 inhabitants, policy implications for a large number of cities falling within this range are unclear. Should further growth be encouraged or discouraged?

Urban Bias

Michael Lipton (1976) was the first to advance the urban bias thesis, which focuses on the advantages that producers and consumers in urban areas (especially capital cities) receive through government policy, and the simultaneous bias against the agricultural sector. According to the urban bias thesis, national policies have been skewed against agricultural development in three ways. Macro-economic policies—especially trade policy—have protected domestic industrial production, while affording little or no protection to agriculture. Simultaneously, public-sector investment has focused on urban infrastructure, despite ample evidence that the social rates of return are higher in rural areas; moreover, many of these urban investment projects have no provisions for recouping costs. Finally, central governments often have compounded these effects by augmenting public employment in cities to a degree unjustified by any conceivable efficiency criterion.

Given this definition, the economist's response is to advocate implementation of a balanced development policy. It is inappropriate for gov-

ernments to designate 'leading sectors', mainly because governments are so likely to be wrong. It is inevitably the case that rapid growth environments will involve considerable growth disparities across sectors, but government should maintain a level playing-field rather than trying to spur growth further by tilting towards those sectors with alleged agglomeration economies, economies of scale, or dynamic comparative advantages. In our view, the extra growth achieved by such favouritism is likely to be more than offset by inaccurate forecasting and inappropriate biases driven by political considerations. This almost certainly has been the record to date. The only biases that may be warranted are, first, those that compensate for clear social externalities or market failures. Also, ostensible biases stemming from a clear recognition of production externalities across sectors will be warranted, but only if the linkages can be clearly established.

Neglect of Agglomeration Economies

Ironically, despite the presence of pervasive urban biases, policy-makers often neglect the agglomeration economies offered by urban areas as they rush to restrain urban growth. Agglomeration economies are cost advantages that accrue to firms that locate in close geographic proximity to one another. Urban economists distinguish between two types of agglomeration economies: localization economies, which benefit only firms within a specific industry, and urbanization economies, which benefit all firms in an urban area. Factors such as better or cheaper access to intermediate inputs or specialized labour pools, and communication among specialists fall under the heading of localization economies. Urbanization economies may involve such factors as better developed insurance and banking services, high-quality public infrastructure (roads, communication links, ports), and well-developed public services (sewerage, electric service).

As already mentioned, governments are unlikely to be very successful in predicting which industries will exhibit large agglomeration economies. On the other hand, the existence of such economies means that governments should not pursue policies which deliberately disperse sectors, and especially manufacturing firms. In this case, agglomeration economies are bound to be lost, and the benefits of dispersion may well be illusory.

Distrust of the Informal Sector

A final, pervasive article of faith in African, South Asian, and many Latin American governments' urban-sector policies has been a distaste for economic activities outside the formal sector. Perhaps because of the difficulty in levying taxes, the small-scale manufacturing and service activities that comprise the informal or 'micro-enterprise' sector are viewed by government officials as being incapable of self-sustaining economic growth. Indeed, to many policy-makers halting urbanization means controlling informal-sector economic activities and the growth of squatter settlements: it rarely means curbing production of luxury housing or public-sector growth, and only infrequently implies restricting modern-sector activities.

In reality, the informal sector enjoys a largely symbiotic relationship with the modern manufacturing and service sectors. The informal sector mainly produces consumer durables, non-durables, and services, and some intermediate goods and repair services for the formal sectors. Consequently, demand for informal products largely depends on formal-sector incomes. Many consumer goods produced in the informal sector, however, compete directly with modern-sector production and with imports. Since local content is much higher in informal-sector production, though, informal production generally utilizes the domestic resource base more efficiently.

In sum, most informal sectors are relatively efficient producers of many goods, including a wide range of import substitutes. They are also a critical source of employment, particularly for those without the skills or connections to obtain a formal-sector job. As its investments are self-financed, informal-sector growth is unlikely to reduce investible resources elsewhere. The sector is also a spawning ground for indigenous entrepreneurial talent.

In view of these considerations, it seems highly inappropriate to suppress or discourage the informal sector. Urban policies such as land-use controls, combined with various licensing requirements and weak small-scale credit markets, do have such a constraining effect. Unreasonably high standards for materials used by government or in government-funded projects also reduce demand for informal-sector goods.

Regional and Population Policies

A variety of policies have been expressly aimed at restricting or redirecting rural–urban migration and reducing population growth. Our objective is to describe briefly (1) the effectiveness of these policies in terms of meeting their objectives and (2) their economic consequences.

Migration and Resettlement Policies

Formal migration restriction has been used only in a few countries. In many formerly socialist countries, internal passport systems controlled population flows, and these worked with varying degrees of success. They are breaking down virtually everywhere today, though, and the consequence is an unleashing of previously restrained population movements, especially towards the largest cities. Since internal labour-market liberalization has not been accompanied by infrastructure development in the destination areas, explosive social problems and rapidly deteriorating access to services have been the consequence. In China, most movement has been from countryside to cities, and inadequate construction has meant widespread homelessness for China's migrants.

Elsewhere, migrant restriction has been infrequent. South Africa under white rule restricted movements, as did Tanzania on occasion. But most countries have sought to curb rural–urban migration by making cities relatively inhospitable—bulldozing squatter settlements, or making it difficult for new migrants to secure property rights to land or access to public services. A small number of countries have actively sought to induce people to migrate from crowded cities or rural areas to more remote, underpopulated zones. Indonesia's *transmigrasi* programme, which moved more than 2 million people between 1980 and 1986 alone (World Bank, 1988), is the most notable example. Its effect should sober advocates of direct population policies: despite considerable expense, the impact on Indonesia's population distribution has been minimal. It is difficult to conclude that direct population policies can have a major impact on the spatial distribution of population, unless ruthlessly enforced by strong authoritarian regimes.

Population Control

A far more effective way to curb urban population growth, especially in countries in which rural–urban migration has waned, is to reduce the natural rate of population growth. Most countries at least in principle support reduced fertility as a social objective. Again, too, determined authoritarian regimes (China is the outstanding example) have been more successful in achieving fertility reductions than countries with weak states.

Depending on the developing country in question, urban fertility accounts for 10 per cent to more than 100 per cent of urban population growth; rural–urban migration accounts for the rest. In countries which are already highly urbanized, urban fertility is more important, since even if rural out-migration rates are high, there are relatively few rural dwellers to migrate. Conversely, a fairly low rural out-migration rate may still translate into a very high urban in-migration rate in countries which have very small urban populations. Thus, in the least developed countries, urbanization will be determined mainly by migration; in more urbanized and more developed countries, natural growth is the primary force.

Because of what is referred to as *demographic momentum*, even dramatic success in reducing fertility will not end urban population growth in the near future. In most developing countries today, declining infant and child mortality rates mean that the number of people in the current 0–20 year cohort who survive to childbearing years will greatly exceed the present childbearing population. For populations to decline, birth-rate declines must more than offset the growth of the childbearing population.

In the case of urban population growth, this momentum is accentuated by a *migration momentum* effect. Rapid growth of younger age cohorts translates into an increase in the overall propensity to migrate to the cities, since young adults are far more likely to migrate than any other age group. Consequently, a bulge in the 0–20 age group today ensures high and possibly

rising rural–urban migration rates for many years, even if particular age-cohort rates are stable (see Becker and Grewe, 1994 and 1996). Indeed, in many African countries, net migration of the elderly is from cities to the countryside.

The objectives behind population and urban growth control policies are often quite similar. Governments fear being unable to provide social services for a surging population; they worry as well about rising young unemployment and resulting political discontent and social disorder. Population and urban growth control policies, however, differ greatly. The most important policies to reduce fertility are to expand contraceptive availability, extend educational opportunities for women, and reduce labour-market discrimination against women. Overall, population control policies have been much more effective, when coherently implemented, than urban growth control strategies.

The impact of population control policies on urban growth and rural–urban migration is unclear. Rapid extension of secondary education, especially to girls, will cause fertility-rates to fall (and have other beneficial effects, as well). But an increase in the number of young adults with secondary education almost certainly will lead to an increase in rural–urban migration: few high-school graduates envision working as peasant farmers as a lifetime occupation. Even direct fertility control through increased contraceptive availability may have unpredictable consequences: women in Asia and Africa may come to imitate their Latin American sisters, who have higher migration rates than men.

Secondary Cities and Market Towns

Several less developed countries have actively promoted secondary cities in order to simultaneously encourage agricultural development and slow the growth of the largest cities. Secondary cities are usually defined as urban centres with populations in excess of 100,000, excluding the nation's largest city. Why this concern with developing secondary cities? The largest city in many less developed countries (LDCs) is often several times larger than the next largest city. With population and economic activity concentrated in one primate city, development possibilities for other regions of the country may be very limited.

Of particular concern has been the negative impact on agriculture of the absence of secondary cities and smaller towns with reasonable infrastructure. Farmers need easy access to storage and processing facilities; without such access, crops may spoil before reaching market. If the nearest market town is very distant, farmers may receive low prices for their crops, since middlemen must recoup higher transport costs. Secondary centres also play a key role in diffusing technological advances to small farmers. Absent secondary cities, farmers may not have access to hybrid seeds, fertilizer, or technical advice.

Nor are secondary cities limited to servicing the needs of surrounding farms. Rondinelli (1983) argues that secondary cities should:

- provide sites for the regional provision of social services;
- provide consumer goods and services to regional consumers;
- act as regional centres for distribution, transfer, storage, brokerage, credit, and financial services;
- provide sources of off-farm employment and supplemental income for rural dwellers; and
- serve as regional transportation and communication centres.

From the point of view of urbanization, secondary cities have the potential to provide attractive alternative destinations for city-bound migrants from rural areas.

The main strategy for promoting secondary cities has been through the provision of infrastructure. However, infrastructure is expensive, and already strained national budgets do not permit the massive expenditures necessary in most countries. At best, governments can shift away from investments in state enterprises to focus on small city and rural infrastructure, and several countries are moving in this direction. One particularly attractive alternative is the devolution of decision-making authority over taxes and spending to local governments.

Government Decentralization

The primacy of many city systems both in less developed and newly industrializing nations reflects in extraordinarily centralized governments. The degree of centralization is often mind-boggling. In Korea and elsewhere, local political leaders are appointed by the national government. In Peru, until very recently, almost all public investment decisions were made by officials of the central government in Lima. Tanzania actually abolished local government in the 1970s. Many countries with the most centralized administrations inherited them from colonial powers who were interested in maintaining centralized control. The francophone countries of West Africa are good examples of this, as are Venezuela, Peru, and to a lesser extent Mexico.

Decentralization of government responsibilities gives decision-making power to those who best know local needs—local residents. To be truly effective, both taxing and spending authority should be given to local governments. If only spending authority is devolved, the local government will continue to depend upon the national government for funds, and consequently will not enjoy true autonomy. Devolution of authority is not without problems, though. Local governments may not have the administrative capacity to design, monitor, and collect taxes; they may not—at least initially—possess the technical, managerial, or administrative capacity to design and implement projects. These problems are especially great in the least developed nations with limited skilled manpower.

A shallow form of government decentralization is the relocation of the capital city. The putative purpose is to promote a more spatially balanced pattern of development. In Latin America, Brazil relocated its capital from Rio de Janeiro to Brasilia; in Africa, Tanzania is moving its capital from Dar-es-Salaam to Dodoma, and Nigeria is moving its capital from Lagos to Abuja; in the former Soviet Union, Kazakhstan is moving its capital from Almaty to Akmola. However, the costs involved are enormous. Moreover, in each of these cases the former capital has remained a more important industrial and commercial centre, so the impetus to regional growth around the new capital has been modest.

Tax and Expenditure Policy

Central governments are often loathe to pursue true administrative decentralization. National politicians and bureaucrats typically want to maintain political power and patronage in their hands. In addition, they may be deeply suspicious of grass-roots democratic processes, and may fear the potential for separatism and national disintegration inherent in the devolution of power. This last fear is quite rational in countries with strong ethnic divisions or with extreme income inequalities across regions.

Given an unwillingness to pursue true decentralization, many governments have opted for a model of 'centralized decentralization'. Two main policies have been used:

- tax breaks to influence the location decisions of private firms; and
- priority provision of infrastructure to previously unfavoured regions.

From 1958 to 1990, for example, Peruvian firms locating outside of Lima were given reductions in their profit tax. Firms locating in the jungle region received the biggest breaks, those locating in the impoverished highland region the next largest, and those locating in coastal towns other than Lima were awarded the lowest tax break. Of course, this tax policy had efficiency costs. A motorcycle assembly plant, for example, located in the middle of the Peruvian jungle in order to benefit from the tax break. Since there were no roads connecting the plant to the rest of Peru, however, the company was forced to ship components and assembled motorcycles by air. The final result was production at negative value added.

Priority provision of infrastructure to poorer regions is another favoured strategy. Brazil spent many millions of dollars constructing the trans-Amazon highway; India has poured funds into projects to divert development away from its largest cities, especially Mumbai (Bombay) and Calcutta; Côte d'Ivoire spent large sums building the port of San Pedro as an alternative export node to Abidjan.

The common thread in these policies is the desire to improve welfare in lagging regions without sacrificing centralized control. Indeed, few countries have not made some efforts to promote the development of 'backward' regions. By and large, the results have been unspectacular.

Sectoral and Macro Policies which Shape Migration Flows

We turn now from policies designed to restrict or channel rural–urban migration to a considera-

tion of the migration consequences of policies aimed at other social and economic variables. The consequences of public policy often extend far beyond the realm for which they were intended, and these unanticipated consequences often dwarf the anticipated effects.

Rural Development Policies

A key policy choice involves deciding whether to keep farm prices (for food and export crops) low, at world levels, or above world levels. *Repressed agricultural prices* reduce rural incomes, and rural out-migration is the immediate result. Ironically, however, agricultural price repression may not lead to sustained urban growth. Low agricultural prices diminish exports, thereby limiting foreign-exchange earnings. Since export earnings are critical for countries dependent on imports for urban capital goods (vehicles, machines) and intermediate inputs (such as unassembled kits or fuel) used by local assembly plants, low agricultural prices may damage urban manufacturing as well as agriculture, and may stall rather than promote city growth in the long run. Whether or not extremely low farm prices deter urban population growth, there is little doubt that they stifle economic growth.

The critical determinant of the effect of agricultural pricing seems to be whether *urban export sectors* are strong. If they are, then agricultural development is inessential for urban development, as export earnings come from urban products, and agricultural inputs can be imported. Low farm prices then simply accelerate rural out-migration. But in very underdeveloped countries, agricultural development is frequently necessary for cities to grow.

Another area in which countries differ markedly is the degree of emphasis on *agricultural exports*, as opposed to *food production*. In some countries, a growing export orientation is accompanied by movement to plantation agriculture. Plantations, in turn, tend to be more capital-intensive than smallholdings and release surplus labour to the cities. But generalizations are difficult. Many plantation crops are also labour-intensive (and conserve on land use relative to food crops): the impact of agricultural commercialization will therefore vary across crops and regions. On the other hand, a growing agricultural export orientation typically implies developing rural infrastructure and marketing services, and hence relatively rapid integration of rural areas into the 'modern' economy.

Few public policies benefit the poor more than those which successfully achieve *rapid income growth in rural areas*. In most cases, rural development means increasing agricultural productivity. Its consequences depend on whether the productivity gains are in export crop production, in the production of foods which compete with imports, or in the production of local foodstuffs. If the productivity gains are in goods which are either exported or imported, and the country in question has a liberal foreign trade policy, then the domestic and world prices will be close, and so an increase in supply will not cause a fall in price. Productivity gains then mean income gains, which in turn induce peasants to remain on the farm. But if the gains come in rural sectors producing only for domestic markets, then supply increases generated by productivity gains imply falling rural prices. Since farmers' revenue equals the value (price times quantity) of output, then supply increases may cause stagnant or falling rural revenue. Furthermore, falling food and industrial crop prices lower the costs of urban industrial production, leading to output and employment expansion, which induces rural–urban migration.

The evidence concerning the effect of rural prosperity on migration is mixed. Clearly, there is net out-migration from destitute areas. But rapid rural development means growing rural incomes and increased demand for services and consumer durables, which tend to be produced in towns and small cities. These towns also serve as processing and marketing centres for prosperous agricultural areas. Moreover, migration to urban areas is more easily financed by prosperous than poor peasants. To the extent that productivity gains raise incomes, which in turn raise demands for rural education, rural out-migration may increase further.

In summary, the effect of rural development on rural–urban migration hinges crucially on the environment in which it takes place. Distress migration of refugees will be driven by very different forces than those which determine the planned migration of young rural dwellers. Furthermore, from a policy perspective, rural development is a far more important objective than urban population control. Finally, the type of urban growth generated by rural prosperity—development of small centres which serve agricultural regions—is generally regarded as desirable by policy-makers.

Finally, *land reform* policies also have important implications for the release of labour to the cities. Redistribution of land from large landowners (or, in formerly socialist countries, the state) to peasants gives the latter group a rural asset which may induce them to stay in the countryside. This is especially the case in countries where land is not transferable (as was the case in Mexico under the *ejido* system), or where land title is sufficiently vague that property cannot easily be reclaimed from renters. In such situations, the beneficiaries of land reform will have little choice but to farm the land themselves, or resort to illegal transactions. Land reform also seems likely to increase the labour-intensity of agriculture, as small farmers typically lack access to capital, have difficulty realizing scale economies, and in some cases produce more labour-intensive crops. To the extent that land reform thus bolsters the demand for rural labour, urban-bound migration will be reduced. But *incomplete rural property rights* almost certainly play a much more critical role in curtailing migration. In fact, as land registration occurs, people will no longer be constrained from selling their plots, and some will sell and then migrate to urban areas.

Macro Policy and Economic Structure

Of all the policies which affect rural–urban migration streams, macro-economic policies are probably the most important. Most developing countries have sharply altered their macro-economic policies over the past decade, with important impacts on rural–urban migration.

Developing countries typically pursued a policy of *import-substitution industrialization* (ISI) after the Second World War. The goal was to reduce dependence on imports from developed countries while simultaneously promoting domestic industrialization. Typical ISI policies included heavy tariff protection of domestic industry, provision of cheap credit to industrial firms, controls on key food prices, and government ownership of heavy industries. While several East Asian countries modified their development strategy—shifting towards a more outward-oriented model—in the 1960s, almost all Latin American and African countries continued the basic ISI policies until the mid-1970s to mid-1980s, when most began either gradual or rapid shifts towards a more outward-oriented strategy.

This policy change was usually involuntary. Many developing countries borrowed heavily in response to the first oil shock in 1973, hoping to maintain economic growth. When the second oil shock of 1979 led to world recession and significantly higher interest rates, adjustment could no longer be postponed. Many countries sought short-term balance of payments loans from the International Monetary Fund; the Fund obliged, but with significant strings attached. These strings were structural adjustment programmes (SAPs), a set of policy prescriptions designed to improve the trade balance and reduce inflationary pressures. Thus, many developing countries' implementation of SAPs reflects a lack of alternatives, plus an understanding that international support will occur only if reasonable adherence to SAPs occurs. But much of the acceptance is also genuine, motivated by the desire to imitate the growth experience of the outward-oriented 'Asian tigers'.

What are the most important policy elements of SAPs? Foremost is a drastic devaluation to improve the balance of trade and permit servicing of foreign debts. Import tariffs and non-tariff barriers are reduced in order to promote more efficient domestic production. Government budget deficits are slashed; many countries, in fact, have run budget surpluses within a few years of implementing SAPs. The growth rate of the money supply is also curtailed. These last two changes are designed to reduce the level of aggregate demand and, consequently, the inflation rate.

These broad policies typically are accompanied by complementary measures. First, the World Bank and other donors have urged that market signals and private producers be given greater roles. Secondly, governments have been pressured to eliminate or shrink direct involvement in production. Thirdly, the slimmed-down public sector has been pushed to give priority to operation and maintenance of essential infrastructure and public services, with increased private-sector participation. These three measures reflect a wish for a leaner, more efficient state. Governments should do what they do well (pro-

vide economic infrastructure and establish clear rules of the economic game) and refrain from doing what they do poorly (attempt to increase welfare by distorting relative prices and engaging directly in production). Finally, international donors and lenders have stressed financial liberalization to generate more domestic savings, and limitations on wage increases in order to combat inflation.

What is the *impact of SAPs* on rural–urban migration? Structural adjustment policies may slow the pace of rural–urban migration significantly in the medium term. *Devaluation* will increase migration to areas and sectors in which export production takes place. Traditional exports from most LDCs are agricultural and mineral goods, although this pattern is rapidly changing, especially in Asia. While some value-added in these activities is produced in cities and towns, a significant proportion is produced in rural areas. The magnitude of the decline in rural–urban migration depends on the degree to which agricultural exports increase in response to a devaluation. In urban areas, the reduction or elimination of import tariffs will reduce output and employment in import-substituting industries, and raise them in import-using industries. If wages are downwardly flexible, potential migrants will be discouraged by both lower wages and a lower probability of finding a job. Even if wages are inflexible, lower employment probabilities will deter potential migrants.

Reductions in government budget deficits also will slow rural–urban migration. The balanced budgets mandated by structural adjustment programmes have reduced education and other social spending. If these reductions keep potential migrants from obtaining education—especially post-primary schooling—many individuals who previously might have migrated may stay at home because they will not have the requisite education to allow them to compete successfully for jobs in urban labour-markets. In addition, to the extent that educational opportunities in urban areas serve as a beacon to migrants, decreases in urban educational opportunities may also reduce migration flows.

Reducing money supply growth similarly slows rural–urban migration. The primary effect of contractionary monetary policy is to increase interest rates. Financial liberalization—primarily eliminating usury laws—will have a similar effect. Formal-sector firms located in urban areas absorb most bank credit, which becomes more expensive as the interest rate rises. In contrast, *credit squeezes and disinflationary monetary policy* have weak contractionary effects on rural producers. For this and other reasons, formal-sector firms are harder hit by structural adjustment than either urban informal enterprises or rural residents. As formal-sector employment growth slows or contracts, rural–urban migration will also slow.

Wage and employment declines due to SAPs will be disproportionately concentrated in urban areas. Export-crop producers gain from SAPs for reasons already outlined, and subsistence farmers will be little affected one way or the other. But urban residents may be hurt significantly. As import-substituting firms face increased foreign competition, they may lay off workers or reduce wages. Since informal-sector firms provide some inputs into formal-sector production, incomes and employment there will suffer as well. More importantly, reductions in incomes earned by formal-sector workers mean that they will have less income to spend on informal-sector goods, further lowering informal incomes and employment.

Finally, reductions in government spending mandated by structural adjustment programmes involve the *tightening of state enterprises' (parastatals') budgets*. Parastatals typically have been managed with objectives other than profit maximization; employment maximization has been a particularly important goal. Parastatals' wage policies also have been designed to curry political favour rather than minimize labour costs. Such wage and employment policies were possible because parastatals traditionally faced soft budget constraints, and workers' demands often were accommodated by additional transfers from the central government. Much of this rent-sharing behaviour has been diminished with the introduction of SAPs. To the extent that above-market salaries lured potential migrants, the attractiveness of parastatal employment has declined as real wages have dropped and employment possibilities worsened during the past fifteen years.

Other forms of *rent-sharing and rent-seeking behaviour* (that is, appropriation of enterprise income as payments to workers and managers in excess of their productivity, based on their political strength and on the firm's ability to pay)

have become less prevalent as well. Urban-biased policies in many countries in the 1960s and 1970s assured the existence of substantial rents in urban areas. One way to think of urban bias is as a transfer of economic rents from rural to urban areas. Rural areas in many LDCs produce the bulk of foreign-exchange earnings; while some value may be added in urban areas, most of the value added from cash crops and mineral exports comes from rural areas. If a nation pursues urban-biased spending policies, however, the bulk of foreign-exchange receipts is spent in urban areas. The availability of rents in the form of public housing, subsidized public education, and electricity, water, and other public services can be a powerful magnet for migrants. SAPs have largely eliminated urban-biased policies in many countries. Excessive protectionism has been dismantled, public-investment decisions are being made with rates of return and cost recovery in mind, and public-sector employment growth has been checked.

Thus, in the medium term SAPs will significantly reduce rural–urban migration. Will they have the same effect in the long run? SAPs seek to restructure LDC economies to expand exports (and also provide immediate debt relief and new loans). Improved access to foreign exchange is urbanizing, as industrial sectors in LDCs often are extremely dependent on intermediate and capital good imports (and urban élites have import-intensive consumption patterns). Consequently, SAPs ultimately may remove a major impediment to sustained urban growth.

What other macro-economic factors affect rural–urban migration? The *structure of aggregate demand* is an important factor. Urban and rural areas produce distinct goods; urban areas produce consumer and capital goods, while rural areas produce food and export crops. Moreover, as incomes rise, the demand for urban-produced goods will rise more rapidly than the demand for rural-produced goods. The magnitude of the difference in growth rates hinges on several factors, starting with the specific goods produced in each region, on the overall GDP growth rate, and on its distribution across income classes. In addition, it also hinges on a country's specific income elasticities of demand for various goods. In particular, the income elasticities of demand for food products will be larger the poorer is the country.

Another prominent economic 'strategy' in developing countries is *populist macro-economic policies*, which, like ISI policies, initially promote rapid rural–urban migration. Populist policies emphasize income redistribution and rapid growth, while placing little emphasis on such basic macro-economic constraints as government budget or trade balances (Dornbusch and Edwards, 1990; and Sachs, 1990). Latin American leaders such as Juan Peron in Argentina during the 1940s and 1950s, Salvador Allende in Chile (1970–3), and Alan Garcia in Peru (1985–90) expended foreign-exchange reserves to expand output. Real wages initially rose in all three cases, making the governments popular. But after a couple years, flaws emerged: overvalued exchange rates produced huge trade deficits, foreign-exchange reserves disappeared, and government budget deficits grew substantially. Perhaps most cruelly, real wages are lower at the end of a populist episode than at the outset, and hyperinflation is also a common outcome. Despite proclaiming concern for the most disadvantaged members of society, populist leaders often end up hurting exactly those groups they promised to help.

In the short run under populist policies, urban real wages rise relative to agricultural wages. Government deficits are used to expand public services to lower and middle classes, and these services are available almost exclusively in urban areas. Both these factors serve to make urban destinations more attractive to potential migrants. The collapse of populist programmes erases these urban advantages, but migrants who were attracted to urban areas are unlikely to return to their origins.

Finally, *anti-poverty policies* affect rural–urban migration as well. To be effective, anti-poverty programmes must be coherently designed and implemented in appropriate locations. Until recently, one could confidently assert that the majority of the desperately poor would be found in rural areas; anti-poverty programmes consequently focused on raising the productivity of rural labour and providing opportunities for both farm and off-farm employment. These anti-poverty programmes will slow rural–urban migration by improving the standard of living of rural residents. While the above characterization of poverty remains accurate for most of Africa and Asia, it has become increasingly inaccurate

in Latin America. As recently as the mid-1980s, rural areas contained significantly more poor people than urban areas. By 1989, however, the majority of Latin America's poor were found in urban areas. This is not to say that the incidence of poverty in urban areas is more severe than in rural areas. In fact, quite the opposite is true: by Morley's (1994) estimate, over 53 per cent of rural vs. 22 per cent of urban residents are poor. But these numbers indicate that anti-poverty programmes in most Latin American countries should contain a significant urban component. To the extent that anti-poverty programmes target urban areas, however, they will tend to increase rural–urban migration.

Urban Labour-Markets, Infrastructure, and Sectoral Policies

The paradox pointed out by Harris and Todaro (1970) is that governments seeking to reduce unemployment via urban job-creation schemes actually may cause unemployment rates to rise. People migrate to cities in anticipation of increasing their earnings. If there is little labour turnover and few new jobs, then there will be few migrants from the countryside. But, with job-creation schemes paying wages greater than rural earnings (even after allowing for differences in living costs), people may find the expected value of urban earnings sufficient to induce migration. Unemployment may actually increase, as not all migrants will be successful in finding jobs.

How important is this phenomenon? In the early development phases in Africa and Asia, following the end of the colonial era, job-creation efforts abounded. But in much of Africa and Latin America, government budget constraints have caused public-works programmes to dwindle or disappear in the past two decades. Indeed, deteriorating prosperity in protected urban sectors since the 1970s has characterized nearly all Africa, along with much of Latin America and South Asia. Real government revenues have fallen; in consequence, so have civil-service earnings. Declining tariff protection, opening markets, and falling government subsidies have been inevitable components of the SAP response to the economic crises, but these steps have not brought prosperity to urban manufacturing enterprises—especially for state-owned enterprises. Rather, real formal-sector wages, and in much of the developing world, real wage gaps between formal- and informal-sector activities have virtually disappeared, when one corrects for differences in skill and experience.

The reduction in gaps between high- and low-wage sectors within urban areas, and between city and countryside, has important implications for rural–urban migration. Expectations among rural African and Latin youth concerning opportunities in the city are far more pessimistic today than they were two decades ago, causing migration rates to stagnate or decline. As Harris–Todaro disequilibrium and urban unemployment are replaced by equilibrium unemployment and underemployment, the expected return to migration has fallen. Falling wage gaps are reinforced by slow job creation, in part because poor agricultural export earnings leave little foreign exchange for imports used in urban activities. Thus, data for four sub-Saharan African nations demonstrate a marked decline in in-migration rates for 20–24 year-olds, who have the highest migration propensity: similar but less pronounced trends exist for other groups as well. Declining in-migration rates are evident everywhere except Gambia, which, with only about 150,000 people in its sole 'metropolitan area', is anomalous, and suffers from major statistical problems in earlier censuses as well (Table 6.1).

TABLE 6.1 Net Urban In-Migration Rates in Structural Adjustment Africa (for 20–24 year-olds)

country	year	male	female
Gambia	1963–73	6.6	2.2
	1973–83	7.7	3.2
Ghana	1960–70	5.8	2.5
	1970–84	2.1	0.5
Zambia	1963–74	8.4	8.1
	1974–80	4.4	3.5
Zimbabwe	1962–9	8.0	6.9
	1969–82	3.3	1.1

Source: Becker and Grewe (1996). Data give net rural–urban 20–24 year-old in-migration, as a percentage of the urban 20–24 year-old population.

Migration and urbanization trends in high-growth Asia are quite distinct. Asia's urban growth typically has been slower than in Africa, which started from a far less urbanized base, has

suffered rural collapse in many cases, and has had high total population growth rates. Nor do Africa's labour-market constraints and slow rates of export and job growth which have caused migration slowdowns characterize most of south-east Asia.

The most extraordinary labour-market effect, however, is taking place in the formerly socialist world. Removal of Draconian constraints on labour- and housing-markets is generating huge equilibrating flows, and hence vast increases in rural–urban migration. In countries such as China and Vietnam, unprecedentedly rapid industrial growth is further fuelling migration.

Among sectoral forces, *education policy* is a critical determinant of urban growth. As Kenyan figures in Becker and Morrison (1993) indicate, rural young adults with secondary education have urban migration propensities two or three times greater than those with primary or no education. Sabot's (1979) job 'queuing' framework provides an explanation: urban jobs go first to those with substantial education, even if such education is not immediately useful. At the same time, it is also difficult to imagine people with high-school education remaining in villages to work in cornfields or rice paddies.

There can be little doubt as well that the *changing age structures, caused in large part by declines in infant and child mortality rates, which are in turn due to the extension of basic medical care*, substantially increase migration rates. Since teenagers and young adults are by far the most sensitive to migration incentives, and have migration rates much higher already than for other groups (Table 6.2), the average migration rate also will increase considerably.

The *size and growth of municipal budgets* affect urban population growth, while the collapse of urban services in Africa almost certainly has forestalled some urban growth. Effective urban growth requires city governments capable of building economic infrastructure to support new private investment. Urban social infrastructure—in roads, cleared sites and/or housing, water, sewerage, and electricity—is also important to accommodate rapid population growth. In the absence of these services, the perceived cost (including non-monetary costs) of moving to cities is much higher. Without adequate social infrastructure, private substitutes for the public sector provide some of the goods, but at very

TABLE 6.2   Net Urban In-Migration Rates, by Region and Age (percentage)

Age group	Sub-Saharan Africa		Latin America	
	male	female	male	female
10–14	2.1	3.4	n.a.	2.2
15–19	4.6	4.6	2.4	3.0
20–24	6.6	4.7	2.1	2.3
25–29	5.1	3.5	1.4	1.1
30–34	2.9	2.2	1.3	1.0
35–39	2.2	2.3	1.1	1.1
40–44	1.9	1.9	1.0	1.1
45–59	1.5	1.3	1.1	1.1
50–54	0.9	1.2	1.1	1.1
55–59	1.1	1.9	1.0	1.1
60–64	0.7	2.1	0.7	1.0
64–69	–0.1	1.8	0.9	1.0

Note: African in-migration rates are unweighted averages of 18 observations from 13 countries between 1960–70 and 1977–88, calculated by Becker and Grewe (1996) using residual inference techniques from census data. Latin American data are similarly calculated unweighted averages from 16 observations covering 13 countries between 1961–71 and 1973–81, as discussed by Becker and Grewe (1994).

high cost. Housing shortages emerge as well, and rents soar—all choking off urban growth.

Urban land markets are critical in this respect. If land markets allocate property to the highest bidder with secure tenure, private housing investment is encouraged. But if recent migrants and private investors lack secure access to land, then fear of arbitrary confiscation (or bulldozing of squatter settlements) is a major deterrent to long-term settlement and investment. Across Africa, one finds highly decentralized, low-density cities, largely because it is risky to build permanent, high-density structures. Instead, the proportion of urban households engaging in some agricultural activities exceeds one-third in many African cities (Becker, Hamer, and Morrison, 1994). But low-density urbanization implies large commuting costs, as well as high costs of providing public services—and in consequence, cities are less efficient, and grow less rapidly.

Housing markets in many LDCs bear little similarity to those in developed countries. Many housing units are constructed either by informal-sector firms or by the residents of the units themselves. In many urban areas, squatter settlements account for much of the housing stock, with units built on land for which the residents originally did not (and perhaps still do not) have legal title. Finally, mortgages are not generally available; households must either save the entire amount needed for the purchase of a house, or they must build their own house incrementally (first the walls, then a roof, then a door, then windows, and so on).

Housing policy in many LDCs has changed markedly in recent years. Traditionally, governments sought to build single-family housing units, in principle for the poor. Such efforts were limited by budget constraints, however, and the number of poor without adequate housing rose in almost every country. Today, many countries have moved to 'sites and services' schemes with far lower unit costs. In these, government usually provides basic infrastructure (sewerage, electricity, and water) and delimits the lots available for settlement. Sometimes residents are offered mortgages. Residents are responsible for the construction of their own dwellings, and often are asked to help install basic or pay for infrastructure. Since residents must contribute with 'sweat equity', the per unit costs of sites and services programmes are far lower than in the more traditional, high-quality approach.

Sites and services schemes unquestionably have been vastly more socially productive than high-quality housing projects. Many also have been quite large: the First Lusaka (Zambia) Upgrading Sites and Services Project, involved the construction of 11,500 new units, and infrastructural and physical improvements to about 20,000 housing units. Thus some 150,000 to 200,000 Lusakans benefited from the project—roughly one-third to one-half of the city's 1974 population.

What has been the impact of these large projects on rural–urban migration and urban growth? Migration theory suggests that, by raising the standard of living in urban areas, they should attract people. Furthermore, because these investments are overwhelmingly located in the largest cities, they primarily should spur growth there. More importantly, were these loans more urbanizing than high-quality housing construction schemes? Apparently not: results from a model of the Indian economy (Becker, Williamson, and Mills, 1992) indicate that the effects on urbanization of changing housing strategy are practically nil.

Transportation infrastructure also has important impacts. Many LDC governments have responded to prolonged economic crisis by neglecting replacement investment in transportation and other infrastructure. Consequently, the international community has responded with many 'rehabilitation' loans. To the extent that these are for intra-urban transport, they clearly lower urban production costs, leading to expanded city growth. On the other hand, removal of port and city transport gridlocks alone will not expand modern sector output and employment, so the effects are unlikely to be large.

Improvements in highways and rail links between urban and rural areas may well be more important. Many rural areas in LDCs are only tenuously linked to the major market centres, limiting the ability of farmers to get crops to market. Greater ease of transport leads to increased incomes in these areas (deterring migration), lower migration costs (increasing migration), and growth of agricultural processing and consumer goods' industries in towns (promoting urban growth). These last effects probably dominate.

Conclusion

We have found that explicit migration and resettlement policies have little impact on rural–urban migration. They are ineffective policy instruments, and often run roughshod over individual liberties. Promotion of secondary cities has the potential to increase agricultural productivity by fostering rural–urban productive links, but the magnitude of spending necessary to provide the requisite infrastructure in these towns makes this a questionable strategy in many cases. More critically, the merit of secondary cities programmes lies in their contribution to rural development and to raising the living standards of the poor in developing countries with weak infrastructure: they are not appropriate measures for limiting growth of a country's largest cities; nor are they likely to be particularly successful. Government decentralization has not been pursued vigorously in most countries, largely because central government officials are loathe to cede power to local decision-makers. 'Centralized decentralization' via tax incentives and infrastructure provision has had only a minor impact on migration flows and urbanization.

Sectoral policies have been more influential. Rural development policies, including land reform, are designed to raise agricultural productivity. However, the links between rural development and rural–urban migration are so complex that it is unclear whether increased productivity will lead to more or less migration. On the one hand, increased productivity means higher incomes and less incentive to migrate. On the other hand, increased incomes will enable more people to afford the costs associated with migration, and spend a higher share of incomes on urban goods (and a lower share for food).

Conditions in urban areas have important impacts on migration flows. Minimal public infrastructure is a prerequisite for the expansion of urban areas. Land registration affects both the growth and nature of cities. The availability of reasonably priced housing will stimulate migration. Improved transportation infrastructure in urban areas will lower production costs, increase productivity and raise output. Improved transportation links between rural and urban areas will also increase rural–urban migration flows.

The most important determinants of migration flows, however, are macro-economic policies, educational policies, and demographic structure. The structural adjustment programmes implemented in many LDCs will significantly slow rural–urban migration in the short-to-medium run; paradoxically, these same programmes may permit the renewal of urban growth in the long run if they remove binding foreign-exchange constraints. Demographic structure is a crucial determinant of migration because the young and better-educated are much more likely to move than older, less-well-educated individuals. Policies which expressly (e.g. population control programmes) or, as an unintended consequence (e.g. reductions in educational expenditures), reduce the size of cohorts most likely to migrate can significantly affect the size of rural–urban migration flows.

In socialist countries undergoing transitions to a more market-oriented regime, such as China and Vietnam, the removal of laws restricting mobility has lead to truly massive increases in rural–urban migration. Clearly, these countries are witnessing decades of desired moves condensed into a very few years. Infrastructure and urban public services are woefully inadequate for the new arrivals, and yet the migration flows continue. The decade of the 1990s in these two countries may produce the most rapid urbanization in human history.

Should rapid rural–urban migration produce alarm? By and large, the answer is a resounding 'No!' In past decades, policy-makers worried about urbanization for inappropriate reasons. Urbanization is unlikely to be excessive if individual migrants are making rational decisions guided by market signals; indeed, internal migration has the potential to increase economic efficiency and increase output and per capita incomes (Morrison, 1993). The irony of past hand-wringing over rural–urban migration is that it was done by the very same policy-makers who pursued urban-biased policies. Migrants are rational decision-makers and will respond to the incentive structure that is presented to them; structural adjustment policies reduce rural–urban migration flows for this very reason. At the same time, it would be irrational to applaud SAPs because they result in slower city growth. They should be applauded if they lay the groundwork for sustained economic growth. Put even more bluntly: explicitly spatial policies make little sense; policy-makers should focus on promoting growth with equity, and let individuals and households decide where to live.

Notes

1. This section borrows heavily from and builds on Becker, Hamer, and Morrison (1994).
2. Data are taken from UNDP (1996) and refer to 1993.

References

Becker, Charles, and Grewe, Christopher (1996), 'Cohort-Specific Rural–Urban Migration in Africa', *Journal of African Economies*, 5 (2): 228–70.
—— —— (1994), 'The Effect of Cohort Shifts on Rural–Urban Migration in Latin America', *Ciencia Economica* (April).
—— and Morrison, Andrew (1993), 'Observational Equivalence in the Modeling of African Labour Markets and Rural–Urban Migration', *World Development*, 21: 535–54.
—— Hamer, Andrew, and Morrison, Andrew (1994), *Beyond Urban Bias in Africa: Urbanization in an Era of Structural Adjustment* (Portsmouth, NH, and London: Heinemann and James Currey).
—— Williamson, Jeffrey, and Mills, Edwin (1992), *Indian Urbanization and Economic Development* (Baltimore: Johns Hopkins University Press).
Dornbusch, Rudiger, and Edwards, Sebastian (1990), 'The Macroeconomics of Populism in Latin America', *Journal of Development Economics*, 32: 247–77.
Harris, John, and Todaro, Michael (1970), 'Migration, Unemployment and Development: A Two-Sector Analysis', *American Economic Review*, 60: 126–42.
Lewis, W. Arthur (1954), 'Economic Development with Unlimited Supplies of Labour', *Manchester School of Economic and Social Studies*, 22: 139–91.
Lipton, Michael (1976), *Why Poor People Stay Poor: Urban Bias in World Development* (Cambridge, Mass.: Harvard University Press).
Mills, Edwin, and Becker, Charles (1986), *Studies in Indian Urban Development* (New York: Oxford University Press).
Morley, Samuel (1994), 'Poverty and Inequality in Latin America: Past Evidence, Future Prospects', Policy Essay No. 13 (Washington, DC: Overseas Development Council).
Morrison, Andrew (1993), 'Unproductive Migration Reconsidered: A Stochastic Frontier Production Function Framework for Analyzing Internal Migration', *Oxford Economic Papers*, 45: 501–18.
Ranis, Gustav, and Fei, John C. H. (1961), 'A Theory

of Economic Development', *American Economic Review*, 51: 533–65.

Rondinelli, Dennis (1983), *Secondary Cities in Developing Countries: Policies for Diffusing Urbanization* (Beverly Hills, Calif.: Sage).

Sabot, Richard (1979), *Economic Development and Urban Migration: Tanzania, 1900–1971* (Oxford: Clarendon Press).

Sachs, Jeffrey (1989), 'Social Conflict and Populist Policies in Latin America', Working Paper No. 2897 (Cambridge, Mass.: National Bureau of Economic Research).

Stark, Oded (1991), *The Migration of Labour* (Oxford: Basil Blackwell).

Todaro, Michael (1969), 'A Model of Labour Migration and Urban Unemployment in Less Developed Countries', *American Economic Review*, 59: 138–48.

UNDP (United Nations Development Programme) (1996), *Human Development Report 1996* (New York: Oxford University Press).

World Bank (1988), *Indonesia: The Transmigration Program in Perspective* (Washington, DC: World Bank).

III Urban Employment Structures

Introduction

RAPID urban growth in developing countries and highly visible poverty in their cities raise the question of how productively these urban populations are employed.[1] In Chapter 7, I explore a notion of 'overurbanization' quite different from that used in Chapter 1. I argue that developing countries tend to be overurbanized because rural–urban migration brings workers to cities that are unable to fully employ their existing labour-force to productive ends. The cost is substantial in two respects: potential rural output is lost, and these additional urban dwellers require more resources than they would in the countryside. Migrants, however, rationally pursue strategies to maximize the welfare of themselves, their families, their kin groups, as we have seen in Part II. The resolution of the seeming paradox derives from the fact that rural–urban migration has a redistributive effect. Rural–urban migrants lay claim to a share in urban income opportunities, and they gain some access to urban amenities. Rural families send their sons and daughters to the city so that they will be able to partake, however little, of its riches.

'Overurbanization Reconsidered' distinguishes three categories of urban workers who are unproductive in a narrower or broader sense. The unemployed are only the most obvious, and probably not the largest, category. Large numbers of underemployed are unproductive to the extent that their aggregate output of products or services could be maintained by fewer workers. Finally, the misemployed produce goods and provide services that find a ready market but may be judged not to contribute to social welfare.

Misemployment is founded in the severe inequalities that characterize most developing countries. People are employed in a wasteful manner because their labour is so cheap—relative to the incomes of the élite, the middle-class, and foreigners. Thus scavengers collect painstakingly what the more affluent discard as worthless.

For three decades now it has been common practice to distinguish between an 'informal' and a 'formal' sector in the urban labour-market. On the basis of research in a low-income neighbourhood in Accra, Ghana, Hart (1973) had emphasized the great variety of both legitimate and illegitimate income opportunities available to the urban poor. The response to his plea that a historical, cross-cultural comparison of urban economies in the development process must grant a place to the analysis of 'informal' as well as 'formal' structures was nothing less than overwhelming: a great deal of research, much of it sponsored by the International Labour Office, has been carried out on the 'informal sector'.[2] It has served to direct attention to a work-force that is typically under-enumerated, commonly characterized as unproductive, and all too often dismissed altogether as making little, if any, contribution to the urban economy.

The analytical value of the simple dichotomy between 'formal' and 'informal' income-earning activities is, however, problematic. Early approaches, in particular the classic statement by the International Labour Office (1972: 6), used multiple criteria to define the sectors. But the labour-market cannot be simply divided into two ideal types: if a multidimensional definition is applied to one sector, the other becomes a quite heterogeneous residual category. More recently, the tendency has been to use a single variable to distinguish the two sectors. Thus Alejandro Portes (1994), while setting illegal activities aside, casts the argument in terms of an 'informal economy' of income-earning activities not regulated by the state. Such an approach can serve to highlight an important variable in the labour-market but fosters the neglect of other variables pertinent to understanding the urban labour-market and addressing its problems.[3] We have already explored the productivity of different types of employment, we are about to address employment relationships, and we will devote a great deal of attention to the issue of differential access to earning opportunities.

As it is commonly used, the concept of the 'informal sector' covers a great variety of activities. Any assessment of the prospects for the 'informal sector', and of policy options, has to be both specific and comprehensive, i.e. it has to focus on particular activities and those engaged in them, and it has to take full account of linkages, with the 'formal sector' in particular (Gilbert and Gugler, 1992 [1982]: 94–100).

Ray Bromley, in Chapter 8, proposes that we conceptualize a continuum of employment relationships ranging from career wage-work to career self-employment. Between these two types of career work he distinguishes four types of casual work that range from short-term wage-work, through disguised wage-work and dependent work, to precarious self-employment. These distinctions serve to make the point that most of those engaged in the least stable and least secure work, while seemingly self-employed, in fact enjoy little autonomy and have rather inflexible working regimes and conditions. 'Disguised wage-work' is paid according to output, like much wage-work—the difference is that it is conducted off-premises. And, 'dependent workers' have contractual obligations that substantially reduce their freedom of action: they have to pay rent for premises and/or equipment, to repay credit, and to purchase or sell at disadvantageous prices.

Bromley presents a survey of street occupations in Cali, Colombia, the most comprehensive survey of its kind ever reported. More than 8 per cent of the city's labour-force is involved in a great variety of street activities. Although some are illegal or illegitimate, most make important contributions to the urban economy: street retailers supply food and manufactured goods to consumers at competitive prices; small-scale transport and a range of other services are provided cheaply; scavengers transform waste into low-cost inputs for artisans and industry. While close to half the street-workers are precariously self-employed, nearly as many are wage-workers in disguise.

Lea Jellinek, in Chapter 9, introduces us to one such self-employed street-worker in Jakarta. Her account of this woman's struggle to support herself and her dependants in Jakarta is unique. Survey research presents a snapshot at one point in time, and even participant observation rarely stretches over more than a couple of years, but

Jellinek has regularly visited Sumira and her daughter for more than two decades. She has thus observed in detail an unfolding chain of events rather than reconstructing a distant past from what informants remember and are prepared to tell. Indeed, Jellinek's experience illustrates the vagaries of such reconstruction even when close rapport has been established.

The story of Sumira and her family demonstrates how revealing a diachronic approach can be. The dramatic ups and downs of Sumira's street sales of prepared food illustrate the precarious nature of even a flourishing trade, and the dramatic impact of economic change and administrative fiat—in the case of Jakarta the repressive measures taken against street-traders and *becaks* and the eradication of low-standard housing. As we follow the life of a remarkable woman, we begin to understand how the changing fortunes of her commerce relate to changes in the composition of her household and affect the very membership of her family, and how the move from the *kampung* to modern apartments transforms the lives of Sumira and those around her.

The urban labour-market in developing countries, like many markets, is fragmented in a variety of ways, i.e. different categories of people enjoy differential access to earning opportunities. Access is usually largely a function of four criteria: education and training, patronage, gender, and age.

'Credentials' are generally accepted as a screening device, even though their relationship to functional job requirements may be quite tenuous. And discrimination on the basis of gender and age is commonly taken for granted. Patronage, in contrast, is usually frowned upon. It is, however, widespread, sustained as it is by strong interests and effective mechanisms. Kin, people of common origin, classmates, people who share the same creed assist each other to secure employment or set up in business. Many employers find it convenient and even advantageous to use 'brokers' who will vouch for new recruits and exert a measure of control over them (ILO, 1972: 509–10).

Because of such particularistic patterns of recruitment, migrants of common origin tend to cluster in certain jobs and trades. Amerindians in Lima provide remarkable examples of immigrants establishing economic niches which offer earning opportunities for newly arrived kin and *paisanos*, as we have seen in Chapter 5. Where people of common origin assist each other in gaining a foothold in the urban economy, work together, quite likely spend leisure time together as well, they establish networks that become a key factor in delineating or even inventing ethnic identities. In many countries such identities are a major organizing principle in political conflict, as we shall see in Part VI.

Women are disproportionately found in the least remunerative and/or lowest status occupations, whether in the streets of Cali or in a shanty-town in Nairobi. Nici Nelson, in Chapter 10, reports how the women living in the shacks of Mathare Valley are much more restricted than men in their choice of economic activity. In the early 1970s, few of the local business establishments were run by women, and almost all women were involved in illegal beer-brewing or sex work. Women were at a disadvantage because they were less educated than men, had fewer skills of commercial value, and supported and cared for children. These handicaps are experienced by women as structural constraints, but these constraints are founded on cultural norms. The norms are an amalgamation of indigenous ideals and European views imported during colonial days: the contemporary consensus defines child care as solely women's work and restricts the skills of most women to the subsistence economy and domestic work.

Nelson repeatedly returned to Nairobi for further research over the last two decades. Some Mathare residents had done very well: those who had been able to invest in companies providing rental housing and others who had been allocated new sites or houses elsewhere. But most Mathare residents were faced with fewer and fewer viable economic options. Women were particularly hard hit: a multinational firm began producing maize-millet beer, and the ban on beer-brewing was so harshly enforced that women had to abandon this their principal economic activity.

Irene Tinker (1994) reports major differences in the position of women in the urban street-food trade across different cultures. In the 'patriarchal belt' that extends from North Africa to West, South, and East Asia few women sell street foods. In sharp contrast, women dominate the street-food trade in parts of Africa South of the Sahara characterized by a pattern of spouses operating in separate spheres and keeping their budgetary affairs apart. South-east Asian countries such as Indonesia and the Philippines present a distinct third pattern: a sizeable proportion of enterprises is run by couples.

Women have come to constitute a large share of the greatly expanded industrial labour-force in several rapidly industrializing countries. East and South-east Asian countries report women holding between two-fifths and more than half of the jobs in manufacturing. In distinct contrast, their employment in manufacturing is extremely low in most of South Asia, West Asia, and Africa. In Latin America and the Caribbean, women hold an intermediate share (ILO, 1995: Table 5A). The rate of mobilization of female labour into industry has been fastest where the rate of growth of industrial employment has been the most rapid, i.e. in the countries that have dramatically increased their exports of manufactures: light industrial consumer goods produced in factories using labour-intensive techniques and employing large numbers of women. A great deal of scholarship has been devoted to women working in export industries in particular (Joekes, 1987; Lim, 1990).

More recently the adverse effects of structural adjustment programmes have become a major concern. The evidence is rather sketchy as yet, but it does suggest that women have often been disproportionately affected. Thus women in Ghana were concentrated in the informal sector which tended to absorb excess labour with a concomitant decline in incomes; in Egypt, where private-sector opportunities for women are limited, the queue for government employment lengthened; and in Chile, public-sector hiring in the early stages of the crisis favoured men. However, in Bolivia much of the employment loss occurred in the male-dominated mining sector (Horton, Kanbur, and Mazumdar, 1994).

Women are subject to serious discrimination in virtually all labour-markets. At its most blatant, women's earnings are less than men's in the same occupation and for the same job. Gender discrimination affects women in a less obvious fashion through job placement. The labour-market is segmented by gender, and women tend to be placed in jobs that involve repetitive, short-cycle tasks that require little or no training: the assembly line is becoming a female institution.

Typically these jobs have no promotion lines leading on to more varied and rewarding work within the enterprise—work that would entail training for higher levels of complexity and that is regarded as more responsible and accordingly better paid. Furthermore, such promotional opportunities as do exist, for example to supervisory positions, tend to be monopolized by men even where women predominate in the labour-force.

Recent research has demonstrated that gender discrimination affects the very organization of work and classification of jobs. John Humphrey, in Chapter 11, shows how gendered occupational and work structures are constructed within the factory and then institutionalized and legitimated through segmented labour-markets. Gender divisions in manufacturing thus do not merely reflect pre-existing gender divisions arising in the domestic sphere: they are actually constructed and reinforced at the work-place.

In his research in Brazilian factories, Humphrey found women concentrated in production jobs, low-grade work, and one or two broadly defined occupations, as well as segregation by department and occupational category. This division of labour resulted from the interplay of several processes: the separation of male and female jobs, the privileged access of men to jobs which require skill and training, the reservation of promotion opportunities for men, the unequal recognition of female and male skills, and the differential 'fine-tuning' of jobs. To explain these processes, Humphrey stands dual labour-market theory on its head. Women were not left at the bottom of job hierarchies because they lacked stability in the job; nor did high turnover keep them from training that would have given them access to better jobs; quite the contrary: women's position at the bottom of job hierarchies was due to their greater stability in the absence of promotion—management saw no need to formally recognize and reward their training, experience, and skills.

The advent of Japanese-style Just-in-Time and Total Quality Management techniques may undermine existing divisions of labour because it tends to integrate tasks, but it is unlikely to bring the end of discrimination against women in the factory. More likely new patterns of segregation will arise, or women will be excluded altogether.

Child labour is common in poor countries, and it is widely perceived as highly problematic. It has, however, received little scholarly attention to date. Children are even more prone than women to dependency in the household and discrimination at work. Arguments that lower wages—relative to adult males—reflect lower productivity are more plausible for children than for women. But for children as well there is often little correspondence between wage and productivity differentials. Indeed, in some cases, the work of children is superior to that of adults, e.g. they are able to tie smaller knots and produce higher-quality carpets. Children are also subject to specific constraints. If women need to remain close to home because of domestic requirements, many children are similarly restricted by concerns for their safety. If women's domestic responsibilities compete with their work for earnings, many children face a similar conflict between school and work. And labour-markets are segmented by age as well as gender. Finally, the gendered organization of work elucidated by Humphrey in Chapter 11 has its parallel in the limited opportunities children find for promotion.

Christiaan Grootaert and Ravi Kanbur, in Chapter 12, discuss the determinants of child labour. They focus on household decisions that shape the supply of child labour, and on aspects of the labour-market and the technology of production that affect demand. The many variables involved change from one context to the next. And the policy implications vary accordingly. Clearly, legislation has to be complemented by economic incentives geared to the specific circumstances. The adoption of legislation, its enforcement, and the provision of incentives require advocacy and mobilization, and ultimately the empowerment of the children and their families.

Notes

1. For reviews of the literature on the urban labour-market, see Berry (1987) and Gilbert (1994).
2. For recent annotated bibliographies on the 'informal sector' in Africa, Latin America, and Asia, see ILO (1991); ILO and PREALC (1991); and Sethuraman (1992).
3. For a recent review of approaches to the 'informal sector', see Rakowski (1994).

References

Berry, Albert (1987), 'The Labour Market and Human Capital in LDCs', in Norman Gemmell (ed.), *Surveys in Development Economics* (Oxford: Basil Blackwell), 205–35.

Gilbert, Alan (1994), 'Third World Cities: Poverty, Employment, Gender Roles and the Environment During a Time of Restructuring', *Urban Studies*, 31: 605–33.

—— and Gugler, Josef (1992 [1982]), *Cities, Poverty, and Development: Urbanization in the Third World*, 2nd edn. (Oxford: Oxford University Press).

Hart, Keith (1973), 'Informal Income Opportunities and Urban Employment in Ghana', *Journal of Modern African Studies*, 11: 61–89.

Horton, Susan, Kanbur, Ravi, and Mazumdar, Dipak (1994), 'Labor Markets in an Era of Adjustment: An Overview', in Susan Horton, Ravi Kanbur, and Dipak Mazumdar (eds.), *Labor Markets in an Era of Adjustment* (Washington, DC: World Bank), 1–59.

ILO (International Labour Office) (1972), *Employment, Incomes and Equality: A Strategy for Increasing Productive Employment in Kenya* (Geneva: ILO).

—— (1991), *The Urban Informal Sector in Africa in Retrospect and Prospect: An Annotated Bibliography* (Geneva: ILO).

—— (1995), *1995 Yearbook of Labour Statistics* (Geneva: ILO).

—— and PREALC (1991), *Retrospectiva del sector informal on urbano en América Latina: Une bibliografía anotada* (Geneva: ILO).

Joekes, Susan P. (1987), *Women in the World Economy: An INSTRAW Study* (Oxford: Oxford University Press).

Lim, Linda Y. C. (1990), 'Women's Work in Export Factories: The Politics of a Cause', in Irene Tinker (ed.), *Persistent Inequalities: Women and World Development* (New York: Oxford University Press), 101–19.

Portes, Alejandro (1994), 'The Informal Economy and Its Paradoxes', in Neil J. Smelser and Richard Swedberg (eds.), *The Handbook of Economic Sociology* (Princeton: Princeton University Press; New York: Russell Sage Foundation), 426–49.

Rakowski, Cathy A. (1994), 'The Informal Sector Debate, Part 2: 1984–1993', in Cathy A. Rakowski (ed.), *Contrapunto: The Informal Sector Debate in Latin America* (Albany, NY: State University of New York Press), 31–50.

Sethuraman, S. V. (1992), *The Urban Informal Sector in Asia: An Annotated Bibliography* (Geneva: ILO).

Tinker, Irene (1994), 'The Urban Street Food Trade: Regional Variations of Women's Involvement', in Esther Ngan-ling Chow and Catherine White Berheide (eds.), *Women, the Family, and Policy: A Global Perspective* (Albany, NY: State University of New York Press), 163–87.

7

Overurbanization Reconsidered

Josef Gugler

THE proposition that developing countries are characterized by overurbanization was widely accepted in the 1950s and into the 1960s. The relationship between level of urbanization and degree of industrialization provided the basis for either a synchronic or a diachronic argument. In cross-section analysis, particular countries were shown to deviate from the general pattern of the relationship; in historical comparison, developing countries were shown to have a degree of industrialization lower than that which characterized developed countries at comparable levels of urbanization in the past. N. V. Sovani's devastating critique of both approaches led to the precipitate retreat of the advocates of the overurbanization thesis; the very notion was all but banned from academic discourse.[1] Unheeded went Sovani's cautionary note that the subject of overurbanization needed to be investigated further. I do not propose to resuscitate comparative arguments but will focus instead on the economic implications of the rapid urban growth that characterizes most developing countries.

Cities in developing countries have substantial surplus labour in various guises. Their labour-force continues nevertheless to increase, swelled not only by natural population growth but also by rural–urban migration that accounts for more than half of urban growth in most developing countries.[2] The process may be labelled 'overurbanization' in so far as

- rural–urban migration leads to a misallocation of labour between the rural and the urban sectors, and
- rural–urban migration increases the cost of providing for a country's growing population.[3]

Most rural–urban migrants correctly assess that they are improving their life-chances. A paradox arises between the rationality of the decision made by individuals or small groups to migrate and the irrationality of the migratory movement when considered at the level of the national economy. This paradox is resolved when the migratory movement is seen as a mechanism that allows some of the disadvantaged rural population to partake in a small measure of the resources disproportionately concentrated in urban areas. Thus, in the absence of effective policies to redistribute productive resources and/or income across the rural–urban divide, rural–urban migration can be argued to contribute to development—if it is defined to include the distributional aspect.

Urban Surplus Labour

Cities in developing countries are characterized by an excess of labour with limited skills. Open unemployment constitutes only one facet of urban surplus labour. A second element is under-

This essay has been revised for *Cities*. It appeared first in *Economic Development and Cultural Change*, 31: 1 (1982), 173–89, and is used here by permission of the University of Chicago Press. Earlier versions were presented to the Migration and Development Seminar at the Center for International Studies, Massachusetts Institute of Technology, Dec. 1978, and to the Conference on Urbanisation in West Bengal at the Centre for Urban Economic Studies, University of Calcutta, Dec. 1980. I wish to thank, without implicating, William G. Flanagan, Alan G. Gilbert, Peter Kilby, and Donald C. Mead for helpful comments.

employment, i.e. labour is underutilized. Finally, many workers, while perhaps fully employed, produce goods or provide services that can be judged to contribute little to social welfare; such labour may be labelled 'misemployed'.

Unemployment

Information on open urban unemployment in developing countries is notoriously problematic. First of all, there is little data. A recent unpublished compilation by the International Labour Office offers estimates of open urban unemployment for seventeen Latin American and Caribbean countries throughout the 1980s. According to these estimates, 6.5 per cent of the urban labour-force were unemployed in 1988. For Africa, estimates are available for only eight countries in the early 1980s. For Asia, the only data on urban unemployment is for China where official sources indicate a decline from 5.4 per cent in 1979 to 1.9 per cent in 1984. Urban unemployment in China used to be limited by controls on rural–urban migration and by 'sending-down' campaigns, most notably the rustication of middle school leavers that took around 17 million youths to the countryside during the Cultural Revolution. However, when these policies were relaxed or reversed altogether, substantial urban unemployment appeared; a low estimate put unemployment at 10 per cent of the labour-force in the non-agricultural sector in 1979.[4]

Secondly, there is good reason to doubt how completely urban populations are covered by censuses, how accurately they are represented in surveys. There is probably a systematic bias in that low-income groups tend to go underreported; in so far as their unemployment rates diverge from the average, the overall unemployment rates reported are affected.

Thirdly, the extent of reported unemployment is very much a matter of definition. Is it restricted to those actively seeking work? Or does it cover all who are available for work, including those who have become discouraged about finding work? The distinction is likely to have a particularly strong effect on the unemployment rate reported for women. This is even more the case for a further issue of definition: are those searching/available for part-time work to be included? Finally, does part-time work disqualify a person from being considered unemployed?

The unemployed are obviously unproductive, but they are usually not representative of the most desperate urban living conditions. In countries where very few qualify for unemployment benefits, it is only the not-so-poor family that can support an unemployed member.[5] If unemployment is frequently reported higher among the urban-born than among immigrants, it is because the families of the urban-born are more likely to be already well established in the urban economy. Given family support, an extended search for a satisfactory job can be a rewarding strategy, especially for those with better qualifications. Higher levels of unemployment among the more educated, a common pattern,[6] thus appear as a function of both the potential rewards for the better educated of an extended job search and the fact that they tend to come from families which are able to support them through a lengthy period of unemployment. In contrast, the poorest, whose relatives and friends cannot help them, and those recent immigrants who have nobody to turn to, are forced to find some livelihood in a hurry; unemployment is a luxury they cannot afford.[7]

Underemployment

I define 'underemployment' here as the underutilization of labour for the purpose at hand.[8] Such underutilization is most conspicuous where labour is idle part of the time. While this is a common pattern in agriculture, it is not unknown in the urban sector where seasonal fluctuations are marked in industries related to the agricultural production cycle, in construction, and in the tourist trade. Underemployment is not limited to these sectors, however, but is much more pervasive.

Underemployment takes three distinct forms. In one guise it is related to fluctuations in economic activity during the day, for example at markets; over the week or month, for example in recreational services; or seasonally, for example in tourism. As activity ebbs, casual labour is laid off and many self-employed are without work. Underemployment takes a second form where workers are so numerous that at all times a substantial proportion are less than fully employed, i.e. a reduction in the number of workers will not decrease aggregate output.[9] In terms of numbers affected, street vendors constitute the most important category in many countries. A third type of underemployment is what may be called 'hidden unemployment': solidary groups continue to employ all their members rather than discharging them when there is insufficient work to keep them fully occupied. Such guaranteed employment is typical of family enterprise, but social ties other than kinship proper, such as common origin or shared religion, can also provide a commitment to maintain every member of the community.[10]

Misemployment

Finally, there is what I call 'misemployment'. Labour may be employed full-time, but the tasks performed contribute little to social welfare. Begging is a clear-cut example.[11] More respectable, but hardly more productive, are the hangers-on to be found in the entourages of the more powerful and affluent. There is also a wide range of illegal activities—whether they all constitute misemployment is a matter for debate: the thief who redistributes resources from the wealthy to his poor family can be argued to perform a service not dissimilar to that of many bureaucrats in a welfare state.[12] And indeed, ultimately the productivity of an activity is social defined.

The notion of unproductive labour dates back at least as far as Adam Smith's *The Wealth of Nations*. It was part of his polemic against the mercantilist state, which redistributed income from its more productive subjects to the sovereign. In this view the political élite, the religious estate, and the cultural and intellectual superstructure had a basically parasitic relationship to the productive classes. Presumably, the larger the surplus income generated by the productive sector, the larger the number of retainers and other parasites that could be supported by the ruling class. Substantial numbers of public administrators in many contemporary societies appear similarly misemployed. As in the courts of yore, their role of hangers-on has become institutionalized.

Much misemployment is based on getting crumbs from the table of the rich. The member of the local élite or middle class, the foreign technical adviser, or the tourist is beseeched for a morsel, or made to maintain a company of sycophants, or has his wallet snatched away. The relationship is vividly portrayed by three activities: the army of domestics that cleans and beautifies the environment of the privileged;[13] the

prostitutes who submit to the demands of those able to pay and thus expose themselves to AIDS;[14] and the scavengers who subsist on what the more affluent have discarded, who literally live on crumbs from the rich man's table.[15]

Of course, most forms of misemployment, unlike begging, make some contribution to social welfare. Waste paper provides a third of the raw material requirements for the paper industry in Cali, Colombia, and some 60 per cent of that waste paper is collected by garbage pickers working on their own.[16] In domestic service, the contribution can be substantial where women with qualifications that are in short supply are released from household work. What is at issue here is that large numbers of people are employed in a wasteful manner because their labour is so cheap—relative to the incomes of the élite, the middle class, and foreigners.[17] Thus what the middle class discards as worthless is painstakingly collected by others so poor that they scratch a livelihood from dirty, damaged, deteriorated 'waste'. Cigarette-butt collecting continues to constitute a major street occupation in Indonesia, a country that has enjoyed a measure of economic development.[18]

Much misemployment serves simply to bolster the status of the more affluent. The point is well demonstrated by the fact that the requirements of middle-class households for domestic help rapidly decrease as domestic wages rise.[19] As Kate Young writes of domestics in Mexico:

Domestic service for these girls is . . . a waste of human resources. It permits little awakening of their latent capacities, and provides no possibility for them to learn a wide range of skills, to value their own work, or to develop an independent and enquiring personality. . . . Domestic service, in reinforcing dependence, does not encourage the questioning of a system in which wealth permits certain categories of people to condemn others to a life of servicing them. Nor does it encourage them to question a system of relations between the genders in which women are essentially seen either as playthings to entertain men or as drudges to service them.[20]

The Opportunity Cost of Rural–Urban Migration

An evaluation of the economic implications of urban surplus labour has to take into account its opportunity cost at the rural end. We now enter the treacherous realm of counter-factual analysis: what would have happened if some of the migrants had stayed in their rural homes?

For a time the assumption was commonly made that the marginal productivity of rural labour is zero in most developing countries.[21] Countries such as Bangladesh, India, Pakistan, Egypt, Ruanda, Burundi, and Jamaica and major regions such as Java in Indonesia were seen to be so severely over-populated that labour was redundant in the rural areas. Elsewhere, particularly in much of Latin America, the mass of the rural population has no access to land, either because of institutional barriers—for example large land holdings controlled by absentee owners frequently are farmed in a rather extensive manner—or because of a lack of resources to open up virgin land. The argument gains in strength if it is remembered that the rural exodus is more than compensated for by natural population growth in nearly every developing country, that is, the rural population continues to increase in absolute numbers in spite of emigration. Admittedly, because of the age selectivity of out-migration, a rural population may increase even while the rural labour-force is reduced. But such instances appear to be exceptional, even at the regional level within countries.

Still, agriculture is characterized by labour bottle-necks during planting and/or harvesting periods. Even where population pressure on land is severe, all hands are needed at certain seasonal peaks of labour requirements.[22] Certainly, where additional land can be brought under cultivation, as is the case in nearly every country in Africa South of the Sahara, the emigration of able-bodied adults implies a loss of potential output. The argument applies in other areas if institutional obstacles to the more intensive farming of land are not taken for granted, or if the opportunities for opening up virgin land are seriously considered.

In countries that are characterized by severe population pressures, these very pressures may encourage changes in agricultural practice that increase output.[23] These can range from an increase in the frequency with which land is cropped, to irrigation, to higher yield crops, and to the use of fertilizer. Recognition that agricultural practice is not static and of the need for innovation, especially where the land/man ratio

is unfavourable, leads to a full appreciation of the opportunity cost of migration that is selective in terms of age and education. The rural areas lose the young, the more educated, and, we may surmise, the most enterprising.

Effective demand for additional agricultural output is manifest in those developing countries that have become dependent on food imports. Certainly, in every country there is scope for improving nutritional standards. Finally, there usually exist opportunities for boosting exports of agricultural products.

We may conclude that rural–urban migration entails a loss of potential agricultural output where uncultivated land is still available, where virgin lands could be developed, and where institutional restraints on the intensification of farming could be overcome. We add that the disproportionate loss of the young, the educated, and the enterprising delays innovation where it is most needed, that is, where population pressure on land appears most severe, given present farming methods.

So far I have restricted my argument to agricultural production, but a large proportion of the rural population is engaged, full-time or part-time, in non-farm activities such as manufacturing, repair, construction, retail and wholesale trade, restauranting, transport, personal services, and salaried employment in teaching, health services, and public administration. Surveys from four Asian and three African countries report that non-farm income generated by rural households ranged from 14 to 45 per cent, averaging 35 per cent.[24] Again, it is the out-migrants who would be best equipped to develop non-farm activities.

R. A. Berry and R. H. Sabot conclude from their review of the available evidence that labour utilization by farm families is high. Where there is a substantial seasonal element in the labour requirements for agriculture, the average annual labour input into agriculture may be low, for example in parts of Africa, but seasonal labour bottle-necks limit expansion of agricultural production under existing technology, and a considerable amount of time is spent on non-farm activities such as crafts and trading during the off-season.[25]

The Relative Cost of Urban Services and Goods

We have seen that rural–urban migration brings workers to cities that are already burdened with surplus labour, and it now further appears that rural–urban migration entails some loss of potential rural output. If it is plausible that rural–urban migration, at current levels, leads to a misallocation of labour between rural and urban areas, we need to address a second issue that appears to be quite clear-cut: how does the resource cost of providing for a rapidly expanding urban population compare with the task of absorbing such additional numbers in rural areas? Housing, transport, garbage and sewage disposal, provision of fuel, and distribution of staple foods stand out as five amenities that are expensive in urban agglomerations but cheap or not required in rural areas. Whether the rural–urban migrant manages to pay for them, urban hosts provide them, or they are subsidized by public authorities, the higher cost of these amenities will usually more than offset the savings that arise in the provision of other services or goods—electricity, pipe-borne water, or health care for example—that come cheaper to urban population concentrations than to a dispersed rural population. This is especially the case if the composition of a low-income budget is taken into account.

The pressure rural–urban migrants exert on existing urban infrastructure in developing countries, and the cost of new infrastructure required by their presence, have received little serious attention.[26] This neglect may be grounded in the tacit assumption that the urban poor remain marginal in that they have neither the political clout nor the market power to make effective demands. Thus, discussions of squatting centre on the diseconomies inherent in such initiatives by the poor to provide housing for themselves but mostly fail to consider the opportunity cost of land, materials, and labour.[27] Even in shanty-towns such costs are substantial. Indeed, the value of the land to be put to alternative uses frequently provides the primary motivation for slum-clearance schemes, opposition to squatting, and demolition of squatter settlements. To the extent that even marginal migrants have access to better amenities than are available in rural areas, their migration has a redistributive effect, an issue I will address in a moment. This redistributive dimension should not obscure the issue at hand: the provision of a subsistence minimum is more costly for additions to urban than to rural populations.[28]

A major part of the costs of providing for increases in the urban population are not borne, in market economies, by those who take decisions affecting the location of jobs and people. In centrally planned economies, in contrast, the higher cost of providing for additions to the populations in urban as opposed to rural settings were readily apparent to policy-makers. They pursued policies designed to economize on the costs of urbanization. Socialist countries used more labour-intensive techniques in agriculture and more capital-intensive techniques in manufacturing than market economies; they encouraged a high level of labour participation of women as well as men; and they allocated a small number of workers to urban services. The urban labour-force required at a given level of industrialization was thus kept relatively low. In addition, substantial numbers of workers were compelled to reside outside the cities and to commute because the provision of housing typically failed to keep up with the expansion in industrial jobs. If the housing shortage made it difficult to establish residence in cities, administrative measures expressly restricted in-migration to major cities such as Moscow. György Konrad and Ivan Szelenyi coined the term 'underurbanization' to characterize this pattern.[29]

In China, rural–urban migration was strictly controlled until recently. Indeed, in the early 1960s, repeated rustication campaigns resulted in actual declines in the urban population. Labour requirements were met by bringing most women into the labour-force and by contract labour: men were recruited from rural areas on temporary contracts; they were housed in dormitories and had to leave their dependants behind.[30]

The Economic Rationale for Rural–Urban Migration

If rural–urban migration entails the loss of potential rural output, if it brings workers to cities that are unable to fully employ their existing labour-force to productive ends, and if the additional urban dwellers require more resources than they would in the countryside, then the conclusion is warranted that rural–urban migration at current rates is inefficient. The label 'overurbanization' seems in order. In the face of this condition rural–urban migration proceeds at a rapid pace. The majority of migrants are ready to enter the labour-force. If they contribute more than half of urban growth, they constitute an even higher proportion of the new entrants into the labour-force.

A substantial body of research on rural–urban migration has accumulated over the last three decades, and the evidence is overwhelming: most people move for economic reasons. When migrants are asked why they moved, they usually cite the better prospects in the urban economy as their chief reason. Also, migration streams between regions have been shown to correspond to income differentials between those regions. And over time, as economic conditions at alternative destinations change, migration streams alternate accordingly. Finally, studies across developing countries report, time and again, that most migrants consider that they have improved their condition and that they are satisfied with their move.

If there is a measure of agreement on this basic proposition—that rural–urban migration is largely motivated by economic reasons—two rather different interpretations have evolved. In Tropical Africa, analysis focused on migrants coming in search of jobs that offered wages and working conditions regulated by legislation and/or collective bargaining. They would spend several months trying to secure such a job but, if unsuccessful, eventually return to the village. In the 1950s and 1960s much urban unemployment in Tropical Africa conformed to this pattern. With independence, urban wages rose substantially in many countries. Rural–urban migration surged, the labour shortages that had plagued colonial governments vanished, and urban unemployment appeared. Much labour migration had been short term, and new arrivals faced little competition from entrenched urban workers and their descendants. Moreover, independence was usually accompanied by a significant expansion in urban employment. In this context, the system of recruiting unskilled labour approximated a random process. Since minimum wages were high relative to rural incomes, even an extended job search was a promising strategy. Joining the urban unemployed, rural–urban migrants tried their luck at the urban job lottery.[31]

In retrospect it is clear that the urban job lottery pattern occurred in exceptional circumstances. More commonly, labour turnover is low, job creation slow, and recruitment anything but random in developing countries. Access to urban earning opportunities is severely restricted—this is the case in particular for the protected sector where wages, working conditions, job security, and social security are guaranteed by unions and/or the state. Employers have established criteria for recruitment into different job categories such as formal educational qualifications, experience, gender, and age. And established workers assist family members, kin, and friends in joining them.[32] The prospects for those who come without the right connections or exceptional qualifications are dim, and an extended search is unlikely to improve their chances. Still, even marginal urban earnings tend to be superior to what the rural areas have to offer.

The Distributional Dimension

Our earlier discussion revolved around a concern with the implications of rural–urban migration for aggregate output. The indications that its contribution to urban output is problematic, that it entails a loss of potential rural output, and that it increases demands on scarce resources led us to conclude that rural–urban migration at current levels is inefficient. Now I have argued that rural–urban migration is a rational response to the economic realities ruralites face, that there are advantages to be gained from the move. The resolution of the seeming paradox derives from the fact that rural–urban migration has a redistributive effect. Rural–urban migrants lay claim to a share in urban income opportunities, they gain some access to urban amenities. Rural families send their sons and daughters to the city so that they will be able to partake, however little, and in whatever demeaning way, of its riches. As Larissa Adler Lomnitz put it in her study of a shanty-town in Mexico City:

The settlers of Cerrada del Cóndor may be compared to the primitive hunters and gatherers of preagricultural societies. They go out every day to hunt for jobs and gather the uncertain elements for survival. The city is their jungle; it is just as alien and challenging. But their livelihood is based on leftovers: leftover jobs, leftover trades, leftover living space, homes built of leftovers.[33]

The condition of those who stay in the rural areas may improve as migrants remit part of their income to their family, provide villagers with access to urban amenities (e.g. education, health care), assist village development, and press village interests with officials at the regional or national level.[34] Or, in some cases, rural conditions may be argued to be at least somewhat better than they would have been if the population pressure on land had become even more severe.

There thus appears to be a trade-off between a misallocation of labour and an improved income distribution.[35] If we define development to include a distributional dimension, an evaluation of rural–urban migration depends on an assessment of its consequences for both output and distribution, and on the respective weighting these are given.[36]

Of Policy and Power

There would appear to be an approach that promises a more efficient allocation of labour between the rural and the urban sector as well as a reduction in the extreme inequalities that characterize most developing countries. It will aim at improving rural living standards by channelling productive resources to the rural areas and/or directing a larger share of income to them.[37] This is not an original policy prescription for a problem that has been recognized for a long time. Julius Nyerere, then the president of Tanzania, raised the issue in the Arusha Declaration in 1967.[38] Michael Lipton coined the term 'urban bias' in 1977.[39] Policy-makers, however, rarely gave more than token recognition to the distortions in the resource allocation between the urban and the rural sector.[40] Only the economic crisis of the 1980s forced many of them to take the issue seriously.[41]

Cities are centres of power and privilege. In every developing country, the urban sector accounts for a disproportionate share of consumption as well as investment.[42] Usually they are even more concentrated in the national capi-

tal.[43] Certainly, industry and modern transport find advantage in spatial concentration. They require urban infrastructure and highly trained specialists. But it would seem that resources for both consumption and investment are invariably apportioned to urban areas beyond the immediate requirements of industrialization. If this proposition is granted, two characteristics of the contemporary setting may be adduced to explain it. On one hand, nearly all those enjoying high standards of consumption are urban-based. Unlike the feudal order of Europe's Middle Ages, and unlike the hacienda system that characterized much of Latin America in the colonial and early post-colonial days, throughout the South today virtually the entire élite is located in cities. Much of the middle class of senior civil servants, professionals, managers, and entrepreneurs also live there, many out of necessity, some by choice. Finally, some sectors of the urban labour-force enjoy a standard of living that is high when compared to the condition of the urban masses. On the other hand, much of the surplus that was appropriated by private interests under *laissez-faire* capitalism, that was drained abroad in the colonial system, is controlled by the state. Its allocation tends to be governed by the interests of decision-makers in improving their immediate environment, in assuring the continued collaboration of the middle class, and in placating strategically placed elements of labour: public resources are disproportionately spent on privileged consumption of the few, and conspicuous investment for the few—in the cities.[44]

Notes

1. N. V. Sovani, 'The Analysis of "Over-Urbanization"', *Economic Development and Cultural Change*, 12 (1964), 113–22. In recent years the notion of overurbanization that defines it in terms of level of urbanization relative to GNP per capita has had a remarkable revival in cross-national analyses, e.g. Bradshaw and Noonan in Ch. 1 of this volume.
2. Sally E. Findley, 'The Third World City: Development Policy and Issues', in John D. Kasarda and Allan M. Parnell (eds.), *Third World Cities: Problems, Policies, and Prospects* (Newbury Park, Calif.: Sage Publications, 1993), 1–31.
3. For our initial reformulation of the notion of

overurbanization, see Josef Gugler and William G. Flanagan, 'On the Political Economy of Urbanization in the Third World: The Case of West Africa', *International Journal of Urban and Regional Research*, 1 (1977), 272–92. David A. Smith, 'Overurbanization Reconceptualized: A Political Economy of the World-System Approach', *Urban Affairs Quarterly*, 23 (1987), 270–94, offers critique of my argument from a dependency perspective.

4. John Philip Emerson, 'Urban School-Leavers and Unemployment in China', *China Quarterly*, 93 (1983), 1–16.

5. R. Albert Berry, in 'Open Unemployment as a Social Problem in Urban Colombia: Myth and Reality', *Economic Development and Cultural Change*, 23 (1975), 276–91, offers a comprehensive discussion of voluntary unemployment and presents data suggesting its importance. His study, as well as Richard H. Sabot's *Economic Development and Urban Migration: Tanzania, 1900–1971* (Oxford: Clarendon Press, 1979) and Dipak Mazumdar's *The Urban Labour Market and Income Distribution: A Study of Malaysia* (New York: Oxford University Press, 1981), indicated that the pool of unemployed comprised predominantly the young and married women. Presumably most of them were dependants.

6. Lynn Squire, *Employment Policy in Developing Countries: A Survey of Issues and Evidence* (New York: Oxford University Press, 1981), 71, summarizes data on rates of unemployment by level of education in ten developing countries. In eight cases the rate was highest, frequently by a large margin, for those with secondary education. Unemployment was lower among those with post-secondary than among those with only secondary education in every country, including countries such as the Philippines, Sri Lanka, and India that are notorious for widespread unemployment among college graduates. Perhaps college students are in a better position to prepare for their transition into the labour-market.

7. For critical discussions of the argument, see Alan T. Udall and Stuart Sinclair, 'The "Luxury Unemployment" Hypothesis: A Review of Recent Evidence', *World Development*, 10 (1982), 49–62; and John Humphrey, 'Are the Unemployed Part of the Urban Poverty Problem in Latin America?', *Journal of Latin American Studies*, 26 (1994), 713–36.

8. Most studies focus on earnings instead, either because the research's prime concern is with urban poverty or because earnings data are more readily available and are assumed to reflect productivity. Of course, the whole thrust of this essay is that such an assumption is most problematic.

9. Such underemployment can be a quite stable feature among self-employed who have low overhead costs and are assured of a minimum number of customers because of personal ties and/or locational advantages.

10. A comparison with the Japanese permanent employment system springs to mind, but the latter precisely fails to provide for workers outside the major firms and indeed the substantial proportion of casual workers in these firms. The analogy of the commitment to full employment in socialist countries is more accurate. In pre-revolutionary Cuba most workers in the sugar fields and the sugar mills were unemployed for a major part of the year. After the revolution they were offered employment throughout the year, and major efforts were directed towards absorbing them in productive activities during the off-season. As in family enterprise, the problem became to what extent full employment, while desirable on equity grounds, only hides unemployment instead of employing workers to productive ends.

11. Ray Bromley, 'Begging in Cali: Image, Reality and Policy', *New Scholar*, 8 (1982), 349–70, provides a unique account. He establishes that the beggars of Cali are relatively few in number and generally severely deprived and disadvantaged individuals.

12. Some illegal activities commonly come to be tolerated as necessary to the very survival of a major part of the urban population. The acceptance of squatting and illegal construction is the most salient example.

13. Domestic workers account for more than 20 per cent of all women in the paid work-force in Latin America and the Caribbean, according to census and labour-force surveys. In addition there are large numbers of women in domestic service who are not counted in the official statistics. See Elsa M. Chaney and Mary Garcia Castro, 'A New Field for Research and Action', in Elsa M. Chaney and Mary Garcia Castro (eds.), *Muchachas No More: Household Workers in Latin America and the Caribbean* (Philadelphia: Temple University Press, 1989), 3–13. The conspicuous use of labour may be argued to be preferable to the conspicuous consumption of imported luxury goods precisely because it provides local employment; in addition, it saves usually scarce foreign exchange. Such an argument takes for granted extreme income inequalities and, in the case of female domestics, sex discrimination in the labour-market. The image of the nursemaid who fulfils her charge's every whim while her own children suffer from neglect is all too disturbing.

14. The judgement that some labour is unproductive because its output is ethically undesirable, e.g.

prostitution, can also be traced back to Adam Smith. A Scottish professor of moral philosophy, he was less willing than his neo-classical successors to accept the legitimacy of given preferences and was prepared to state that some preferences are better than others.

Many urban women derive their livelihood from prostitution in most developing countries. Estimates of the number of women in sex work are, however, usually problematic, as Gavin W. Jones, 'Economic Growth and Changing Female Employment Structure in the Cities of Southeast and East Asia', in Gavin W. Jones (ed.), *Women in the Urban and Industrial Workforce: Southeast and East Asia*, Development Studies Centre, Monograph 33 (Canberra: Australian National University, 1984), 17–59 shows. He suggests that prostitution-related occupations constitute 7 to 9% of female employment in Bangkok and Manila, cities notorious for attracting sex tourism.

15. The examples of misemployment presented here are drawn from the service sector, but the notion of misemployment can also be applied to labour that produces goods of dubious value, e.g. the tobacco industry.

16. Chris Birkbeck, 'Garbage, Industry and the "Vultures" of Cali, Colombia', in Ray Bromley and Chris Gerry (eds.), *Casual Work and Poverty in Third World Cities* (Chichester: John Wiley, 1979), 161–83.

17. Indigenous labour is inevitably cheap relative to the incomes of foreign tourists. The internationalization of prostitution is a particularly revolting consequence, the Thai sex industry being the most blatant example.

18. Daniel T. Sicular, 'Pockets of Peasants in Indonesian Cities: The Case of Scavengers', *World Development*, 19 (1991), 137–61.

19. Robert Fiala, 'Inequality and the Service Sector in Less Developed Countries: A Reanalysis and Respecification', *American Sociological Review*, 48 (1983), 421–8, shows that inequality is a significant factor in expanding the service sector. His analysis of 23 developing countries found a concentration of income in the top 20% of the population in 1960 to be associated with an expanded service sector in 1970.

20. Kate Young, 'The Creation of a Relative Surplus Population: A Case Study from Mexico', in Lourdes Benería (ed.), *Women and Development: The Sexual Division of Labor in Rural Societies* (New York: Praeger, 1982).

21. Thus the Lewis–Fei–Ranis model assumed a rural sector characterized by labour that had zero or very low productivity (Arthur W. Lewis, 'Economic Development with Unlimited Supplies of Labour', *Manchester School of Economic and Social Studies*, 22 (1954), 139–91: Gustav Ranis and John C. H. Fei, 'A Theory of Economic Development', *American Economic Review*, 51 (1961), 533–65).

22. Seasonal migration, where it is geared to the agricultural production cycle, appears optimal from this angle.

23. Ester Boserup launched the contemporary debate on the relationship between population growth and food supply in *The Conditions of Agricultural Growth: The Economics of Agrarian Change under Population Pressure* (London: Allen & Unwin, 1965). She argued that population growth is the prime cause of agricultural change and focused on the more frequent cropping of the available land as the principal mechanism. In *Population and Technological Change: A Study of Long-Term Trends* (Chicago: University of Chicago Press, 1981), Boserup emphasizes that investments in rural infrastructure are a prerequisite for the intensification of agriculture.

24. Thomas P. Tomich, Peter Kilby, and Bruce F. Johnston, *Transforming Agrarian Economies: Opportunities Seized, Opportunities Missed* (Ithaca, NY: Cornell University Press, 1995), 44–6. Apart from the difficulties encountered in defining rural areas, it should be noted that non-farm income includes the earnings of rural residents commuting to urban jobs. With improved communication and rising incomes such commuting is becoming ever more common. See e.g. Graeme Hugo, 'Urbanization in Indonesia: City and Countryside Linked', in Josef Gugler (ed.), *The Urban Transformation of the Developing World* (Oxford: Oxford University Press, 1996), 133–83.

25. R. Albert Berry and Richard H. Sabot, 'Labour Market Performance in Developing Countries', *World Development*, 6 (1978), 1199–1242.

26. For a review of research that bears on the cost of garbage and sewage disposal, electricity, pipe-borne water, and health care in urban as compared with rural areas, see Johannes F. Linn, 'The Costs of Urbanization in Developing Countries', *Economic Development and Cultural Change*, 30 (1982), 625–48.

27. For a review of the debate on 'self-help housing', see Ch. 18 in this volume.

28. Harry W. Richardson, 'National Urban Policies and the Costs and Benefits of Urbanization', in Dominick Salvatore (ed.), *World Population Trends and their Impact on Economic Development* (New York: Greenwood Press, 1988), 95–106, suggests that absorption costs per capita are 2.5 to 4 times higher in urban than in rural areas.

29. György Konrád and Ivan Szelényi, 'Social Conflicts of Underurbanization', in Alan A.

Brown, Joseph A. Licaro, and Egon Neuberger (eds.), *Urban and Social Economics in Market and Planned Economies*, i. *Policy, Planning and Development* (New York: Praeger, 1974), 206–26.

30. For a recent discussion of the various measures adopted in China in the 1960s and 1970s to 'economize' on urbanization while industrializing rapidly, see Kam Wing Chan, *Cities with Invisible Walls: Reinterpreting Urbanization in Post-1949 China* (Hong Kong: Oxford University Press), 68–96.

31. For a detailed discussion of rural–urban migration, see Alan Gilbert and Josef Gugler, *Cities, Poverty, and Development: Urbanization in the Third World*, 2nd edn. (Oxford: Oxford University Press, 1992), 62–86.

32. For an account of Amerindian migrants controlling economic niches in Lima, see Ch. 5 in this volume.

33. Larissa Adler Lomnitz, *Networks and Marginality: Life in a Mexican Shantytown* (New York: Academic Press, 1977), 208; first published as *Como sobreviven los marginados* (Mexico City: Siglo Veintiuno Editores, 1975).

34. For an account of the efforts of urban residents on behalf of rural development in south-eastern Nigeria, see Ch. 4 in this volume.

35. See the reaction to our initial reformulation of the notion of overurbanization by Peter Marris in 'The Political Economy of Urbanization: A Comment', *International Journal of Urban and Regional Research*, 2 (1978), 171–3.

36. For a discussion of poverty-weighted indices of development, see Montek S. Ahluwalia and Hollis Chenery, 'The Economic Framework', in Hollis Chenery, Montek S. Ahluwalia, C. L. Bell, John H. Duloy, and Richard Jolly, *Redistribution with Growth* (London: Oxford University Press, 1974).

37. The policy options are myriad, from rural infrastructure to subsidized fertilizer, to credit, to agricultural research, to extension services, to farm mechanization, to small-scale industries; from reducing the tax load of the peasantry, to raising prices that have been depressed by price controls or an inflated foreign-exchange rate, to giving rural areas a larger share of health, educational, and other services. The effect of different approaches on the rate of rural–urban migration requires careful enquiry, e.g. mechanization may displace labour and encourage out-migration; the structure of increased demand for agricultural inputs and for consumption goods and services will affect the growth of rural non-farm employment.

38. Julius K. Nyerere, 'The Arusha Declaration', in *Ujamaa—Essays on Socialism* (London: Oxford University Press, 1968); first published in 1967 in Swahili by TANU, the party organization.

39. Michael Lipton, *Why Poor People Stay Poor: Urban Bias in World Development* (London: Temple Smith; Cambridge, Mass.: Harvard University Press, 1977).

40. Severe distortions in the resource allocation between the urban and the rural sector were avoided by the 'Four Little Tigers': peasants did rather well in Taiwan and South Korea after successful land reforms that came in the wake of the Kuomintang's defeat on the mainland and the Korean War, respectively; Singapore jettisoned its hinterland by separating from the Federation of Malaysia in 1965; and a colonial border cut Hong Kong off from its hinterland until 1997.

41. For a review of the academic debate over 'urban bias' and policy reversals, see Ch. 2 in this volume.

42. Even socialist countries, while reducing income inequality within the urban and the rural sector, respectively, found it difficult to deal with inter-sector inequality. In China, the urban–rural ratio of nominal consumption, as high as 2.9 to 1 in 1978, declining to 2.2 to 1 in 1985, reached 3.1 to 1 in 1990. Hidden subsidies further boosted the consumption of urban residents. All in all, urban–rural inequality remained as high in China as in other poor countries in Asia, at times it was even more pronounced (Xiangming Chen and William L. Parish, 'Urbanization in China: Reassessing an Evolving Model', in Josef Gugler (ed.), *The Urban Transformation of the Developing World* (Oxford: Oxford University Press, 1996), 61–90.

43. For evidence that spatial patterns of government expenditures promote the growth of a country's largest city, and in particular the capital city, see Samuel H. Preston, 'Urban Growth in Developing Countries: A Demographic Reappraisal', in Josef Gugler (ed.), *The Urbanization of the Third World* (Oxford: Oxford University Press, 1988), 11–31; an earlier version appeared in *Population and Development Review*, 5 (1979), 195–215.

44. The consumption and investment in cities of resources produced elsewhere is particularly striking in oil-rich countries where it is facilitated by direct state control over oil revenues.

8

Working in the Streets of Cali, Colombia: Survival Strategy, Necessity, or Unavoidable Evil?

Ray Bromley

THIS essay reviews some of the theoretical, moral, and policy issues associated with the low-income service occupations found in the streets, plazas, and other public places of most African, Asian, and Latin American cities.[1] These 'street occupations' range from barrow-pushing to begging, from street-trading to night-watching, and from typing documents to theft. They are often grouped together in occupational classifications, and they are generally held in low esteem. The street occupations are frequently described by academics and civil servants as 'parasitic', 'disguised unemployment', and 'unproductive', and they are conventionally included within such categories as 'the traditional sector', ·'the bazaar economy', 'the unorganized sector', 'the informal sector', 'the underemployed', and 'sub-proletarian occupations'. It seems as if everyone has an image and a classificatory term for the occupations in question, and yet their low status and apparent lack of developmental significance prevent them from attracting much research or government support.

This discussion of 'street occupations' is based mainly on 1976–8 research in Cali, then Colombia's third largest city with about 1.1 million inhabitants. The occupations studied in the streets and other public places of Cali are remarkably diverse, but they can be crudely described under nine major headings:

Retailing: the street-trading of foodstuffs and manufactured goods, including newspaper distribution.

Small-scale transport: moving cargo and a few passengers for payment, using *motocarros* (three-wheel motorcycles), horse-drawn carts, bicycles, tricycles, handcarts, or direct human effort as porters.

Personal services: shoe-shining, shoe repair, watch repair, the typing of documents, etc.

Security services: night-watchmen, car-parking attendants, etc.

Gambling services: the sale of tickets for lotteries and *chance*, a betting game based on guessing the last three digits of the number winning an official lottery.

Recuperation: door-to-door collection of old newspapers, bottles, etc., 'scavenging' for similar products in dustbins, rubbish heaps, and the municipal tip, and the bulking of recuperated products.

Prostitution: or, to be more precise, soliciting for clients.

Begging.

This essay has been revised and expanded for *Cities*. A shorter version appeared first in Alan Gilbert, Jorge Hardoy, and Ronaldo Ramírez (eds.), *Urbanization in Contemporary Latin America* (Chichester: John Wiley, 1982), 59–77. By permission of John Wiley & Sons Ltd.

Property crimes: the illegal appropriation of movable objects with the intention of realizing at least part of their value through sale, barter, or direct use. This appropriation can be by the use of stealth (theft), by the threat or use of violence (robbery), or by deception ('conning').

Of these nine categories, 'retailing' was the largest in 1977, accounting for about 33 per cent of the work-force in the street occupations, followed by small-scale transport and gambling services, each accounting for about 16 per cent of the work-force. The six remaining categories each accounted for 2–10 per cent of the work-force in street occupations.

With the exception of small-scale public transport, all these occupations can be conducted in private locations as well as in public places, and private locations are generally considered more prestigious. Private premises give a business an aura of stability and security which is not available to most businesses conducted in streets, plazas, and other areas of public land. Those who work in public places may try to obtain a degree of stability by claiming a fixed pitch, and by building a structure there to give them some protection, but their tenure is almost always precarious, and their investment in 'premises' is likely to be very limited. Thus, the street occupations are classically viewed as 'marginal occupations', as examples of how the poor 'make out', or as the 'coping responses' of the urban poor to the shortage of alternative work opportunities and the lack of capital necessary to buy or rent suitable premises and to set up business on private property.

Even though they are an integral part of the street environment and interact strongly with the street occupations, four major groups of economic activities are not considered as street occupations *per se*: the off-street private shops, supermarkets, market stalls, etc., which open on to the street; government and company employees who are responsible for building, maintaining, and cleaning the streets; the police and soldiers who are responsible for law and order on the streets; and the operators of larger scale public transport vehicles such as buses, trucks, and taxis. These occupations are considered in relation to the street occupations, but not as part of the street occupations. All of them have a strong off-street base, most having working regimes and relationships rather different from those prevailing in the street occupations, and many have much higher levels of capital investment in premises, equipment, or merchandise.[2]

Because they are neither practised in conventional (off-street) establishments, nor in the homes of the workers, the street occupations are usually severely under-represented or excluded altogether in statistics based on sample surveys of establishments or households. In spite of their under-representation in most official statistics, however, their highly public location ensures them a prominent place in the urban environment and popular consciousness. Even if they have no direct dealings with those who work in the street occupations, the general public can hardly fail to be aware of their existence.

The streets of the city serve a wide variety of interrelated purposes: as axes for the movement of people, goods, and vehicles; as public areas separating enclosed private spaces and providing the essential spatial frame of reference for the city as a whole; as areas for recreation, social interaction, the diffusion of information, waiting, resting, and, occasionally, for 'down-and-outs and street urchins', sleeping; and as locations for economic activities, particularly the 'street occupations' (Anderson, 1978: 1–11 and 267–307). Within the functional complexity of the street environment, the street occupations are both strongly influenced by changes in other environmental factors, and also contributors to general environmental conditions. Thus, for example, street-traders and small-scale transporters depend upon the direction, density, velocity, and flexibility of potential customers' movements, and are immediately affected by changes in traffic flows and consumer behaviour. At the same time, they influence patterns of movement and overall levels of congestion.

Average incomes per worker in the street occupations in Cali in 1977 were only equivalent to about US$3 per day, and were roughly comparable to the sums paid to unskilled urban wage-workers in casual employment (i.e. without long- or indefinite-term contracts). The distribution of incomes in the street occupations was highly skewed, with the majority of workers having incomes below US$3 per day, but with a minority having incomes far in excess of this mean value. On a 'good day', a prosperous street-trader with a large capital stock of merchandise could make US$30 or more in profit, and a skilful street-thief might make much more than that. The success of a few of the most prosperous, skilful, or simply lucky people working in the street occupations, however, should not blind us to the bare economic subsistence of the great majority of the participants in these occupations. Most street occupations are intensely competitive, and improvements in profit levels and incomes tend to lead to additional workers entering, and/or to increased competition from larger, more capital-intensive enterprises based in off-street locations, forcing average incomes down again to their previous level.

As well as low status and low average incomes, the street occupations are characterized by relatively low inputs of capital in relation to labour, and by low 'formal' educational requirements. In 1976–8 most street enterprises operated with a total working capital equivalent to less than US$100 in terms of equipment and/or merchandise. Many porters, watchmen, scavengers, beggars, and thieves incurred no monetary expenditures in order to be able to work, beyond the costs of their normal clothing, food, and transport to a work-place. Even the most elegant and well-stocked street stall was unlikely to have a value of more than US$1,200, and the most expensive small-scale public transport vehicle, the *motocarro*, had a maximum capital cost of about US$560. Basic literacy and numeracy are generally useful to participants in the street occupations, but even these relatively low educational standards are rarely required. Instead, the occupations are characterized by skills learned outside the government educational system, such as hard bargaining, quick-wittedness, manual dexterity, good memory, an engaging personality, and physical endurance. On-the-job experience and effective utilization of social networks are particularly important in the street occupations, together with such difficult-to-alter variables as 'an honest face', physical toughness, or beauty.

The most basic need of the urban poor is an income in goods and/or money to provide for food, drink, housing, clothing, and other necessities. An income may come from government or private charitable institutions, from investments, moneylending and renting, from windfall gains, or from work. For the poor, work is the normal way to obtain an income, and for around 8.4 per cent of Cali's working population in 1977, roughly 29,000 people, some or all of their work

was in the street occupations (Bromley and Birkbeck, 1984: 187–90).[3] Ideally, work should be both enjoyable and rewarding, yielding an income and a sense of personal achievement and satisfaction. Instead, to most people, including those working in the street, work is boring and exhausting, and even dangerous or degrading. Furthermore, work opportunities are usually scarce and insecure, and work is often inadequately remunerated, leading to poverty and deprivation. All of the occupations under consideration here have some of these negative characteristics of work, and together the street occupations form a complex of low-status, poorly remunerated, insecure forms of work. Although many who work on the streets comment that their occupations are less exhausting than heavy manual labour like cutting sugarcane and carrying bricks on construction sites, the street occupations usually require long hours and uncomfortable conditions. The 'curse of Adam' weighs heavily on the urban poor, and most perceived advantages in street occupations reflect even worse conditions in alternative occupations.

Work, Illegality, and Informality

Work is defined here as 'any activity where time and effort are expended in the pursuit of monetary gain, or of material gain derived from other persons in exchange for the worker's labour or the products of such labour'. In other words, work is the labour involved in producing goods and services for exchange, and it is 'income-generating'. The category of 'work', thus defined, excludes the equally important category of 'expenditure-reducing' activities which can be described collectively as 'subsistence labour', for example growing food for household consumption, self-help house construction and repair, unremunerated housework and child-minding, voluntary unpaid help given to friends and neighbours, and walking or cycling to places of work or recreation so as to avoid paying transport fares. Any form of work which is regularly performed by a given person may be described as an occupation.

Under these definitions, such classic lumpenproletarian occupations as begging, prostitution, and theft can all be viewed as work, and hence can be analysed together with the remaining street occupations. The presence of these illegal, disreputable, or public nuisance activities in the category of 'street occupations' serves to emphasize the high degree of differentiation which exists within this category. Of course, illegality and disrepute extend much further than begging, prostitution, and theft, as a few traders deal in contraband, stolen, or falsified merchandise, various forms of street gambling are illegal, and substantial numbers of persons working in transport and gambling services and the overwhelming majority of street-traders do not possess the licences and documents required by official regulations.

Illegality is widespread in the street occupations, and serious problems and suffering may be associated with theft, robbery, conning, and prostitution. Most cases of illegality in the street occupations, however, are either trivial or common to many other groups across the social spectrum. The most frequent complaints against street-traders, for example, come from the owners and managers of large shops and department stores: that they spoil the appearance of the street; that they cause congestion in downtown areas; and, that they pose unfair competition to indoor businesses because very few of them charge sales taxes to their customers and virtually all either drastically underestimate their earnings in income-tax declarations or do not make any declarations at all.

Many street enterprises provide useful services, but do not have some or all of the permits they are required to have, or fail to meet official specifications on receipting, taxation, equipment, uniforms, hygiene, etc. Just like big business corporations, wealthy taxpayers, and real-estate dealers, workers in the street occupations often cut corners and break a few rules here and there in order to make a living or a little extra profit. Complying with each and every official regulation can be costly and time-consuming. For minors and for adults who have lost or never obtained key identity papers, compliance may be impossible. In addition, many of the regulations are rarely enforced, and some officials will accept a bribe to stop enforcement.

Hernando de Soto and his collaborators (de Soto, 1989) have developed a simple, clear concept of 'informality' to explain the socio-legal status of most street occupations, numerous other small enterprises, and self-help housing through squatting and illegal subdivision. Their concept of informality has no relation to the International Labour Office's seven-criterion definition for the formal and informal sectors (ILO, 1972: 6; Sethuraman, 1976), an approach which presumes numerous intercorrelations which have never been empirically demonstrated. Instead, de Soto and his colleagues go back to the Age of Enlightenment and the works of Adam Ferguson, Adam Smith, and Thomas Paine, making a clear and very sharp distinction between 'natural laws' (moral principles that are socially necessary, right, and just) and 'formal laws' (the artificial creations of governments). Because governments can be inefficient, repressive, exploitative, over-intrusive, or simply 'too big', their formal laws are not necessarily socially useful, right, or just. Individuals have rights and responsibilities—to work, to do business, to support their dependents, etc.—and de Soto, following Paine (Bromley, 1994: 134–5), believes that they have a moral duty to break unjust and unreasonable formal laws which prevent them from doing those things. In de Soto's view, most people working in the street occupations are part of a grass-roots uprising against unjust and excessive regulations, starting and developing their own businesses outside the framework of 'formal law'.

Using a simple means/ends criterion, the de Soto approach divides economic activities into three basic groups: formal, where both means and ends are legal; informal, where ends are legal but means are nominally illegal; and, illegitimate, where both ends and means are illegal and/or anti-social. Formal enterprises obey the spirit and the letter of the law. Informal enterprises obey the spirit of natural law but not the letter of formal law; they perform useful functions and provide necessary services, but do not obey every official regulation applying to their activities. Illegitimate activities contravene the principles of natural law—they are anti-social and/or criminal, whether or not they are officially proscribed by formal law. The means/ends criterion is reinforced by a second criterion, social utility, that the people involved and the society as a whole are better off if the law on these activities is broken than if it is obeyed. Hence, 'an activity is informal when it neither produces a deterioration in the social situation nor an antisocial result when the law and the regulations applicable to it are disobeyed' (Ghersi, 1991: 46). This second criterion excludes such morally questionable economic activities as prostitution, child labour, and begging from the category of informality.

De Soto portrays informal activities as manifestations of the vitality and entrepreneurial dynamism of the poor. They break a few rules here and there, but only to support themselves and their dependants, to make a living, and to avoid crime or destitution. He argues that they contribute massively to the provision of services, to the economy as a whole, and to capital investment in housing and commercial real estate. In his view, underdevelopment results from a bloated and overly bureaucratic governmental apparatus and from entrenched mercantilist interests which restrain competition and entrepreneurship. His writings tend to focus on 'informality' by the poor, but the logic of his analysis suggests that the rich and big business may break 'formal' laws just as much as the poor do.

The socio-legal problems that de Soto focuses on are societal problems of enormous importance: the role of the state, the rights and responsibilities of the citizen, and the nature of the legal system. Whether or not we agree with all his views, they do provide some key pointers for the analysis of street occupations and the formulation of policies affecting those occupations. The street occupations are highly visible and often scapegoated, but many of the problems they pose and face are posed and faced by less visible and notorious off-street activities. They are societal problems which should be addressed systematically rather than by scapegoating one occupational group.

Employment Relationships

In this analysis, the term 'work' has a different meaning from the term 'employment'. 'Employment' is used to denote a relationship between two parties, an 'employer' and an 'employee', the former paying the latter to work on the former's behalf for a significant period of time (at least a working day), or for lesser periods on a regular basis. When there is a direct two-tier employer–employee relationship based on some form of contract (an oral or written agreement), there are two main forms of working relationship: 'on-premises working', when the employee works at a site owned, rented, or operated by the employer; and, 'outworking', when the employee works away from the employer, usually in his own home, in the streets, or in some door-to-door operation. An employee may be paid wages per unit of time worked, per unit of 'output', or by some combination of the two. When work is remunerated wholly or partly by the unit of time worked, whether as 'on-premises working' or 'outworking', it is generally recognized as a form of employment. When it is remunerated solely per unit of 'output', however, it is usually only viewed as employment if it is conducted 'on premises'. When conducted off-premises, piece-work is conventionally viewed as a form of self-employment, and this conception is embodied in Colombian labour legislation and in the perceptions of most middle- and upper-class Colombians. In contrast to such views, and in keeping with our definition of employment, 'outwork on a piece-work basis' is viewed here as a form of employment remunerated by a piece-wage. Because it is not officially or widely recognized as such, it may be viewed as 'disguised wage-working', as distinct from the more widely recognized forms of wage-working which take place 'on-premises' or which involve off-premises work remunerated per unit of time worked.

When a worker is not employed by someone else either overtly or disguised through off-premises piece-working, two alternative working relationships are commonly found; 'dependent working' and 'self-employment' (Bromley and Gerry, 1979: 5–11). Dependent workers are not employees and have no fixed margins and commissions, but they have obligations which take a contractual form and which substantially reduce their freedom of action. These obligations are associated with the need to rent premises, to rent equipment, or to obtain credit, in order to be able to work. Although the appropriation of part of the product of the worker's labour is not as clear and direct in the dependent-working case as in the disguised wage-working case, there is normally an appropriation process through the payment of rent, the repayment of credit, or purchases and sales at prices which are disadvantageous to the dependent party in the relationship. In contrast, true self-employment has no such relationships; the workers work on behalf of themselves and other persons that they choose to support. Self-employed workers must, of course, rely on inputs provided by others, on the receipt of outputs by others, and on a system of payment. However, the bases of their self-employment are that they have a considerable and relatively free choice of suppliers and outlets, and also that they are the owners of their means of production. They are dependent upon general socio-economic conditions and on the supply and demand conditions for their products but they are not dependent upon specific firms for the means to obtain their livelihoods.

The employment relationships described above form a continuum ranging from wage-work to self-employment. Two major variables, 'relative stability and security of work opportunities', and 'relative autonomy and flexibility of working regimes and conditions', can be used to divide up the continuum into six major categories whose relationships and characteristics are set out in Figure 8.1. The two extremes of the continuum, described as 'career wage-work' and 'career self-employment', have relatively high levels of sta-

bility and security, and they are not encountered among the street occupations of Cali. The four intermediate relationships have relatively low levels of stability and security, can be described collectively as 'casual work', and are found in varying proportions among the street occupations. The six categories of the continuum are not commonly distinguished, and ambiguous cases will arise as in all classifications, but they do provide a much richer and more appropriate focus for studies of contractual relationships, economic linkages, and changes in employment structure than the more conventional 'dualist' distinction between wage-work and self-employment with no intermediate categories.

In the streets of Cali, self-employment is a much less common phenomenon than might at first glance seem apparent, and there are signs that it is diminishing in significance in the face of the expansion of disguised wage-working and dependent working (Birkbeck, 1978a, 1979; Bromley, 1978; Gerry, 1985; Gerry and Birkbeck, 1981). According to our estimates, only 40–45 per cent of those working in the street occupations were in precarious self-employment,

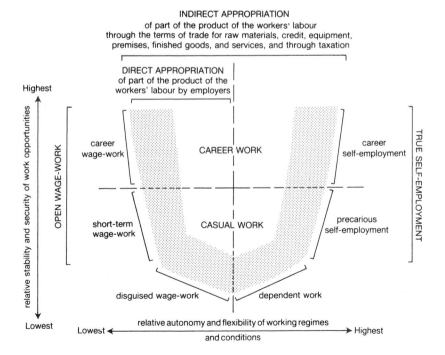

FIG. 8.1 The continuum of employment relationships

39–43 per cent were disguised wage-workers, 12–15 per cent were dependent workers, and only about 3 per cent were short-term wage-workers.[4]

Disguised wage-workers and dependent workers have a variety of obligations to employers, contractors, suppliers, property-owners, and usurers, yet do not have the employees' rights specified in government labour legislation. They are 'disenfranchised workers'. Thus, for example, *motocarro* drivers notionally own their vehicles, but a substantial minority are buying them on hire-purchase, or have borrowed money for the full purchase, so that they are really dependent workers. Similarly, most night-watchmen who work on the streets watch over a single property, or a group of neighbouring properties, guarding the property of the same owner(s) night after night. Though they are officially viewed as self-employed, they are paid a fixed sum by the owner(s) every day, week, or month, and their working relationship is effectively one of disguised wage-working. As a further example, those street-traders who have kiosks or fairly sophisticated mobile stalls usually rent the units or buy them on some form of credit (dependent working), or, as in the case of soft drinks company kiosks and ice-cream company carts, they have been lent these units by companies on the condition that they sell company products on a commission basis (disguised wage-working). Even in the street-trading operations requiring least capital and yielding the lowest incomes, such as door-to-door newspaper selling by small boys and small-scale fruit and vegetable selling in the street markets, dependent working relationships are formed when, because of their own property, the sellers have to obtain the merchandise on credit from other sellers or wholesalers. The condition of this credit is that they must sell that merchandise and pay for it before they can have another lot of merchandise on credit.

The low status and economic dependence of many of the people working in street occupations, combined with the intensely competitive, individualist mentality associated with most of these occupations, contributes to a lack of political and economic organization. The street occupations are moreover fractionalized by the high socio-economic differentiation within and between occupations, low general levels of education, insecurity, instability, official ignorance and persecution, and conflicting loyalties to a variety of different 'masters'. Although trade unions and co-operatives exist, these associations are usually small, unstable, and relatively ineffective. Worse still, they are frequently corrupt and/or personalistic, there are often several different rival organizations within a single occupation, and only a small proportion of the total work-force belongs to any organization. Even among lottery-ticket sellers, the most unionized group in Cali's street occupations, only half the total workers were paid-up members of trade unions or occupational associations in 1977, and they were divided between five different organizations. Only about one-eighth of street-traders were in unions and associations, and they were split between six different organizations (Bromley, 1978: 1167). Among most of the other street occupations, the proportion of workers in such organizations was even lower, and several occupations had either never had a 'formal' organization, for example theft, begging, and knife-grinding, or had only defunct organizations which no longer held meetings or collected subscriptions, for example shoe-shining and garbage-scavenging. Viewing the street occupations as a whole, worker solidarity is ephemeral or non-existent, and group interests tend to be subordinated to individual concerns and ambitions.

The Diversity of Street Occupations

The typology of street occupations in Cali shows considerable variety of types of work. Further variety is emphasized by consideration of the nature of work, and the 'rag-bag' character of the general descriptive category of services.

In classifying work and economic activities, it has become customary to identify three major economic sectors (Fisher, 1933, 1939): 'extraction' (primary production), 'manufacturing' (secondary production), and 'services' (occasionally described as tertiary production). Of these three sectors, the third is by far the most heterogeneous, and many attempts have been made to subdivide it into smaller categories (e.g. Foote and Hatt, 1953; Gottmann, 1970). All the street occupations fit loosely into the service sector, and the diversity of occupations found in the streets reflects some of the diversity in this sector.

The term 'services' ranges from the highest levels of government, education, and research, to such low-status workers as watchmen, porters, and scavengers. The category of services also includes work in such diverse institutions as churches, the fire brigade, the police, and the armed forces, so that the general character of the service sector is really that of the 'everything else sector' after extraction and manufacturing have been separated out from the total range of economic activity. Within the broad category of services, I include begging and property crime, occupations which are usually viewed as 'parasitic' in that the worker serves himself at the expense of the victim (Hirst, 1975: 225). We should not forget, however, that the beggar may provide a service to donors by giving them some moral satisfaction, and that the thief may serve 'fences' by supplying them with cheap stolen goods.

Most people's attitudes to services tend to be coloured by whether or not they consider services to be 'productive', and as a related issue, how they define 'productive'. The word 'productive' has a good connotation, whilst the word 'unproductive' is little short of insulting. In some circles, productive is used only to describe the extraction or manufacture of goods, and in this sense, services are clearly not productive. In other circles, productive is used to describe any activity which 'creates surplus-value' or even simply 'adds value', and there is no distinction between primary, secondary, and tertiary sectors in terms of their innate productivity or non-productivity.[5] In a sense, this is merely a terminological debate. Whether or not services can be produced like goods, some services must be provided, if only to sustain the output of the primary and secondary sectors. No economy could function for long without transport or commercial distribution, and some services are clearly 'essential', whether or not they are defined as productive. In contrast, the production of such superfluous goods as electric toothbrushes and plastic beads can hardly be described as satisfying any basic human need, or as making any major contribution to the economy.

Most street occupations provide some wanted services, as evidenced by the fact that the public is prepared to pay for these services. Of course, we may feel that some of these services, such as prostitution and the sale of gambling opportunities, should not be offered to the public, but we should recognize that they are available off-street as well as on-street, and that even worse social problems may result from their indiscriminate repression. The lessons of the Prohibition Era in the United States are as valid today as they were in the 1920s. Furthermore, of course, the repression of any income opportunity may endanger the livelihoods of its workers, and also the income flows and accumulation patterns associated with it. Thus, for example, a complex set of moral dilemmas faces anyone seeking to formulate policies towards the sale of lottery and *chance* tickets. Not only is it necessary to bear in mind the welfare of the ticket sellers and their dependants, but also the facts that the lotteries are government-organized to generate revenues to finance part of the social-welfare system, and that *chance* is organized by capitalist enterprises paying taxes to local government (Gerry, 1985). Further dilemmas arise when one tries to assess the overall effects of gambling on the gambler and his/her household, and the degree to which the 'ethic of the windfall' implicit in gambling may strengthen individualistic and competitive traits among the poor, reducing the potential for worker solidarity and class conflict.

For those services which are offered both on-street and off-street, the street occupations play an important role by increasing levels of competition in the economy, and hence reducing the likelihood of the formation of oligopolies working against the interests of the consumer. In two price surveys in Cali, for example, street-vendors were shown to sell most basic foodstuffs in small quantities at prices significantly below those of supermarkets, with the sellers in the street markets generally recording the lowest prices in the whole city.[6] Thus, some street occupations contribute towards lowering the cost of living in the city, and particularly towards holding down the prices of foodstuffs for the urban poor, who buy a greater proportion of their food in the streets than better-off social groups. By holding down the cost of living for the urban poor, these street occupations contribute to holding down the costs of wage-labour (see Oliveira, 1985; Williams and Tumusiime-Mutebile, 1978). Street-retailing also plays a role in encouraging consumption, both by selling at relatively low prices and by making items available in a wider range of locations and for longer periods of time on each day of the week. Further encouragement to consumption is given by the cheapness of many transport services, including such small-scale transporters as *motocarros* and handcarts, which enable the consumers to get their purchases back home. It should be evident, therefore, that most of the street occupations are eminently functional to the socio-economic system as a whole, rather than simply 'marginal occupations' of no social or economic significance beyond providing subsistence to those who work in them.

Three street occupations stand out from the others as not clearly offering a service to the general public: recuperation, begging, and property crime. Recuperation as a street occupation is distinct from the municipal collection of refuse, and is oriented mainly towards the recovery of paper, metal, glass, bone, cloth, and other products which can be sold to artisans and industrial firms. As an occupation, it serves manufacturing industry directly by collecting (or, in a sense, extracting) useful materials, and by reducing industrial costs and national imports. The benefits of recuperation can potentially be passed back to the consumer in terms of cheaper manufactured goods and lead even to increased manufacturing employment. Recuperation is not parasitic, but rather symbiotic, to some extent benefiting all parties involved, though providing the scavenger with only a low-income job and rather degrading and frequently unhealthy working conditions. Begging is a very different case in that it is genuinely parasitic, but as there is no victim, it cannot be considered seriously anti-social. Property crime is yet again a different case, being parasitic, having a clear victim, and hence, given that almost everyone is at risk, being decidedly anti-social.

Underemployment: Concept and Reality

In Colombia, any discussion with an academic or civil servant on the characteristics of the street occupations is likely to be cut short by a kindly, 'Ah, yes, you mean underemployment'. All work in low-income services, and often also the work of artisans and peasant farmers, is lumped together as underemployment, and the street occupations are considered the classic examples. The implicit assumption is that the workers are

doing something, but not much, and that they are in some sense 'less employed' that those who work in government, public services, factories, or capitalist agriculture. When one asks what underemployment means, answers are usually tautoiogous or contradictory, revealing the different meanings attached to the popular usage of the word 'employment' as well as general confusion about underemployment. Four main approaches are taken individually or in combinations to the definition of underemployment: that the workers work less hours than they wish to or ought to according to some élite-defined norm; that the workers have very low productivity in terms of the amount of work completed per unit time; that the workers are inadequately remunerated for their labour; and that the workers do not have a 'normal relationship' with an employer, in other words, they are not on-premises wage-employees. The variables embodied in the four approaches are not necessarily mutually correlated, and none applies exclusively to those occupations which are usually described as underemployment.

Underemployment is a decidedly derogatory term mainly used by the upper and middle classes to describe the work of the poor. Those who work on the streets do not usually describe themselves as underemployed, but rather stress the long hours that they have to work to earn a subsistence income, their lack of capital, and the hard, tedious, or dangerous nature of their work. If underemployment simply means low incomes, clearly the street occupations reflect underemployment. If underemployment means low productivity per worker because of low capital investment and intense competition between workers, then again the street occupations reflect underemployment. However, these two criteria for defining underemployment are simply the criteria for defining poverty: low incomes and little or no capital. If, instead, we look at how much those who work in the streets actually work, there is no evidence that they work less than those in most other occupations. Most people working in the streets work rather long hours and a seven-day week. Although some sellers of goods and services spend substantial periods between sales waiting for the next customer, such slack periods are also found in most government and private offices and in many off-street commercial enterprises.

One remarkable feature of many street occupations is, discounting the cost of labour, the high return they can give on very small amounts of capital invested. US$10 loaned to a retailer at the beginning of a day can yield US$15–20 at the end of the day, enabling the retailer to pay back the loan with an exorbitant 5–10 per cent interest, and to keep the balance for his/her own subsistence. Thus, large numbers of workers can make an income above the national minimum wage (which was equivalent to US$2 per day at the end of 1977) with an extraordinarily low capital investment. Capital is used very efficiently, though of course this efficiency is conditional on the low cost and high availability of labour, and upon the donor of credit limiting the number and size of loans so as to be able to supervise the recipients closely and avoid frequent non-repayment.

On-premises wage-working governed by written contracts and official labour and social-security legislation is only available to a minority of the labour-force in Colombia and other Latin American countries. However desirable regulated, protected, on-premises wage-employment and its associated bonus payments, holiday pay, redundancy pay, family benefits, and social security may be, it is necessary for governments to spread their attention more widely so as to cover short-term wage-workers, disguised wage-workers, dependent workers, and the self-employed. Simply to consign all of these forms of working away as underemployment which will eventually disappear is to ignore the majority of the working population and the worst forms of poverty and exploitation associated with work.

The urban poor are loosely tied into a vicious circle of low capital, low training, shortage of remunerative work opportunities, and low incomes. Only a minority with considerable luck, talent, or initiative can break out of this situation, and the success of a minority is often conditional on the relative stagnation of a great majority. The vicious nature of the circle within which the urban poor usually work is accentuated by a variety of exploitative contractual relationships, and by the lack of effective organizations and participatory structures. The application of the term 'underemployment' to a wide variety of the occupations of the urban poor can have serious negative effects, both in lowering the status of the occupations concerned, and in tending to throw the blame for low incomes and productivity caused by poverty upon the poor themselves—in effect, blaming the poor for being poor.

Regulation and Persecution of the Street Occupations

It is not surprising, and in total accordance with the legal system, that clearly illegal activities such as property crimes, trading in contraband goods, and the sale of prohibited gambling opportunities, are persecuted occupations in Cali. Indeed, many complain that these occupations are not persecuted enough. It is also hardly surprising that such occupations as prostitution and begging, viewed as 'disreputable' by most of the population, are officially regulated and suffer from periodic police harassment (see Bromley, 1982). In the case of prostitution, however, official attitudes are decidedly ambiguous, and there are many complaints that organizers of prostitution and upper-class prostitutes are free from harassment, while lower-class prostitutes are frequently persecuted.

The intervention of the authorities in occupations which are not clearly criminal, immoral, or anti-social according to conventional, élite-defined standards, is much more complex and diverse. In general, the concern is to regulate activities by introducing checks and controls on prices, standards, and locations, and by limiting entry to the occupations. Government personnel are appointed to enforce these regulations, and penalties are prescribed for offending workers. Thus, for many street occupations, e.g. the operation of a street stall, the commercial use of a *motocarro*, horse-drawn cart, or a handcart, and shoe-shining, registration procedures have been introduced, and regulations have been made as to when, where, and how these occupations should be practised. Hundreds of pages of official regulations (MSC, 1971; GV, 1976) specify the municipal, departmental, and national government's regulations on street occupations, and substantial bodies of police and municipal officials are expected to administer these regulations. In reality, however, these regulations are excessively complex, little known, and ineffectively administered, resulting in widespread evasion, confusion, and corruption.

An example of corruption associated with the regulation of the street occupations is the case of the Calle Trece-Bis street-market, close to the main shopping area of central Cali (Bromley, 1978: 1170). This street-market functioned without official permission and against the wishes of virtually all higher level municipal officials from 1972 until July 1978, and it was active again for several years after October 1978. When the market was functioning, the traders made a daily collection so as to gather together the necessary funds to pay off each shift of police patrolling the area, a routine bribe to enable the market to continue its activity. In the period from July to October 1978, the street-market was temporarily 'eradicated' in a very determined municipal campaign to control street activities in that section of the city, an area bisected by the inner-city ring-road, and highly visible to passing tourists and rich *Caleños*. After a change of municipal administration, however, the political will for 'eradication' was sharply reduced, and despite the hardships caused by the recent persecutions, most of the traders returned to their previous locations.

In general, those who administer law and order on the streets complain that there are so many people working in the street occupations, and that there is such widespread ignorance and disrespect for the official regulations, that controls must be very selective. The main objective is usually 'containment': to hold down the numbers of people working in the streets of such priority areas as the central business district, the upper-class shopping-centres and residential zones, and the main tourist zones.

Although there are occasional cases of assistance by the authorities to street occupations, as when help was given in improving street stalls and providing uniforms for street-vendors and shoe-shiners at the time of the Pan-American Games in Cali in 1971, official intervention in the street occupations is essentially negative and restrictive, rather than supportive. The basis of government policy is that 'off-street occupations' should be supported in the hope that they will absorb labour from the street occupations. However, this objective has not been achieved because insufficient investment funds have been mobilized, and because investment has concentrated in areas which generate relatively few work opportunities. In the mean time, the street occupations have often been persecuted, and opportunities to improve working conditions in these occupations have generally been neglected.

Street occupations conflict strongly with the prevailing approaches to urban planning. Although Cali has a warm, dry, and congenial climate for economic activities and social interaction in the open air, city planners have usually been concerned with reserving the streets for motorized transport and short-distance pedestrian movement, and concentrating economic activities into buildings. In general, no special provision has been made for street occupations, and restrictions on *motocarros*, non-motorized transport, and the sale of goods and services in the streets have been imposed to reduce traffic congestion, road accidents, and the incidence of these occupations. Cali is officially twinned with Miami and has strong links with other North American cities. Urbanistically, Cali is being planned along North American lines as a sprawling automobile-dependent metropolis, and the street occupations are, from the planners' point of view, an unfortunate embarrassment.

Women and Children in Street Occupations

Among the urban poor, conventional official definitions of 'labour-force' and 'economically active population', based upon the idea that neither children nor housewives earn an income, are simply irrelevant. When personal incomes are low, when the membership of households is often unstable, and when many households with children are headed by a single, separated, divorced, or widowed woman, there is a strong pressure on all household members to seek work opportunities. Work is a form of personal security as well as a contribution to the household budget. Instability and insecurity of work and income opportunities are endemic among the urban poor (McGee, 1979; Rusque-Alcaino and Bromley, 1979), and reliance on only one breadwinner increases the risk of disaster. An adult male breadwinner may be the victim of theft, arbitrary arrest, or the eradication of job opportunities (Cohen, 1974), or he may choose to abandon family responsibilities and to spend heavily on tobacco, alcohol, drugs, or gambling to escape a depressing reality.

Despite all these pressures for female involvement in the labour-force, about 70 per cent of those working in the street occupations in Cali are males, at least three-quarters of these falling into the 18–55 age-range. Around 15 per cent of those working in the street occupations are aged under 18, about three-quarters of these being boys.[7] Some occupations, for example virtually all work in transport and security services, shoe-shining, and most forms of street theft, are almost exclusively male preserves. In general, therefore, males and adults predominate in the street occupations, though females and children are numerically quite significant. Only prostitution is almost exclusively a female preserve, though women predominate in many forms of retailing, particularly the sale of fruit, vegetables, and cooked foodstuffs. Child workers are mainly concentrated in scavenging, newspaper-selling and other small-scale retailing, shoe-shining, and petty theft.

In general, women and children are especially concentrated in the least remunerative or lowest-status street occupations, and have less access to capital than men. Particular occupations are age- and sex-specific, and although this division of labour may at times be convenient, it mainly acts to reduce the range of work opportunities and the potential income available to women and children. Women and children working on the streets are much more liable than adult males to be subjected to physical threats and sexual harassment, and this factor sharply reduces the number of girls and young women in the street occupations. Children working in the streets have great difficulty in obtaining and keeping any significant capital equipment or merchandise for their occupation, and most young women without access to significant capital are aware that prostitution is potentially their most remunerative form of work.

Conclusions

This rapid summary of the characteristics of street occupations, based on the example of Cali, has emphasized the diversity of these occupations and the impossibility of applying a uniform set of policies to all street occupations. These occupations deserve a greater degree of attention and respect than they have received, and it is necessary to convert the predominance of negative policies to a predominance of more positive measures. Most potential improvements to the working conditions in street occupations are relatively inexpensive, and, as de Soto (1989) has argued for Peru, some would actually save government money by reducing the number of regulations and the costs of enforcing these regulations. There is no reason, therefore, why a 'humanization' of the street occupations should hold up vital investments in agriculture, manufacturing, and public services. An improvement of the working conditions in the street occupations will not greatly increase the number of workers in these occupations and lead to accelerated rural–urban migration if appropriate investments are made in all regions of the country, involving small cities and rural areas as well as the major metropolises.

For substantial numbers of the urban poor, working in the streets is a survival strategy. Their occupations fit de Soto's profile of 'informality'—performing useful functions and providing necessary services, but contravening numerous official regulations in the process. In many cases, such street occupations are the most viable alternatives to parasitic or anti-social occupations, or to destitution. Their negative features are reflections of much wider social *malaises* which cannot be resolved simply by regulating and persecuting the street occupations. Gross poverty and social inequality are institutionalized in Colombia, and it is unreasonable to blame the poor for their own situation and to fail to tackle the conditions which underlie their poverty.

There is a growing awareness that our concern to increase production should be tempered by periodic consideration of what we are producing, and for whose benefit. Thus, for example, the work of slaves or serfs, or even of low-paid wage-workers, to enrich their employers, is not necessarily productive for society as a whole, but only for a privileged minority. So-called unproductive work may be desirable if it contributes to the welfare of members of society other than the worker, or even if it simply contributes to the welfare of the worker without prejudicing the welfare of others.

Many of the street occupations, and particularly those concerned with public transport and food retailing, are important to the functioning of the urban socio-economic system. Indeed, in some cases where significant capital investments are required or where official controls are exercised over-zealously, a shortage of service provision may lead to increased costs and inconvenience for the consumers of those services. In Cali, for example, there are serious deficits of public transport and food-retailing facilities in some sectors of the city, and these deficits raise living costs and reduce the numbers of work opportunities available. More generally: the removal of street-traders would encourage price speculation in the shops and supermarkets and might reduce the sales of some agricultural and manufacturing enterprises; the removal of street newspaper-sellers would severely damage the sales of the press and would bankrupt some of the newspaper companies; the abolition of street sales of lottery tickets would reduce government revenues and lead to the closure of several social-welfare institutions; and the elimination of recuperation would lead to increased imports and higher costs for Colombian manufacturers.

A few street occupations, of course, are decidedly anti-social and/or parasitic, and are obviously undesirable. For the moment, however, hardly anyone would claim to be able to eliminate criminal or immoral activities, and some presence of these occupations is probably an unavoidable evil. Anti-social and/or parasitic occupations can be controlled, and efficient control may reduce their incidence, but there is no evidence that they can be eliminated. Any hope of their disappearance must depend on general improvements in social conditions and the creation and improvement of alternative work opportunities, and not simply on the repression of those involved.

In summary, therefore, working in the streets encompasses a very wide range of service occupations, almost all of which can be described as survival strategies for those who work in them, many of which can be described as necessities for those they serve, and for the efficient functioning of the contemporary national economy, and a few of which can be described as unavoidable evils. In policy terms, the most urgent requirement is to generate more work opportunities both outside the street occupations, and in the more necessary street occupations, so as to divert manpower from anti-social and/or parasitic occupations, and to improve the general range of income opportunities available to the urban poor. It is also necessary to adopt a more positive series of policies towards those street occupations which are not clearly anti-social or parasitic, simplifying rules and regulations and administering them more equitably, providing work-places and sources of credit and training for workers, and encouraging the formation of workers' organizations without co-opting them into the web of governmental paternalism. Unfortunately, of course, there can be no assurance, either in Colombia or in most other Third World countries, that the sort of government which would adopt such policies will assume power in the near future. In such circumstances, increased worker consciousness and the mobilization of those working in the street occupations are crucial to press for more favourable government policies and to develop alternatives to the present dependent links with employers, contractors, suppliers, usurers, and the owners of sites and equipment.

Notes

1. The author is indebted to Chris Birkbeck and Chris Gerry for critical comments on an earlier draft. The research was financed by the UK of Overseas Development and was conducted in association with the Servicio Nacional de Aprendizaje, Regional Cali. The views expressed here, however, are exclusively those of the author and do not imply the agreement of any British or Colombian institution.

2. In public transport, for example, there is a sharp distinction between the vehicles used in small-scale transport, none of which cost more than US$560 new in 1977, and those used in larger scale transport, none of which cost less than $4,000 new, and most of which cost over $20,000 new. This distinction is paralleled by the much heavier involvement of government regulation, subsidies, and credits, and of capitalist companies employing wage-labour in the larger scale forms of transport (Birkbeck, 1978*b*).

3. Estimates of the numbers working in the street occupations are based on a combination of different sources and research methods: official registers, particularly of transport vehicles and *chance* sellers; street counts at different times of the day, week, and year in all concentrations of people working on the streets, and in sample residential neighbourhoods; estimates given by knowledgeable individu-

als, particularly the more experienced workers, union leaders, administrators, and policemen; and, calculations derived from estimates of supply, demand, and/or turnover. Estimates were made for each occupation, and then added together to produce a general total for the city.

4. These estimates are based on the same sources as the estimates of numbers of persons working in the street occupations, though they have a lower level of accuracy. Most of those workers who are in true self-employment work on their own, but a few are involved in partnerships and a very small number actually employ others as wage-workers.

5. For a useful review of these issues, see Hirst's (1975: 221–30) discussion of Marx (1969: 155–75, 387–8, and 399–401); see also Birkbeck (1982).

6. The main survey was conducted on Friday, 3 Sept. 1977, and the other survey was held on the following day. A 'shopping basket' of basic foodstuffs was purchased in a wide variety of different sectors of the city and types of establishment, according to a prearranged sampling scheme. All purchases were made by the same person, Carmen Rosario Asprilla, and the author is indebted to her for her assistance with the survey. She is a relatively poor black person who has been resident in Cali for several years. Once the purchases had been made, all products were weighed and details of quality were noted.

7. Estimates of the proportions of males, females, adults, and children in the street occupations are based on the results of the street counts mentioned in n. 4 and on general observations at points of concentration of the street occupations. The main count was made on the afternoon of Tuesday, 13 Sept. 1977, but further sample counts were made at different times of the day and night, on all of the different days of the week, and in all of the months of the year.

References

Anderson, Stanford (1978), *On Streets* (Cambridge, Mass.: MIT Press).

Birkbeck, Chris (1978*a*), 'Self-Employed Proletarians in an Informal Factory: The Case of Cali's Garbage Dump', *World Development*, 6: 1173–85.

—— (1978*b*), 'Small-Scale Transport and Urban Growth in Cali, Colombia', in William M. Denevan (ed.), *The Role of Geographical Research in Latin America*, Publication 7 (Muncie, Ind.: Conference of Latin Americanist Geographers), 27–40.

—— (1979), 'Garbage, Industry, and the "Vultures" of Cali, Colombia', in Ray Bromley and Chris Gerry (eds.), *Casual Work and Poverty in Third World Cities* (Chichester: John Wiley), 161–83.

—— (1982), 'Property-Crime and the Poor', in Colin Sumner (ed.), *Crime, Justice and Underdevelopment* (London: Heinemann), 162–91.

Bromley, Ray (1978), 'Organization, Regulation and Exploitation in the so-called "Urban Informal Sector": The Street Traders of Cali, Colombia', *World Development*, 6: 1161–71.

—— (1982), 'Begging in Cali: Image, Reality and Policy', *New Scholar*, 8: 349–70.

—— (1994), 'Informality, de Soto Style: From Concept to Policy', in Cathy A. Rakowski (ed.), *Contrapunto: The Informal Sector Debate in Latin America* (Albany, NY: SUNY Press), 131–51.

—— and Birkbeck, Chris (1984), 'Researching Street Occupations of Cali: The Rationale and Methods of What Many would Call an "Informal Sector Study"', *Regional Development Dialogue*, 5 (2): 184–203.

—— and Gerry, Chris (1979), 'Who are the Casual Poor?', in Ray Bromley and Chris Gerry (eds.), *Casual Work and Poverty in Third World Cities* (Chichester: John Wiley), 3–23.

Cohen, D. J. (1974), 'The People Who Get in the Way', *Politics*, 9: 1–9.

de Soto, Hernando (1989), *The Other Path: The Invisible Revolution in the Third World* (New York: Harper & Row); English translation of *El Otro Sendero* (Lima, 1986).

Fisher, A. G. B. (1933), 'Capital and the Growth of Knowledge', *Economic Journal*, 43: 379–89.

—— (1939), 'Production, Primary, Secondary and Tertiary', *Economic Record* (June): 24–38.

Foote, N. N., and Hatt, P. K. (1953), 'Social Mobility and Economic Advancement', *American Economic Review*, 43: 364–78.

Gerry, Chris (1985), 'Wagers and Wage-Working: Selling Gambling Opportunities in Cali, Colombia', in Ray Bromley (ed.), *Planning for Small Enterprises in Third World Cities* (Oxford: Pergamon), 155–69.

—— and Birkbeck, Chris (1981), 'The Petty Commodity Producer in Third World Cities: Petit Bourgeois or Disguised Proletarian?', in Brian Elliott and Frank Bechhofer (eds.), *The Petite Bourgeoisie: Comparative Studies of the Uneasy Stratum* (London: Macmillan), 121–54.

Ghersi, Enrique (1991), 'El otro sendero o la revolución de los informales', in Gustavo Marquez and Carmen Portela (eds.), *Economía Informal* (Caracas: Ediciones IESA), 43–64.

Gottmann, Jean (1970), 'Urban Centrality and the Interweaving of Quaternary Activities', *Ekistics*, 29 (174): 321–31.

GV (Gobernación del Valle) (1976), *Código de Policía del Valle del Cauca* (Cali: Imprenta Departamental).

Hirst, Paul Q. (1975), 'Marx and Engels on Law, Crime and Morality', in Ian Taylor, Paul Walton, and Jock Young (eds.), *Critical Criminology* (London: Routledge & Kegan Paul), 203–32.

ILO (International Labour Office) (1972), *Employment, Incomes and Equality: A Strategy for Increasing Productive Employment in Kenya* (Geneva: ILO).

McGee, T. G. (1979), 'The Poverty Syndrome: Making Out in the South-East Asian City', in Ray Bromley and Chris Gerry (eds.), *Casual Work and Poverty in Third World Cities* (Chichester: John Wiley), 45–68.

Marx, Karl (1969), *Theories of Surplus Value*, i (Moscow: Foreign Languages Publishing House).

MSC (Municipio de Santiago de Cali) (1971), *Compilación de Disposiciones Legales sobre Urbanismo, Saneamiento y Otras Materias* (Cali: MSC).

Oliveira, Francisco de (1985), 'A Critique of Dualist Reason: The Brazilian Economy since 1930', in Ray Bromley (ed.), *Planning for Small Enterprises in Third World Cities* (Oxford: Pergamon), 65–95.

Rusque-Alcaino, Juan, and Bromley, Ray (1979), 'The Bottle Buyer: An Occupational Autobiography', in Ray Bromley and Chris Gerry (eds.), *Casual Work and Poverty in Third World Cities* (Chichester: John Wiley), 185–215.

Sethuraman, S. V. (1976), 'The Urban Informal Sector: Concepts, Measurement and Policy', *International Labour Review*, 114 (1): 69–81.

Williams, Gavin, and Tumusiime-Mutebile, Emmanuel (1978), 'Capitalist and Petty Commodity Production in Nigeria: A Note', *World Development*, 6: 1103–4.

9

Displaced by Modernity: The Saga of a Jakarta Street-Trader's Family from the 1940s to the 1990s

Lea Jellinek

A PASSION for hot lemon-juice and fried prawn-crackers brought me by chance to Sumira's stall.[1] I loved sitting there deep into the night, watching the pedestrians go by. She parked her stall in the heart of the city, beside a water fountain opposite Merdeka (Freedom) Square. In the mid-1960s and early-1970s there was nothing unusual about street-stalls and traders in the centre of Jakarta. They were patronized by most of the city's population, rich and poor. The streets were alive with people passing to and fro. Most went on foot. Many travelled by bicycle or *becak*.[2] Only government employees and the very rich went by car.

Sumira's Flourishing Trade

Sumira started her day's work at 7 a.m. with a walk to a nearby market. She bought the foods she needed for her stall from the same few traders every day. Her survival as a trader depended on her ability to manage her modest finances very carefully. Her prices had to be competitive and profit margins were small. Most of what she earned from one day's trading had to be used to buy raw materials for the next day. The more she spent on herself, the less she could

buy for her stall. There was little room for error. Sumira watched the market-traders keenly. She could accurately assess the price of her purchases before they were weighed. She began by offering a price she knew to be too low. Traders responded with prices they knew to be too high. On most occasions a compromise was soon reached and both would seem happy with the deal. Sometimes Sumira would hold out for a price which a trader refused to accept, or he would demand a price which she thought excessive. Sumira would move on to the next market-trader, who would have witnessed the unsuccessful negotiations, and start bargaining all over again. No one harboured ill feelings, though it was all in deadly earnest. Like Sumira, the market-traders also depended on the daily earnings for their next day's survival.

Sumira returned from the market in a *becak* laden down with fresh fruit, live chickens, and the other goods she had bought. The household was soon busily engaged in food preparation. They chopped and peeled and fried for the next few hours. Sumira's tiny house was taken over by delicious odours and baskets of banana fritters and fried chicken. The cramped and primitive kitchen did not seem to limit the output.

Cooking was completed by 3 or 4 o'clock in

This account grew out of my relationship with Sumira, whom I first met in 1972. She introduced me to her neighbours, and my research of her community evolved gradually.

the afternoon. The food was carefully stacked into Sumira's cart which, loaded to the brim with the day's cooking, was then pushed to Sumira's trading site. Sumira operated her stall from 5 o'clock in the evening until the early hours of the morning, serving her prepared food or cooking afresh as required. After 1 a.m. Merdeka Square was reduced to the glow of kerosene lamps from the few night-traders, the bell-sounds of passing *becaks*, and an occasional pedestrian. Before dawn Sumira's tarpaulin was taken down and neatly folded and any left-over food carefully put aside for the next night's trade. With the help of her husband and another assistant the cart was packed up and pushed back home for storage near Sumira's house. By 4 a.m. she was asleep. Three hours later she would have to awaken to start the new day's work. Without her two-hour siesta in the early afternoon, while Tati the second wife did the cooking, Sumira would never have been able to endure these long hours of work.

Like many of her neighbours Sumira was a migrant to Jakarta. When she first arrived in the city in 1948, the population was about 800,000. Twenty years later the population had increased fivefold to 4 million. Migrants poured into the city at a rate of 100,000 per year.[3] Sumira came from Semarang, a small town on the north coast of Central Java. Her father was a clerk in the railways, her mother the daughter of a local headman.[4] Sumira was lucky to be able to attend school for six years and learned to read and write. Her introduction to trading came during the tumultuous years of the Japanese occupation (1942–5).[5] Sumira's father had lost his job and her mother set up a stall in their home to earn money to feed the family. Sumira married in Semarang and then with her husband, Yanto, migrated to Jakarta, where relatives had assured them that there were greater opportunities.

From his wartime contacts with the Japanese army and the Allied forces Yanto had acquired many new skills. In Jakarta he tried his hand at *becak* repair and, when that did not do well, he turned to repairing radios. In between he tried a variety of jobs, including casual labour and trade. At the time they were living with relatives, and Sumira once again assisted in the running of the household's food-stall. By the early 1950s they were able to rent a room of their own. Sumira left the family stall and took work as a

ward assistant in a hospital. After a few months she found a better-paid position weighing ingredients for prescriptions in a privately run pharmacy.

After moving several times in the early 1960s, Sumira and Yanto finally obtained a home of their own. It was a small bamboo hut with two rooms and a thatched roof. During the dry season it had an earthen floor but that soon turned to mud in the monsoon rains. Wire-mesh windows helped to keep out the hands of passing strangers but not the noise and cooking smells from neighbouring households.[6]

The amenities of the *kampung*[7] were primitive. The narrow winding paths which separated the tightly packed shanties from one another turned to quagmires in the wet season. Rubbish accumulated in the doorways. The only toilet facility was the nearby Cideng canal which flooded several times in each monsoon with subsequent outbreaks of disease. Two wells originally provided drinking-water but later they became polluted and drinking-water had to be brought into the *kampung*. Vendors squeezed along the narrow pathways carrying cans on long shoulder-poles.

A few years after moving to the new house, Yanto failed to come home every night, explaining his absences in one way or another. Sumira eventually discovered that he had taken a second wife. She was devastated and began to feel insecure. To become self-supporting, she decided to start trading by herself. Her business prospered over the next year or two and she was in need of additional manpower. By then Yanto was womanizing again. Tati, the second wife, came to Sumira in sorrow and anger not knowing what to do. Sumira invited Tati to live with her and help her trade. Feeling desolate and betrayed Tati accepted gratefully and a seemingly improbable partnership ensued.[8]

Yanto liked to have many wives. They provided security and gave him prestige in the eyes of other men. When one wife did not satisfy his needs, he found another and then another, even if this jeopardized the stability of his families. In this way he had a choice of sex partners, accommodations, and food. His wives worked and gave him support. Women by contrast, such as Sumira and Tati, relied on their children or neighbours for support in old age. From a young age, they were brought up to be more responsible and hard-working than men.

Sumira's and Tati's enterprise thrived. In 1968 Yanto, who temporarily had been brought to heel, built them a push-cart. This enabled Sumira and Tati to offer not only the cooked foods they were now selling but a variety of new items including fresh bananas, cigarettes, beer, and of course the hot lemon-juice that attracted me. The business went so well that by 1974 Sumira was able to rebuild her house. She was one of the first in her neighbourhood to do so. Sumira's renovations transformed the house into a two-storey structure with solid brick walls, a tiled floor and roof, glazed windows, and a large, carved wooden door. Her old house had always been open but Sumira's new door was frequently slammed shut.[9] A water-pump was installed so that the members of Sumira's household no longer had to queue with the neighbours at the communal well. As Sumira prospered, the members of her household began to withdraw from the surrounding community.

By 1975 the house was well stocked with consumer goods. The front room housed a battery-powered television set and a large radio. Four plastic chairs and a small coffee-table seemed to take up most of the remaining space. Along one wall there was a new sideboard, decorated with plastic flowers and plastic fruit and full of crockery. Toys were scattered everywhere. Sumira's kitchen contained a variety of recently acquired pots and pans and a new kerosene stove. An imported Chinese kerosene pressure lamp replaced Sumira's old, locally made one. Electricity had entered the *kampung* and by paying a small fee Sumira was able to obtain an illegal connection from a neighbour. This enabled her to install a fluorescent light in the bedroom. There was an ice-chest, a thermos, a cupboard, and two wardrobes full of valuable batik cloths. Upstairs, overlooking the roofs of neighbouring households, family members now slept on beds with mattresses instead of the bamboo mats they had formerly rolled out on the floor.

As Sumira prospered, new members were progressively added to the household. Yanto surreptitiously took a third wife, Nani, who lived in a village 30 kilometres from Jakarta. Sumira, whose two children had died when still young,[10] agreed to the marriage when she was promised a child of that union. A year later, Nani duly handed Sumira her first child. Sumira called the baby Rahmini and brought her up as her own child, although the adoption was never legally formalized. Sumira's mother came to live in the house to look after Rahmini, while Sumira and Tati continued to trade. The two of them were now often working up to 20 hours a day, so Sumira invited people from the third wife's village to assist in preparing food for the stall.

Where once only three people (Sumira, Tati, and Yanto) lived in the house running the stall, there were now eight. Yanto, who was frequently away in the third wife's village, increasingly felt that pushing the cooked-food cart was below his dignity. He persuaded Sumira to search for a neighbour to do the job. Sumira recruited a young man called Achmad. Yanto seldom helped in Sumira's enterprise after Achmad was hired. The fact that he spent an increasing amount of his time in the third wife's village annoyed Sumira greatly. As the senior wife, Sumira felt that he should spend more time with her. Yanto nevertheless continued to draw an income from the stall, even in his absence.

Nani's fellow-villagers viewed Sumira as a rich townswoman and she felt compelled to live up to that image. She held a large festivity in the village as Rahmini's birth approached and another after she was born. The celebrations cost her at least two months' earnings but they were not the only occasions on which she financed festivities in the village. Once, for example, she hired a film and movie projector. She marked another celebration by contributing a live goat and festive meal.

The growing demands of kinsfolk, family, and friends, both in the village and in Jakarta, imposed a considerable financial burden on Sumira. During each Lebaran (the major Islamic celebration after the fasting month of Ramadhan) she bought members of her household new clothes and cooked elaborate meals. During Idul Adha (a festival which emphasizes communion with the poor) she donated substantial sums of money to the local mosque and contributed to the cost of a goat that was slaughtered in the festivities. Rahmini's snacks cost as much in a day as the entire household spent on a meal. Previously, Sumira's household had modestly eaten the left-over food from her stall, but now Sumira bought extra delicacies such as bread, fresh fruit, cakes, and ice-cream. New luxuries such as soap, toothbrushes, toothpaste, and shampoo appeared beside the water-pump. Dishes were no longer cleaned with the ash used by *kampung* dwellers but with detergent. Sumira and Tati had always sewn their own clothes, but now Sumira hired a woman from the *kampung* to make clothes for them. Each festive season she paid a neighbour to repaint her house.

Declining Fortunes

As Sumira's aspirations and living expenses began to soar, the future of her stall, and thus of her income, became increasingly uncertain. Signs that her trade was doomed became more evident. The central city area was rapidly being transformed by new, wide, and ever-busier roads and the construction of numerous multi-storey offices, banks, and hotels. Sumira's community prospered from this building boom. Many labourers in the neighbourhood worked on construction sites, while other members of the community sold food or provided transport to workers engaged on the projects. But the community was gradually engulfed by the city it helped to create. In 1974 Sumira was directly affected, for the footbridge across the canal to Jalan Thamrin, the main highway running north–south through the city, was dismantled and a high steel fence erected in its place. *Kampung* dwellers now had to make a long detour to get to the city centre.[11] Sumira and her assistants had to find back routes away from the traffic to get to their trade location. On several occasions Sumira's cart was struck by speeding cars. The vacant plot of land beside Jalan Thamrin where Sumira and many other traders had stored their carts was fenced off for the construction of a multi-storey office block. As available land at the city centre became scarce, it became increasingly difficult and expensive for Sumira to find a place to store her cart.

Step by step, Sumira's stall was forced out of the central business district. She had started her trade in front of Sarinah department store less than five minutes from her home. After the first anti-trader campaigns in 1970,[12] she had moved north along Jalan Thamrin to Merdeka Fountain, where I first met her. To avoid encountering the police clearance teams, she

became a night-trader. Her main clients were the soldiers and policemen who guarded the nearby offices, government buildings, and banks. Some of her customers were involved in the trader-clearance campaigns and they warned Sumira when it would be unsafe to set up her stall at Merdeka Fountain. As she provided a good service at a time and place where it was needed, her clients protected her against harassment. She in turn gave them generous helpings of food. While other traders went bankrupt, Sumira was able to survive and even thrive.

This happy state of affairs could not last. By 1975 the anti-trader campaign was being so firmly enforced that not even Sumira dared venture out to her normal location near the fountain. She had to be content with a more secluded spot along a back street between a number of banks and a military depot. As a result of her move she lost many of her valued customers. However, the move was not the only factor undermining her trade. Sumira began to find herself trapped in a cost–price squeeze from several directions. Her rising transport costs are a good illustration of this dilemma.

Sumira had always used a *becak* to ferry her from the market-place to her home when she was loaded down with her early morning shopping. Later she also used a *becak* to ferry her to her stall in style. Sumira used the same *becak*-driver each day. He was content to be paid with a meal from her stall. But just as the government was pursuing its anti-trader campaign, it was also implementing a ban on *becaks* in the streets of central Jakarta.[13] As these policies came into effect, Sumira's driver could no longer operate in safety. Instead Sumira had to transport her goods from the market by *bajaj* (motorized *becak*) which were twice as expensive as *becaks*. Moreover, their drivers would not accept payment in kind.

Throughout Jakarta, the old markets of crowded muddy passages and makeshift stalls were being replaced by multi-storey concrete buildings. The stall-holders in the old markets had paid only nominal site rates, usually illegal taxes pocketed by local officials. They could hold their prices down because their overheads were minimal. As the markets were organized and rebuilt, additional fees were levied for the rent of kiosks, rubbish collection, the maintenance of law and order, and market administration.[14]

These additional charges were passed on to buyers like Sumira, who in turn should have raised their prices. As the anti-trader campaigns progressively eliminated traders from large parts of Jakarta, those who were displaced re-clustered together in a few locations, near markets, major crossroads, and railway stations until they were dispersed again. Competition became intense. Despite rising costs, traders undercut one another in a desperate bid to attract customers.

By 1976, Sumira and her fellow street-traders were also competing with an increasing number of office canteens and small cafeterias which could buy in bulk and undercut their prices. Sumira's customers were increasingly reluctant to pay for a rickety seat at her roadside stall when, for a little more, they could dine elsewhere in greater comfort. Most of the offices which lined Jalan Thamrin set up their own canteens, so that employees no longer had to seek cheap food on the street. The government even launched a campaign in the newspapers and over television and radio to advise people against eating from street-stalls which were said to be unhygienic. They did not conform with the city's 'modern' appearance and were accused of cluttering up the streets and hampering the flow of traffic. The military men who still patronized Sumira's stall were forbidden to eat there after a canteen was set up in their depot. Sumira believed that this directive was given to ensure the success of the army canteen which was run by the wife of one of the senior officers.

Competition from off-street cooked-food outlets was reinforced by the changing tastes of a growing middle class, which benefited from the economic prosperity of the late 1960s and early 1970s. Soldiers and policemen who had once been content to eat at stalls like Sumira's became reluctant to do so. To be seen eating on the street was an embarrassment. Kentucky Fried Chicken, McDonald's, and more modest Indonesian-style cafés were increasingly preferred to roadside stalls.

In an almost self-defeating effort to maintain her trade Sumira held her prices constant but lowered the quality and quantity of the food. She offered more rice, less meat, and fewer vegetables. She no longer used the best quality 'Saigon rice'. Her customers noted the decline in quality of the food and complained. But there was little

Sumira could do. Between 1972 and 1976 the number of her customers dropped sharply. In 1972 she had served over 100 military men, police, drivers, and lower-level civil servants each day. By 1976 she served no more than 50.

In the face of ever-increasing difficulties, Sumira somehow continued to trade. In late 1976, however, Tati, the second wife, ran away. It was a major blow. The quality of the food at Sumira's stall was almost entirely due to Tati's skills as a cook. And Tati was not just a skilful cook. She was also by far Sumira's most industrious worker. She was responsible for most of the food preparation. After spending some hours in Sumira's hot, steamy kitchen, she helped Achmad push the loaded cart to its trading-site while Sumira travelled to the site by *becak*. Tati then crouched on the ground behind the stall preparing food or washing dishes, while Sumira chatted to customers and handled the money.

Stalls like Sumira's rely on cheap family labour. Tati had received food, clothes, and accommodation but hardly any pay. As Sumira's stall prospered Tati felt embittered by her lack of a share in the profits. When Sumira hired Achmad to help Tati push the cart to its trade location the two of them conspired to run away and set up a stall of their own. Before their plans could be brought to fruition, however, Achmad was expelled from the household for flirting with Tati.[15] Less than a year later, Tati fled. Sumira thereby lost the backbone of her enterprise. Sumira's aged and arthritic mother took over Tati's role as cook and Sumira looked for another man to help push the cart. She hired a young man who was recommended by a neighbour. A month later he absconded, taking Sumira's radio and kerosene lamp with him. Sumira could not push her cart to her trading-site unaided. Without an assistant, preferably a man, she could not work.

After searching for many days, Sumira recruited another young man, Narto, whom she found sleeping on the streets. He had come to Jakarta from a village near Semarang, Sumira's birthplace, but he had no work or place to stay. He was delighted with the prospect of work and more than satisfied with the accommodation provided. Narto proved an able and willing worker, and Sumira started to trade again, full of hope that this time her luck would hold. She must have known, however, that her problems were far from over.

Almost a year earlier, some of Sumira's military protectors from the nearby army depot had warned her that their base was to be moved to the outskirts of Jakarta to make way for a highway. They suggested Sumira move to their new location, for they were sure she would fall victim to anti-trader raids if she stayed in the central city area. Sumira toyed with the idea of moving but decided to struggle on with the difficulties she knew rather than face the risks of the unknown. Within the year the soldiers moved to their new base. Sumira realized almost immediately how vulnerable she had become, for she was ordered to move within a few days after the soldiers had left. When she tried to defy the order, she was threatened with a fine and the loss of her cups and plates. A week after restarting her trade, she was again ordered to leave and realized that this time she would have to go.

The next few days were a nightmare. She and Narto pushed the food-stall along endless roads in a vain search for a new trading-site. All the sites they came to were either occupied or subject to frequent anti-trader raids and too dangerous to use. Their search seemed hopeless. Sumira sat down and wept. They moved on till finally they found a vacant site. Sumira dared not leave her cart for fear that it would fall prey to an anti-trader raid. The two of them worked all day and slept on the roadside by their cart at night. Narto was used to life in the open but Sumira found it a strain. They had no toilet or washing-facilities. But they worked on. One day Sumira's mother and Rahmini came to ask for some housekeeping money. Sumira burst into tears. In the past 24 hours she had only earned the equivalent of 8 US cents.[16] That was as much as Rahmini had spent on a single snack in their better days. When I first met Sumira six years earlier she had earned one hundred times that amount in a day.

Sumira's managerial skills seemed to have deserted her.[17] The food she sold catered to nobody's tastes. She was serving an odd mixture of coffee, eggs, and rice. Eggs and coffee were too expensive for the poor, who could only afford rice, vegetables, and tea. Rice and eggs were too modest for the better-off who wanted meat and fruit. Potential customers passing by her stall were not impressed by the range of food and simply moved on to another vendor.

As her trade and livelihood disintegrated, Sumira seemed to lose her hold on life. While her stall went well, she thought nothing of working 20 or more hours a day. Now as her plight became increasingly desperate she lost the will to work at all. Everything seemed futile and hopeless. Whatever she did went wrong. She forgot to pay the monthly interest on some batik she had deposited with a pawnbroker and forfeited US$11 worth of cloth.

Sumira's neighbours, whom she had shunned in her better days, rallied to her cause. Some, who had weathered the collapse of their own stalls, advised Sumira to dispose of her cart and become a mobile trader. Carts were hard to store, difficult to push and almost impossible to move out of the way of a police raid. With the small sales Sumira was now making, it seemed pointless retaining her cart and incurring a charge for storage. Regretfully Sumira sold her cart to a neighbour for US$40 and obtained more than enough capital to establish herself as a mobile trader.[18]

With lots of advice and help from her neighbours, Sumira set about cooking food that would find a ready demand and that she and Narto could carry. On her first day she slipped and fell and her basket of freshly cooked food was upturned in mud. It was an inauspicious beginning. But Sumira and Narto persisted. Rahmini realized that something was seriously wrong, even though she was only 6 years old at the time. She began to feel insecure and insisted on accompanying Sumira as she worked. One day as Sumira set down her load on the edge of a little alleyway near Hotel Indonesia, her plates slipped and rolled down an embankment into a murky canal. Sumira clambered down to retrieve her plates. Rahmini began to cry hysterically, apparently fearing that Sumira would drown.

Sumira had been unable to pay Narto for some weeks. Narto realized that it was becoming increasingly hard for Sumira even to feed him. Rahmini could carry the pot of tea or extra baskets of food that Sumira needed. Sadly he said goodbye and promised he would return to his parents' village near Semarang. A year later Sumira heard he had gone back to sleeping on the streets. He was too embarrassed to return home destitute. Sumira never learnt if he had found another job, returned home, or perhaps, died destitute in Jakarta. Achmad, her former assistant, was found dead under one of Jakarta's bridges.[19] Sumira was told he had become ill and desperately poor, though no one was sure what caused his death.

Despite her setbacks Sumira continued to trade. Without her cart she was certainly more mobile. She went from street to street seeking a place where she could stand with her child and sell her few wares. But wherever she went, other traders told her to move on. Every location it seemed was already taken. Traders were *sympathetic to Sumira's plight*, but it was a cut-throat world and each trader had to look after his/her own interests. Some had known her in years gone by as 'the successful trader by the fountain'. They suggested other places to try, sending her from one street corner to another. There was said to be a potential market of labourers here, or drivers there. Eventually Sumira found a location on a small street to the north of Hotel Indonesia. She found out what to cook and when to sell by asking the mini-car drivers and labourers who worked nearby. They warned Sumira that her predecessors had been cleared from the very site that she was proposing to use. She realized she would have to exercise great care. Sumira put considerable effort into her cooking and her trade started to develop. She could not afford to relax, however, for the threat of an anti-trader raid was ever present.

Sumira did not know any of the police at her new site. She had lost the network of informers who could warn her of an impending raid. The local police still tried to extract bribes, often demanding a packet of the most expensive cigarettes she stocked. There was now nothing more costly in her wares that they could demand. Even the loss of a packet of cigarettes was more than she could afford. She realized that, if she gave in, they would return again and again. Their friendship did not seem worth cultivating even if she could have afforded it, for she felt the police would not help her when she was in need. Instead, she had to rely on her own cunning. She soon learned that the area around Hotel Indonesia was cleared whenever there was an important conference. All the drivers and traders in the area took a keen interest in any forthcoming conference and speedily passed on any information they acquired. Sumira and her fellow-traders stayed away whenever an important meeting was scheduled

so they would not tarnish Jakarta's image in the eyes of any foreign dignitary who happened to speed past in his chauffeur-driven Volvo or Mercedes-Benz. If she did encounter the odd policeman, she soon learned to exploit her daughter's presence quite shamelessly. She would explain that she was only trying to support her little girl. Rahmini, in turn, soon became quite adept at bursting into tears as soon as Sumira gave her the appropriate cue. They became an accomplished duo.

Sumira earned a greatly reduced income, about US$47 gross a month,[20] from her food selling and learned to live frugally again. She battled to make some headway and counted every rupiah she spent. It became increasingly and painfully apparent that Yanto and his third wife, Nani, were an expense she could no longer sustain. From the beginning Sumira was critical of Nani and on several occasions demanded that she leave Jakarta and return to her village.[21] But life in Nani's village had become impossible. It had been a bad year and there was not enough to eat. There was no money in the village to pay for such luxuries as Yanto's radio repairs or Nani's washing of clothes. To make matters worse, Yanto was ill. He was coughing up blood and had lost weight. He had tuberculosis, though of course he had not seen a doctor. Even if he had found the money for a consultation, he would not have been able to pay for any medicines that might be prescribed. Nani had given birth to a second child but she realized she could not afford to keep it. Yanto and Nani begged Sumira to adopt their child but Sumira sadly explained that she was in no position to do so. The baby died, presumably of malnutrition, less than 12 months later.

Sumira felt trapped. She understood why Yanto and Nani were spending more and more time with her. When she had earned well, she had welcomed Yanto's visits. Now it was a question of her survival. But how could she convince her husband and his third wife that she could no longer support them? Sumira decided that her only course was to cease trading and dispose of all her possessions. Yanto and Nani would then realize that Sumira's position was as desperate as theirs.

In June 1979 Sumira stopped trading. She progressively sold off her possessions and her household lived off the proceeds. By word of mouth it became known through *kampung* brokers that she had items to sell. Offers were made through intermediaries for her wardrobes, sideboard, cooking utensils, and eventually, even her television set, which she was most reluctant to let go. Yanto, of course, demanded a share. Behind her back he tried to sell some of her possessions and pocket the money.

Sumira's household economies were far-reaching. She rented out half her house, severed her electricity supply, and reverted to the use of kerosene. To save on fuel she rarely cooked and used only one lamp, so the room was dimly lit. Her house fell into disrepair. The well gave forth just a trickle of murky water. Its pipes had rusted but Sumira could not afford to replace them. Sumira no longer asked the vendor or a servant to bring drinking- and cooking-water to her house. She collected it herself from a kilometre away where, if she queued, she could get it free of charge. All these measures enabled Sumira to survive without working for about six months. The household became spartan by any standards, let alone those of her recent past. Gradually Sumira's strategy began to work and Yanto and Nani's visits became a rarity.

Less than four years earlier Sumira had been one of the richest people in her community, now she was one of the poorest. Her neighbours felt pity for her. One regularly brought Sumira's mother and daughter porridge from the hospital where she worked. Another provided them with the flavoured iced-water Sumira so loved to drink. A third delivered rice and tidbits of food left over from her cooked-food trade. Another invited Sumira, Rahmini, and the grandmother to watch television. A spiced-vegetable trader paid Sumira's mother a small sum for babysitting, while she went out to trade. When she had money Sumira had alienated many of her neighbours by her individualism, consumerism, and snobbery. But now in her difficulties, she sought comfort and support from those around her and they responded to her needs. But Sumira found it hard to admit to others just how far her fortunes had declined. Once as she was struggling to carry her household's daily water supply a *kampung* dweller who had known Sumira in her better days asked what had become of her servant. Sumira replied she had given her temporary leave to return to her family in the village.

Like Sumira's neighbours I found it difficult to

be critical of her when her fortunes were down. When she was doing well, she was sometimes strong, arrogant, and mean. She made use of the people around her for her own benefit. Even if her personality basically remained the same intense suffering had mellowed her and brought out the softer, gentler sides to her character. She was more introspective, more sensitive to other people's hardships and sorrows. As she was no longer busy trading 20 hours a day, she had more time to sit and talk to her neighbours and find out about their problems.

Her months away from trading and the constant fear of a raid started Sumira thinking. If she took up trading again she could, at most, look forward to a modest income if everything went well. She could earn as much if she had a regular job and she would then be spared the uncertainties and insecurity of her past existence. Sumira asked me to help her find a job.

Even though she was not a good cook and had never worked in a Western household, I was able to find her a job amongst my expatriate friends. At first she was overwhelmed by the prosperity of the household where she worked. She could not get over the spaciousness of the house or the size of its garden. Her entire *kampung* neighbourhood of several hundred people would have fit in the garden. Yet only six people lived there. Sumira was amazed by the enormous refrigerator and the pantry stocked with tins of imported food for months in advance. Like other *kampung* dwellers Sumira had always bought just enough food to last for the day. Despite her urban background, she had never used a telephone or seen a shower, a Western-style toilet, or a swimming-pool. The unlimited supply of hot and cold running water in the three bathrooms and kitchen was a novelty. The automatically opening car-port door was a source of astonishment. Her employer's noisy teenage children gulped down bottles of milk and gorged themselves on uncounted mouthfuls of chocolate cake. Yet, she observed, when a little sugar or spice was missing from the pantry there was inevitably an uproar. More than anything else, Sumira was amazed to see the family's children shout at their parents. The family was wealthy and secure, but Sumira noted that did not seem to make them any happier or more contented with their lot.

Though Sumira had at last found a measure of

security at least for as long as her employer remained in Jakarta, she felt imprisoned behind the high walls of the expatriate compound. She missed the ceaseless activity of her *kampung*'s crowded lanes, the familiar faces, the gossip, the smell of *kampung* food, and above all, her family. Sumira worried about leaving her aged mother and young daughter to fend for themselves. Her job required her to live where she worked. She was given one day off a week and that was her only opportunity to see her family. Sumira's mother was in her late sixties and had become accident-prone. She had fallen down the steep rickety ladder in their two-storey home and broken an arm. On another occasion she had spilt boiling cooking-oil over her leg.

Rahmini, now 8 years old, was left to care for her grandmother. When Sumira had first left home to work as a cook, Rahmini had cried bitterly. Without telling anyone she fled alone to the third wife's village to find her father. Then she played truant from school and was rumoured to be roaming the streets. Her grandmother could not control her. Sumira was concerned that Rahmini could not yet read or write. Once a week when Sumira returned home, she tried to tutor her daughter but her efforts were in vain. Sumira worried about Rahmini's future. She feared it would be worse than her own. While her trade had prospered, she had dreamed of her daughter becoming a doctor. Now her aspirations were much more modest: 'If only Rahmini could learn to type or sew'. Sumira felt sadly that Rahmini was unlikely to get as good an education as she herself had had 40 years earlier under the colonial administration. Sumira would not be able to rely on Rahmini to support her when she was old.

As a domestic servant in 1979, Sumira was paid in cash about one-fifth of what she had earned as a food-trader in 1972, if inflation is taken into account.[22] If was just enough to feed herself, her daughter, and her mother. She did obtain medical care, clothing, accommodation, and some food for herself and occasional presents for her daughter and mother, but she was cut off from her family and her roots, and forced to live in an alien and lonely world. She longed for the interaction with people, her freedom, and the independence of petty trade. The job as a cook had its attractions, however. It was less arduous than trading. Sumira was nearly 50

years old and she had to think of her future. Her new position offered her a sense of security. She would have a job so long as she did her work properly. She was no longer at the mercy of soldiers and the police, of rising prices, and a diminishing clientele. Despite the drawbacks of her new job, the lack of freedom, and the realization that it offered little scope for material improvement, life as a domestic gave her economic security. For Sumira that was worth a lot.

Involuntary Resettlement

Sumira's peace of mind did not last long. In 1981 the government announced that her neighbourhood was to be demolished. The government was proposing to replace the houses with apartments, which would be made available to the inhabitants of the *kampung*. Was this the same government that had banned their *becaks* and outlawed their petty trade? The *kampung* was full of households like Sumira's who had lost their livelihoods as a direct result of government policy. Was the government now going to help them? There were families in the *kampung* who had been the subject of earlier government demolition programmes. Sumira herself had once been forced to move as a result. Everyone knew that the government forcibly evicted *kampung* dwellers and either claimed that the *kampung* was illegally sited—and thus not entitled to compensation—or made such paltry payments that they scarcely covered the costs of transporting one's belongings to a new site, let alone buying another home. Fear and rumour swept the *kampung*.

For Sumira now isolated from her community the news was especially alarming. Once a week on her day off she made her way to her home wondering each time what she would find. Would her house still be standing? How would her infirm mother or her 9-year-old daughter face up to officials when they came? But the officials seldom came and the neighbourhood had little firm information to go on. Each time Sumira returned there was new speculation and new fears. The government held a public meeting at which it outlined its proposals. But the various officials contradicted each other. The brochures they distributed did not seem to agree with what they had said. The confusion and the apprehension increased.

The government proposals were in fact confused. The government announced its intentions before it had finalized its plans or properly defined its aims. The government planned to re-house Sumira's neighbourhood in subsidized apartments. The earliest announcements, however, mentioned prices for the proposed apartments that were beyond the financial capacity of all but a handful of the inhabitants. Moreover, many *kampung* dwellers used their houses to store and prepare materials for their livelihood, just as Sumira had prepared and cooked food in her home whilst she ran her stall. But the government was going to prohibit its apartments being used for commercial purposes. To its credit, the government did substantially modify its proposals to try to accommodate the objections of the *kampung* dwellers. It offered generous compensation and alternatives for those who did not want to accept an apartment.[23] Unfortunately, it never overcame the suspicion of the community. People observed that the proposals continued to change no matter what was announced and believed that the junior officials with whom they had to negotiate were enriching themselves at the *kampung* dwellers' expense.

I encouraged Sumira to take a government apartment. It seemed too good an opportunity to miss. The apartments would be located in the central-city area where Sumira's house now stood. Those who lived in the neighbourhood could buy them at a subsidized rate. Repayments on the units could be by monthly instalments at very low interest rates over a period of 20 years. Each unit would have running water, electricity, and gas. When Sumira's employer departed for overseas, she would readily find domestic work within walking distance of her home in the élite suburbs of Menteng or Kebayoran. If she carefully placed the compensation she received for her home in a bank, she could pay off the apartment with the interest she earned.

Sumira did not see things my way. She had never entered a bank, let alone used one. They and the officials working within them were impersonal and intimidating. Most importantly, interest payments were against the teachings of Islam. According to Sumira, the usurer would be strangled by snakes in hell. Sumira longed to use the compensation for her house to buy the con-

sumer items—radio, television, and sideboard—she had been forced to sell. Rahmini wanted a bicycle like the one promised to the children next door upon their parents' receipt of compensation. Sumira's mother wanted a gold ring like the one Sumira had borrowed and pawned when her trade was going badly. Yanto insisted on his rightful share, saying that he planned to better equip his radio repair business in the village. Sumira had long-standing debts which needed repayment. For example she still owed US$76 on the renovations made to her home in 1974.

Sumira agonized over her options. How would she pay for the monthly charges for water, gas, and electricity? Previously she had paid for kerosene and water daily as the need arose and she had money to spare. When times were bad, she simply cut down on the quantities she used, or accepted these items on credit until she could pay. But the government was not offering such flexibility. Failure to pay on time would entail a fine. Minimum charges would be levied even if no water, electricity, or gas were used. She wondered how the instalments for her apartment would be paid if she became too old or too ill. Would she be thrown out of her unit? Would the government carry out repairs free of charge as it promised or would Sumira be forced to pay for them? If Sumira and her household were assigned units on the fourth floor how would her rheumatic mother, or even Sumira when she was older, clamber up so many flights of stairs?

But Sumira made her wager. She overcame her Islamic scruples about interest payments by making a donation to an orphanage. With my help and encouragement she banked most of her compensation payment. The interest on this was almost enough to cover her repayments on the apartment. Some of her neighbours were not so wise and embarked on an orgy of consumption. They bought television sets, motorcycles, and clothes. Vast fortunes were spent in a few days. In other instances, men ran off with the money, leaving their wives and children penniless.

After a long and difficult transition period in appalling temporary accommodation, Sumira, her mother, and Rahmini moved into a government apartment in April 1984. Although Sumira had asked for a ground or first-floor unit, she was allocated an apartment on the fourth floor. Yet the household was so relieved to move out of the temporary accommodation that not even

Sumira's elderly mother complained of the four flights of stairs. As Sumira had feared, the bills for electricity, gas, and water soon began to prove a burden. Even though the apartments were subsidized, the mortgage repayments and the charges for services represented a threefold increase over Sumira's expenses prior to demolition. They consumed half of her income. On her modest and fixed wage it was going to be hard enough meeting her monthly commitments. What would happen if she had an unexpected medical bill or if Rahmini needed a new school uniform? Sumira could only pray that she would manage somehow and that they would not have to move again.

Yanto never saw the new apartment. Sumira received word of his death just as her old home was demolished. It took her two years to save enough money for a trip back to the village to see Yanto's grave and a further two years before she could pay for his funeral rites.

Sumira's mother died less than a year after the family moved to the apartment. The loss affected Sumira badly. She felt intensely sad and isolated. Now she and Rahmini were alone in the world. Her mother had stood by her when Sumira was in great difficulty. She could have gone to stay with her other daughters who were all comfortably off but she chose to stay by Sumira who, she felt, needed her moral support through marital difficulties and financial hardship. She had remained when the household moved to temporary accommodation while they waited for over two years for their apartment. The three of them had to squash together in a cubicle eight metres square. The overcrowding, lack of amenities, and dust must have been a nightmare.

After her mother's death, Sumira left her job to look after Rahmini. She felt she could not leave Rahmini unsupervised throughout the week whilst she lived and worked elsewhere. Rahmini was now 12 years old and more frequently on the streets than at school. She was terrified of staying in the apartment alone and kept running away to neighbouring houses.

Instead of earning an income from domestic work, Sumira felt the two of them could live off some of the money in the bank. But Sumira gradually became more and more depressed. She was now spending all day in her apartment. Most of her neighbours had managed to cover the bare concrete walls and floors of their new accommo-dation with colourful tiles and paint. They had bought new furnishings, curtains, and blinds. Sumira felt embarrassed. Only she and Rahmini lived in what looked very much like a gloomy, empty prison cell.

Sumira decided to withdraw all her compensation money and renovate her house. She had the walls plastered and tastefully painted in creamy white. The floors in the kitchen, bathroom, and guest-room were tiled. I was aghast, wondering how she would survive. Sumira was no longer working, yet she had spent all the money I had persuaded her to bank on renovating her apartment. How would she pay for her gas, electricity, water, and the instalments for the apartment? How would she feed herself and her daughter? I deeply regretted that I had ever meddled in her life. How could I, a secure middle-class Australian, hope to understand the values, the dreams, the pressures, and especially the unpredictability of the life of a Jakarta *kampung* dweller?

But as my doubts grew, Sumira became more convinced that she had made the right choice. She was delighted with her renovated apartment and, from the beginning, had appreciated the convenience of running water, drainage, and electricity. She explained she was now able to cover all her overhead costs by subletting a room. She argued that she would never have been able to attract a sufficiently high rent without renovating. Subletting was illegal but Sumira—and her many neighbours who were making ends meet in the same way—simply registered her tenant as a member of her household.

Apartment Life

Ibu Sumira died at the age of 56, only two years after moving into the apartment. First her toes were cut off and then her leg was amputated at the knee. For two years she had been complaining of hot flushes in the head and unpleasant numbness and pins and needles in her feet. She had difficulty sleeping and throughout the night poured water over her head and legs to stop the numbness and pain. I had assumed that the previous decade of hardship—the loss of her trade-stall, demolition of her *kampung* home, death of her mother, and premarital pregnancy of her daughter—had taken its psychological toll. But, she was physically ill and nobody was telling, because diabetes is viewed by *kampung* dwellers as cancer is viewed in our society.

Rahmini admitted that when Sumira was in the hospital, she was about to give birth and could not bear to be close to Sumira because of the stench coming from her gangrene. Only Mimin, a bold and determined *kampung* neighbour, visited Sumira regularly before she died, alone in the hospital. The hospital had refused to treat Sumira, until I sent money. Only Mimin was prepared to hunt for Sumira in the Morgue when her dead body was left amongst many others. Mimin rummaged through the bodies, turning them over and pushing them aside to find Sumira. A lesser person could not have done what Mimin did—retrieve a body from amongst the dead. She washed and buried Sumira without her leg. A day earlier Rahmini had been asked to take the leg home and bury it, but could not bring herself to do so. Eight years later Rahmini's main concern was that Sumira's grave was still only marked by a piece of wood with 'Sumira' written across it. The proper Islamic ceremonies guaranteeing Sum's safe passage into the future had not been performed and the grave was likely to be dug up and used for somebody else's burial.

Still childlike, Rahmini also looked the bosomy mum as she lovingly fondled her 2-year-old son, Gugun. After giving birth, she had put on weight and used less make-up. She still had smooth velvety skin, a boyish face, and short-cropped hair. At times she returned to being the helpless child with shy downcast eyes and weak baby voice. She wore an attractive fawn jump suit with mauve jacket, gold earrings, necklace, and bracelet. We sat in the airless room at the back of Sumira's apartment on an imitation Persian carpet, next to the colour television set and fan. Gugun's plastic car stood in the doorway. Rahmini had decorated the walls of her room with some dolls, ornaments, and artificial oranges. The family clothes were neatly stored behind a zip-up plastic wardrobe.

The four yellowish-white walls of the apartment stared down at us. Rahmini and I had little to say to each other. Life in the apartments seemed empty compared to Sumira's life in the *kampung* of two decades earlier, when Rahmini

was born. I remembered the colour, and the hustle and bustle, with neighbours beavering away at many different activities. People walked, talked, and sat about. Dramas were constantly played out on the streets. There was never a dull moment. People in every household seemed to have something to do—dresses to sew, food to cook, ice-cream to make, water and kerosene to deliver, herbal medicines to prepare and sell. In the apartments, by contrast, people sat about with blank faces and looked bored. But Rahmini did not want to return to the *kampung* of her childhood. She was still haunted by the fear of fire which had burnt one of the houses. She did not want to return to carting water each day as Sumira had done, or having to use kerosene for lighting and cooking, or going to the toilet over the canal. She was a proud apartment-owner with only two more years to pay off the instalments on her unit.

When Sumira was taken to hospital, the local headman had taken all the papers relating to Sumira's apartment. Two of Sumira's friends—Mimin and Babe—forced the local headman to return these documents to Rahmini. They advised Rahmini to hold on to Sumira's apartment and view it as a valuable resource. They helped her find tenants and taught her how to organize her money. Maybe she had also learnt something about finance from watching Sumira trade.

Half of the *kampung* dwellers who had accepted apartments along with Sumira in 1981 had been forced out of them by 1991. They had ended up selling their units to Golden Truly, a powerful supermarket chain run by one of Jakarta's wealthy conglomerates. It was not clear why the supermarket chain wanted to get hold of the apartments—perhaps to house its own employees and then to knock them down and build a multi-storey car park, office blocks, and luxury apartments. Despite their feelings of hostility and fear towards the supermarket, many *kampung* dwellers patronized its shelves for their daily needs and when they had difficulties with payments for their units, they were enticed to sell to Golden Truly.

When Wawan was temporarily out of work, Rahmini tried to trade in the apartments. At Golden Truly, she bought chicken pieces laid out on neat plastic trays covered in glad wrap with computerized price tags. It was a far cry from the days when Sumira went to the open air market, bargained energetically with her regular suppliers, and eventually bought a couple of live chickens which she carted home by *becak*, killed, washed, plucked, cooked, and sold from her semi-mobile stall on the streets of Jakarta. Rahmini liked the convenience of the Golden Truly chicken pieces. She prepared what looked like Kentucky Fried Chicken for the neighbouring children. Like Sumira, she liked to have an entourage and recruited children to do the cooking, selling, and eating. They ran in and out of Rahmini's apartment all day. The trouble was that she spent US$5 buying the chicken and only received US$3 from her sales. Subsequently, Rahmini tried, every so often, to trade different types of cooked food but she was never successful.

Wawan had turned out to be a loyal and loving husband. When Rahmini first met him he had been a wild street boy. Rahmini now complained that he was too loyal. Every night he returned from work and fondled their baby. Wawan's mother's attempts to wean her son off Rahmini and find him another partner had failed. Even before Rahmini's marriage, Sumira had not liked Rahmini's mother-in-law because she refused to contribute any money for the marriage. The mother-in-law had dreamt of better prospects for her son, but in 1986, just before her death, Sumira had insisted that Rahmini marry Wawan because he had made her pregnant. According to Javanese tradition, marrying, having the baby, and then divorcing gave Rahmini and the child sufficient respectability. Even though Sumira felt unwell and was short of funds, she managed to raise enough cash to organize a modest wedding. Eventually the mother-in-law came good with US$60 for the celebration. Sumira died ten days before the birth of Rahmini's baby and the baby died shortly after birth from a navel-cord infection.

For a number of years after Sumira's death, Rahmini lived with her mother-in-law while renting out her flat. She felt that these years, from 1987 to 1991, were the worst years of her life. For the first time in her life, Rahmini had to cook, clean, and wash clothes. Rahmini was lonely and wanted a child but her mother-in-law insisted that as long as Wawan was not working, they should not have another baby. She forced Rahmini to use contraception. Wawan, a tall, thin, handsome boy with naïve face and flowing

black hair had dropped out of school. His father, a driver for a Japanese firm, promised to find him work. In the meantime, Rahmini and Wawan tried to earn some income by selling boiled vegetables in peanut sauce. In 1991, Rahmini stopped using the spiral and became pregnant. In that year, Wawan's father at last found his son a job as a driver with a Japanese firm.

As soon as they could afford it, Rahmini and Wawan moved out of the mother-in-law's house and back to Sumira's apartment. Rahmini continued to rent out individual rooms while living in the apartment. Babe, an elderly handyman and former admirer of Sumira who lived across the corridor, rebuilt Rahmini's 42-square-metre apartment so that it looked like a dormitory. A narrow corridor ran through the middle with partitioned rooms to either side. The kitchen, bathroom, and toilet at the back were shared. Rahmini rented two rooms to four young men who worked at the Hilton Hotel. She took the back room for her family. Seven people, and later ten, lived in the 42-square-metre unit. Rahmini partitioned the back room where she slept with her husband and baby with a wardrobe, providing a small space for a mattress for her real mother, Nani, her younger brother Ramli, and her sister Sana. At the time, I thought she was generous letting her mother, brother, and sister stay there.

Nani had had seven children by Yanto, four of whom survived. Rahmini was the oldest. Another had been adopted by Nani's younger sister in Parung. The two youngest lived with Nani in Rahmini's apartment. They were bright-eyed, eager, intelligent children, longing to learn. They had a soft, shy seriousness that reflected a difficult past. They had never been spoiled like Rahmini even though they shared the same mother and father. Rahmini had spent her first six years amid the economic boom of Jakarta with four adults—Sumira, Tati, Sumira's mother, and Yanto—who gave her everything. By contrast, her younger brother and sister had been brought up 30 kilometres from Jakarta in a bamboo hut without any luxuries. While they were small, Yanto was aged and sick. In 1982 he died of tuberculosis. Nani earned a bare livelihood as a washerwoman, begging work from relatives and neighbours and occasionally trading snacks or cigarettes.

Uncaring Relatives

Rahmini had only found out that she had been adopted at six years of age when Nani had visited Sumira in the *kampung* and occasionally worked for her: neighbours had told Rahmini 'that's your real mother'. Rahmini adopted Sumira's view of Nani as a stupid village woman who did the menial chores. Nani's only resources were her hands. Although she should have been paid US$2.50, the people in the flats were only prepared to pay her US$0.50 for an hour's massage. In Rahmini's apartment, she scrubbed clothes and floors. She prepared food. She went next door to earn extra money washing clothes. On one occasion, she had washed clothes for a neighbour for months and then suddenly the neighbour left without paying. She had been hoping to obtain money for her two younger children's schooling. Nani could earn US$15–25 a month from washing clothes and giving massages.

In August 1994 Nani complained of difficulty breathing, headaches, and dizzy spells. Over the upper part of her breast was a large hot red swelling. For some time she had known she was ill but had not been able to do anything about it. She had gone for a chest X-ray but could not pay the necessary US$12 and been turned away. I gave Nani the money for an X-ray and she was diagnosed as having a breast tumour. The tumour had to come out and the operation would cost US$200. Rahmini said she did not have the funds. Rahmini blamed Nani for not having treatment earlier when the swelling was smaller and the treatment would have cost less. Rahmini's new colour television set, fan, and carpet could have been sold to help pay for the operation, but Rahmini insisted that she was still paying for these items on credit and they could not be sold. The television set cost US$223 and the money had been borrowed from her mother-in-law. Each month Rahmini had to pay her mother-in-law an instalment of US$20. Rahmini worried that if she failed to pay, her mother-in-law would try to cause problems with her marriage. The carpet was also being paid off at US$9 per month and the money for that had been borrowed from Babe. When Rahmini left the room, Nani commented, 'She values her possessions more than my health'.

To see if she could get support for the operation from her seven brothers and sisters, Nani went back to her home village. She returned to Rahmini's apartment saying that everything had been arranged. Each sibling would contribute US$15 which would make up the necessary US$100 for treatment by a traditional healer in the village. Nani planned to return to the village to convalesce and gain strength for the operation. She and her children would live with one of her siblings and Rahmini had agreed to pay a monthly allowance of US$15 for food and the children's education.

Two weeks later, Rahmini and I went to check on how Nani was doing. We were returning to the village where Rahmini had been born and circumcised and where Sumira had held several festivities in the early 1970s. Sumira had been the rich town-trader coming with gifts from the city. Rahmini, dressed in an elegant outfit, playing the part of the prosperous apartment-owning urban-dweller reminded me of Sumira, but she came empty-handed.

The differences between town and country had almost disappeared. All the way from Jakarta to Parung we travelled on highways and main roads, amid hooting traffic—cars and buses churning out hot grey fumes. Cement and brick houses lined both sides of the road. Gone were the empty spaces, the fruit trees, villages, rustic market with traders selling fruits and vegetables. The village had grown in all directions over paddy-fields and through bamboo thickets. Away from the main roads, houses built of brick or bamboo, or both, were crammed one on top of the other along narrow paths. It was hard to find the huge holy tree where I had accompanied Sumira in 1973 to pray for a new two-storey house, and again in 1976 to pray for a restaurant.[24]

Nani was nowhere to be seen. Each relative laughed at the suggestion that they were going to pitch together to help with Nani's treatment. The richest sister's house was decorated with a white terrace and Italianate columns. A new luxurious sofa with built-in pillows filled the guest room. The sofa cost US$250 and was being paid off at US$7 per month. The sister pointed to her seven children, most of whom were still at school and costing her money. She could not give her ill sister money but could offer her and her two children the occasional meal. In a neighbouring house, Nani's brother and his wife complained that they too were short of funds. He lacked capital for his goat trade. They could offer Nani a room to sleep in but not cash. The same story was repeated in each of the relatives' houses.

Nani had been turned away by each of her relatives and was living in a bamboo hut on the edge of the village facing a large overgrown Chinese grave. The relative who offered accommodation, had offered it at a rent of US$10 per month. Nani preferred to rent a room for US$7 per month in a modest bamboo hut from a poor unrelated widow. She preferred the kindness of a stranger and bamboo walls that breathed rather than the solid brick houses of her more prosperous relatives. She was offended by their refusal to help with her medical treatment and therefore refused to accept their offer of food or accommodation. Her children had not enrolled in the village school because they lacked the US$5 entry fee. Nani's son went to the goat market each day and tried to raise US$0.25–0.50 per day for food.

The poorest but kindest of Nani's relatives, her younger sister who had married one of Nani's husbands and adopted Nani's third daughter, allowed Nani and her two children to seek refuge in her house during the day but could not provide any financial assistance. Her husband had another wife and six children in Depok, a suburb on Jakarta's outskirts, but every two days visited Nani's sister—his second wife—in Parung where he rented a house for her.

Two years later in 1996 I found Nani lying on the floor of Rahmini's flat in Jakarta barely able to breath. She had arrived from the village an hour before, hardly managing to climb off the bus and make her way from the crowded bus station across the polluted city and up the four flights of stairs to Rahmini's flat. Upon arrival, she collapsed on the floor with an aching head, burning stomach, and hot swollen chest. Nani had come to Rahmini to ask for money for medical treatment. Despite what Rahmini had told me the relatives in the village were not providing Nani with the medical care she needed. The last batch of injections they had bought her were rotten and had to be thrown out.

Over the past two years Nani had survived in the village from washing clothes for neighbours when she felt sufficiently strong and from her son's sale of plastic bags in the market. Her daughter, Sana, worked as a baby sitter for a family in the village where she received accommodation and food but no income. Even though Nani was seriously ill, she still worked. She had

no option. She rented the same old bamboo house in the village for US$20 per month and needed another US$15 per month for food.[25] Rahmini claimed that she regularly sent her mother money, but when she left the room, Nani admitted that Rahmini's monthly payments arrived only sporadically.

In one small corner of Rahmini's flat—where Nani and her two young children had formerly slept—was a gleaming large new white fridge, a shining new rice storage unit, and an elegant dappled blue ceramic water dispenser. Nani said she had been forced out of Rahmini's apartment by these possessions. Rahmini explained that 'the refrigerator was part of Wawan's inheritance. Wawan's mother liked shopping and had given each of her children their share of their father's superannuation in the form of household appliances. She liked to give them consumer goods so that their houses were full—not empty—and thus not an embarrassment in front of their neighbours'. When she was ill two years earlier Nani believed she had been forced to leave Rahmini's flat because Rahmini, and perhaps Wawan's family, feared that she would lay claim to the flat as the oldest surviving member of Sumira and Yanto's family. Nani and Rahmini's relationship, the broken thread between mother and daughter, was symbolic of what was happening to relationships throughout the city.

Like the poorest in the city and the village, Nani was willing to share what little she had. When I gave Nani a few tablets to assist with her many pains she asked whether she could put some tablets aside for her youngest sister in the village who was suffering from the same illness—tuberculosis. The younger sister had probably caught the illness from Nani by sharing the same plate. Nani's stomach looked unnaturally bloated. In between difficult breaths, she explained that she ate fattening pills and applied make-up so that she did not look so unwell. Her illness represents the Indonesian society around her—desperately ill but covering it up: form concealing reality.

Notes

1. In earlier accounts I referred to her as 'Bud'. Sumira was her real name, but I had not dared use

it when writing her story for fear that the author-ities might harm her for being my friend and telling me about life in the *kampung*. Now that she is dead, she can no longer be hurt, and I feel that she deserves her own name, not the silly 'unJavanese' name I gave her. Also, I used to refer to Tati as Nanti, to Yanto as Santo, to Semarang as Tegal, and to Nani as Ade, while I referred to Rahmini, Achmad, and Narto by their proper names.

2. The *becak* is a light tricycle peddled from behind with seating for two passengers in front under a collapsible roof.

3. See Jakarta City Government, *Gita Jaya* (Jakarta, 1977), 97; Planned Community Development Ltd, *Kampung Improvement Programme Jakarta, Indonesia* (Washington, DC: IBRD and IDA, 1973), Annex 1, 5; J. Heeren, 'The urbanization of Djakarta', *Ekonomi Dan Keuangan Indonesia* (Jakarta: University of Indonesia) 8: 11 (1955), 700.

4. Some details of Sumira's life changed as I came to know her better. Earlier I had understood her father was a policeman. See L. Jellinek, 'The Life of a Jakarta Street Trader', in J. Abu-Lughod and R. Hay, Jr. (eds.), *Third World Urbanization* (Chicago: Maroufa Press, 1977), 252. Later, she told me he was a clerk in the railways. Sumira's mother had a number of husbands and it is possi-ble that one worked as a policeman, another as a clerk.

5. There is a discrepancy in my data with regard to Sumira's first trading experiences. For some years I thought her initial trading experiences were in Jakarta (*Third World Urbanization*, 252), but later I learned that she had helped her mother trade in Semarang during the Japanese occupation.

6. For more details on Sumira's house and furnish-ings and how I came to be invited to her home, see *Third World Urbanization*, 244–7.

7. *Kampung* here means 'urban village'.

8. For Tati's background and role, and Sumira's treatment of her, see *Third World Urbanization*, 253–4.

9. For a more detailed account of the old and new house, see L. Jellinek, 'The Life of a Jakarta Street Trader—Two Years Later', Working Paper 13 (Centre of South-East Asian Studies; Melbourne Monash University, 1976), 2–3.

10. Initially, I thought Sumira had only had one baby ('Life of a Jakarta Street Trader', 4), but later she told me she had had two, both of whom died.

11. For evidence of other *kampungs* stifled by the expanding modern metropolis, see G. H. Krausse, 'The Kampungs of Jakarta, Indonesia: A Study of Spatial Patterns in Urban Poverty', Ph.D. diss. (University of Pittsburgh, 1975), 57–8.

12. See G. J. Hugo, *Population Mobility in West Java* (Yogyakarta: Gadjah Mada University Press, 1981), 211–12; G. F. Papanek, 'The Poor of Jakarta', *Economic Development and Cultural Change*, 24: 1 (1975), 9–12; L. Jellinek, *The Wheel of Fortune* (Sydney Allen and Unwin, 1991), 74–9.

13. See Hugo, *Population Mobility*, 211; Papanek, 'The Poor of Jakarta', 9–12; R. Critchfield, 'Desperation Grows in a Jakarta Slum', *Christian Science Monitor*, 12 Sept. 1973, 9; *Wheel of Fortune*, 61–5.

14. See T. G. McGee and Y. M. Yeung, *Hawkers in South-East Asian Cities: Planning for the Bazaar Economy* (Ottawa: IDRC, 1977), 50–2; *Laporan Hasil Survey Profile Pedagang Kaki Lima Di DKI Jaya* (Jakarta: University of Indonesia, 1976), 28; *Jakarta Post*, 21 Dec. 1983, 2; *Jakarta Post*, Jan. 1984, 2; Praginanto, 'Pak Parto Pengusaha Warung Tegal', *Galang* (Lembaga Studi Pembangunan, Jakarta), 1: 1 (1983), 44–7.

15. For a more detailed description of the relationship between Tati and Achmad, see 'Life of a Jakarta Street Trader', 7.

16. US dollars are used throughout even though all transactions amongst *kampung* dwellers in Jakarta are performed in Indonesian rupiah. Comparisons of prices and incomes over time are easier in US dollars because the dollar was subject to much less inflation than the Indonesian rupiah. Also, most readers will be more familiar with the inflation of the US dollar. One US dollar bought Rp 420 in 1971, Rp 1,000 in 1983, and Rp 2,000 in 1994.

17. For Sumira's remarkable managerial skills, see *Third World Urbanization*, 249–51, 255–6; and 'Life of a Jakarta Street Trader', 11–12.

18. Having watched *kampung* dwellers and traders for longer, I now wonder whether there was so much sympathy for Sumira when she was forced away from her trade location and eventually went bank-rupt? In the mid- to late 1970s life may have been gentler or, I may have been seeing things on the surface—seeing the harmony rather than the con-flict. In the 1990s the more brutal aspects of life stand out. The sympathy of traders and poor people for one another's plight has been replaced by competition and some pleasure at seeing another person fall. A bankrupt trader means that there are new opportunities—new trade loca-tions—for other traders. Many traders are inter-ested in buying the stalls of bankrupt traders cheaply. The quality of most traders' food has declined dramatically from the days when Sumira served delicious meals. Most cooked-food sellers have suffered like Sumira.

19. Increasing evidence during the 1990s suggests that many of the urban poor die—unnoticed—in this way.

20. It is extremely difficult to calculate the earnings of food-traders. They seldom distinguish between their business and their domestic budgets. Their sales generate an income which has to provide for their purchases for the next day's sales. But it must also cover any costs that arise from their trade, such as a tax or the repair of a cart, as well as any household expenses, such as children's schooling or the repayment of a loan. Traders appreciate that the more they spend on themselves, the less stock they will be able to buy and the less they will earn. The household's food is usually obtained from the stall, frequently from its left-overs, or by barter from other food-traders.

21. For more details on Nani's relationship to Sumira, see 'Life of a Jakarta Street Trader', 4–5.

22. I estimate that Sumira was earning Rp 120,000 or US$285 a month net at the peak of her trade in 1972. Seven years later she was earning Rp 40,000 or US$63.

23. In subsequent years, *kampung* dwellers were treated much worse when their homes were demolished. In 1981, the flats were new and the government appeared to be concerned and cautious. It tried to negotiate with the residents of Sumira's community, offering alternative housing to those who found flats unacceptable. It provided relatively good rates of compensation and kept the level of instalment payments for the flats low. Ten years later, demand for land by government and developers has increased, but rates of compensation have not kept pace with the rapidly rising land prices, while instalment payments are closer to commercial rates. The government has become much less generous, and *kampung* dwellers are forced to move in a variety of ways. The government instructs the *kampung* dwellers to move. Brokers working for government and private developers spread rumours to pressure people to leave the area. 'Accidental' fires break out in areas targeted for demolition. Canals which flow past these communities, helping to drain them, suddenly become blocked by soil from 'infra-structure works' and flooding occurs for six months instead of just one or two days a year. Services such as water, electricity, and garbage collection, that had been introduced into the community by the Kampung Improvement Programme, disappear. Local government calls on the military police to force the *kampung* dwellers to move.

24. For an account of our expeditions to the sacred tree, see 'Life of a Jakarta Street Trader', 27–8.

25. The Indonesian poverty line is Rp 150,000 (US$75) compared to Nani's income of Rp 70,000 per month (US$35). Clearly Nani had no resources for medical treatment. Rahmini and Wawan, by contrast, earned about Rp 400,000 (US$200). Over a quarter of this went on flat instalments, gas, electricity, water, and rubbish collection; one half was spent on food and Gugun's schooling; and the remainder went on credit payments for consumer goods, Wawan's needs for drink, cigarettes, and food while driving, and on occasion, some may have been sent to Nani and her children in the village. Ramli, Nani's youngest child, went to the city to request/pick up these funds if and when they were available.

10

How Women and Men Got By and Still Get By (Only Not So Well): The Gender Division of Labour in a Nairobi Shanty-Town

Nici Nelson

In the original version of this essay I examined the gender division of labour in informal-sector activities as I recorded them in the first half of the 1970s in a Nairobi shanty-town, Mathare Valley.[1] Twenty years have passed since the data on which that original essay was based was collected. During that period a number of short return visits to Nairobi allowed me to update my research.[2] This new data adds an historical perspective to the original analysis and allows me to make additional observations on women and men in the informal sector as the processes of urbanization and the penetration of the Kenyan economy by capitalism proceed.

The Setting

At the time of my initial research, Kenya had a population of 12 million people. Though at that time only one in ten Kenyans lived in urban areas, these areas were growing very rapidly. Today Kenya has 28 million people and about one in four live in urban areas. Cities continue to grow rapidly, though there is some evidence that

Nairobi's rate of growth is slowing down and smaller towns are expanding more rapidly. High population growth has fostered high differentials in income between rural and urban economies, and a shortage of arable land has made urban migration attractive to many throughout the post-independence era. Though just after independence the majority of urban residents were male, over the last 30 years this sexual imbalance has been reduced. In the post-colonial period, increasing numbers of women have been migrating to town, many of them single women (Nelson, 1991).

Kenya in the 1970s was a relatively successful mixed capitalist economy. Major earners of foreign exchange such as tea, coffee, pyrethrum, and tourism were completely or partially in the hands of national parastatal bodies, though private enterprise was encouraged, including multinational investment. The government also had controlling interest in strategic industries such as banking. Leys (1975) described the contradictions in the government's economic nationalism and its continuing dependence on foreign capital and questioned the capability of Kenya's economy to grow under the conditions of what he

This essay has been revised and updated for *Cities*. It first appeared in Ray Bromley and Chris Gerry (eds.), *Casual Work and Poverty in Third World Cities* (Chichester: John Wiley, 1979), 283–302. By permission of John Wiley & Sons.

termed 'the political economy of neo-colonial-ism'. His analysis has been widely accepted, though it must be noted that he later somewhat revised his analysis by recognizing the strength of the Kenyan bourgeoisie in capital accumulation (Leys, 1978). The growth in GNP has varied since independence: high in the late 1960s to early 1970s, falling drastically after the oil crisis of 1973 and returning to reasonable growth in the mid-1980s. However, it is clear that what economic growth there has been has not resulted in a commensurate growth in jobs. Unemployment and underemployment in the urban economy have increased steadily since the early 1970s. As a result, urban migrants have been forced, in ever-increasing numbers, to support themselves by petty commodity production in self-built shanty areas. It is estimated that as many as one-third of Nairobi's population live in such communities.

Mathare is one of these shanty-towns. In the early 1970s, it housed one-fifth of Nairobi's population. After the lifting of restrictions on Gikuyu movement at the end of the Mau Mau Emergency in 1962, migrants coming to Nairobi began building on the slopes of the Mathare River Valley, a 15-minute drive from what became the site of the Nairobi Hilton. They were, in the beginning, landless Highland Bantu (Gikuyu, Meru, Embu, Kamba). While it was difficult in 1970 to establish the exact break-down of population by sex, it was clear that women outnumbered men and about two-thirds of them were heads of household. Less than 10 per cent of women were employed in the formal sector, as opposed to 20 per cent of the men (Etherton, 1973).

By 1970 the Nairobi City Council estimated that 70,000 people were accommodated in either individually built mud-and-wattle houses with cardboard roofs, or company-built wooden houses with tin roofing. By the early 1980s, the population was estimated at 100,000. This number has remained stable until the present day.

Informal-Sector Activity in Mathare Valley

From its inception, Mathare Valley could have been described as an informal-sector commun-ity.[3] Certainly in the 1970s the housing was all 'informal' and largely self-built; and, according to Ross (1973), 80 per cent of the total popula-tion (and by my calculations, 90 per cent of the women) made their living in the informal sector within the confines of the valley itself. I shall discuss briefly the organization, the costs and revenues, and the gender division of labour of the following informal-sector activities in Mathare: (1) the entertainment industry, (2) self-built housing for sale, (3) shops, (4) other small businesses, and (5) hawking, and how these have changed in the last 20 years.

The Entertainment Industry

Mathare is ideally located to provide the services which Ross (1973) refers to as the 'entertainment industry'. The valley was, and still is, surrounded by areas of large numbers of men living alone without wives and families (who may be living in the rural areas). These areas include the staff quarters of the Mathare Mental Hospital, the Police Lines, the Army Camp, and the Kenya Air Force. The neighbourhoods to the west and south are dormitory areas for men working in Nairobi, many sharing rented accommodation with other men. In the early 1970s the 'enter-tainment industry' provided reasonably priced alcohol, companionship, and sex. During their leisure hours, men flooded into the valley look-ing for beer and companionship. Men were there all day till the early hours of the morning roam-ing the alleyways looking for prettier faces, fresher beer, livelier talk, or a woman willing to let them stay the night (Nelson, 1977).

Buzaa (maize beer) brewing was so prevalent in Mathare it could almost be called a local industry. It was the product for which the area was known throughout Nairobi during the 20 years after independence. Eighty per cent of the women in both my samples brewed beer; this included all the independent heads of households without formal wage-work and many of the mar-ried women. The beer business in Mathare was largely in the hands of women in this period; women brewed and sold out of the rooms they lived in. Thus each woman's room was used for domestic purposes as well as to manufacture beer and to serve it to customers drinking on the spot. I have described the process of brewing and selling of *buzaa* elsewhere (Nelson, 1978, 1979).

While it was lucrative, it was also uncertain and dangerous, principally because it was, and still is, illegal;[4] women ran the risk of being arrested or having their beer, raw materials, and/or equipment destroyed in a raid, or of being forced to bribe the police. Although women worked as individual entrepreneurs, they evolved complex networks of co-operation between patron and clients, neighbours and friends, to cope with these difficulties as well as the economic uncertainties of giving credit to men over whom they had no sanctions.

Despite the uncertainties, *buzaa* brewing was at that time a rational choice for many women in Mathare. Compared to the employment open to these relatively uneducated or undereducated women (e.g. house servant and barmaid), beer brewing involved shorter hours, better remuneration, and permitted women to combine business and domestic work.[5]

Buzaa brewing required little training and relatively little start-up capital. In the mid-1970s the ingredients for a brew of 16 gallons came to Ksh 30. When commencing brewing, women could borrow the simple equipment necessary for the brewing. For retailing beer, a newcomer only needed a room from which to operate and vessels to drink from, usually cleaned half-litre oil tins purchased from men who recycled them from the dumps.

Though *buzaa* brewing and selling was primarily women's work in Mathare, there were subsidiary activities which were conducted by men. There were a number of men who sold firewood which they brought from the rural areas to be used for the frying of the fermented maize flour. Certain men assisted with the arduous work of straining the maize/millet residues from the standard 16-gallon batches of *buzaa*, though they were usually alcoholics content to be paid in drink. .

Limited quantities of gin (locally called *changaa* or Nubian gin) was distilled and sold in Mathare Valley. It was usually made in more remote and wooded areas mainly by men. It was very lucrative. Important people in Nairobi were said to finance it for the profits. Because of health risks, it has always been a very serious legal offence, attracting heavier penalties than brewing. In 1974 the fine for gin-distilling was over Ksh 1,000 as compared to Ksh 100–200 for *buzaa* brewing. To escape the attention of the police, distillers chose remote rural venues for their manufacturing and thus needed a vehicle for transport. The only woman distiller in Mathare had a boyfriend who provided a car and protection.

Gin was regularly retailed by women, distributed to them in plastic bleach bottles by quick-witted boys, sure-footed enough to elude the police. In the 1970s two men distilled Nubian gin and also ran unlicensed 'bars' furnished with tables, benches, and portable record-players blaring out the latest hit tunes, modelled on the Western-style bars elsewhere in Nairobi. This development intensified in the 1980s.

When I returned to Mathare in the 1980s and early 1990s, I found that the brewing of *buzaa* beer and its retailing out of individual women's rooms was practically defunct. In 1974, 90 per cent of my sample earned their income, or substantially supplemented their income, through the production and/or sale of beer. In the mid-1980s, it became apparent that a shift in informal-sector activity had occurred. The women told me that while in the 1970s they had 'fed and educated their children well' through the production and sale of beer, this had now become impossible to do.

What had happened was that the production of maize beer had been taken over by a capitalist enterprise. In South Africa, a firm had discovered how to halt the fermentation of maize beer through additives, thus making its production and sale on a large scale economically feasible. In 1982 a multinational firm which was already retailing this type of beer in southern Africa, opened a factory for the production of *buzaa* under the brand name of *Chibuku* in Kenya. In the following year, the President of Kenya decreed a ban on the production of so-called 'native alcohol'. The decree was enforced throughout Kenya. After the Presidential decree, the brewing of *buzaa* became illegal under any and all circumstances. This should have been no problem, since it had always been illegal for Mathare women to brew. However, now police pressure had increased. In 1974 fines were Ksh 200 and custodial sentences for persistent offenders were three months long while police bribes were Ksh 20. In 1985 the fines had risen to Ksh 1,000 or 1,500 and custodial sentences could be as long as one year. Police bribes had escalated to Ksh 500. Women claimed that the frequency and ferocity of police raids had increased. In

1985 all the women I contacted had been directly affected by this government campaign. A fine, a period of imprisonment, or a couple of police bribes were enough to convince a woman that the brewing of beer was no longer a viable option.

Men still came to Mathare to drink *buzaa*, though now not in the homes of individual brewers/sellers but at unlicensed bars run by men. They could do this with impunity because they only sold commercially brewed *Chibuku* and because they paid large bribes to the police. I calculated that while in 1974 there were approximately 1,000 women selling *buzaa* to small groups of men in Village II alone, in 1985 there were ten bars selling commercially produced *Chibuku* and almost no women selling *buzaa*.

What are women doing instead? A small number of my previous informants had obtained houses in New Mathare and lived on rents, no longer needing to carry on informal-sector activity, such as brewing or selling beer. Of those left in Old Mathare, some now sold only gin and marijuana because they were easier to conceal from the police. Others had moved into various forms of trading or informal-sector employment such as house servants and barmaids in the male-owned bars. There had also been an increase in high-volume, low-return sex work. Women were forced to take more casual customers at lower rates to sustain a reasonable standard of living.

Sex work has always been part of the entertainment industry in Mathare Valley; indeed, the link between sex and alcohol is common throughout the world. In 1974 Mathare's independent women made the greater part of their income retailing *buzaa* and Nubian gin: sex work had been merely a way to supplement their income. Even married women were forced to do this when the husband did not bring home enough money. Sex work took various forms. Women charged Ksh 5 for a short 'bed ride', or 'Quick Service' as it was then called. Staying the night cost from Ksh 5 to 20, depending on how close the end of the month payday was, and how much a woman needed money or liked the customer. I could only estimate roughly how much women depended on casual sex work, but it was clear that most women resorted to it in an emergency. Few depended on it as their only source of income. The general opinion was that women who did go were lazy and ran the risk of being beaten or robbed by unscrupulous, vicious, or drunken customers.

Most women favoured longer term relationships with lovers or with *Bwana ya town* (town husbands). Lovers did not live with their women, but merely visited; 'town husbands' lived with their 'town wives' in a conjugal unit of limited duration. The woman rented the room and retained rights to it after the man had left and the man did not have any responsibility to or rights over children born of the union. Because both these relationships had a strong monetary component, I have classified them as part of a continuum of sex work. Women explicitly discussed these relationships in instrumental and monetary terms, saying: '*Bwana* should help with the rent and give money for food. What else is their use?' or 'Any *bwana* who doesn't help me with money when I need it would have to leave'. There was no direct payment for sexual services as there was with casual sex; the economic exchange was couched in terms of gifts, help, or 'loans' which were rarely repaid. Though women valued sexual attraction, affection, and companionship, it was clear in their gossip sessions comparing the relative merits of their *bwana* that the most important criterion of a good man was his generosity with money. Women manipulated the various kinds of sexual relationship to maximize their economic security.

In 1990 I returned for another period of fieldwork. I found that this particular strategy was still practised by single women raising children alone in Nairobi. However, in the intervening 20 years the strategy had been altered in two ways. The first change has been the long downward slide in the economy. The buoyant economic growth of the 1960s and early 1970s halted during the oil crisis of 1973, and the economy went into a recession which continued until the mid-1980s. Though the economy has improved since then, it has been the type of economic development which privileged the wealthier and a few particular groups such as the coffee producers in the periods of high coffee prices. In 1990 the cumulative effects of years of high inflation, relatively little economic growth, structural adjustment, and increased unemployment had severely affected the free-wheeling leisure life-style of the majority of Nairobi men. The effect on the women of Mathare and their daughters, many of whom no longer live in Mathare, is that their potential lovers and customers have less money to spend on women outside their marriages. Women who depended on all or part of their income from friends or lovers could no longer depend on one man. The economic strategy of sex work has become less certain and lucrative.

Sex work has also become more risky. The increasing prevalence of HIV/AIDS, which in Africa is largely transmitted heterosexually, makes casual sex of any kind dangerous for both parties.[6] It was the assertion of the women I talked to in 1990 that some men are limiting their extra-marital sexual encounters. To my eyes, though, men drinking in the bars and chatting up single women seemed to be as numerous as they ever were.

Housing for Rental

When Mathare valley was repopulated after the lifting of the Emergency prior to Independence in 1963, many Gikuyu migrants flocked into the valley to build mud-and-wattle housing for occupation and renting. In 1968 a number of the wealthier local inhabitants in collaboration with outside investors incorporated themselves as companies, bought land on the upper edges of the valley from the Indian owners, and built large blocks of concrete-floored, tin-roofed wooden houses. In 1969, after a period of confrontation with the Nairobi City Council in which the fate of Mathare hung in the balance, a moratorium was declared on further house-building. This ruling was enforced by the District Officer and the police under his command.

House-building was a very lucrative financial proposition at this time. In the early period most people built mud-and-wattle housing with cardboard roofs. In the 1960s the mud housing units cost approximately Ksh 400 to 500 to build—with monthly rents at Ksh 30 in 1969 and Ksh 45 in 1973, the investment could be paid off in less than a year and a half. Some women mobilized work parties of friends and lowered the costs to Ksh 250. After the original investment was recouped, rents on houses were pure profit. Owners paid no rates, taxes, water, or electricity bills, and repairs were limited to remudding after a severe rainy season or patching the roof with a few more flattened cardboard boxes. The demand for rooms in Nairobi was much higher than the supply, and Mathare rooms were rarely empty for more than a day or so, regardless of their condition. Investing in housing yielded a quick and steady return.

Investment through a registered company was a more costly and complex procedure. Builders of the early mud houses merely had to apply to the local leaders for permission to build. Mathare, by 1970, was divided into ten named and spatially distinct 'villages', each with its self-appointed leaders, usually older men and women who arrived early in the settlement period. In contrast, to join a company one had to pay a membership fee of Ksh 1,000. After this, one could buy as many shares in the company as one could afford, each costing Ksh 1,000. The number of shares determined the number of rooms listed under the member's name. Contractors built these housing units approximating Nairobi City Council building standards, though they lacked water, electricity, and drainage. If costs of joining a company were high, the returns were good. At an average rent of Ksh 80 per room, an investment could be recouped in much the same time as for a mud house.

The numbers of male and female house owners presented a fascinating puzzle for the researcher. In the period after the struggle for survival with the City Council, each village committee of KANU, the only political party in Kenya at the time, conducted a survey of all individually owned housing. This register of owners was to be the basis for an equitable allocation of 800 low-income houses and site-and-service plots built in 1973 by the City Council and the National Christian Council of Churches to rehouse owners of Mathare's mud houses—rather a futile exercise, since there were 10,000 such owners in Mathare.

I was only able to examine the register for Village II and that only briefly. In that village, 66 per cent of registered owners were women, though they owned only 50 of the units registered. The Secretary of Village I confirmed that the proportion of women owners was much the same in his village; but the Secretary of Village III, a village settled later than I and II and with a majority of Luo residents, claimed that the proportion of women owners was much lower. It was indeed my impression that the proportion of women was lower in the areas of the valley settled after 1963–4. Just after the Emergency, large numbers of older, independent Gikuyu women, many widowed or separated during the Emergency, lived and worked in Nairobi—it had been easier for Gikuyu women than for Gikuyu men to stay in Nairobi during the Emergency. The surge of building in Mathare came at just the right time to release the small capital savings of these women.

Women members constituted only a third of the companies' membership. These companies required higher capital investment and consequently attracted many prominent businessmen and politicians from outside the valley. Fewer women could raise the minimum Ksh 2,000 subscription.

During my returns to Mathare over the last 20 years, three trends emerged among the house owners of Old Mathare. For those few fortunate men and women who received a New Mathare house or site in 1973 or a unit in the World Bank low-cost housing schemes built between 1978–85 in eastern Nairobi, their economic future in the city was secured. Only a few very poor aged widows, who obtained units by virtue of being clients of important KANU members, lacked the business acumen, the family support, or the luck to maintain the payments on their houses or plots. They forfeited their plots and were back in Old Mathare renting accommodations.

Those Mathare residents who were members of the companies have also flourished. Nairobi has expanded greatly during the last 20 years and most of the land within the city boundaries has been developed. The demand for housing remains high and developers have become interested in the hilly, rocky Mathare Valley once regarded as marginal. During the first half of the 1980s, many of the companies tore down their original, basic, barracks-style housing in the upper part of the valley and erected two-storey, stone-built structures with courtyards and gates, which command higher rents than the old company housing. Mathare Valley looks very odd in these redeveloped areas, as if chunks of the nearby lower-middle class neighbourhoods had been uprooted randomly and plunked down in the middle of a shanty-town. These houses are rented to office workers and teachers.

Thirdly, those men and women who still live in the valley, whether in the old-style company housing or the mud houses, experience deteriorating conditions with less and less hope that any

official help will be forthcoming to improve their standard of living. When many companies went through a period of brutal often violent (utilizing arson and bully boys) redevelopment in the early 1980s, the displaced tenants drifted to the eastern end of the valley where there were still empty areas of government land. There they squatted throwing up provisional mud houses. The housing stock has become increasingly run-down, reflecting the residents' mood of hopelessness and fear about the future. Those left behind in the valley have fewer and fewer viable economic options. Women's options appear even more circumscribed than those of men.

Shops

The results of a 1973 survey of shops and other businesses located in two of the 'villages' of Mathare, with ownership broken down by sex, are presented in Table 10.1.

In the category of shops I included *duka*, i.e. small general stores, small restaurants, butcheries, charcoal shops, vegetable sellers, and wholesalers of *kimera* and maize flour. These businesses required a minimal establishment, if only a rudimentary roof. Needless to say, all these businesses were unlicensed and technically illegal.

Duka were usually converted rooms with an enlarged front window and a counter area. There were a great many of these small establishments in Mathare. In Villages II and III alone, there were 44 to serve a population of approximately 7,000. Most were small and stocked a limited range of domestic goods and non-perishable foodstuffs. A well-stocked *duka* might also contain such articles as beans, rice, sugar, tea, flour, cocoa, cooking fat, salt, chilli and curry powder, jam, maize meal, milk, eggs, potatoes, onions, matches, kerosene, cigarettes, needles, thread, safety-pins, aspirin, rubber baby-bottle teats, soap, wicks for lamps, and wirewool. Sales volume was low, since locals bought supplies only in small quantities. The *duka* were run by family labour and made up for a low volume of trade by staying open 12–14 hours a day. Credit was granted only to well-known customers living close by, a means of insurance against a creditor's 'midnight flight'.

Few, if any, *duka* owners kept accounts or had a precise grasp of cash flow or the costs of a

TABLE 10.1   Survey of 204 Small Businesses in Villages II and III in Mathare Valley, 1974

A. Shops

	General stores	*Hoteli* (small restaurants)	Butchers	Charcoal-sellers	Vegetable-sellers	*Kimera* and maize flour-sellers
Male	40	11	19	24	8	16
Female	4	1	1	5	15	9
TOTAL	44	12	20	29	23	25

B. Other small businesses

	Tailors	Carpenters	Shoemakers	*Dobi* (washermen)	Water-sellers (Village II)	Others (bicycle repairers, carver, charcoal stove maker)
Male	12	7	6	6	4	10
Female	3	0	0	0	0	3
TOTAL	15	7	6	6	4	13

shop's stock. One woman informant said she spent Ksh 1,500 to stock a shop for her son. A typical response to my questions about gross or net profits was 'Money just comes in and goes out'. Owners casually used money and food from the shop for their families' daily living expenses, further complicating attempts to estimate income. One guessed that he made a gross profit of Ksh 1,000 monthly, out of which he paid Ksh 100 in rent for house and shop. He was more businesslike than many others and his was a particularly large and well-stocked store. However, even he was vague on the monthly cost of restocking. It seemed to be a matter of whether he had cash when the wholesaler vans called. These vans normally traded with legal stores in neighbouring areas and might go into Mathare at irregular intervals, stopping at the larger *duka* there. Smaller shopkeepers merely walked to the neighbouring Eastleigh and bought from shops there. This inevitably raised costs. One woman would buy a dozen pints of milk a day, paying 75 cents each and charging her customers 80 cents from her small *duka*.

Forty of the 44 *duka* were owned by men, though many of them were helped by their wives. The same is true of small restaurants. Owners without wives available to work, hired women

workers, young boys, or old men. In central Kenya, cooking is the one household task which is unnegotiably female, the last task which men willingly take on. This may explain why only young or very old men would be willing to do it for a living in Mathare.

Fare in such a small restaurant would be limited to tea, white bread, eggs, fermented porridge, and various forms of *irio* (boiled and mashed maize, beans, and potatoes), and stews In the mid-1970s the prices were reasonable, e.g. 50 Kenyan cents for a large plate of *irio*. Overheads were minimal. Food stuffs would be bought each day from the Municipal Market. I obtained no exact figures on profit margins; one owner estimated that he made Ksh 1,000 in a good month. Janet Bujra (1975), working in neighbouring Pumwani slightly earlier than I, estimated that some of the restaurants might have grossed Ksh 3,000 a month—these restaurants were larger than their Mathare equivalents, a reflection of Pumwani's greater prosperity.

Butcheries were very much in evidence in Mathare in the 1970s, 20 in Villages II and III. These outlets primarily catered for visitors who enjoyed eating roasted meat with their beer, since the local population rarely could afford meat. Butcheries did a roaring trade in the early part of the month, when workers were paid. During the periods of slow business at the end of the month, butchers cut down on the meat they ordered. One owner who sold on average 240 kilos of meat per month claimed that he could sell 100 kilos of meat in the first week of the month and only 30 kilos in the last. Meat cost Ksh 2 per kilo at the Kenyan Meat Commission, and he retailed it at the government-controlled price of Ksh 6 a kilo. Thus his gross profit was approximately Ksh 960, his net profit Ksh 760 after deducting his cook's salary and his rent, but not counting some loss in bones and scraps.

Butcheries not only sold raw meat, but roasted meat on the spot. Usually a couple of shaded benches were provided for clients to wait while the fatty meat roasted appetizingly over the coals. Some butchers also provided a soup made of bones and scraps.

All but one of Mathare's butcheries were owned and operated by men. The woman owner had hired a male manager. This possibly related to an old Gikuyu belief that women should have nothing to do with the slaughtering of cattle or the portioning out of meat.

By the late 1980s, there were few butcheries left in Old Mathare. Most roasted goat-meat selling was now carried out in an area known as Kwa Michael, a cluster of shanties on a dirt road near New Mathare. Nearly 60 outlets served a huge clientel roast goat and bottled beer. Most of the customers drove there in their cars. All owners and employers were men. The flight of butcheries from Old Mathare is directly related to the disappearance of *buzaa* retailing there.

Charcoal shops were usually the most rudimentary of the establishments, often consisting only of an awning to protect the charcoal from the rain. They were supplied by lorries which transported charcoal into Nairobi from the rural areas where it is made. Most charcoal sellers had an arrangement with a rural charcoal-burner, often a relative or a friend from their home area. The charcoal was delivered in large burlap sacks at regular intervals, and was sold by the measure, a large bowl or a tin. Profit margins were high. A bag of charcoal from the rural wholesaler cost Ksh 7.50 to 11.00. When a bag was divided into 36 units at Ksh 1 each, this yielded a profit of Ksh 25 to 28. Most charcoal retailers retailed a bag a day. Thus the small charcoal seller, e.g. one who retailed bags of charcoal in small quantities

rather than selling whole bags, could earn Ksh 700 a month; a larger operator who also sold entire bags of charcoal could earn more. Rental costs, if any, were low. Most operators merely stored the coal in their own room, dragging it out to sit under their awning each day on a piece of waste ground near by, often near a well-frequented road.

It is surprising that only 5 out of 29 charcoal sellers were women. Capital investment and skill needs were low, and there seemed to be no Gikuyu ideas about the gender division of labour to create barriers to women selling charcoal or for customers patronizing a woman charcoal seller. On the contrary, Kenyan women have always been fuel providers for the household. It was certainly regarded as a dirty job: the faces and hands of charcoal sellers seem permanently engrained with coal dust. Perhaps women, who are intimately concerned with household cleanliness and clothes washing, resented the constant cleaning such work necessitated.

By the mid-1980s the charcoal trade had become much less profitable. First, supplies were difficult and expensive to obtain. The deforestation of Kenya had increased apace, and the only available trees were now far from Nairobi. And because of higher cost and lower incomes, Mathare people bought less and less charcoal. It was palpably obvious that eating habits had changed drastically over the last 20 years. Tea was seldom made. Visitors and thirsty children were now offered Fanta instead of tea. Quick-cooking maize meal *ugali*, with a spinach relish had almost universally replaced the long, slow-cooking *iro*.

The ideal image and actual role of Gikuyu women as food providers undoubtedly contributed to the fact that two-thirds of the vegetable sellers were women. The low capital investment necessary to sell vegetables is also a factor. Vegetable sellers need no business place except for a plastic sheet spread on the ground. The more substantial establishments might have an awning, or even tables to display the goods. This was not a very lucrative form of self-employment. Mathare sellers went very early to the Municipal Market to lurk on the periphery illegally purchasing small amounts of vegetables from the licensed dealers. Such marginal purchasers could be harrassed by the police. Periodically the City Council would order a clean-up campaign to arrest and fine such luck-less individuals on the grounds of obstructing the public streets. Since these small-scale dealers paid more for their 'wholesale' vegetables than licensed dealers, profit margins were quite small. The perishability of the produce also forced deal-ers at the end of the day to clear easily spoilable stock such as greens at cost. One woman claimed that she could barely make enough to feed her four children by this daily activity. Vegetable sell-ing was certainly the most transient type of busi-ness. Women frequently took a short breather from brewing beer after the police raids or drunken customers became too much for them and switched to hawking vegetables. However, women who started selling vegetables to escape the rigours of *buzaa* brewing, soon complained about the roughness of the City Council police, the long hours spent sitting in the sun, and the low financial returns. Eventually they would return to the more lucrative option of brewing *buzaa*.

Sadly in the 1980s and 1990s, the number of women engaged in various forms of selling and hawking, including vegetables, had greatly increased. Left with no other option, women sat by the road on pieces of sacking selling a hand-ful of tomatoes or 6 stalks of spinach. Either that or they walked endlessly from door to door with small quantities of produce for sale. One Mathare woman grumbled to me in 1990 that it was hardly worthwhile selling in this way, the returns were so minuscule.

The numerous brewers involved in *buzaa* man-ufacture in the 1970s meant that there existed a demand for the raw materials: maize flour and *kimera*, a type of millet flour. Most women did not have the time or transport facilities to buy these ingredients themselves in neighbouring areas. A single brewing would use 45 lb. of maize flour and 10 lb. of *kimera*. Also a woman carry-ing a 45 lb. bag of flour on her back would be particularly conspicuous and vulnerable to police harassment. For these reasons brewers preferred to buy from wholesalers who transported the materials by truck into the valley.

The retailing of these products was lucrative. Two women wholesalers alternated travelling to the country. At each trip they would buy seven 180 lb. bags of *kimera* and transport them back to Nairobi, all the while hoping that they would not be stopped by the police checking for the authorization required to transport food in bulk.

They estimated they made a net profit of Ksh 400 per trip, and they averaged three such trips a month.

One-third of the flour dealers in Mathare were women. Most of these were older women who had already made a good deal of money brewing *buzaa*. They may have wished to shift to a less arduous and risky business, one which involved less energy and mental wear and tear. It was a natural transition for them to begin providing materials for the brewing process. Many of these women had been brewers on a larger scale than average, brewing perhaps three or four times a week instead of the more normal once or twice a week. As a result they were reliable wholesale suppliers of beer and had established networks of female clients who bought beer from them on a regular basis. When they set themselves up in the flour wholesaling business, their former *buzaa* clients switched their trade to them, saying that they knew these women would buy only the best ingredients. In the 1990s this opportunity no longer existed.

Other Small Businesses

This category includes entrepreneurs producing a product—tailors, carpenters, shoemakers—as well as those providing a service—*dobi*, i.e. wash-ermen, water-sellers, traditional doctors, circum-cisors, and hawkers—who have no place of business. I have less than complete data on some of these activities.

Only 3 of 15 tailors in the two villages sur-veyed in 1973 were women. Two factors limited the entry of women to this business: the high cost of the treadle-operated sewing-machine (which was more than Ksh 1,500) and women's limited access to tailoring training in polytechnics or informal apprenticeships. In the early 1970s, one woman had paid Ksh 2,000 for two year's part-time instruction in tailoring at a local school.

All the carpenters and shoemakers in the two villages were men. Many of the carpenters had learned their skills as casual labourers on build-ing sites or as apprentices to other small carpen-ters Most shoemakers had been apprentices to other shoemakers; one had learned his craft at a mission trade centre. There seemed to be no clear pattern for apprenticing to a trade. In most trades, apprentices would work for several years for little or no pay doing increasingly responsible tasks. Since so many young men were eager to enter the lucrative and prestigious car mechanics trade, car mechanics could ask for payment from their apprentices.

Dobi were few in Mathare. Paying to have someone do one's laundry was a luxury for most inhabitants. The *dobi* working in Mathare washed the clothes of single male office workers who needed a smart appearance at work. They were all male, even though clothes washing is very much a woman's task in the domestic realm. This appears to be a colonial legacy. Most Europeans hired male house servants for washing and ironing. Women's tasks gained prestige when incorporated into the monetary economy, as Bujra (1990) showed in her life history of a Tanzanian man who was a cook for a European household. The *dobi* in Mathare were old house servants who now made a poor living at a low-status job.

Water-selling was another low-status male occupation, arduous work for little return. The four water-carriers in Village II were young single men. All that was needed to set up in business was a cart costing Ksh 30, a number of metal drums, and a strong back. One such vendor made only Ksh 120 a month.

In 1974 there was one male *mkanga* or doctor and one female circumcisor operating in Villages I and II. Both were reputed to be rich. People maintained that indigenous doctors were invariably male. As well as treating illnesses with herbal medicines, these 'doctors' made love potions and good luck charms. They also had magic which could reveal the identity and whereabouts of thieves.

All female circumcisions had to be performed by women. The Gikuyu form of circumcision entails the removal of the tip of the clitoris only. The female circumcisor boasted of being very 'modern' in her techniques and claimed to have a certificate in circumcision from a hospital. She eschewed the usual 'dirty and dull knives', as she put it, using instead a double-edge razor. She dressed the wounds with sanitary towels and administered aspirin afterwards. 'Just like a real doctor', as she proudly described it.

Hawkers were not included in the business survey of Mathare because of difficulties in locating and counting them. Hawkers continue to be numerous in Mathare, especially on the first days of each month, when wage-workers have recently been paid. They hawk vegetables, used and new clothes, medicines recycled from trash bins, trinkets, and the cleaned half-litre oil tins which women use to serve *buzaa*. Some vegetable hawkers are women, but in general hawkers are predominantly male.

Conclusion: How Women and Men Get by in Different Ways

I will conclude with a number of observations on the gender division of labour I found in informal-sector activities in Mathare in the mid-1970s and some of the changes I have observed over the last 20 years.

When examining the list of activities practised by men and women, two conclusions about the gender division of labour become obvious. First, men, in the main, carry out all the activities which require a relatively high capital investment: running a *duka*, tailoring, or owning a taxi. This is as true now as it was in the 1970s. In the late 1980s, it was men who were able to take advantage of the opportunity to build formal bars for the sale of commercially manufactured *chibuku* beer, obtaining it wholesale and bribing the police to operate without licences. Men in the informal sector are neither the sole support of their dependants nor the caregivers to their children. Most Kenyan men have two options. They may keep their family on a small farm in the country where their wives farm and perhaps work casually as agricultural labour to supplement the family income. If the family is landless and lives in town, men expect their wives to brew beer or hawk vegetables to feed the family and to assist in their businesses. In Mathare, only women married to men in the formal economy or highly successful in self-employment, do not work. A man struggling in a highly competitive urban economy has the advantage of being able to channel a significant proportion of his income into capital investment. An independent woman in the informal economy is at a disadvantage if she has children. Indeed a high proportion of the so-called 'big women' of Mathare's informal economy were childless or had expanded their economic activities after their children were grown and independent. Often older children contributed to a joint household (Nelson, 1979).

Most women with children spent their income on upkeep and school fees, unlike a man in the same situation. While men see their financial future in building up a business and accumulating capital, women do see it in terms of educating their children. This is not to say that men did not educate their children, but that men in the informal economy can expect financial contributions from their wives.

Second, men do the work which involves skills of the industrialized sector. They learn in school or on the fringes of the modern industrialized economy to be tailors, carpenters, shoemakers, car mechanics, and taxi drivers. Women have not acquired these skills. Their skills have remained overwhelmingly those of the subsistence economy and the domestic sector. Women in Mathare were and continue to be more restricted than men in their choice of economic activity. At the end of my research in 1974, I conducted a survey of 10 per cent of the women in Mathare Village II: almost all, whether married or unmarried, worked in the informal sector brewing beer or practising commercialized sex. In a survey of all informal businesses other than those of beer brewers/sellers in Villages II and III, I noted that there were 41 women entrepreneurs and 163 men. Women were concentrated in three out of eleven categories of activities: selling vegetables, *kimera* and maize flour, and charcoal (Table 10.1). Men who were employed or self-employed in the informal sector were involved in a much wider range of activities in Mathare, and even more so elsewhere in Nairobi. Men could be motor mechanics, small-scale building contractors, house repairers, glaziers, plumbers, electricians, casual workers on building sites or in industry, drivers of legal or illegal taxis, fare-boys on buses, delivery boys, thieves and muggers, fences for stolen goods, barbers, hawkers of anything portable, street sweepers, dustbin scavengers, bicycle repairers, makers of shoes from old tyres, carvers of wooden utensils or tourists curios, makers of aluminium cooking pots, charcoal stoves, and on and on (Hake, 1977).

Women, on the other hand, were clustered in activities which relate to women's roles in the domestic realm: provision of food, beer, child care, and sexual companionship. In both the informal sector and the formal sector throughout the world women are mainly found in domestic service, nursing, teaching, cooking, and sex work, using the skills they had been taught for use as organizers of the domestic domain (Arizpe, 1977). Educational systems and cultural expectations frequently reinforce this division of labour. It is also obvious, looking at the data, that women seek forms of employment which fit in with child-care responsibilities, sometimes choosing less lucrative work which permits them to integrate their day-to-day domestic work with their income-generating activity, something noted in many labour markets of the world (Afshar, 1985).

Sex work as a form of informal-sector work deserves particular consideration. It is almost entirely the work of women.[7] It is an issue which raises a great deal of hostility either because it is seen as an aberration from the 'normal' emotionally intimate act between man and woman—an attitude fraught with moral connotations—or because it is seen as a form of exploitation. Should sex work be seen as a moral issue? Certainly early writers such as Hellman (1948), describing women doing sex work in South Africa, saw it as a part of the breakdown of community morals attendant on African migration into urban areas. Is sex work perhaps an exploitation of women by men in which women are degraded victims? Or is it merely a service little different from other commercialized services, such as cooking food, performed for reasons of duty in the domestic sphere? It is certain that Mathare women took this pragmatic, morally neutral view. They referred to commercialized sex as 'selling their kiosk', i.e. selling a commodity like any other sold at a kiosk, and at their frankest often made an explicit comparison between marital and commercial sex. One woman told me that while I accepted food and rent for sex from one man, my husband, she accepted cash for sex from many men. 'What is the difference?', she asked. When sex work takes place in a social situation without pimps or madames, as was and still is the case in Mathare and Eastlands, I think it is possible to accept Mathare women's views that it is an economic option yielding reasonable returns for relatively little work, even if it involves some short- or long-term physical risk. Bujra (1975: 215) commenting on women in Pumwani, writes that 'prostitution allowed them the independence and freedom from exploitation that would not have been possible to them had they chosen any of the other socioeconomic roles open to them.' More work must be done to devise models to deal with commercialized sexual relationships which avoid both pseudo-scientific moralizing and a simple assumption of social deviance. An approach which relates commercial sex to women's domestic labour, as White (1990) has done for sex workers elsewhere in Nairobi, could well prove fruitful.

The reasons for the narrow repertoire of women's activities are structural and cultural. Women's education and skills levels are lower than their male counterparts'. Since colonial times they have had limited access to skills with commercial value and this educational policy changed slowly in the post-independence era. In 1985 I contacted several Central Province Technical Training colleges. The only courses open to women were domestic science and tailoring. When I queried why plumbing, carpentry, car mechanics, accountancy, etc. were not open to women, I was told that women would not sign up for such courses. This circular argument neatly absolves the college administrators from the need for action and from blame. The structural constraint is founded on definitions of female and male and the proper gender division of labour. This consensus seems to be an amalgamation of indigenous ideas and imported European views. Women are probably constrained from being butchers because of an indigenous convention that men hunted wild animals or slaughtered domestic cattle and divided the meat. Women are not carpenters, plumbers, and car mechanics because Europeans imported their views of these jobs as the 'natural' province of men.

Child care poses another constraint which is structural in nature but cultural in origin. In the Central Province, as elsewhere in Kenya and most of the world, views of the domestic division of labour defines child care as solely women's work. Therefore, married women are constrained in their search for a productive role in the city by child-care responsibilities. Child care doubly constrains unmarried mothers: they have to both care for their children and support them. Female relatives, in the rural areas or in Nairobi, lighten the load of many single mothers operating in the informal sector in Mathare (Nelson, 1987). Few women have older relatives nearby to provide daily care or access to alternative provisions, such as crèches or nursery schools. Unless they can call on their mothers or sisters to foster their children, they are tied to the house or must pay for mother-substitutes.

The way that Mathare people perceived informal economic activities changed over the last two decades. In 1974 it seemed to me that differences in perception were determined by a combination of sex and age. Young men in Mathare viewed their particular activity in the informal economy as a stop-gap measure. They always talked about getting away and finding a 'real job'. They were rationalizing and salving their pride while doing a job considered demeaning, both by themselves and by others. Women, regardless of age, and older men had fewer options, so they regarded the informal sector as a way of life. Whether this led to greater commitment in skills and investment is not clear. Young men did appear to spend their earnings on fashionable clothes and 'high-status' consumer goods rather than on expanding their enterprises. But it is difficult to tell whether this was a manifestation of their youth or of their relative high level of education which led them to see a higher standard of living as theirs by right.

At the time, I noted that both women and older men had lower educational standards but did not develop this factor as a significant one in explaining their different perception of their future in the informal-sector economy. In 1990 I interviewed a large number of women who had been young girls at the time of my initial research about their economic survival strategies. What struck me then was that while many were involved in various forms of informal-sector activity such as sex work, smuggling cheap second-hand clothes purchased in Tanzania into Kenya, working as barmaids or house servants, all but two of the forty, unmarried, young women stated uncategorically that they were looking for 'real work'. The two who were content hawking vegetables had almost no education. The remaining 38 women all had Standard VII (primary school) or above and many had had further training, e.g. typing. It is clear that when women attain higher educational levels, their aspirations also rise above a career in the informal sector. These young women were noticeably better dressed than the previous generation of Mathare women. However, all but two had children to care for and this meant that they had to spend most of their earnings on domestic expenses rather than on consumer goods. The parlous state of the Kenyan economy probably means that most of these women will be condemned to a 'career' in the informal sector, just like their mothers.

This look at two decades of change in the Mathare informal sector highlights an issue relating to the globalization of the economy. Women's loss of their most lucrative option in the informal sector, i.e. beer brewing, was directly related to the involvement of a multinational firm in the production of maize-millet beer which led to the tightening of anti-brewing legislation. The penetration of capital into this form of petty commodity production further marginalized women.

The informal sector provides many women in poor countries with the only viable alternative to marriage. Men do not have to choose between economic independence and marriage. Women often have to, or indeed do not have a choice. In Africa at any rate, the city is often the only option available to rural women who have little chance of getting married, those who are widowed, divorced, or who have children out of wedlock. Women with few 'modern' skills and little education can market their domestic skills in a socio-economic setting which allows them to care for their children. Without question the women in this sector are exploited, poor, and overworked. Yet, faced with major structural and cultural constraints, unable to either secure employment at the lower end of the formal-sector job hierarchy, or to marry someone with a relatively well-paid formal occupation, women have little choice but to pursue such income-generating opportunities as are available to them.

Notes

1. I am an anthropologist by training. The original data used in this essay was collected between 1971 and 1974 in Mathare Valley for a Ph.D. degree. I returned to Mathare in 1978, 1984, 1985, 1990, and 1994.

   The methods used in the initials research were participant observation, interviews, and surveys. Participant observation was carried out in three of Mathare Valley's ten 'villages', but I visited the other villages as frequently as possible in order to ascertain whether what I observed in Villages I, II,

and III was similar to what seemed to be happening elsewhere.

   This data was supplemented by a number of long, structured, open-ended interviews with 89 selected respondents. Due to the tense situation in Mathare *vis-à-vis* city and national authorities, it was difficult to interview people at random. I tried to stratify the sample on the basis of age, education, and status as owner or tenant. At the end of my period of research, I conducted a random sample survey collecting basic demographic data. I used the results of this survey to compare my intensive interview sample with the total universe of women in Mathare. My sample was younger and better educated than the norm, possibly they understood more clearly what I was trying to do. In addition, there was a larger number of house owners in my sample than was the norm. It is possible that better-off women had more time to devote to this curious stranger with her interminable questions.

2. Most of the subsequent periods of research were too short to permit more than ethnographic interviewing and limited participant observation in the form of visits. In 1984 I spent three months contacting and re-interviewing nearly 60 of my original informants. In 1990 I spent another three months in Nairobi. Though my research purpose in 1990 was to contact younger women outside of Mathare, most of those were daughters of women I had previously known in Mathare. I also spent periods of time contacting, visiting, and talking to previous informants.

3. See the ILO report on employment in Kenya (1972) and Hart (1973) for early developments of the concept of the informal sector. Moser (1978) provides a comprehensive overview of the pros and cons of informal-sector concepts. Feldman (1991) presents a good overview of the evolution of the concept over nearly twenty years of use and argument about its use.

   In the original version of this essay, I questioned why gender analysis figured so little in informal-sector descriptions. While analysis of the informal sector has moved on from urban-centric micro-studies in Third World cities to more structuralist analyses of global economic restructuring and the informalization of production, the concepts of these debates are gender-neutral. Little has changed in the understanding of women's position in the informal sector. Feldman (1991: 6) questions 'why women were not present in early research on the informal sector and why, even today, they remain almost invisible in analyses and theorization on the informal sector debate'.

   One of the first problems is the lack of rigorous, detailed primary data. It continues to be the case that men are badly enumerated in the informal sector, and women are virtually invisible.

4. The previous legal status of *buzaa* production was complicated. Distilled alcohol was entirely banned for obvious reasons, but in Nairobi it had been possible to brew beer for personal reasons, such as a wedding. The brewer had to obtain a licence from the District Officer's office—not an easy procedure—which allowed her or him to brew 16 gallons of *buzaa* on a one-off basis. In smaller municipalities in Kenya, local bars could obtain licences to retail *buzaa*, but individual women could not brew and sell it legally in their homes.

5. Both barmaids and house servants earned on average Ksh 100 a month. In 1974 the minimum wage was Ksh 200 a month, and there were Ksh 7.14 to the US dollar. Women brewing beer in Mathare in 1974 could earn anything from Ksh 250 to 400. In addition, the work was done in their own homes, which enabled them to care for their children while they brewed and sold. By contrast a woman working as a barmaid or house servant had to pay for someone to care for her children in her absence, which could come to as much as Ksh 30 a month. Afshar (1985) has demonstrated that women will choose less lucrative work, if it allows them to combine productive and reproductive roles.

6. A report of the Panos Institute (1988) makes it clear that HIV/AIDS is heterosexually transmitted in Africa. In fact it may well be that women are more at risk than men. This is especially true of women sex workers. A longitudinal study of a sample of 1,500 sex workers in Nairobi revealed that in six years the percentage of those who were HIV-positive had risen from one-third to nearly 90% of the sample.

7. In Africa, relatively few males make their living selling sexual services, except in a few beach resorts, such as the Gambia and the Kenya coast which attract European tourists. Little accurate, in-depth information exists on the incidence and organization of sex work in Third World cities. There are no models to deal with it as a form of informal-sector activity.

References

Afshar, Haleh (1985) (ed.), *Women, Work and Ideology in the Third World* (London: Tavistock).

Arizpe, Lourdes (1977), 'Women in the Informal Sector: The Case of Mexico City', *Signs: Journal of Women in Culture and Society*, 3: 25–37.

Bujra, Janet (1975), 'Women Entrepreneurs', *Canadian Journal of African Studies*, 9: 213–31.

—— (1990), 'Men at Work in the Tanzanian Home: How did They Ever Learn', in Karen Hansen (ed.),

African Encounters with Domesticity (New Brunswick, NJ: Rutgers University Press), 242–65.

Etherton, David (1973), *Mathare Valley* (Nairobi: University of Nairobi, Housing Research and Development Unit).

Feldman, Shelley (1991), 'Still Invisible: Women in the Informal Sector', in *Women and International Development Annual* (Boulder, Colo.: Westview Press), 183–201.

Hake, Andre (1977), *African Metropolis: Nairobi's Self-Help City* (Brighton: Sussex University Press).

Hart, Keith (1973), 'Informal Income Opportunities and Urban Employment in Ghana', *Journal of Modern African Studies*, 11: 61–89.

Hellman, Ellen (1948), *Rooiyard: A Sociological Survey of an Urban Native Slum Yard*, Rhodes Livingstone Papers, No. 13 (Cape Town: Oxford University Press).

ILO (International Labour Office) (1972), *Employment, Incomes and Equality: A Strategy for Increasing Productive Employment in Kenya* (Geneva: ILO).

Leys, Colin (1975), *Underdevelopment in Kenya* (London: Heinemann).

—— (1978), 'Capitalist Accumulation, Class Formation and Dependency: The Significance of the Kenyan Case', *Socialist Register*, 241–66.

Moser, Caroline (1978), 'Informal Sector or Petty Commodity Production: Dualism or Dependence in Urban Development', *World Development*, 6: 1041–64.

Nelson, Nici (1977), 'Dependence and Independence: Female Household Heads in Mathare Valley, a Squatter Community in Nairobi, Kenya', Ph.D. thesis (Univ. of London, School of Oriental and African Studies).

—— (1978), 'Female Centred Families: Changing Patterns of Marriage and Family among *Buzaa* Brewers', *African Urban Studies*, NS Special Issue on East Africa, 3 (Winter): 85–104.

—— (1979), 'Women Must Help Each Other: The Operation of Personal Networks among *Buzaa* Brewers in Mathare Valley, Kenya', in P. Caplan and J. Bujra (eds.), *Women United: Women Divided* (London: Tavistock Publication), 77–98.

—— (1987a), 'Rural–Urban Child Fostering in Kenya: Migration, Kinship, Ideology and Class', in Jeremy Eades (ed.), *Migrants, Workers, and the Social Order*, Association of Social Anthropologists of the Commonwealth 26 (London: Tavistock Publishers), 181–98.

—— (1987b), ' "Selling Her Kiosk": Kikuyu Notions of Sexuality and Sex for Sale in Mathare Valley, Kenya', in Pat Caplan (ed.), *The Cultural Construction of Sexuality* (London: Tavistock), 217–39.

Nelson, Nici (1991), 'The Women Who Have Left and Those Who Have Stayed Behind: Migration in Central Kenya', in S. Chant (ed.), *Gender and Migration in Developing Countries* (London: Belhaven Press), 109–38.

Obbo, Christine (1975), 'Women's Careers in Low Income Areas as Indicators of Country and Town Dynamics', in David Parkin (ed.), *Town and Country in Central and Eastern Africa* (London: International African Institute), 288–93.

—— (1980), *African Women: Their Struggle for Economic Independence* (London: Zed Press).

Panos Institute (1988), *Aids and the Third World* (London: Panos Institute).

Ross, Marc (1973), *The Political Integration of Urban Squatters* (Evanston, Ill.: Northwestern University Press).

White, Luise (1990), *The Comforts of Home: Prostitution in Colonial Nairobi* (Chicago: University of Chicago Press).

11

Gender Divisions in Brazilian Industry

John Humphrey

IN Brazil as in many other developing countries, women have been drawn into the labour-force in increasing numbers in the course of the past 30 to 40 years. Female participation rates have risen sharply, and the female share of total employment has continued to rise. Even during the 1980s, often referred to as the 'lost decade' in Brazil, the female share of total employment continued to rise. In the São Paulo Metropolitan Region, women accounted for 33 per cent of all those in work in 1979. By 1990 this figure had risen to 38 per cent (IBGE, 1979, 1990).

The continuing entry of women into the labour-force has not in any way undermined the gender division of labour. Gender segregation and gender inequalities remain as great as ever. The key question is not whether women are disadvantaged, but why. In this essay, the factors structuring the gender division of labour will be examined for the case of one particular sector of employment—the formal sector of manufacturing industry. It will be argued that while the segregation of male and female labour appears to support the dual labour-market hypothesis, the patterns of employment observed for female and male factory workers can best be explained by considering the way in which factories are defined as gendered spaces. Gendered occupational and work structures are constructed within the factory and then institutionalized and legitimated through segmented labour-markets.

Gender divisions in manufacturing do not merely reflect patterns of gender relations in society. Gender divisions specific to factory situations are constructed in them. This means that in manufacturing gender divisions are constructed and reconstructed as patterns of work and employment change. Segmented labour-markets play an important role in creating and maintaining gender divisions, not as causes of such divisions, but as mechanisms for ensuring their persistence. This analysis will then be used to explore the possible consequences of the restructuring of labour process arising from 'Japanese' methods such as Just-in-Time and Total Quality Control.

This study is based on research into the sexual division of labour among hourly-paid workers in Brazilian industry undertaken in the early 1980s. Reference will be made to seven factories: three electrical plants, two factories making motor components, one making pharmaceuticals, and one making toiletries. The factories varied in size from 400 to 3,000 employees, with the proportion of male workers in the hourly-paid labour-force varying from 25 to 72 per cent. Clearly, the factories are not a representative sample of firms in Brazilian industry. They were chosen to represent large, modern firms in non-traditional industries which would use relatively sophisticated personnel management policies.[1]

Gender Divisions in Brazilian Manufacturing

Many women found work in manufacturing in the 1970s and 1980s. Their problem was not marginalization from industrial employment as such, but rather the nature of their incorporation into it. As one might expect on the basis of experience in many other countries, men held a virtual monopoly on managerial and technical occupations. Female managers were rare, and very few women were found in engineering and technical occupations. According to a survey of establishments in the city of São Paulo, there were twenty male technical workers employed for every one woman in the late 1970s.[2] Even among the workers who are the focus of this study, blue-collar manual workers, there was a well-defined sexual division of labour.

The SENAI survey found that women workers were concentrated in the lower reaches of the job hierarchy, while men were spread more evenly across it. Men outnumbered women by almost eight to one in those jobs SENAI classified as skilled. Women were overwhelmingly concentrated in semi-skilled jobs, defined as 'requiring dexterity', 'repetitive', and needing 'concentration'. In effect, the category includes any work with machines not requiring specialized training, and a lot of semi-skilled work is paid at the same rate as the lowest SENAI category, 'labouring'. Taking just the three manual categories (skilled, semi-skilled, and labouring), 83 per cent of women were employed in semi-skilled work. Male manual workers were distributed much more evenly across the skill categories: 24 per cent skilled, 60 per cent semi-skilled, and 17 per cent in labouring jobs.

The segregation of female and male workers is far from confined to the issue of access to work defined as skilled. Data from the factories studied show that in addition to the exclusion of women from many functional areas—such as stores and maintenance—and the concentration of women workers at the bottom of job hierarchies, a considerable segregation of the male and female labour-forces along occupational and departmental lines was also evident. The depth of the division between male and female within manufacturing establishments, which was evident in all of the factories visited, can be illustrated by a description of the situation in one of the automotive plants. The plant had an hourly-paid labour-force of 578 workers in September 1978, of whom 72 per cent were male and 28 per cent female. Its occupational structure incorporated a sexual division of labour which ran along functional, hierarchical, and occupational lines.

As might be expected from the results of the SENAI survey, male workers in the plant monopolized almost all of the skilled occupations within it. There were virtually no women working in machine-setting, tool-room, and maintenance functions, and men also monopolized store-keeping work. In these four areas there were just 3 unskilled women employed, compared to 221 men—53 per cent of the male work-force in the plant. In effect women were confined to jobs in production, packaging, and quality control: 98 per cent of all women working in the plant were in such functions, compared to just 44 per cent of men.

In common with many factories in Brazil, the plant had a wide variety of occupational titles which were grouped into a number of grades. There were 22 different grades for hourly-paid workers as a whole and 14 for production workers. Although wages for each grade overlapped, each grade had a starting wage and increments according to merit and length of service. As might be expected, the exclusion of women from many areas of skilled work meant that a much greater proportion of men than women held higher grade jobs. Of all male hourly-paid workers 73 per cent had jobs on grade eight or higher, compared to only 2 per cent of the women. Women were concentrated at the bottom of the job hierarchy: 97 per cent were employed on the bottom four grades, compared to only 13 per cent of men.

This concentration of women workers in the lower grades was not just the result of their exclusion from such skilled work as maintenance and machine-setting. Even in the production, packing, and quality-control jobs, women were concentrated in lower grade occupations, while men were distributed across a much wider range of grades. In production jobs, almost half the men were on grade eight or higher, while all the women were on grade four or below. The highest grade a female quality-control worker could attain was the lowest grade at which male quality controllers were employed. Even within the production and quality-control areas, therefore, women were concentrated on the bottom rungs of the occupational hierarchy.

The job titles for women and men were also segregated. The 578 hourly-paid workers in the plant were grouped into job titles which were almost totally single-sex. One male worker was classified as 'Machine Operator I', which was otherwise totally female; three women and sixteen men shared the title 'Quality Control Inspector I'; five cleaners (four men and one woman) shared the same job title. All other job titles were single-sex. Men and women were inserted into two segregated and differentiated occupational structures within the plant, each with its own jobs, wage levels, and patterns of mobility. Male workers in low-grade occupations were segregated from female workers in occupations of similar status, and such male workers had access to male-only higher paid jobs. Even in those few cases where women gained access to higher grade jobs, it was through specifically 'female' mobility chains into 'female' jobs.

Not all plants were like this automotive plant, but it was by no means exceptional in its occupational structure, sexual division of labour, and segregation of female and male workers. In the other automative and electrical plants quite similar occupational structures were found. They all showed a concentration of women in production jobs, low-grade work, and one or two broadly defined occupations, together with distinct promotion patterns for men and women and segregation by department and occupational category. In one electrical factory, for example, five of the twelve production departments were overwhelmingly female (185 women and 2 men—excluding the male supervisors), and five departments were entirely male except for two women secretaries. The only area in the factories studied where it was common to see men and women working together was quality control. In factories in pharmaceuticals and toiletries, the nature of the work results in a much less elaborate structure of occupations and grades, but the overall structure of employment was the same.

Not surprisingly, it was extremely rare to find male and female workers doing similar work with the same job title. In the seven plants studied, just two exceptions to this rule were found. First, in the toiletries plant men and women were employed on the same work of filling tubes and bottles with toiletry preparations. However, the male workers were paid a higher wage for this.[3] Secondly, men and women commonly worked together in quality-control departments in the automotive and electrical industries, and there was often an overlap between male and female job titles. The highest grade female quality-control workers shared the same grade as the lowest grade men, as was seen in the automotive plant.

These patterns of occupational hierarchy and segregation had clear implications for the wages and earnings of female and male workers. Differences in rates of pay between high-grade and low-grade jobs are considerable in Brazilian industry. In one of the electrical plants, for example, a (male) carpenter working on the production of television and speaker cabinets could earn two and a half times as much as a (female) assembler. As a result, differences in the grading of male and female occupations led to big differences in wages for men and women. Hourly wage rates for male workers were much higher than those for women. In the case of the electrical plant, for example, average hourly wage rates for hourly-paid men were 46 per cent higher than for women. Even among production workers the average hourly rate was 41 per cent higher for men than for women, and for quality-control workers the difference was 29 per cent. Data from five plastic firms revealed a roughly similar situation: overall, male hourly wage rates were 74 per cent higher than the average female rate, and in production and quality control the differences were 32 per cent and 43 per cent, respectively.

Wage differentials for male and female workers were reinforced by patterns of promotion for male and female workers. In the electrical factory, average hourly wage rates for newly hired workers (less than four months in the firm) were 19 per cent higher for men than for women. For workers with between one and three years' employment, the differential rose to 34 per cent, and for workers with more than six years' employment, the differential in average hourly wage rates for male and female production workers reached 45 per cent. Such widening wage differentials were not just the result of men entering the kinds of occupations which offered chances for training and advancement. Even when men and women were recruited to the same occupations, as could happen in the plastics industry, wage differentials widened over time. In mixed-sex occupations men and women employed for only a short time in the plants had roughly similar wage levels, but as length of employment rose, differences emerged. Female wage rates remained roughly constant, while male rates rose. Even in low-grade labouring and 'production assistant' work, female wages rose less than male as length of service increased. Men gained in two ways, therefore. Within a given occupation, their chances of advancement up incremental scales were greater, and they also had a better chance of promotion to higher grade occupations.

Segmented Labour-Markets

The analysis of occupational structures shows that men have jobs with promotion prospects, while women remain in dead-end, low-wage, and presumably unskilled jobs. One well-known theory which purports to explain this pattern of employment is that of dual or segmented labour-markets. Widely applied to analyses of both the division of labour by gender and discrimination against women workers, the notion of a dual or segmented labour-market has been summarized by Barron and Norris (1976: 49):

1. There is a more or less pronounced division into higher-paying and lower-paying sectors;
2. Mobility across the boundaries of these sectors is restricted;
3. Higher-paying jobs are tied into promotional and career ladders, while lower-paying jobs offer few opportunities for vertical movement;
4. Higher-paying jobs are relatively stable, while lower-paying jobs are unstable.

Jobs in the higher paying sector, which also offer opportunities for advancement and stability of employment, are called the 'primary labour-market', while the low-wage, unstable, and dead-end jobs are called the 'secondary labour-market'.

Segmented labour-market theorists suggest that primary labour-markets in industrial establishments are constituted by patterns of internal promotion and training. Workers move up and along 'mobility chains' which link jobs together in a more or less natural sequence of experience and training on the job. It is also suggested that workers' productivity can be increased by the experience of work itself, with the result that once workers gain access to certain kinds of jobs, their advantages relative to other workers are reinforced by the very fact of having gained access. Internal promotion systems within large enterprises create primary labour-markets, formed by clusters of jobs which offer workers relatively high wages, opportunities for training

and promotion, and stability of employment. Outside of this primary labour-market employment conditions are much poorer.

For both orthodox theorists (Piore, 1975: 147) and the radicals (Reich, Gordon, and Edwards, 1980: 239; Edwards, 1979: 179), the productive structure of an enterprise more or less determines the number of primary sector and secondary sector jobs, chances of promotion, and mobility chains. For orthodox theorists, this structure is determined by technology, which creates particular job structures. For radical theorists, segmented labour-markets grew up as a form of control over labour, but they do correspond to differences in work content. In both cases wages and promotion reflect productivity and learning.

Segmented labour-market analyses of female employment address themselves to the question of why men tend to occupy the higher rungs of the occupational ladder and move up the mobility chains that give access to them, while women tend to be left firmly at the bottom, occupying the low-wage, dead-end jobs. The answer usually given is that women lack access to the productivity-enhancing opportunities that would allow them the chance to raise their wages and occupational status. As Lloyd and Niemi (1979: 12) have argued, 'Productivity-enhancing job experience and differential access to such experience are the keys to the vicious circle of constrained opportunities in which the woman worker is still trapped.' Initial differences in endowments between men and women are reinforced by employment itself.

In both the orthodox and radical versions of segmented labour-market theory, the specification of a particular job is seen to determine the worker's productivity, his/her skill, the training opportunities which might arise, and the chances of access to better occupations. Hence, the structure of primary and secondary labour-markets appears to be determined by technology and work content.

The effect of this is to leave occupational structures largely outside the discussion of women's oppression. The factors which determine the characteristics of jobs and the allocation of individuals to them are not seen as gendered. When dual labour-market theory specifically addresses the issue of the sexual division of labour, it can only explain why women have certain attributes (or lack others) which would, if they were possessed by men, also deny them access to the primary labour-market. By and large, workers are placed in the jobs for which they are suited.

In taking this approach, segmented labour-market theories (in common with neo-classical and Marxist ones) suppose that the labour-market itself operates in accordance with non-gendered principles. Walby (1986: 81) makes the point with force:

A major problem with Barron and Norris' analysis is ironically, their lack of appreciation of, and analysis of, patriarchal structures in the labour market. Despite their emphasis on the importance of the labour market much of their article is taken up with merely a description of the characteristics that women bring, or are believed to bring, to the labour market. They describe the structuring of the labour market into two sectors in non-gender-specific terms and, I would suggest, mistakenly ignore the structuring of the labour market itself by sexual divisions. They treat sexual differentiation as determined largely outside the labour market by the sexual division of labour in the household. This sexual differentiation is then incorrectly treated as a given which is unmodified by the workings of the labour market.

It has become increasingly obvious that gender permeates all social institutions, and that the supposedly objective economic laws of market competition work through and within gendered structures. The market does not value male and female labour independently of gender. This point has been stressed in studies of sex and skill. Labour-markets operate not only to exclude women from skilled jobs, but also to downgrade jobs when they are performed by women. As Phillips and Taylor have argued: 'It is the sex of those who do the work, rather than its content, which leads to its identification as skilled or unskilled' (1980: 85). The importance of this point should not be underestimated. It is a fundamental challenge to most accounts of the operation of labour-markets and the causes of sex discrimination in work. This approach suggests that occupational structures do not exist independently of the characteristics of those who fill them. Technology does not define particular jobs, it defines tasks which can be grouped into jobs (collections of tasks) in different ways. These jobs can then be structured into promotion chains in different ways. This approach further implies that there is not necessarily a strong link

between pay and skill or productivity. One of the key arguments put forward by Phillips and Taylor is precisely that similar work can be classified as skilled or unskilled (and paid accordingly) according to the sex of those performing it.

In this perspective, a number of complex processes structure gender-divided labour-markets in factories. They include the separation of male and female jobs, the privileged access of men to jobs which require skill and training, the reservation of promotion opportunities for men, the unequal recognition of female and male skills, and the differential 'fine-tuning' of jobs (decisions about which peripheral tasks are attached to or separated from the core activities of any particular job). These overlapping processes are all driven forward by the same overall tendency for factory work to be divided along gender lines.

From this perspective, dual labour-markets do not simply reflect pre-existing gender divisions arising in the domestic sphere. Nor can they be seen as a mechanism for allocating men and women to pre-defined good and bad jobs, as is suggested by Hartmann (1979). Rather, they are the means by which gender divisions are made viable in situations in which they might otherwise enter into shock with egalitarian principles such as equal pay for equal work or promotion according to merit.

An example from one of the Brazilian electrical plants illustrates some of the mechanisms which allocate women to low-skilled jobs, undervalues their skills, and reserve promotion for men. The plant produced silicon chips, and most of its work consisted of making, testing, mounting, and sealing them. The women workers were concentrated in production and quality control, and on the lowest rungs of the occupational ladder. Seventy-six per cent of all the hourly-paid workers were female, and in production and quality control there were 35 men and 374 women. Ninety per cent of the women were classified as 'production assistants', the lowest paid job in the plant. They earned even less than unskilled workers in the canteens and kitchens, and their wages were very close to the minimum payable in the industry.

Some of the women production workers did simple but exacting jobs, such as cutting and breaking wafers into single chips, sorting chips into approved and rejected, and wiring a small number of connections. Such work required a degree of manual dexterity, good eyesight, and considerable concentration, but would be classified as semi-skilled by SENAI's criteria. However, there were other jobs which needed considerably more training and experience, chief among which were those related to the production of the wafers themselves. These were made in a special sterile area, sealed off from the rest of the plant. The wafer department was the foundation of the plant's production, and great care was essential if acceptably low levels of wastage were to be achieved. All the non-supervisory staff in the department were women, and they used sophisticated machinery in all the stages of production—photo-engraving, etching, and depositing of particles. Each woman was expected to be able to carry out any of the tasks in the sterile area. Although the machinery was highly automated and set up by technicians, the women were obliged to work with extreme care and attention, and to have experience and responsibility. In the photo-engraving stage, for example, the worker had to take a batch of wafers, check that they were the batch specified on the instruction card, check the kind of chip being made and the layer being applied, and then select the correct mask and align it over the first wafer in the batch in the right position, using a microscope to check alignment at a number of points.

The inspection of wafers which followed the engraving was even more complicated. This involved a visual inspection by microscope at 9 reference points to check for approximately 20 basic defects. According to the male supervisor of the department, the best young female workers, chosen after having had experience in other jobs in the plant, required between 4 and 6 months' intensive training and constant supervision in order to be able to spot the main defects that might arise. The women working in the department represented a considerable expenditure in terms of training for the company, and also an asset in terms of experience and knowledge. However, in spite of this, and in spite of the need for high standards of care, cleanliness, attention, and discipline, every female worker in the section was classified as an unskilled 'production assistant' and paid the lowest wage in the plant.

Had male workers been employed there, the occupational structure would have looked very different. The plant's manager observed that prior to the construction of the new factory and the introduction of more advanced technology, both men and women had been employed on the basic production processes. In his opinion, while the quality of the work and the productivity of the male workers had been equal to that of the female, the relatively high rate of male turnover had been a problem. Male workers had tended to leave in search of better opportunities, and the manager attributed this to dissatisfaction with low wages and poor promotion prospects.[4] Women workers, he argued, presented no such problems: they remained in the firm for relatively long periods, in spite of being confined to low-wage, dead-end jobs. As Brazilian managers so often remark, women workers seem to be more stable within a given occupational category, while male workers either push for promotion or leave in search of opportunities elsewhere.

This pattern of promotion was seen in other plants, and it is particularly interesting because of the relationship it implies between promotion and turnover. Generally speaking, managements neither provided clear lines of promotion for women nor expected women workers to demand it. An experienced senior personnel manager in one of the automotive plants expressed a common managerial view: 'Women are more stable within a given function . . . They are capable of staying in the same job for much longer [than men], as long as it is well-paid . . . Women are more stable from a job point of view.' This factory paid the lowest wage rates among those studied, but in spite of this, management could still rely on women to show acceptable (to management) rates of turnover, even while working in dead-end jobs. Another manager in the same plant described the firm's lack of interest in promoting women:

There's a certain under-valuation, complacency [in relation to women's work]. After four years a woman can still be doing the same job. There might be a certain difficult job, which needs a certain dexterity. So the foreman prefers to keep her in that job. That's what tends to happen in production. And the women put up with it. They are educated to be like that.

Managements were not under pressure to promote women because they did not present the problems seen with the men. In other words, low wages for women are the consequence of their low rates of turnover in low-wage jobs. Male turnover in such work is a problem which would have to be resolved through paying higher wages and providing promotion opportunities. The exception to this rule would be dead-end male jobs such as cleaning. In such cases, managers pick the men very carefully, choosing those without career prospects and without family prospects (marriage, having children) which would lead to pressure for higher wages.[5] Women are paid low wages because they have low rates of turnover, not because they have high rates, as is usually argued in segmented labour-market theories.

It is not difficult to imagine what would happen in the silicon chip plant were, for some reason, it to become impossible to employ women. The firm would have to employ men, and even if the pattern of work and definition of tasks were to remain unaltered, there would be drastic changes in employment policies. The firm would increase wage rates as a first step to reducing male turnover, but this need only be done for selective jobs. One cost-efficient way of reducing turnover would be to recognize that certain jobs require more training, responsibility, and experience, and reward these more highly. Mobility chains would be constructed linking the less skilled occupations to the more skilled, and this would provide an incentive for the lower-paid and less-skilled workers to stay in the plant and try for promotion. The jobs in the sterile department would be re-graded, and the skills involved recognized. Male workers would be offered the incentive and prestige of promotion to more skilled tasks, and management would both stabilize experienced workers and also be able to use promotion as a means of discipline and control. All this could be achieved without any substantive changes in the work actually carried out in the plant, although it is quite likely that the distribution of tasks would be altered. Men might be encouraged to gain knowledge about how machines work—to adjust them or provide information to maintenance workers on faults—whereas women are often discouraged from 'meddling' with machinery which is regarded by men and managers as beyond their abilities to understand. Better wages and promotion opportunities would be provided by creating a typical internal labour-market, but the internal labour-market would be the result of the need to provide

the former, not its cause. The actual work performed in the plant would remain the same, even if it was distributed across different occupations.

Dividing the Male and the Female

Certain stereotypes which characterize men and women's work are very widespread in industry. As Game and Pringle suggest in their study of the white goods industry in Australia, these stereotypes usually take the form of a series of polarities defining masculine and feminine: skilled/unskilled, heavy/light, dirty/clean, dangerous/less dangerous, interesting/boring, mobile/immobile. In each case, the first part of the couple is considered natural to men, the second typically female (Game and Pringle, 1983: 28–9). These categories could just as easily be drawn from Guilbert's study of the Parisian metalworking industries. One might add two more couples found in the latter's study: responsible/ not responsible, which denotes the tendency for men to be employed on costly machinery or products, and machine ability/lack of machine ability, to denote the belief of managers (and of male workers) that women lack mechanical sense (Guilbert, 1966: 135).[6]

The pairings listed above form the lines along which distinct male and female identities are constructed in factories. They are not straightforward descriptions of the types of work women and men actually do, nor even ideal types (in the sociological sense) which define the general characteristics of female and male work, or female and male workers. First, they are designed to establish a difference, a divide, between men and women. They specifically discount the overlap in attributes and aptitudes between the female and male populations by suggesting that women and men are suited to very different types of work. Specifically, in the manufacturing setting, the pairings select particular traits which define a masculine attribute and leave the feminine as merely its negative. While female work could, in theory, be described as 'continuous', 'requiring concentration', 'requiring dexterity', and so on, with the male equivalents being 'intermittent', 'requires less concentration', and 'not requiring dexterity', in practice, female attributes are not seen as giving value to work and are not recognized. They are not seen as properly industrial attributes. From the outset then, the elements used to establish gender identities define women as inferior and inadequate—lacking in masculinity, in other words. Secondly, the pairings are used selectively. Attributes are given great importance in some circumstances and not in others. In one factory situation, emphasis might be placed upon the difference between heavy and light work, while in another this factor may be irrelevant and the line of cleavage between the masculine and the feminine constructed along the axis of, say, mechanical ability.

The characterization of jobs as light or heavy, clean or dirty, can also be highly selective. Dirt in the factory context means grease, oil, paint, and so on. The men who consider that women should not be exposed to dirt do not usually mind if women clean the toilets. Similarly, a general association of femaleness and clean/light work does not prevent women working in poor conditions in particular plants, subject to dangerous chemicals or poisonous fumes. The elements of gender division are used selectively to create and justify differences. The fact that women do a job may even 'feminize' its perceived characteristics.

The construction of masculine and feminine identities in a manner which devalues and subordinates women can be seen clearly in the treatment of the issues of physical force and danger. In many cases the ideology of masculinity encourages male workers to submit themselves to poor working conditions, taking pride in their capacity to cope. Hirata's (1989) study of men in the Brazilian glass industry points to a culture of risk and courage which motivates men to take on dangerous work. This reinforces their own sense of masculinity, while effectively barring women's access. Managers approve of such divisions. Jobs requiring strength or involving exposure to danger in the factories studied were deemed to fall outside women's competence. Even when women might prove in practice that they could do certain 'male' jobs, managers remained concerned about what was right and proper for women workers:

The women in the firm where I worked even did the sanding down in the paint section. They choked with the dust. The only thing they didn't do was work with heavy machinery. I don't think women should do sanding work because they have more delicate fingernails than men, and the chemical gluing processes are not advisable for women either. One would have to see that it did not endanger a subsequent pregnancy. One would need medical advice. (Manager, woodworking department, electrical plant)

In the factory setting, concern expressed about women's health (or, in many cases, the health of potential children) is used to construct a barrier to women's labour-force participation. For men, such concerns are either not expressed—as with chemicals which might cause cancer or sterility, for example—or seen as problems to be overcome by technical measures. They are not used to exclude men from certain areas of work. In the case of women, medical problems are, once again, often used as a demonstration of women's unfitness for many types of industrial work. Similarly, lifting heavy weights can be used to exclude women from certain types of work, even though ways to get around such lifting may well be devised to help smaller or older men. The strength of such divisions lies in the fact that male and female workers share these perceptions. As Jenson (1989: 150) notes, women and men seek to do jobs defined as feminine and masculine, respectively.

Labour-markets do much more than merely reflect differences in the quality of labour which is available for hire. They are an element of the construction of a gendered work environment which actively separates and unequally values male and female labour. Without these structures, the unequal access of men and women to jobs and the unequal valuation of male and female work could not be sustained. As Cockburn (1990: 5) notes, the gendering of work is a continuing process. It has to be revised in the face of changing labour processes and it has to continue to convince those involved that it is just. This is why segregation and separate mobility chains are so important. They prevent comparisons of work and worth. When segregation is relaxed, then male privilege becomes too clearly apparent and, for women at least, much less tolerable. The case of women quality controllers in one of the electrical factories in Brazil provides a clear illustration. They were resentful because younger, less experienced men were promoted while they were not. Sex differentiation constantly has to cope with this tension. It is based on ideas of difference which can be undermined by the reality of work. Labour-market segmentation prevents comparisons and shores up gender stereotypes.

The differences in treatment of male and female workers are by no means confined to employment and training. The way in which work is defined also varies between the sexes. In-depth study of plants in Brazil shows that women's time and space is much more controlled than that of men. Women are much more likely to carry out work involving short job cycles and repetitive work. Movement—to fetch materials or tools, for example—might be built into repetitive male jobs, while women working at tables assembling components would have materials and tools brought to them. Women often referred to the pressure to work continuously. Women workers could whisper to each other, for example, but only if they kept their hands moving (Humphrey, 1987: 131). Women constantly referred to management in terms of 'looking' and 'keeping watch'. Female departments were more often organized on open-plan lines with the manager having a clear view from his (no female managers were found) office over the work area. This control over time and space extended into other areas. Women were not allowed to go to the washroom with the same freedom as men. It was common for each group of women to have a metal tag which they had to take with them to the washroom, thus ensuring that only one woman could go at any one time. Similarly, some factories provided special smoking areas for women, but not for men. Even over apparently trivial questions, control over women was enforced more rigidly than for men. For example, in some plants men were allowed to wear sweaters over their overalls in cold weather, but women were not. The plants had two distinct gendered spaces, with different rules applying in each.

Gender Divisions in Just-in-Time Production

In the past few years, Brazilian firms have sought to improve productivity through the application of so-called 'Japanese' management techniques—Just-in-Time (JIT) and Total Quality Management (TQM). Significant improvements in productivity and quality have been achieved by firms using such methods (Fleury and Humphrey, 1993). The introduction of JIT/TQM

may have a major impact on the division of labour in manufacturing and the jobs taken by women and men. Equally, pre-existing gender divisions may influence the way firms put JIT/TQM into practice.

JIT/TQM may undermine existing divisions of labour because it tends to integrate tasks. TQM puts increased emphasis on the role of production workers in producing products to the correct specification. Quality is monitored by those making the product. The checking and control functions formerly given to quality-control workers are increasingly passed to production workers. JIT emphasizes the continual flow of products through the factory and the organization of work into closely linked units of production dedicated to particular ranges of products. Within each unit of production, workers may move between different jobs, or operate a number of machines, or work in a team. In addition, the production workers may take on tasks related to machine preparation and preventive maintenance in order to increase the flexibility of production units and reduce the amount of time taken for stoppages.

To the extent that JIT/TQM leads to the reorganization of work along these lines, the horizontal and vertical division of labour is undermined. Workers move between more tasks of the same kind (working at different points in an assembly operation, for example) and take on new tasks of a different kind (typically, inspection, measurement, minor adjustment to equipment, etc.). To what extent, then, do these changes in the way tasks are grouped into jobs pose a threat to women's employment? If men continue to monopolize manufacturing skills, might women become even more marginalized in manufacturing work? Cockburn argues that whenever the content of work changes, the division between technical competence and women's work remains. On the basis of studies of technical change in a number of different work situations she concludes:

New technology had arrived, the old machines had been scrapped, but when the dust had settled it became clear that the high level and interactive technical skills remained just as before in men's hands. . . . Particularly striking, however, in the detailed evidence gathered from women and men were the coherent representations by both sexes of the affinity between masculinity, technological knowledge and sustained careers, and an even more complete association of femininity with technical incompetence. (Cockburn, 1990: 4)

If technical competence is attached to a wider range of occupations with JIT/TQM, what will happen to women's access to factory work? Jenson puts the point directly when considering women's position in flexible specialization, which is a characterization of many aspects of the JIT/TQM system: 'The ethos of the all-rounder—which comes closest to the ideal-typical worker in discussions of flexible specialization—is a masculine one. It involves not only certain abilities to deal with the machinery but also a sense of camaraderie which is highly social and which depends on being accepted by the group' (Jenson, 1989: 151). Or, to put it another way, if women were incorporated into Third World manufacturing on a large scale in the 1970s and 1980s on the basis of performing fragmented, routinized, repetitive tasks, what prospects do the possible reintegration of tasks hold out for them?

On the basis of the analysis so far, it is clear what will not happen. The process of reorganizing work will not be one of simply redefining tasks in accordance with technical norms which are defined independently of the characteristics of actual or potential labour-forces and then deciding which workers, male or female, are most suited to the new work. First of all, there are bound to be differences within JIT/TQM. Secondly, new gender divisions of labour will have to be constructed in specific settings. This will involve struggles over access to work. Thirdly, the elements involved in new jobs and the suitability of women and men to carry them out could be evaluated in different ways. Will, for example, the 'male camaraderie' mentioned by Jenson be crucial for group formation—in other words, will men exclude women for being female? Or will women's ability to work well in groups and their better communication skills give them an advantage for teamworking? Will this advantage be recognized? Fourthly, new gender divisions could well be created within JIT/TQM.

Even a cursory examination of Japan reveals enormous gender divisions and differences in firms using JIT/TQM. One extreme example of this is analysed by Hirata (1992). In a Japanese yoghurt factory, tasks were indeed integrated. Women on the assembly lines carried out qual-ity-control functions, data collection, and routine maintenance. These activities did not lead to higher status. These women workers worked one hour less per day than the full-time workers and were classified as part-time. This meant they were denied holidays, social security, pensions or the right to join a union. In spite of enormous importance given to polyvalence and quality throughout the plant's operations, only a minority of women workers enjoyed the rights associated with life-time workers.[7]

There are also indications that JIT/TQM is implemented in different ways in different industries. It is apparent that Japanese firms transfer 'Japanese' practices to the West to a much greater extent in the motor industry than in the consumer electrical industry (Kenney and Florida, 1995). In the motor industry, teamworking, job rotation, quality circles, etc. are common, but in most of the electrical industry, Japanese firms do not use these practices, preferring the usual division of labour. A large number of studies on Japanese firms in the electrical industries of California (Milkman, 1992), Mexico (Kenney and Florida, 1992; Shaiken and Browne, 1991), and the United Kingdom (Sewell and Wilkinson, 1992) have shown little use of quality circles, improvement groups, teamworking, or systematic rotation of labour. This does not mean that such plants lack either TQM or JIT. Rather, this is achieved without some of the features associated with such practices in the motor industry. In simple operations, particularly in assembly, low-stock systems with a certain degree of flexibility and considerably emphasis on quality can be constructed without abolishing the established division of labour.

This contrast between the electrical and motor industries could be the result of Japanese firms transplanting only simple processes in the electrical industry—processes which in Japan might either be automated or performed in low-wage subordinate enterprises. However, the hypothesis that this is a distinct pattern of JIT/TQM being implanted in the consumer electrical/electronics industry merits further investigation: the fact that women often work in the industry might be a key explanatory factor.

A further possibility is that firms create different types of multi-skilled jobs for female and male workers. This is the situation described in some Argentine factories by Roldán (1993).

When managements introduced multi-tasking (workers doing different tasks of the same skill) and multi-skilling (workers performing tasks involving different skills) they created distinct categories for women and men. In one plant, skilled male workers combined machine preparation and monitoring with peripheral tasks such as cleaning, while the women combined machine operation only with other low-grade tasks (1993: 48–9). The main divides remain technical knowledge and repetitive assembly work. Women have no access to the former, while management does not consider men suitable for the latter.

Roldán suggests that when multi-skilling is introduced there is a tendency for women to be replaced by men. The dual system of different multi-skilled jobs for men and women will be phased out as the women leave the plant. Employment protection legislation in Argentina means that the firms cannot afford to dismiss the women and replace them immediately by men. A slow process of substitution will take place.

In the Brazilian factories studied by Fleury and Humphrey (1993), various patterns of segregation and integration were found, including cases where extensive teamworking involved mixed-sex teams. In one plant producing headlight and tail-light assemblies for cars, work had been reorganized into cells, each with a team. The plant management planned to make the team members competent to perform different tasks in the cell and to be able to carry out quality checking, statistical process control (SPC), and routine maintenance. All the workers in the plant, male and female, were being trained in basic skills—basic mathematics, SPC, rapid tool-changes, safety, and basic maintenance. In 1993 every worker in the plant was to receive 56 hours basic training in these areas of competence. The firm was clearly pursuing a strategy of multi-skilling and multi-tasking right across the plant.

From management's point of view, the process of investing in labour was gender-neutral. Training was being applied across the board, with the aim of raising the abilities of all workers in the plant. Women might in this situation receive the same training and have the same opportunities as men. They might gain access to technical knowledge, and the redefinition of work would allow women to move out of narrowly defined 'female work' and into a broader range of tasks.

But, where workers and managers have no commitment to equal opportunities, the outcome may be very different. Women might be gradually displaced through the hiring of men; or work teams might become single-sex, with male teams being given more training; or hierarchies might emerge within teams, with men gaining access to higher levels of skill.[8] There is no certainty as to which of these differing outcomes will occur. It depends on management perceptions, labour-market conditions, and the demands made by female and male workers. All that can be said is that the processes which will define the future gender division of labour in the plant will be strongly influenced by representations of femininity and masculinity and the interplay of gender interests and the pursuit of efficiency. Managements will seek to use labour in an effective manner, but this aim will be pursued within the context of their own perceptions of gender differences as well as the perceptions of male and female workers. Women may come to recognize that they have the abilities to take on tasks previously barred to them. Alternatively, they may come to regard such work as 'unfeminine'. Similarly, men may seek to exclude women from the new multi-skilled jobs, or leave some of them entirely to women, or work alongside women.

Exclusion seems more likely when men enjoy employment protection, as in Argentina, while in Brazil, male workers are unlikely to achieve women's exclusion. New patterns of segregation may arise, however. Only time will tell, but if a new gender division of labour is established, it will be strengthened and justified by the occupational structures which form around it.

Notes

1. For further information on the sample, see Humphrey (1987: 202–6).
2. This survey, by the Serviço Nacional de Aprendizagem Industrial (SENAI), was carried out in a large number of establishments in the city of São Paulo in the late 1970s. The author is grateful to the Research Department of SENAI for allowing access to the results.
3. Men were paid higher wages, even though the more demanding work labelling high-value products was given to women because they were considered to be better at it.
4. The manager was referring to the 1970s, when a

booming industry created a scarcity of male labour (Humphrey, 1984).

5. In one plant, a manager described one of the male cleaners in terms of his age (over 50) and his recent arrival in São Paulo from the countryside. Being absolute without good job prospects in the city, he was 'suitable' for a dead-end job.

6. These examples are taken from European-dominated work cultures. In Asia and Africa, expectations of women's and men's work might be different.

7. Hirata (1984) has analysed elsewhere the link between employment and family structures in Japan, arguing that Japanese employment practices for both female and male workers depend upon a particular pattern of domestic relations.

8. For example, for each task or skill, the factory defined four levels of competence—no knowledge, rudimentary understanding, full understanding, and able to train others. Will men and women in teams reach similar levels of competence? Will women train men?

References

Barron, R. D., and Norris, G. M. (1976), 'Sexual Divisions and the Dual Labour Market', in D. L. Barker and S. Allen (eds.), *Dependence and Exploitation in Work and Marriage* (London: Longman), 47–69.

Cockburn, Cynthia (1990), 'Technical Competence, Gender Identity and Women's Autonomy', paper presented to World Congress of Sociology, Madrid, July.

Edwards, Richard (1979), *Contested Terrain* (New York: Basic Books).

Fleury, Afonso, and Humphrey, John (1993), *Human Resources and the Diffusion and Adaptation of New Quality Methods in Brazilian Manufacturing* (Brighton: IDS Research Report, No. 24).

Game, Ann, and Pringle, Rosemary (1983), *Gender at Work* (Sydney: George, Allen and Unwin).

Guilbert, Madeleine (1966), *Les Fonctions des femmes dans l'industrie* (The Hague: Mouton).

Hartmann, Heidi (1979), 'Capitalist Patriarchy and Job Segregation by Sex', in Zillah Eisenstein (ed.), *Capitalist Patriarchy and the Case for Socialist Feminism* (New York: Monthly Review), 206–47.

Hirata, Helena (1984), 'Vie reproductive et production: Famille et entreprise au Japon', in Marie-Agnès Barrère-Maurisson *et al.* (eds.), *Le Sexe du travail* (Grenoble: Presses Universitaires de Grenoble), 191–205.

—— (1989), 'Du souffle au "float": Transferts et changements technologiques dans l'industrie du verre' (Paris: GEDISST-IRESCO, mimeo).

—— (1992), 'Organização do trabalho e qualidade industrial: Notas a partir do caso Japonês', *Estudos Avançados, Coleção Documentos* (São Paulo: IEA–USP).

—— and Humphrey, John (1986), 'Division sexuelle du travail dans l'industrie Brésilienne', in Nicole Aubert, Eugène Enriquez, and Vincent de Gaulejac (eds.), *Le Sexe du pouvoir* (Paris: Epi), 175–89.

Humphrey, John (1984), 'The Growth of Female Employment in Brazilian Manufacturing Industry in the Nineteen Seventies', *Journal of Development Studies*, 20 (4): 224–47.

—— (1985), 'Gender, Pay and Skill, Manual Workers in Brazilian Industry', in Haleh Afshar (ed.), *Women, Work and Ideology in the Third World* (London: Tavistock), 214–31.

—— (1987), *Gender and Work in the Third World* (London: Tavistock).

IBGE (Instituto Brasileiro de Geografia e Estatística) (1979), *Pesquisa Nacional por Amostra de Domicílios, Região Metropolitana de São Paulo, 1979* (Rio de Janeiro: Instituto Brasileiro de Geografia e Estatística).

—— (1990), *Pesquisa Nacional por Amostra de Domicílios, Região Metropolitana de São Paulo, 1990* (Rio de Janeiro: Instituto Brasileiro de Geografia e Estatística).

Jenson, Jane (1989), 'The Talents of Women, the Skills of Men: Flexible Specialization and Women', in Stephen Wood (ed.), *The Transformation of Work?* (London: Unwin Hyman), 141–55.

Kenney, Martin, and Florida, Richard (1992), *Japanese Maquiladoras* (University of California–Davis, Program in East Asian Business and Development: Working Paper, No. 44).

—— —— (1995), 'The Transfer of Japanese Management Styles in 2 US Transplant Industries—Autos and Electronics', *Journal of Management Studies*, 32 (6): 789–802.

Lloyd, Beth, and Niemi, Cynthia (1979), *The Economics of Sex Differentials* (New York: Columbia University Press).

Milkman, Ruth (1992), 'The Impact of Foreign Investment on US Industrial Relations: The Case of California's Japanese-Owned Plants', *Economic and Industrial Democracy*, 13 (2): 151–82.

Phillips, Anne, and Taylor, Barbara (1980), 'Sex and Skill, Notes towards a Feminist Economics', *Feminist Review*, 6: 79–88.

Piore, Michael (1975), 'Notes for a Theory of Labour Market Stratification', in Richard Edwards, Michael Reich, and David Gordon (eds.), *Labour Market Segmentation* (Lexington, KY: D. C. Heath), 125–50.

Reich, Michael, Gordon, David, and Edwards, Richard (1980), 'A Theory of Labour Market Segmentation', in A. Amsden (ed.), *The Economics of Women and Work* (Harmondsworth: Penguin), 232–41.

Roldán, Martha (1993), 'Industrial Restructuring, Deregulation and New JIT Labour Processes in Argentina: Towards a Gender-Aware Perspective?', *IDS Bulletin*, 24 (2): 42–52.

Sewell, Graham, and Wilkinson, Barry (1992), '"Someone to Watch Over Me": Surveillance, Discipline and the Just-in-Time Labour Process', *Sociology*, 26 (2): 271–89.

Shaiken, Harley, and Browne, Harry (1991), 'Japanese Work Organization in Mexico', in G. Székely (ed.), *Manufacturing across Oceans and Borders* (San Diego: University of California, Center for U.S.–Mexican Studies, Monograph Series, 36), 25–50.

Walby, Sylvia (1986), *Patriarchy at Work* (Cambridge: Polity Press).

12

Confronting Child Labour: A Gradualist Approach

Christiaan Grootaert and Ravi Kanbur

CHILD labour is upsetting. The popular images in the industrialized world are drawn from Dickens and the 'dark, satanic mills' of the Industrial Revolution, on the one hand, and the sweatshops and street children of the cities of the developing world, on the other. A common, and natural enough, reaction in developing countries has been legislation to ban child labour, following the historical lead of the now developed world when it emerged from its period of industrialization. Trade sanctions are being recommended in some developed countries against the exports of developing countries which use child labour. Many, including developing country governments, see this as a thinly disguised protectionist device. Others argue that child-labour legislation, even if it could be enforced, is not the only way, or necessarily the best way, of tackling the problem of child labour.

Against this background, this essay presents an overview of the recent literature on child labour with a view to determine consistent and feasible policies to deal with the problem. We take an economic perspective and focus on demand and supply factors. We set up a framework for discussing the incentives that lead to child labour and that can be used as means for combating it. We first consider the conceptual and empirical problems in defining child labour, and discuss some recent estimates of its extent. We then move to a discussion of the determinants of child labour, focusing first on supply, then on demand. Subsequently, we have set out the welfare economics framework within which policy interventions can be analysed. Finally, we assess a range of policy interventions, including legislation.

There seems to be an emerging consensus in the literature that policies to deal with child labour will have to vary depending upon which types of child labour and accompanying arrangements are prevalent, and depending upon the institutional and administrative capacity of the country in question. While child-labour legislation is an important component of the policy package suggested here, by itself it is neither necessary nor sufficient for making a rapid and significant dent in the problem. It has to be accompanied by a range of incentives, for schooling, for example, and a range of targeted interventions. This, together with equitable economic growth, it is argued here, is what will eventually reduce child labour to levels that can be addressed by legislation.

This is a condensed and edited version of a background paper 'Child Labor: A Review' prepared for the World Bank's *World Development Report 1995*. An earlier version appeared under the title 'Child Labour: An Economic Perspective' in the *International Labour Review*, 134: 2 (1995), 187–203 and is used by permission of the International Labour Organization. Any views expressed in this essay are those of the authors only and should not be attributed to the World Bank or its affiliated organizations.

The Nature and Magnitude of Child Labour

How much child labour is there in the world? The answer to this question depends, of course, on what one means by child labour. To begin with, it is not clear how to define 'child'. In the West, it is customary to do so by age, but in many societies cultural and social factors enter as well (Rodgers and Standing, 1981). The evolution from childhood to adulthood passes through biologically and socially defined life phases, during which the degree of dependence and the need for protection of the child gradually decline. For example, in many societies an apprentice even if only 8 or 9 years old is often not considered a child—a determination based on social status rather than age (Morice, 1981). In that sense too, many societies, especially poor rural ones, do not view child work as 'bad'. Rather, it is part of the socialization process which gradually introduces the child into work activities and teaches the child survival skills. This view is present in many African countries (Bekombo, 1981; Agiobu-Kemmer, 1992, Amin, 1994).

The concept of 'work' is equally problematic to apply to the range of activities in which children are engaged. They can range from help with domestic work, to work in the household enterprise or farm, to wage-work. It can be light artisanal work, trading, or heavy physical work.[1] For the purpose of defining a policy towards child labour, both the nature of the work and the nature of the relationship between the child and the employer must be considered. A key element is whether the arrangement is 'exploitative'. In the extreme, this can take the form of bonded labour, quasi-slavery, or a feudal relationship. A debt incurred by the parents can be the 'bond' whereby a child is forced to work. Several million bonded child labourers are thought to be working in South Asia (ILO, 1992). It can also be considered exploitative when a child starts full-time work at too early an age, or works too many hours, or when the work imposes excessive physical, social, and/or psychological strains which hamper the child's development (UNICEF, 1986; ILO, 1992). Let us illustrate the variety of child-labour situations by describing three prototypical urban child workers.[2]

Ade, street-trader in Lagos
Ade is in the last class of primary school, and like more than one-third of his class-mates he hawks goods on the streets after school hours. The children are exposed to air pollution, vehicle accidents, abduction, and sexual exploitation and assault. Their income is a significant contribution to household income, and helps in the purchase of school books and other supplies.

Juan, quarry-worker in Bogotá
Most of the quarries in Bogotá operate without a permit from the city authorities. Children use shovels and sieves to separate rocks according to size. Juan, at age 13, is among the older children workers in the quarries. In spite of the hard work, 50 per cent of them attend school, often in the evening at a community centre. Most of the children live in squatter settlements at the outskirts of Bogotá. Their parents have little or no education, and most have no stable employment.

Taeng, masseuse in Bangkok
Taeng left her native village in Northern Thailand when she was 17 to come to Bangkok, with only a few years of primary eduction. An agent, who lent her parents money to build a house, arranged her placement. She sends money home and sees herself as the family breadwinner. Many girls like her return home after a number of years, to a mix of sympathy and derision from their fellow-villagers.

The International Labour Office (ILO) has recently produced statistics on child labour based on a uniform definition—the economically active population under the age of 15 (Ashagrie, 1993). That attempt highlighted the difficulties that arise in terms of data availability. Eventually, a number of sources had to be used, including a set of specially designed questionnaires sent to 200 countries and territories—the response rate was uneven across regions. On the basis of returns from 124 countries, the ILO estimates that there were about 78.5 million economically active children under 15 years of age in 1990. The figure for children aged 10 to 14 was 70.9 million, a participation rate of 13.7 per cent.[3]

By contrast to the ILO total, UNICEF (1991) estimated that there were 80 million children aged 10 to 14 who undertook work so long or onerous that it interfered with their normal development. The differences in the estimates result from differing definitions and methodologies. How dramatic the differences can be even within one country is demonstrated by the estimates Weiner (1991: 20–1) reports from India where 82 million children are not in school: the 1981 census reported 13.6 million children in the work-force, the official National Sample Survey of 1983 17.4 million child labourers, and a study by the Operations Research Group of Baroda, sponsored by the Labour Ministry, concluded that the child labour-force was 44 million, including children paid in kind as well as in cash. More recent estimates reported by Gulrajani (1994) put the number of child workers in India between 20 and 100 million, suggesting that the divergence in estimates has not diminished.

Determinants of Child Labour

The absence of systematic data collection on the incidence of child labour affects the research done on its determinants. Most work is based on case studies covering a sub-national area, often one or a few villages, at best a province or region. Much of it dates back to the period 1978–85, perhaps motivated by the United Nations' declaration of 1979 as the Year of the Child.[4]

Our discussion of the determinants of child labour starts with the literature on fertility and time allocation within the household. Obviously, the number of children in the household determines the potential supply of child workers; hence fertility behaviour is a determinant of the supply of child labour. Also on the supply side, the role of risk management in the household affects the extent of child labour. On the demand side, the two main determinants of child labour are the structure of the labour-market and the prevailing production technology.

Household Size and Time Allocation

In any household, a child's non-leisure time is available for schooling, home production, or income-earning work in the market. The way the household allocates the child's time depends, *inter alia*, on household size and structure, the productive potential of the child and its parents (mainly its mother) in home and market work,

and the degree to which children can take over work from parents (again, mainly the mother). The allocation of time depends on the potential income from child labour. This potential in turn affects the desired household size (see e.g. Nakamura and Nakamura, 1992; Hotz and Miller, 1988; and Rivera-Batiz, 1985). It encompasses both the income from their work as children and their income transfers to parents in old age.[5]

A recent review of the evidence on this relationship from developing countries suggests that larger household size reduces children's educational participation and progress in school, and reduces parents' investment in schooling. Both factors make it likely that larger household size increases the probability that a child will work.[6] Lloyd (1994) finds that the magnitude of this effect is determined by at least four factors:

- the level of socio-economic development: the effect of household size is larger in urban or more developed areas;
- the level of social expenditure by the state: the effect of household size is weaker if state expenditures are high;
- family culture: the effect of household size is weaker where extended family systems exist, e.g. through the practice of child fostering;
- the phase of demographic transition: the effect of household size is larger in later phases.

One implication is that the empirically observed magnitude of the effect of household size on child labour varies enormously from place to place, depending upon a combination of factors (Cochrane, Kozel, and Alderman, 1990). Let us consider some of these issues in greater detail.

An econometric study for the Philippines found that the relationship between household size and child work is not the same for market and for domestic work, and that it depends on the gender and the birth order of the child (DeGraff, Bilsborrow, and Herrin, 1993). For example, the presence of older siblings decreases the likelihood of market work by a child, especially if it is of the same sex, suggesting that older siblings free younger ones from market work, presumably because older children are likely to earn more. However, such a substitution effect was found to be absent for domestic work This study as well as others (see Lloyd, 1993) document gender roles in child labour: in many settings boys are more likely to be engaged in market work and girls are more likely to be engaged in domestic or farm work.

The degree to which boys or girls, or all children equally, are affected by household size is also determined by cultural factors. In Malaysia, for example, Chinese girls appear disadvantaged in larger households but their brothers are not (Shreeniwas, 1993). In India, families from urban slums in Tamil Nadu discriminate in order to provide a few children, mainly boys, with 'quality' private education. Families from Uttar Pradesh try to provide all children equally with less expensive public education (Basu, 1993).

Although school attendance cannot be considered the 'inverse' of child labour, the literature on the determinants of school enrolment has established two pertinent effects. First, there is a trade-off between the schooling of girls and the labour-force participation of mothers. When mothers go to work in the market, girls stay at home. In this sense, the opportunity cost of girls' schooling is not their foregone wages, but those of their mothers. Secondly, the most important determinants of school enrolment are parents' education (especially mothers' education) and household income level. Further, increased household income due to mothers' earnings will at some point establish a preference for 'quality' children. The implications for child labour are that in poor households, when mothers need to enter the labour-force, child labour will increase because girls in particular will be pulled out of school to take over domestic work or their entry into school will be delayed. However, as income increases, the effect of the mother's work on household income will become more important and child labour will decrease. This process is likely to be affected by the same societal factors just identified, namely the level of development, the level of social expenditure, cultural factors, and the phase of demographic transition. Most case studies of child labour do indeed identify poverty of the household and a low level of parental education as important factors in determining child labour (ILO, 1992).

It would stand to reason that the overall condition of the education system can be a powerful factor on the supply of child labour. Bonnet (1993) argues that the failure of the education system in Africa has led many parents to view child labour as the preferred option for their children. Education is no longer a road towards obtaining a diploma assuring a modern-sector job. In an economic environment where survival depends on work in the informal sector, many parents conclude that taking children out of school and putting them to work is the most sensible solution for survival and the education method which offers the best prospects for the future. As one African commentator put it, 'Education broadens your mind but it does not teach you how to survive' (Agiobu-Kemmer, 1992).

The extent to which the state can influence a household's decision with respect to child labour, and counteract the effect of fertility and large household size, is highlighted in a study of Malaysia (Shreeniwas, 1993). As part of its policy to reduce ethnic inequalities, the Government of Malaysia systematically favours education of Malays through scholarships and other subsidies. As a result, no negative effects from household size emerged for these households in contrast to Chinese and Indian households, which did not benefit from government subsidies and among which a strong negative effect of household size on schooling was observed.

The final supply-side determinant of child labour considered here relates to the labour-market itself: the wage level in the market—both the wages of children and those of adults. Evidence from Egypt (Levy, 1985) and rural India (Rosenzweig, 1981) suggests that the supply of child labour increases when wages increase, especially so for younger children. However, the supply of child labour for wages is reduced when the wages of mothers increase. The data from Egypt show that this effect is strongest for younger children. The data from India show that an increase in women's wage rates decreases girls' labour-force participation, but has no effect on boys' participation. The opposite is true for men's wage rates: increases reduce the labour supply of boys, but have no effect on girls' labour supply.

Household Responses Towards Risk

Households send children to work in order to augment household income but also to manage better the income risk they face. Child labour can be part of a strategy to minimize the risk of interruption of the household's income stream, and hence to reduce the potential impact of job loss by a family member, etc. (Cain and Mozumder, 1980). The impact of such a loss is more severe for poor households. Where the level of income is very low, any interruption can be life threatening, particularly in the absence of savings, liquid assets, or ability to borrow (Mendelievich, 1979). Therefore the risk argument provides a further explanation as to why child labour is more prevalent among poor households. This income-insurance argument is not the same as that of old-age insurance: parents want to have children so that they will take care of them in their old age. The latter argument explains a high demand for children, but not child labour. In fact, it should induce parents to send children to school, since that would increase their earnings and hence the potential transfers they can provide for their parents in old age.

The policy implication of the income-insurance argument is that in settings where household risk management is an important reason for child labour, attempts at its forced abolition, e.g. by legislation on child labour or compulsory schooling are likely to fail, since they would threaten household survival. Such attempts at abolition would need to be accompanied by mechanisms providing households with insurance against income fluctuations in other ways, e.g. by short-term credit which does not require collateral.

The Structure of the Labour-Market

The economic value of children, and its implications for reproductive behaviour, cannot be properly assessed without reference to the structure of the labour-market. The latter determines the level of wages, which in turn determines the contribution of children to household income. A key factor is the flexibility of wages. In competitive markets where wages are flexible, children can substitute for adults in the market-place. Where minimum wages are established, whether due to legislation or to collective bargaining, employers will prefer adult workers—assuming their productivity is higher than that of children. Effective minimum wages can thus in principle deter child labour, although in practice one must ask whether minimum wage legislation is any more likely to be effectively enforced than legislation banning child labour.

There is little hard evidence on the level of remuneration of children and the extent of wage discrimination against them. In a review of case studies, Bequele and Boyden (1988) conclude that children's earnings are consistently lower than those of adults, even where the two groups are engaged in the same tasks. Jomo (1992) reaches the same conclusion based on several case studies on Malaysia. Even in the industry of hand-knotted carpets, where child workers are seen to have a skill advantage over adults—with more nimble fingers they can tie smaller knots, evidence for India indicates that children are systematically being paid lower wages than adults (Gulrajani, 1994). Where the demand for child labour in the market is concentrated in the hands of a few firms, children's wages will be depressed more. Such conditions occur often in developing countries due to concentrated ownership of land, credit and product monopolies, share-cropping arrangements, imposed or natural restrictions on labour mobility, or a lack of alternative employment possibilities (Cain and Mozumder, 1980).

The relative importance of the formal sector in the economy and the degree of segmentation between it and the informal sector also affect the demand for child labour. In general, the evidence suggests that the extent of child labour in the formal economy is small, with the possible exception of plantations (see e.g. Bonnet, 1993, for Africa; Goonesekere, 1993, for Sri Lanka). However, in many countries there is a tendency towards an 'informalization' of production methods with formal enterprises either breaking up into smaller units or sub-contracting to households or informal enterprises—mainly to escape social legislation and charges which add to the cost of labour. In such conditions, the demand for child labour may well increase.

An important aspect of the issue of child labour and remuneration is the apprenticeship system, which ties a child to a small enterprise, usually for many years, in principle to learn a trade. In practice, especially in the early years of the apprenticeship, the child often serves 'the master', and only later will there be any actual learning (Mendelievich, 1979). Nevertheless, an apprenticeship can contribute to a process of socialization while transferring know-how, and some have argued that it would therefore be a mistake to view it as exploitation of child labour (Bonnet, 1993).

The Role of Technology

The second major factor underlying the demand for child labour is the technology of production, since it affects the extent to which children have an advantage over adult workers. Many of the cases where this factor plays a crucial role are those that incite reports in the press and by voluntary agencies. Examples are the boys who work in mines or as chimney sweeps because the tunnels/chimneys are too small for adults to crawl through, the girls who weed and pick cotton, the children who weave carpets. In these cases, children are 'superior' workers relative to adults, and require less capital input.

Changes in technology can have a profound impact on the incidence of child labour. In the Industrial Revolution, the mechanization of spinning and weaving led to a reduction in the demand for child labour (Galbi, 1994). The Green Revolution in India led to reduced child labour and increased school attendance (Rosenzweig, 1981). The mechanization of Egyptian agriculture, especially the expanding use of tractors and irrigation pumps, reduced the demand for child labour in tasks such as picking cotton, driving animals to power waterwheels, and hauling freight (Levy, 1985). In the Philippines, the introduction of electricity in the community reduced the amount of market labour for children, and electricity in the home reduced the amount of home production by children (DeGraff, Bilsborrow, and Herrin, 1993). Even quite elementary technological change can have a major impact: in the quarries of Bogotá, the introduction of wheelbarrows displaced children who previously carried rocks piece by piece (Salazar, 1988).

Today's technology can have contradictory effects on the demand for child work. Miniaturization and assembly-line production in the electronics and electrical appliance industries has once again led to demand for 'nimble fingers'. In garment production, the advent of fairly cheap multi-function sewing machines has once again made possible home production which often involves girls. Empirical assessments of the implications of technological change are lacking and would be needed before the importance of technology relative to other demand factors can be assessed.

The Welfare Economics of Child Labour

Now that we have reviewed some of the key determinants of child labour, we are in a position to discuss policy interventions to reduce child labour. However, a decision to seek to reduce or even ban child labour must be based on a conviction that there is 'too much' of it relative to the social optimum. Whether or not there is too much or too little of an activity can be examined systematically using the framework of conventional welfare economics—such an analysis may also provide insights into appropriate interventions. This section outlines a few of the issues and possible interventions.

The basic analytical framework applied here is that of household decision-making in the allocation of children's time between labour and non-labour activities, together with an assessment of private and social returns to each activity. Each household will seek to allocate the time of its children to wherever the perceived private return, the gain to the household, is highest. The crucial question is whether the private return matches the social return, suitably defined. Conceptually, there are three sets of issues. First, there are those to do with pure efficiency, where no distributional questions are raised. Secondly, there are issues involving intra-household distributional considerations. Thirdly, there are the issues involving inter-household distribution. Let us consider each of these in turn.

Suppose, for example, there is a failure in the market for education, whereby the social returns to primary education are higher than the private returns, e.g. primary education for girls leads to lower fertility and this is desirable from the social point of view. This means that in the social optimum more children would be at school than is the case on the basis of household decisions alone. What should be the nature of the intervention? Basic welfare economics suggests that it is best to attack market failures in the very markets in which they occur—it is second-best to intervene in related markets. Thus, in this situation it would be second-best to attack the problem by taxing or banning child labour, thereby inducing households to use that time in alternative ways. Rather, policy should focus on raising the private rate of return to education to bring it closer to the social return. The same argument applies in cases where incomplete markets for risk spreading lead to the use of child labour as a diversification device. The best solution is to encourage the development of credit and risk markets.

Where there is discrimination against children within the household, child labour may be a problem even in the absence of market failures. Such discrimination often affects female children. The appropriate intervention will depend on the intra-household allocation process. In the unitary model of the household where the head of the household is the sole decision-maker, the issue is how to rearrange incentives for the head of household so that he does 'the right thing' from a societal perspective. This can be done either by 'taxing' or banning child labour or by subsidizing education. There is no longer an obvious ranking of these two alternatives. If there is intra-household bargaining—perhaps between the father and the mother–child dyad, directly altering the bargaining power of the mother is a possible instrument. Increasing wages, even to the child, can then be seen as strengthening the mother–child dyad.

Households differ in their wealth and capabilities such that an aggregation of households leads to a distribution of welfare outcomes. In what sense might there now be 'too much' child labour? The answer is that it may be the case that child labour is associated with low-income households, in which case a reduction in household poverty would lead to a reduction in child labour. In this situation, child labour should be used as a targeting device. Interventions which transfer resources, e.g. nutrition, to child labourers can be effective, since this is a way to transfer resources to poor households. Of course, such interventions may well lead to an increase in child labour, but that should not necessarily worry us from this perspective, and, in the short run, if the objective is to help poor households and if the poverty alleviation effects outweigh possible increases in child labour.

Within this framework, we can now examine the consequences of a popular policy intervention: banning child labour. If the ban is enforced, it means that children will no longer be found in the labour-market and will spend more time in family work or schooling. If there was previously an inefficiency in the eduction market, so that there were too few children in school from the point of view of the social optimum, then an effective ban will move us closer to that optimum. However, suppose that child welfare depends on the cash income of the child, e.g. because it strengthens the bargaining power of the mother–child dyad. Then, of course, banning child labour may leave the child worse off after the intra-household bargaining is completed. Likewise, where child labour is a manifestation of poverty, a ban on child labour makes the poor household worse off, since it is a restriction of its income opportunities.

If, as is quite likely, the ban is not enforced, it will create rents in the system. Quite simply, if employers are the ones who would be fined, they bribe the policemen or others so that they can let them continue to employ child labour. But this bribe is an extra cost to them and will reduce the demand and/or the wages for child labour. To the extent that more children would then go to school, this will move the system closer to efficiency if the market failure is in the market for education. But again, negative outcomes for the child are possible if the bargaining power of the mother–child dyad is reduced, or if poor households have lower income from child labour.

The welfare economics of child labour thus leads to a complex and densely textured analysis which does not suggest a single, or even a dominant, way of approaching the issue. An array of policy instruments is likely to be required, addressing different aspects of failures arising from efficiency or distributional considerations. In particular, it seems clear that legislation, even if it can be enforced, is at best only one instrument in an array that has to be deployed.

There is, however, a major counter to the welfare economics perspective on child labour. This is the non-welfarist framework where certain rights are self-evident, natural, and given. A good example is slavery. It is generally accepted that no one has the right to sell himself or herself into slavery—irrespective of whether this would be welfare-improving in terms of efficiency or even distribution. It can be argued that child labour falls into this category, that child labour violates a basic human right, and should therefore be banned. From such a basic rights perspective a focus on international conventions and legislation emerges. Not only is the signing of conventions and passing of legislation symbolic—a clear expression of the acceptance of the right in question—but vigorous attempts at enforcing such legislation are seen as furthering a basic right. From this perspective, the fact that attempts to enforce legislation may in some sense hurt the very group whose right is being protected is besides the point.

Policy Intervention: The Need for a Diversified Approach

It is now well understood that a major cause of child labour is quite simply the poverty of the household. General economic development, equitably distributed, is the best and most sustainable way of reducing child labour. Beyond this, our review has shown that special emphasis should be given to fertility reduction and to reducing the costs of school attendance, if the objective is to reduce child labour. Also, the impact of technology in determining the demand for child labour has to be a matter of special concern. In terms of specific project interventions, programmes in Manila (Gunn and Ostos, 1992) and elsewhere show the value of an integrated approach covering all of these factors as well as the provision of alternative income sources for children and improved employment opportunities for their parents. While there is general agreement on such policy interventions, there is considerable debate on the role of legislation in addressing the issue of child labour.

Historically, the single most important and common approach to the problem of child labour has been the adoption of legislation. The ILO has sponsored many Conventions and Recommendations banning child labour, and most countries now have some form of legislation or regulation prohibiting the employment of children below a certain age and specifying the conditions under which minors may work (Bequele and Boyden, 1988). The age threshold and the scope of the legislation vary. In the majority of countries the minimum age for employment is 14 or 15, but there are about 30 countries where it is only 12 or 13 years (ILO, 1992). In many countries higher minimum ages apply for hazardous work. Almost all legislation exempts work in household enterprises, but some limits the scope further by excluding also domestic service and agriculture.

A fundamental problem with enacting and enforcing legislation banning child labour is that there are few interest groups to support it: the government often considers it embarrassing to admit the existence of child labour, the employers of children will be hostile to legislation, and so, quite likely, will be the children themselves and their parents (Morice, 1981). Legislation is therefore likely to be effective only where there is a capable administration determined to implement the laws, where there is considerable difficulty in hiding child labour, and where relatively little advantage is to be gained from child work (Rodgers and Standing, 1981). The employment of children in the cotton mills of eighteenth-century Europe began to fall a half century before the introduction of legislation, mainly because it was to the advantage of factory owners to switch to adult workers (Galbi, 1994). Legislation to ban child labour thus fitted in with an ongoing economic trend.

Documentation on effective enforcement of child-labour legislation is spotty at best. Most information pertains to labour inspectors describing their degree of understaffing and their difficulties to visit factories and enforcing penalties before the courts. There is little question that in many countries labour inspectorates are seriously understaffed. In the Philippines, for example, there are fewer than 200 inspectors nationwide for almost 400,000 employers (Bequele and Boyden, 1988).[7] The ongoing trend towards informalization in the labour-market makes inspection even more difficult.

In putting legislation in place an argument can be made for using a graduated approach. As Rodgers and Standing (1981: 39) point out, 'it is one of the ironies of child labour that, where it is prohibited by law, the law is likely to leave child workers unprotected, since legally they do not exist.' Legislation on working conditions, e.g. safety, working hours, needs to be made to apply to all workers, including children. It may be easier, in a first phase, to force employers to limit children's workdays and to provide adequate lighting or safety equipment, rather than to force them to forego child labour altogether. Compulsory schooling laws can be seen in this context. The record of enforcement seems better here than with legislation banning child labour (Weiner, 1991). Part-time work might be allowed, at least initially, in combination with school attendance. Potentially, the local community has a crucial role in this, since community-based monitoring of school enrolment and attendance is likely to be more effective than occasional visits from an inspector of the Ministry of Education.

Ultimately then, child labour is best addressed through a combination of legislation and economic incentives (Myers, 1991; Goonesekere, 1993). This is now well accepted by international agencies addressing the problem, such as UNICEF and ILO (see e.g. UNICEF, 1986; ILO, 1992). It is unlikely that any one approach will succeed everywhere and the balance between legal and economic measures needs to be adapted to the incidence of child labour, the type of relevant work and work arrangements, and market conditions.

To illustrate, Brazil and India have both followed approaches which combine legal action with economic incentives. In India, the legal framework is provided by the 1986 Child Labour Act, prohibiting the employment of children below 14 in hazardous occupations. The economic framework is in the National Policy on Child Labour, which targets education for all children up to age 14, and proposes an extensive system of non-formal education combined with employment and income-generating schemes in areas with a high incidence of child labour. A series of pilot projects has been established (Narayan, 1988). In Brazil, the focus is on street children. The government has recognized that its customary bureaucratic procedures fail to address this problem effectively. Instead, a community-based strategy was pursued, and the government's role limited to providing technical support. By 1986, local volunteer 'commissions' existed in most major urban areas which were able to mobilize community-based resources (Myers, 1988).[8] To illustrate different approaches to child labour, we revisit the three prototypical child workers we described earlier, and suggest ways to help them and their families.

Ade, street-trader in Lagos

Since Ade is not yet 14 years old, his street-trading activity is illegal according to Nigerian law. The police make some effort to enforce the law. In 1988, almost 7,000 street-traders were arrested, and most were convicted. However, a new wave of kids immediately took their place. Street-trading makes an important contribution

to household income and many children combine it with educational achievement. The latter are jeopardized when children work too many hours. Rather than banning the activity, the law might focus on restricting trading hours. In order to protect the children, some of their trading places might be consolidated in informal open air markets. Also, a system could be instituted to monitor the health condition of the children.

Juan, quarry-worker in Bogotá

Juan has to work in the quarry because his father has no education and can obtain employment only irregularly. Juan's work is part of the family's survival strategy. Moreover, his parents do not perceive education to have a high pay-off in Colombia's economic situation characterized by high unemployment. Many of the children in the quarry do go to school, and school meals and free shoes and clothing would be likely to increase attendance. The introduction of wheelbarrows has already reduced the demand for child labour, and the appearance of dump trucks has accelerated the trend. The government could promote the adoption of these innovations, perhaps by small subsidies. This would enhance the demand for adult workers from the same squatting areas.

Taeng, masseuse in Bangkok

Taeng, and many girls in her situation, come from poor farming villages with insufficient economic opportunities. Creating more local income opportunities is the prime challenge to reduce the supply of girls being sent to Bangkok. Improved agricultural extension, double cropping, new cash crops, all will help. Opening up opportunities outside agriculture would both increase income and reduce income volatility. Improving access to credit for the farmers from another source than the 'agent' would avoid child prostitution as the means to pay off a loan. On the legislative side, the problem is quite obviously enforcement, since prostitution is already illegal in Thailand. The difference between the law and the reality is such that enforcement needs to be focused on child prostitutes. In view of growing health problems (AIDS, drug use) among prostitutes, free medical services to those below 18 years of age would be a case in point where the illegal activity can be used as a targeting device.

If economic incentives and legislation are the two pillars on which efforts to help working children need to be based, their effectiveness will be greatly enhanced by simultaneous efforts at advocacy and mobilization, and by empowerment of the children and their families. Because many of the most exploited and endangered working children go unnoticed, their situation must be brought forcefully to the attention of government and the public so as to mobilize a constituency to defend them. Advocacy entails combating ignorance but also prejudice, fear, and the denigration of working children. There is irony in such negative views, because working children often assume great responsibility to help themselves and their families (Myers, 1991). Many working children and their families operate within very limited economic and social options—effective empowerment can expand those options and give the children a larger share of society's possibilities and benefits. Myers (1991) discusses ways to achieve this, most of which are characterized by the strong involvement of local communities, aided and supported by national or international organizations.

A stigma is attached to child labour, certainly among government officials. Sometimes a more positive view can open a road to helping working children more effectively, especially by extending certain benefits to child workers. This is especially appropriate where child labour is a manifestation of poverty. Many working children are inadequately fed and do not go to school. The government could provide factory meals to working children, just as it sometimes provides school meals. Similarly, primary health-care workers could be mandated to visit and provide free care to child workers. Employers could be required to restrict children's workday to five or six hours so that they can attend school at least part-time. Such measures would benefit children who are confronted by economic difficulties which are not likely to improve in the short run. Admittedly, the added benefits may be perceived as an increase in the children's real wage and increase the supply of child labour, but this is not a reason for not helping child workers and their families.

Such programmes, of course, require rather profound shifts in many governments' current attitude of denial and/or 'all-or-nothing' bans. We think that such gradual solutions, which recognize an economic reality and devise policy accordingly, will contribute to the elimination of child labour in the longer term.

Conclusion

On 30 September 1990, the first World Summit for Children promised to reduce child mortality and malnutrition, and set targets to be reached by the year 2000. There were no explicit goals relating to child labour, but the targets included basic education for all children and completion of primary education by at least 80 per cent of children. These goals, if met, will reduce child labour.

The evidence reviewed in this essay shows that education interventions play a key role in the reduction and eventual abolition of child labour. Other measures are needed as well, such as fertility reduction, the modification of technology, improvements in employment opportunities for adults, labour-market policies, and legislative action. Such measures need to be accompanied with efforts at advocacy for working children, and empowerment of the children and their families.

A consensus that effective action must first of all aim at protecting the children and improving their living and working conditions is emerging in the literature as well as in the policies of international agencies concerned with child labour. This implies a less stigmatized view of child labour and a gradualist approach in policy intervention. The condition of child labour can itself be used as a targeting device to help the children with schooling, health care, and nutrition. The long-term objective to eliminate child labour needs to be approached through a package of legislative action and economic and social incentives that takes into account the circumstances of child labour as well as institutional and administrative capacity.

Notes

1. For a useful typology of children's activities, see Rodgers and Standing (1981).
2. These cases are our composite constructions based on case studies in Oloko (1991); Salazar (1988); Phongpaichit (1982); Weiner (1991); and Myers (1991).
3. As this essay went to press, ILO announced an increase in its estimate of 5 to 14 year old child workers to 250 million.
4. A very useful bibliography covering the literature of this period has been provided by the ILO (1986).

A more recent bibliography is found in the Nov. 1994 issue of *Labour, Capital and Society*, 27 (2). Recent case studies are reported by Bequele and Boyden (1988); Myers (1991); Jomo (1992); and Goonesekere (1993).

5. Formal presentations of models of the household economy which explicitly take into account the economic contributions of children can be found, *inter alia*, in Levy (1985); Rivera-Batiz (1985); and Sharif (1994). Much of this work is based on Rosenzweig and Evenson (1977).

6. Education and work are not the only factors that affect child welfare. There exists a vast literature which has demonstrated that large family size adversely affects many aspects of child welfare: health, intelligence, physical development, etc. (see the review by King, 1987).

7. A detailed and interesting description of enforcement problems and issues in the case of Sri Lanka is provided by Goonesekere (1993).

8. For recent information on these cases, see Boyden and Myers (1995). For further discussion on street children, see UNICEF (1986) and Taçon (1991).

References

Agiobu-Kemmer, I. S. (1992), 'Child Survival and Child Development in Africa', Bernard van Leer Foundation Studies and Evaluation Papers, No. 6 (The Hague).

Amin, A. A. (1994), "The Socio-Economic Impact of Child Labour in Cameroon', *Labour, Capital and Society*, 27 (2): 234–48.

Ashagrie, K. (1993), 'Statistics on Child Labour: A Brief Report', *Bulletin of Labour Statistics*, 1993–3, ILO.

Basu, A. M. (1993), 'Family Size and Child Welfare in an Urban Slum: Some Disadvantages of Being Poor but "Modern" ', in Lloyd (1993), 375–414.

Bekombo, M. (1981), 'The Child in Africa: Socialization, Education and Work', in Rodgers and Standing (1981), 113–30.

Bequele, A. and Boyden, J. (1988) (eds.), *Combating Child Labour* (Geneva: ILO).

Bonnet, M. (1993), 'Child Labour in Africa', *International Labour Review*, 132 (3): 371–89.

Boyden, J., and Myers, W. (1995), 'Exploring Alternative Approaches to Combating Child Labour: Case Studies from Developing Countries', Innocenti Occasional Papers, Child Rights Series, No. 8 (Florence: UNICEF).

Cain, M., and Mozumder, A. (1980), 'Labour Market Structure, Child Employment and Reproductive Behavior in Rural South Asia', World Employment Programme Research; Population and Labour Policies Working Paper, No. 89 (Geneva: ILO).

Cochrane, S., Kozel, V., and Alderman, H. (1990),

'Household Consequences of High Fertility in Pakistan', World Bank Discussion Paper, No. 111 (Washington, DC).

DeGraff, D. S., Bilsborrow, R. E., and Herrin, A. N. (1993), 'The Implications of High Fertility for Children's Time Use in the Philippines', in Lloyd (1993), 297–330.

Galbi, D. (1994), 'Child Labour and the Division of Labour', mimeo (Cambridge: Centre for History and Economics, King's College).

Goonesekere, S. W. E. (1993), *Child Labour in Sri Lanka—Learning from the Past* (Geneva: ILO).

Gulrajani, M. (1994), 'Child Labour and the Export Sector: A Case Study of the Indian Carpet Industry', *Labour, Capital and Society*, 27 (2): 192–214.

Gunn, S. E., and Ostos, Z. (1992), 'Dilemmas in Tackling Child Labour: The Case of Scavenger Children in the Philippines', *International Labour Review*, 131 (6): 629–46.

Hotz, V. J., and Miller, R. A. (1988), 'An Empirical Analysis of Life Cycle Fertility and Female Labor Supply', *Econometrica*, 56: 91–118.

ILO (International Labour Office) (1986), *Annotated Bibliography on Child Labour* (Geneva: ILO).

—— (1992), *World Labour Report 1992* (Geneva).

Johnson, D. G., and Lee, R. D. (1987) (eds.), *Population Growth and Economic Development: Issues and Evidence* (Madison: University of Wisconsin Press).

Jomo, K. S. (1992) (ed.), *Child Labour in Malaysia* (Kuala Lumpur: Varlin Press).

King, E. M. (1987), 'The Effect of Family Size on Family Welfare: What do We Know?', in Johnson and Lee (1987), 373–411.

Levy, V. (1985), 'Cropping Pattern, Mechanization, Child Labor and Fertility Behavior in a Farming Economy: Rural Egypt', *Economic Development and Cultural Change*, 33: 777–91.

Lloyd, C. B. (1993) (ed.), 'Fertility, Family Size and Structure—Consequences for Families and Children', Proceedings of a Population Council Seminar, New York, 9–10 June 1992 (New York: Population Council).

—— (1994), 'Investing in the Next Generation: The Implication of High Fertility at the Level of the Family', Research Division Working Paper, No. 63 (New York: Population Council).

Mendelievich, E. (1979) (ed.), *Children at Work* (Geneva: ILO).

Morice, A. (1981), 'The Exploitation of Children in the "Informal Sector" ', in Rodgers and Standing (1981), 131–58.

Myers, W. (1988), 'Alternative Services for Street Children: The Brazilian Approach', in Bequele and Boyden (1988), 125–43.

—— (1991) (ed.), *Protecting Working Children* (London: Zed Books Ltd).

Nakamura, A., and Nakamura, M. (1992), 'The Econometrics of Female Labor Supply and Children', *Econometric Reviews*, 11 (1): 1–71.

Narayan, A. (1988), 'Child Labour Policies and Programmes: The Indian Experience', in Bequele and Boyden (1988), 145–60.

Oloko, B. A. (1991), 'Children's Work in Urban Nigeria: A Case Study of Young Lagos Street Traders', in Myers (1991), 11–23.

Phongpaichit, P. (1982), 'From Peasant Girls to Bangkok Masseuses', in *Women, Work and Development Series*, 2 (Geneva: ILO), 1–80.

Rivera-Batiz, F. L. (1985), 'Child Labor Patterns and Legislation in Relation to Fertility', mimeo (Bloomington: Department of Economics, Indiana University).

Rodgers, G., and Standing, G. (1981) (eds.), *Child Work, Poverty and Underdevelopment* (Geneva: ILO).

Rosenzweig, M. R. (1981), 'Household and Non-Household Activities of Youths: Issues of Modeling, Data and Estimation Strategies', in Rodgers and Standing (1981), 215–44.

—— and Evenson, R. (1977), 'Fertility, Schooling, and the Economic Contribution of Children in Rural India: An Economic Analysis', *Econometrica*, 45 (5): 1065–79.

Salazar, M. C. (1988), 'Child Labour in Colombia: Bogotá's Quarries and Brickyards', in Bequele and Boyden (1988), 49–60.

Sharif, M. (1994), 'Child Participation, Nature of Work, and Fertility Demand: A Theoretical Analysis', *Indian Economic Journal*, 40 (4): 30–48.

Shreeniwas, S. (1993), 'Family Size, Structure, and Children's Education: Ethnic Differentials over Time in Peninsular Malaysia', in Lloyd (1993), 331–72.

Taçon, P. (1991), 'A Global Overview of Social Mobilization on Behalf of Street Children', in Myers (1991), 87–98.

UNICEF (1986), 'Exploitation of Working Children and Street Children', UNICEF, Executive Board, 1986 Session, Report E/ICEF/1986/CRP.3 (New York).

—— (1991), *The State of the World's Children 1991* (Oxford: Oxford University Press).

Weiner, M. (1991), *The Child and the State in India: Child Labor and Education Policy in Comparative Perspective* (Princeton: Princeton University Press).

IV   Social Organization in the City

Introduction

THE city presents problems of social integration and social control. The individual or family recently arrived from a village, from another town, or from another part of the city may be surrounded by unfamiliar faces in the neighbourhood and at work. In a crisis there may be nobody to turn to for help. At the same time, anonymity or superficial acquaintance limit the control that can be exerted collectively as well as in transactions among individuals. Even parents have little control over what their older children do much of the time.

Actually, most urban dwellers are well integrated. For some rural–urban migrants, the village continues to provide social integration as well as economic security, as we have seen in Part II. And most migrants move to a location where they expect to be received by relatives or friends. Some find housing in the same neighbourhood, or a job with the same employer. But even when not so clustered, people of common origin frequently maintain close ties. The associations or 'clubs' of Amerindians in Lima are famous. Moßbrucker shows, in Chapter 5, how they meet the migrants' socio-cultural needs, provide access to economic opportunities, and maintain and support relations with the village or region of origin. The urban-born, in turn, grow up in contexts of family, neighbourhood, school, and work which effectively integrate most of them.

Women usually do most of what Ursula Sharma (1986) calls 'household service work': creating and maintaining ties with kin, neighbours, and friends who are sources of information and aid. Larissa Lomnitz (1994) draws on her own research on urban upper, middle, and poor strata in Mexico and Chile to demonstrate how in each case a system of informal exchange is embedded in long-term social relations—primarily with kin, but also with fictive kin, neighbours, friends, and schoolmates. These networks constitute 'social capital' that can be converted into economic assets: security, status, and power. For the poor the foremost problem is sheer phys-

ical survival; for the middle class it is the maintenance of a perpetually threatened economic status; while among the very wealthy it is the retention of their economic privilege. Women are the principal agents in creating and maintaining these social networks as they draw on the specific roles assigned to them in the rites of sociability, in the forms of economic exchange, and in the cultural links which determine trust.

Mutual aid among the urban poor was the focus of Lomnitz's classic study in Cerrada del Cóndor, a shanty-town in Mexico City, presented in Chapter 13. She argues that the urban poor in Latin America find their ultimate source of livelihood in market exchange, but cannot survive individually: the market fails to provide any security, and the poor are not in a position to accumulate savings. They survive by complementing market exchange with a system based on resources of kinship and friendship, a system that follows the rules of reciprocity embedded in a fabric of continuing social relationships among equals. In Cerrada del Cóndor, the patterns of individualism and mistrust prevalent in rural Mexico are superseded by powerful tendencies towards integration, mutual assistance, and co-operation. Recent arrivals are housed, sheltered, and fed by their relatives in the shanty-town; the men are taught a trade and oriented towards available urban jobs, in direct competition with their city kin. The migrants thus become integrated into local networks of reciprocity. How common such support networks are in Mexico is a matter of debate (Willis, 1993).

The economic crisis of the 1980s brought an increase in mutual-help networks of relatives and friends in Mexico and elsewhere in Latin America. Sharing of housing became more common, as Kusnetzoff reports for Chile in Chapter 19, and communal kitchens appeared. Such household and community survival strategies generated tensions which led to the break-up of households and community fragmentation. As the economic crisis continued, so the possibilities of obtaining help from family and friends diminished, whilst state-provided services also deteriorated. Unorganized forms of protest, such as the collective looting of shops and shopping malls, became more common. And there was an increase in violence, robbery, alcoholism, drug consumption, and prostitution (Oliveira and Roberts, 1996).

The role of kin varies greatly across cultures, across socio-economic strata, and with political and economic conjunctures. Rugh (1984: 244) reports from Cairo that families lacking a broad set of kin conveniently nearby compensate by placing greater reliance on nuclear family members and those relatives who happen to be close at hand. Friends, office colleagues, and neighbours come to be treated as sister, brother, or other kin. The practice is most common among first-generation urban dwellers who are poor and particularly dependent on support networks. Even in China, where the state provides a subsistence floor for most, people still call on kin when the extraordinary emergency does occur and money or assistance with an ill person is needed. Since the increase in unemployment in the late 1970s, kin have been called on to help with job contacts. Kin working in shops or elsewhere in the commercial network help find scarce goods. And when kin live in the nearby countryside, there may be frequent visiting back and forth—the urbanite bringing gifts of consumer goods that cannot be found in the countryside and the villager bringing gifts of food not available in the city (Whyte and Parish, 1984: 335–6).

Unlike the typical rural setting, the city gives a measure of choice as to how closely one associates with various relatives, whether to discontinue relationships with some of them altogether. In Cairo, the established middle class, which has long since made its adjustments to city living, weighs the costs and benefits of relationships, and the competition for educational achievement and for status tends to cut if off from those below (Rugh, 1984: 244). A. L. Epstein (1981) describes the pattern in the 1950s on the Copperbelt in what has since become Zambia. The corporate kin group of the village is transformed into a social network of urban kin, a network maintained and developed selectively. At the same time, the range of recognized kin may be extended. A minimum level of support may be

expected from all people sharing a common origin.

Household composition has been commonly assumed to change from the extended family to the nuclear family in the course of industrialization. In fact, pre-industrial patterns varied widely. The nuclear family, comprising parents and their unmarried children, was the norm in many pre-industrial societies. And even where the extended family—married siblings co-residing with their parents—was the ideal, it was realized only by a minority, for example by some but by no means all wealthy families in China (Baker, 1979: 1–25).

In contemporary cities, joint families, i.e. married siblings maintaining a common household, are the exception. Stem families, that is, three-generation households, are more common. In Calcutta, a survey found about one in five households to be constituted of stem families, about one in eight of joint families in 1973–4 (Chakrabarti, 1985). A study of urban China suggests that half or more of men as well as women over the age of 60 lived with a married son or, rather rarely, a married daughter in the 1970s (Whyte and Parish, 1984: 157, 144). Among Taiwanese couples married more than ten years, living in large cities, and with the husband's parents available, 27 per cent lived in stem families and another 4 per cent in extended families, in 1986 (Weinstein et al., 1994).

Stem families are usually based on the elders' control over resources—over land or cattle in peasant societies, over family enterprise in the urban setting. Where they do not hold such power over productive resources, urban elders often control precious housing and assist with child care and domestic work. Such assistance is crucial for mothers of young children who work away from home and lack access to adequate child-care facilities. Lee, Parish, and Willis (1994) apply a 'power and bargaining model' to the former situation, a 'mutual aid model' to the latter. They conclude, however, that an 'altruism/corporate group model' is most appropriate to explain the results of a 1989 survey of married children's financial support for their parents in Taiwan. This model suggests both an effective communality of interests between generations and an assumption that contracts can be enforced across time. Such long-term patterns of intergenerational loans and insurance are grounded in filial piety—a norm that continues to be effective not only in Taiwan.

Children are socialized into obligation, guilt, and gratitude towards their parents in most cultures. Some governments explicitly support filial piety, e.g. in their choice of school textbooks—and rely on it to the extent that they provide very little old-age support. Where economic growth has been rapid, as in Taiwan and the other three 'Little Tigers', Hong Kong, Singapore, and South Korea, most children are more prosperous than their parents, and the duty towards parents is a general generational responsibility. This may be expected to change once the rate of economic growth slows and the economic gap between children and parents narrows.

Survey data from 24 developing countries support the proposition that in nearly all these countries men and women marry later in urban than in rural areas (Morgan, 1993). Delayed marriage may be in the interests of parents who continue to receive a large share of their children's income, but it also weakens their control: they are likely to be less involved in the selection of the marriage partner, and the young couple are less likely to reside with parents. Marrying at a later age is also likely to strengthen the position of the wife vis-à-vis her spouse and his parents.

Increasing numbers of women work outside the home. They are subject to discrimination in the labour-market, as we have seen in Part III. But their experiences in the work-place and the income they earn also affect their domestic position. Still, Ramu (1989: 54–70) reports from a 1979 survey of married industrial workers in Karnataka, India, that, even though almost half would have preferred to stay at home, the vast majority had to legitimize their outside work by assuming primary responsibility for familial tasks.

Hale Cihan Bolak, in Chapter 14, analyses the negotiation of conjugal rights and responsibilities in dual-earner working-class households in Istanbul. She describes five distinct marital relationships within this narrowly circumscribed universe. Traditional patriarchal constructions of gender, responsibility, and male authority, and the limitations on women's choices posed by the threat of violence and lack of alternatives continue to constitute powerful determinants of marital relationships. But they interact with the economic reality of women's contribution to family income as well as the emotional dynamics of the couple's relationship.

Bolak emphasizes that changes in kin relations in the urban setting profoundly affect marital relationships. As the support and control of kin weaken, marital negotiations become more flexible. Emotional factors such as understanding and attention gain in importance. The negotiating power of women increases, at the same time the social constraints compelling men to discharge their familial responsibilities diminish.

In the extreme case men abdicate all responsibility, abandoning their family. Other men contribute so little to the household, while demanding so much, that women prefer to raise their children on their own (Chant, 1985). Not all single mothers are left without any support from the fathers of their children, but women desperately struggling to fend for their children and themselves are all too common.[1] We encountered some such women in Nelson's account of Mathare Valley in Chapter 10. But even where co-resident parents both contribute to household income, women typically devote a larger share of their earnings to their families' subsistence— 'maternal altruism' may be re-enforced by 'maternal self-interest' for women who are more likely than men to become dependent on support from their children. Bolak speaks of gendered spending prerogatives and priorities. And Roldán's (1988) account of different patterns of income allocation in working-class households in Mexico City demonstrates the disproportionate share of their earnings men keep for their personal expenses.

Some children have lost just about all parental support. Street children are the most conspicuous example. They have been a common sight in Latin America for many years. They are increasingly found in Asia and Africa as well. Some estimates put the number of street children in developing countries at 30 million. They have caught the attention of journalists, charitable organizations, and policy-makers, but academic researchers appear singularly oblivious to their plight.

Janet Salaff, in Chapter 15, shows how government policies have transformed social organization in general, and the family in particular, in Singapore. These policies were designed to assist rapid economic growth which in turn provided the resources for ambitious social policies. An expanding system of quality education and training established a highly skilled labour-force. A pension scheme ensured a high savings rate. Central city slums and village settlements were replaced by high-rise buildings, and the apartments were increasingly owner-occupied. The population thus was drawn ever more tightly into the market economy, and a growing proportion of women joined the labour-force. Singaporeans became less dependent on the support of kin and neighbours. Residential and occupational mobility further weakened community ties. The assurance of economic security in old age, the employment of women, and anti-natal policies led to a precipitous drop in fertility. Within less than a generation, the state had 'modernized' the family. This transformation greatly reduced differences in family styles, even while it reproduced class differences.

There are some parallels between the changes Salaff reports from Singapore and the changes Jellinek, in Chapter 9, describes in Jakarta. In both cases petty enterprises were eliminated, low-standard housing eradicated. Both authors comment on the ascendance of market exchange and the decline of community. Singapore, however, was exceptionally successful in promoting economic growth: the island's entire labour-force was engaged in increasingly productive tasks. Men were thus assured of employment at rising wages even while women increasingly joined the labour-force. Apartments in high-rise buildings became available to most people. In distinct contrast, Jellinek observed men losing their jobs, women no longer able to find earning opportunities, and the resettled unable to hold on to their apartments.

Urban life was profoundly transformed in the world's largest country, China, after the Communists came to power in 1949. Deborah Davis, in Chapter 16, reviews the series of dramatic changes Chinese cities underwent over close to half a century. Industrial enterprises became the organizing feature of the city. The housing, consumer goods, and various services they provided to their workers divided the city into enterprise villages striving for uniformity and self-sufficiency. Davis emphasizes the extent to which the ideals and practice of the Chinese Communist Party were articulated for the nation as a whole and implemented with remarkable consistency across a wide diversity of settings. In terms of wages, benefits,

and housing, an exceptional degree of equality was established both within cities and across cities and towns. The rural population, however, had to make do with a much lower standard of living, while its access to the city was severely restricted. Estimates put the urban–rural ratio of nominal consumption as high as 2.9 to 1 in 1978, declining to 2.2 to 1 in 1985, and reaching 3.1 to 1 in 1990 (Chen and Parish, 1996).

The most far-reaching transformation was effected during the Cultural Revolution that began in 1966 and continued until the death of Mao Zedong in 1976. Martin King Whyte and William Parish (1984: 358) describe the model of urbanism that emerged during that period. Among the basic structural elements of this model were strict migration controls and minimal urbanization, in spite of considerable economic development; a penetrating system of residential and work-unit organization; a highly developed bureaucratic allocation system; an emphasis on production rather than consumption; a relatively egalitarian distribution system; a rejection of schools as the basic mechanism for sorting talent; much stress on citizen involvement in public health, social control, and other realms; and rigid taboos on all forms of dress, expression, ritual life, and communication that did not conform to the official ideology. These structural elements contributed to a number of distinctive social features: high stability in jobs and residences, involvement and familiarity with neighbours and workmates, minimal differentiation of consumption patterns and life-styles, low divorce, high female work-participation rates and rapid changes in fertility and religious customs.

Whyte and Parish (1984: 368) observe how entrenched certain urban characteristics, such as bureaucratism and the urban prestige hierarchy, proved to be. They further emphasize that when aspects of social organization were effectively changed, the results were not always as expected. The pursuit of equality through class struggle and an emphasis on class labels alienated many, failed to eliminate awareness of occupational rank and privilege, and destroyed much of the former unity in the struggle for national strength and economic growth. The pursuit of comradely relations produced more interpersonal knowledge and concern, but also a certain wariness in personal relations and weak attachment to middle-level community and work organizations.

After the death of Mao Zedong, the new ideology of 'market-socialism' initiated a far-reaching transformation. Davis shows how markets, decentralization, and foreign investment increased the economic autonomy of cities and strengthened their ties with the hinterland. Pro-market policies reunited rural and urban China and produced massive rural–urban migration. Among new entrants to the urban work-force, there was a steady tilt towards employment in the private sector. Working over-time or holding a second job became common. Income inequalities grew between cities and within cities. Yet the legacy of the socialist transformation persists: urban residents continue to enjoy equitable access to basic goods and services, and they support redistributive income policies.

Note

1. Single mothers and their children need to be conceptualized separately from the category of 'women-headed households'. For a detailed discussion that elucidates the conceptual issues and problematizes widely used estimates, see Varley (1996).

References

Baker, Hugh D. R. (1979), *Chinese Family and Kinship* (New York: Columbia University Press).

Chakrabarti, Prafulla (1985), 'Some Aspects of "Complex" Families in Calcutta', *Journal of Comparative Family Studies*, 16: 377–86.

Chant, Sylvia (1985), 'Single-Parent Families: Choice or Constraint? The Formation of Female-Headed Households in Mexican Shanty Towns', *Development and Change*, 16: 635–56.

Chen, Xiangming, and Parish, William L. (1996), 'Urbanization in China: Reassessing an Evolving Model', in Josef Gugler (ed.), *The Urban Transformation of the Developing World* (Oxford: Oxford University Press), 60–90.

Epstein, A. L. (1981), *Urbanization and Kinship: The Domestic Domain on the Copperbelt of Zambia* (London: Academic Press).

Lee, Yean-Ju, Parish, William L., and Willis, Robert J. (1994), 'Sons, Daughters, and Intergenerational Support in Taiwan', *American Journal of Sociology*, 99: 1010–41.

Lomnitz, Larissa (1994), 'Urban Women's Work in Three Social Strata: The Informal Economy of Social Networks and Social Capital', in Gay Young and Bette J. Dickerson (eds.), *Color, Class & Country: Experiences of Gender* (London: Zed Books), 53–69.

Morgan, Philip S. (1993), 'Third World Urbanization, Migration, and Family Adaptation', in John D. Kasarda and Allan M. Parnell (eds.), *Third World Cities: Problems, Policies, and Prospects* (Newbury Park, Calif.: Sage), 235–54.

Oliveira, Orlandina de, and Roberts, Bryan (1996), 'Urban Development and Social Inequality in Latin America', in Josef Gugler (ed.), *The Urban Transformation of the Developing World* (Oxford: Oxford University Press), 252–314.

Ramu, G. N. (1989), *Women, Work and Marriage in Urban India: A Study of Dual- and Single-Earner Couples* (New Delhi: Sage).

Roldán, Martha (1988), 'Renegotiating the Marital Contract: Intrahousehold Patterns of Money Allocation and Women's Subordination among Domestic Outworkers in Mexico City', in Daisy Dwyer and Judith Bruce (eds.), *A Home Divided: Women and Income in the Third World* (Stanford, Calif.: Stanford University Press), 229–47.

Rugh, Andrea B. (1984), *Family in Contemporary Egypt* (Syracuse, NY: Syracuse University Press).

Sharma, Ursula (1986), *Women's Work, Class, and the Urban Household: A Study of Shimla, North India* (London: Tavistock).

Varley, Ann (1996), 'Women Heading Households: Some more Equal than Others?', *World Development*, 24: 505–20.

Weinstein, M., Sun, T. H., Chang, M. C., and Freedman, R. (1994), 'Co-Residence and Other Ties Linking Couples and their Parents', in Arland Thornton and Hui-Sheng Lin (eds.), *Social Change and the Family in Taiwan* (University of Chicago Press), 305–34.

Willis, Katie (1993), 'Women's Work and Social Network Use in Oaxaca, Mexico', *Bulletin of Latin American Research*, 12: 65–82.

Whyte, Martin King, and Parish, William L. (1984), *Urban Life in Contemporary China* (Chicago and London: University of Chicago Press).

13

The Social and Economic Organization of a Mexican Shanty-Town

Larissa Lomnitz

A COMMON prejudice found in the sociological literature on poverty consists in portraying the urban poor as people bedevilled by a wide range of social pathologies, amounting to a supposed incapacity to respond adequately to social and economic incentives. More social scientists have directed their attention towards the material and cultural deprivation that meets the eye than towards the socio-cultural defence mechanisms which the urban poor have devised. My work in a Mexican shanty-town, as summarized in the present chapter, deals with a basic question: how do millions of Latin Americans manage to survive in shanty-towns, without savings or saleable skills, largely disowned by organized systems of social security?

The fact that such a large population can subsist and grow under conditions of extreme deprivation in Latin American cities has important theoretical implications. Obviously, the members of such a group can hardly be described as 'unfit' for urban life in any meaningful sense. On the contrary, the proliferation of shanty-towns throughout Latin America indicates that these forms of urban settlement are successful and respond to some sort of objective social need (Mangin, 1967; Turner and Mangin, 1968). My

own work in Mexico City tends to support this view, by providing evidence that shanty-towns are actually breeding-grounds for a new form of social organization which is adaptive to the socio-economic requisites of survival in the city. In this essay, I show that the networks of reciprocal exchange among shanty-town dwellers constitute an effective stand-by mechanism, whose purpose is to provide a minimum of economic security under conditions of chronic underemployment.

This study is the result of two years of field-work in a shanty-town. I have used participant observation and unstructured interviews in the anthropological tradition. However, I have also made a number of quantitative surveys covering the totality of households in the shanty-town, in order to substantiate the major conclusions. This eclectic approach works better in the urban environment, where the homogeneity of the unit of study cannot be taken for granted. Statistical results are not presented for their own sake; rather, they are used to fill out the conceptual model based on direct observation. For a comprehensive account of my study, see Lomnitz (1975).

This essay has been revised for *Cities*. It appeared first in Wayne A. Cornelius and Felicity M. Trueblood (eds.), *Anthropological Perspectives on Latin American Urbanization* (Latin American Urban Research, 4; Beverly Hills, Calif.: Sage Publications, 1974), 135–55. © Larissa Lomnitz

The Research Site

The shanty-town of Cerrada del Cóndor sprawls over a ravine in the southern part of Mexico City, facing a cemetery on the opposite slope. The ravine represents the natural boundary between two residential middle-class neighbour-hoods of fairly recent development. This area makes up the hilly outskirts of the ancient town-ship of Mixcoac, a part of urban Mexico City since about 1940. Prior to that time, a few small entrepreneurs raised flowers and tree seedlings on the hills and worked the sand-pits in the ravine.

The earliest settler bought a tract of barren land about 1930, at the location of the present shanty-town. There he settled with his family and began to manufacture adobe bricks. He was later joined by a caretaker of the sand-pits. Within ten years, there were about a dozen families living in Cerrada del Cóndor, all of them workers in the adobe industry. Around that time the owner decided to sell fifteen small lots to new settlers, who immediately started to build their own homes.

By the end of the 1940s, the whole surround-ing area was being urbanized. The southward growth of the city had begun to swallow up the towns of Mixcoac and San Angel. The shanty-town, however, was unfavourably located on the slopes of the ravine and was bypassed by devel-opers. During the 1950s, 30 new families arrived. These included workers in the sand-pits, the adobe works, and in the housing projects of the neighbouring hills, particularly the suburb of Las Aguilas. A considerable number of relatives of the original settlers also came directly from rural areas. After 1960, the shanty-town began to grow very rapidly: 111 families arrived during this period, plus around 25 families who left during the period of the study (1969–71) and are there-fore not included in the survey.

As both the adobe factory and the sand-pits closed down, their owners became the slumlords of the shanty-town; yet they did not move away. The settlers pay rent for their houses or for plots of land on which they erect houses of their own. At the time of the study Cerrada del Cóndor had about 177 households, most of which were sur-veyed.

Origins of the Settlers

A full 70 per cent of the heads of families and their spouses (hereafter referred as the 'settlers') are of rural origin, having migrated to the Mexico city metropolitan area from localities of less than 5,000 inhabitants. The remaining 30 per cent were born in the Federal District, either as sons or daughters of rural migrants, or as inhab-itants of the small towns which are now part of the southern residential area of the city.

Eighty-six per cent of the rural migrants moved directly to Mexico City, without interme-diate stops. This high proportion applies to all age-groups. In fact, about 70 per cent of all migrants moved in family groups, and only 30 per cent were single. The rural migrants came from the most impoverished sectors of the Mexican peasantry. Eighteen states are repre-sented in the shanty-town, but the states of Guanajuato, México, and San Luis Potosí account for 56.6 per cent of all migrants. Veracruz, Zacatecas, and Hidalgo come next with about 6 to 7 per cent each. In discussing their reasons for migrating, nearly all migrants declared that they had been landless field-work-ers, or that their landholdings had been too poor for subsistence.

Thirty-five per cent of the migrant heads of families and their spouses were illiterate; another 9 per cent had never been to school but knew the rudiments of reading and writing. Another 33 per cent had had one to three years of schooling. It is probably fair to say that more than half the settlers of rural origin were functionally illiterate at the time when they reached the Federal District. They had neither savings nor skills of any value in the urban labour-market.

Among those born in the Federal District, the illiteracy rate was significantly lower. Only 17 per cent had never been to school, and nearly half of those had taught themselves the rudiments of reading and writing. We shall see later that there is a significant correlation between schooling and economic status, as measured by income and material possessions.

When migrants reach the city, they normally move in with relatives. The presence of a relative in the city is perhaps the most consistent element within the migration process. The role of this rel-ative determines the circumstances of the migrant family's new life in the city, including place of settlement within the metropolitan area, initial economic status, and type of work. There is no escaping the economic imperative of living near some set of relatives: the initial term of stay with a given kinship set may be variable, but subsequent moves tend to be made with reference to pre-existing groups of relatives elsewhere. Unattached nuclear families soon attract other relatives to the neighbourhood.

The Villela Group: An Example of Kin-Mediated Migration

Among the 30-odd households from the state of San Luis Potosí, 25 came from the *hacienda-ejido* Villela near Santa María del Río. These families are related through consanguinal and marriage ties. The experience of the Villela group will serve as an example of the process of kin-mediated migration as observed in Cerrada del Cóndor.

The settlement of migrants from Villela goes back to the early 1950s, when two young men from the village decided to try their luck in Mexico City. They found work in the adobe factory and settled in Cerrada del Cóndor. One year later, one of them brought a sister and two nieces with their offspring to Mexico City. These nieces later brought their mother and brothers, and other relatives in successive waves. Two other Villela families also migrated to Cerrada del Cóndor and became related to the first family group by marriage or *compadrazgo* (fictive kinship).

After working at various trades, one of the migrants was fortunate to find work as a carpet-layer. Later migrants were lodged, fed, and counselled by those among their kin already in residence, with the result that practically all men now work in the carpeting trade. This pattern can be observed quite generally among family networks and is by no means unique to the Villela network. Thus, all the men in one network polish tombstones; in another, they work as bakers; still others are members of bricklayer crews, and so on.

Villela settlers in Cerrada del Cóndor maintain a closely knit community within the shanty-town. They founded the oldest functioning local association, the Villela football club, with three teams in constant training which participate in league tournaments in Mexico City. Social contact among Villela families is intense, and there is a great deal of mutual assistance among them. All migrants express satisfaction at the positive results of their move to Mexico City, and not even the grandmothers show any nostalgia for Villela, where, they say, 'we were starving'.

Moves Within the City

It has sometimes been assumed that migrants to Mexico City initially tend to gravitate towards the crowded tenements in the old downtown area (Turner and Mangin, 1968). Our research in Cerrada del Cóndor does not confirm this hypothesis. Instead, the place of initial residence is determined by the residence of pre-extant cores of relatives in the city. In general, the migrants continue to move within the city, but always in the same general sector. Thus, the settlers of Cerrada del Cóndor were born in or migrated initially to the southern part of the metropolitan area, and few of them have more than a very superficial acquaintance with other parts of the city, including the downtown area. Few of the men venture farther into the city than their jobs require. Women and children barely know anything of the city beyond a church, a market, or the home of some relative.

The mechanism of moves within the urban area was studied in some detail. Families moved, on average, once every five years for the first ten or fifteen years of married life; some families never seemed to settle down. Moves seemed to be caused largely by displacement due to the southward growth of the city, coupled with a desire to seek better work opportunities and more congenial kin. There is an important turnover in Cerrada del Cóndor: during the period of my study, about 25 families moved away, and some 40 new families moved in. As previously noted, Cerrada del Cóndor has been bypassed by developers and represents an area of refuge for those displaced by urban growth.

Most of the new settlers in Cerrada del Cóndor, however, merely follow the pull of relatives who already live in the shanty-town. These relatives have told them about cheap available housing and have offered the exchange of mutual help, without which life in a shanty-town is extremely difficult. Thus, kinship is also the determining factor in the process of residential mobility within the city. When job opportunities seem sufficiently bright for a nuclear family to move into a new area, they soon bring in other relatives from nearby areas or directly from the countryside. New migrants may be recruited during trips to the village, as migrant families keep in regular contact with their place of origin. Visits occur normally on festive occasions, such as holidays and celebrations.

In conclusion, it may be said that each migrant helps several new migrants to settle in the city, or to move from another part of the city to this shanty-town. This they do by providing temporary or permanent lodging, food, information, assistance in job-hunting, moral support, and the basis for a more permanent form of exchange to be discussed later.

Economics of Shanty-Town Life

The general economic setting of Cerrada del Cóndor is one of extreme poverty. A typical dwelling consists of a single room measuring 10 by 12 ft., containing one or two beds shared by members of the family. There may also be a table, a chair, a gas or petroleum stove, and sometimes a television set—33 per cent of all households own one. There are three public water faucets in the shanty-town, which are used by most of the population (a few clusters of dwellings have a faucet of their own). There is little public sanitation and drainage; more than four-fifths of the population use the bottom of the gully for a latrine. Sanitary conditions are made worse by the presence of a large public garbage dump next to the shanty-town. There is no regular electric service; power is obtained by illegal hook-ups to the power lines. There are no paved streets, only alleys and gutters left between residential units.

Families residing in Cerrada del Cóndor can be classified according to the following four levels of living:

Level A. Three or more rooms, running water, bathroom or privy, brick construction, cement or tile floor, dining-room furniture, living-room, electric appliances such as sewing-machine, washer or refrigerator, gas stove.

Level B. Two rooms, cement floor, no running water, some furniture such as a trunk or closet, table and several chairs, some electrical appliance(s), gas stove.

Level C. The same as level B but no electrical appliances (except for radio or television), lower-quality furniture, petroleum-burner for cooking.

Level D. One room with or without small lean-to for cooking, no furniture (except beds and an occasional table or chair), clothing kept in boxes or under the bed, no appliances (except for radio or television), petroleum cooking only.

Analysis by means of contingency tables showed that the four criteria used (housing, furniture, electrical appliances, and type of cooking) were highly intercorrelated.

The distribution of levels of living within Cerrada del Cóndor was as follows:

Level A	7.8%
Level B	8.9%
Level C	23.8%
Level D	59.5%

The level of living was highly correlated with other economic indicators, particularly the occupational status of the breadwinner. Table 13.1 shows that the total of unskilled labourers, journeymen, servants, and petty traders comes close to the total of families classified in levels C and D.

Unskilled labourers or apprentices include hod-carriers and other construction workers (foremen excepted), house painters, sand-pit workers, brickmakers, bankers' helpers, truckers' helpers, carpet-layers, electricians, gardeners, and other unskilled labourers paid by the day who earn the minimum legal wage or less. Semi-skilled or skilled journeymen or craftsmen include independent or freelance workers such as bakers, carpet-layer foremen, construction foremen, electrician foremen, truck-drivers, tombstone-polishers, carpenters, cobblers, blacksmiths, potters, and so on. These workers may earn higher wages, but their job security is usually as low as that of the unskilled workers. Some of them have developed a steady clientele and work with their own assistants, usually relatives. Industrial workers are those who work in an industrial plant, usually with the lowest wages and qualifications: watchmen, car-washers, janitors, and unskilled labourers. The service

TABLE 13.1   Occupations of Heads of Households in Cerrada del Cóndor

Occupation	Men		Women	
	n	%	n	%
Unskilled labourers or apprentices	51	32.9	1	4.5
Semi-skilled or skilled journeymen or craftsmen	48	31.0	—	—
Industrial workers	16	10.3	—	—
Service workers	5	3.2	12	54.5
Traders	7	4.5	4	18.2
Employees	8	5.2	1	4.5
Landlords	5	3.2	1	4.5
Unemployed	15	9.7	—	—
Housewives	—	—	3	13.6
TOTAL	155	100.0	22	100.0

workers include waiters, water-carriers, watch-men, icemen, and domestic servants. Traders include all kinds of street-vendors. None of these has a steady income or social security. The employees are unskilled workers who earn fixed salaries: municipal workers (street-sweepers, garbagemen), and a few similarly employed with private corporations. These workers have relatively high job security and other benefits. Finally, there are six households whose income is mainly derived from rentals of property in the shanty-town.

About 10 per cent of the household heads were out of work at the time of the survey. However, more than 60 per cent of those who said they were working consider intermittent joblessness for variable periods of time to be normal. Thus, the majority of the working population in the shanty-town are underemployed ('eventuales'), and have no job security, no social security, and no fixed income. They exist from day to day, as urban 'hunters and gatherers'. Members of level-of-living D belong to this group. More than half the settlers in level D were illiterate. None owned both his home and the lot it was built on; nearly two-thirds paid rent for both. All lived in a housing unit consisting of a single room. The average number of people per room was 5.4, if the cooking was done inside, and 6.2 if there was a lean-to for cooking. Of all acknowledged cases of problem drinkers, more than 75 per cent belonged to level D.

In contrast, members of level A were practically all owners of their homes and lots. Most of them were either born in the Federal District or had lived in town for many years. There was practically no illiteracy, and most settlers had completed third grade. They tended to belong to the upper types of occupations: landlords, employees, traders, and industrial workers; their most distinctive trait was job security. More than half the households in this group included two or more breadwinners. The men were either abstemious or moderate drinkers.

Levels B and C are intermediate, but they can be sharply distinguished. Level B is urban in type of dwelling, furniture, and life-style, while Level C is still rural in most of these respects. Households of either type can usually be recognized by a glance at their belongings. The transition from C to B is not determined so much by gross income as by the degree of cultural assimilation to urban life: hence, time of residence in the city is a significant factor. Even the highest income in Cerrada del Cóndor would easily be used up by a single heavy drinker. Wives receive a weekly allowance and have no direct knowledge of their husbands' income. Working wives contribute their total income to household expenses; likewise, sons and daughters hand their earnings to the mother. The husband's contribution to raising the economic level of the household is largely limited to major appliances which are purchased on the instalment plan. However, if the husband enjoys a steady income and has a tolerant view of a working wife, the economic improvement of the family may be rapid by shanty-town standards. Nevertheless, the transition from level C to level B is rarely accomplished before a household has completed ten years of residence in the city.

Social Organization

The pattern of social organization which prevails in the shanty-town can be described as follows. Most nuclear families initially lodge with kin, either in the same residential unit (47 per cent), or in a compound arrangement (27 per cent). Compounds are groups of neighbouring residential units which share a common outdoor area for washing, cooking, playing of children, and so on. Each nuclear family in such a cluster forms a separate economic unit. Families in the compound are related through either consanguinity or marriage ties; each compound contains at least two nuclear families.

Extended families, e.g. two brothers with their wives and children, may share the same residential unit temporarily; in the case of newly married couples with the parents of either husband or wife, the arrangement may be more permanent. Any room or group of rooms having a single private entrance is defined as a residential unit: this excludes tenements of the 'vecindad' type, consisting of a series of rooms opening on an alley with a public entrance gate, which may contain several independent family groups. Extended families contain at least two nuclear families; these share the rent expenses or own the property in common. Sometimes they also share living expenses.

Extended households are less stable than compounds. Nuclear families in an extended household tend to move into a nearby room of their own, or to join a different set of relatives elsewhere. However, those who move away in search of independence and privacy eventually return for security or assistance. In the case of compounds the rate of desertion is much lower. Of 44 nuclear families who joined a compound arrangement since the beginning of married life, only seven moved away in an attempt to form an independent household.

Thirteen couples began their married lives as independent households within the Federal District; these represent the major exception to the pattern of social organization described above. The heads of these households had all been born in Mexico City or had lived there for many years. Yet even these households do not remain independent for long, since they tend to attract other kin who join them in an extended or compound arrangement. The complete data for household types is summarized in Table 13.2.

TABLE 13.2 Types of Households in Cerrada del Cóndor

Extended families	29
Nuclear families in a compound-type arrangement	68
Independent nuclear families:	
(a) without kin	30
(b) with kin in Cerrada del Cóndor	28
Other unknown	7
TOTAL HOUSEHOLDS IN SURVEY	162

Independent nuclear families are in the minority; those who live within walking distance of relatives are usually waiting for a vacancy to move into a compound-type arrangement. In this case, there is much visiting, mutual assistance, and other types of interaction even though the related nuclear families are not yet fully integrated into a compound. The term 'nuclear family' is used in a broad sense here, as each nuclear family may include one or more individually attached kin. Most of them are older persons or young children of unmarried recent arrivals from the country. Nuclear families may also include the offspring of a previous union of the mother; in two such cases, there was no offspring from the present union.

Thus, the social organization of the shanty-town may be described as a collection of family networks which assemble and disband through a dynamic process. There is no official community structure; there are no local authorities or mechanisms of internal control. Co-operation within the family networks is the basic pattern of social interaction. There is a pattern of movement from the extended family towards the compound type, as illustrated by a survey tracing the moves of each household over the past years. The results show that there has been an increase of 29 per cent in compound arrangements, against a decrease of 46.7 per cent in extended family arrangements, as compared with the initial state of residence.

This pattern can be viewed as the outcome of a dynamic process, which depends on economic circumstances, the stage in the life cycle, the availability of housing vacancies, personal relationships with relatives, etc. The initial choice of moving in with the family of either spouse is usually an economic one. Since young husbands or wives often do not get along with their in-laws and conditions in an extended family may be very crowded, the couple tends to move out. However, new circumstances, such as the arrival of children, desertion of the husband, loss of employment, and so on frequently compel the family to return to the shelter of relatives. The preferred arrangement is the compound which combines proximity of kin with an adequate amount of independence and privacy.

Kinship Relationships Outside the Shanty-Town

A pattern of residential moves such as we have described, sometimes over large distances, implies a substantial amount of contact among kin extending beyond the physical boundaries of the shanty-town. The existence of such contact was confirmed initially through personal observation and later by means of a kinship census covering all households in the shanty-town.

Contact with relatives within the Federal District depends on kinship distance and on physical distance. Informants tend to list first their nuclear family of orientation, then other relatives by order of spatial proximity: first those who live in the shanty-town, then those who live in a nearby shanty-town such as Puente Colorado, and so on. If a relative is not particularly close and lives as much as two hours away by bus from Cerrada del Cóndor the contact is unlikely to be significant, and may be lost after a generation. The mother is often the only nexus between such relatives, and contact vanishes after her death. Of course, if there exits a true closeness of relationship each set of relatives will exert a great deal of 'pull' on the other, in order to encourage them to move into as close a neighbourhood as possible.

Changes in socio-economic status become a factor which influences the intensity of contact between kin. A female informant commented that she hardly ever saw her sisters, who were married to skilled industrial workers: 'To tell the truth, I don't like to go and see them because they can dress very nicely and I can't afford to and so . . . I feel ashamed'. The informant is the daughter of a skilled worker, who married a man who 'never finished grade school and is worse off' than her sisters' husbands. Her parents were opposed to the match because her husband had no skills, 'not even that of a truck-driver, a barber, or a carpenter', but she was in love and they went to live with his parents. At first they lived in the same residential unit in Cerrada del Cóndor; now they have a room near by, because 'each on his own is better'.

Contact between migrant families and their relatives in the countryside usually takes the form of visits to the village on festive occasions, such as Mother's Day, All Saints' Day, and the festival of the patron saint of the village. Unmarried migrants often return to the city accompanied by a smaller brother, sister, or cousin whom they help out until they find a job. Married migrants frequently maintain a share in a small plot of land which they own jointly with a brother, and they time their occasional visits to coincide with harvest time, and so on. Most migrants send money home to their parents or close relatives. Through word of mouth or correspondence, they keep up with village gossip; eventually they are instrumental in promoting the migration of their close kin from the village. For years there is a steady stream of relatives from the country, who are lodged and fed for indeterminate periods of time depending on resources and needs.

Contact with the village gradually wanes over the years. About one-fourth of all informants said they had relatives in the country but had lost all touch with them: 'I haven't seen my folks since I got married and moved to the city eighteen years ago' . . . 'I never went back home since my mother died—my father married again and I don't get along with my stepmother' . . . 'My brothers have moved to the city; I used to visit them and bring them money; now I don't go any more' . . . 'I never went back since my grandparents and parents died'. Other migrants, however, said they maintained significant contact through visits to and from the village; through economic interests (land owned in common); through correspondence; through remittances of money; through sentimental pilgrimages (visiting mother's grave on All Saints' Day), and so on.

Local Groups

The shanty-town is not organized around central institutions of any kind. Instead, there are several types of groupings, of unequal importance: (a) the family network; (b) football teams; (c) the medical centre; (d) temporary associations.

The family networks will be discussed in greater detail below, as it is our thesis that they represent the effective community for the individual in the shanty-town. They are composed of members of an extended family, or a compound, but may include neighbours who are assimilated through fictive kinship. We shall see how these networks have developed into systems of reciprocal exchange of assistance, which are key to the survival of large numbers of people under the severe economic handicaps of shanty-town life.

Other forms of organization at the community level are relatively rudimentary. There are four football teams in Cerrada del Cóndor. Three of these teams belong in effect to a single large family network, the Villela network described above. The fourth is a more recent team whose membership is recruited among young people of the shanty-town irrespective of family origins. Football teams represent one of the few vehicles of social contact between men of Cerrada del Cóndor and men who live in other parts of town. After a game, there are drinking sessions which reinforce the team spirit and friendship among members of the team.

The shanty-town's medical centre was organized and financed by a group of middle-class ladies from the neighbouring residential district, with some assistance from a nearby church. Later, the national Children's Hospital agreed to staff the centre, but this help has recently been withdrawn. In spite of the modest assistance offered, the centre has become an important part of shanty-town life. It is a place where children are welcome during most hours of the day, and where many girls and women receive guidance from an understanding social worker.

There is no local organization for solving the common problems of shanty-town life. Groups of neighbours may band together for specific issues; this has happened three or four times in the existence of Cerrada del Cóndor. The first time was to request the installation of a public water outlet. Another time, a group of women jointly requested an audience with the First Lady, in order to lodge a complaint about spillage of oil from a refinery that was causing brush fires in the ravine. These exceptional instances of co-operation merely serve to highlight the absence of any organized effort to solve community problems.

The residents of Cerrada del Cóndor have little contact with city-wide or national organizations. Articulation with Mexican urban culture occurs mainly through work and through mass media such as radio and television. School is, of course, very important for the children. Adult reading is limited to sports sheets, comics, and photo-romance magazines. Only about one-tenth of the men belong to the social-security system. About 5 per cent are union members. In general, extremely few people belong to any organized group on a national level, such as political parties, religious organizations, and so on.

Social Networks

According to Barnes (1954), a network is a social field made up of relations between people. These relations are defined by criteria underlying the field. While Barnes saw a network as essentially unbounded, Mayer (1962) showed how certain types of migrants encapsulate themselves in a bounded network of personal relationships. In the case of Cerrada del Cóndor, we find networks defined by criteria of neighbourhood, social distance, and exchange of goods and services.

Each network is constituted of nuclear families, not individuals. Initially we shall use an operational definition of networks as clusters of neighbouring nuclear families who practise continuous reciprocal exchange of goods and services. A total of 45 such networks are identified in Cerrada del Cóndor. The social relationships which form the basis of these networks are as follows: 30 are networks based on consanguineal and marriage ties, 7 are based on kinship but included also one or more families not related by kinship, and 8 are formed by families not related by kinship. The number of nuclear families per network is shown in Table 13.3. The average network contains four nuclear families. Of course, the number of families in a network is not static but changes with time. Initially, the network may

TABLE 13.3 Number of Nuclear Families per Network

Nuclear families	Cases
2	9
3	13
4	10
5	6
6	5
uncertain	2
TOTAL	45

be composed of two or three families; it may grow until a part of the network is split off because of lack of room or facilities. Table 13.3 does not include unattached nuclear families (estimated at less than ten).

It is possible for several networks to be interrelated through kinship. Thus, the Villela macro-network includes about 25 nuclear families grouped into five networks. Each of these networks internally displays a high degree of reciprocal exchange of goods and services on a day-to-day basis. The resources of the macro-network are used more on ritual occasions, in important matters such as job placement, in the expression of kin solidarity (football teams), and drinking. Reciprocal exchange does occur between families belonging to different networks within a macro-network system, but the recurrence of such exchange is less frequent, because no single nuclear family in the shanty-town has enough resources to maintain a generalized day-to-day exchange with such a large group of families.

All nuclear families in a network practise reciprocal exchange with each other on an equal footing. In addition, a nuclear family may maintain dyadic exchange relations with families outside the network or belonging to other networks. These dyadic ties are important because they provide the mechanism through which an outside family can be attracted to join the network, or by which a network which has outgrown its optimal size may split.

A network constituted as an extended family practises a generalized exchange of goods and services, which includes the informal pooling of resources for rent and entertainment, joint use of cooking facilities, communal child care, and many others. Each nuclear family contributes according to its ability and receives according to

the availability of resources within the network. There is no accounting of any kind among the members of such a network. In a compound, on the other hand, each nuclear family has a roof and an economy of its own; yet there is an intense exchange of goods and services in the form of daily borrowings of food, tools, and money. Reciprocity here is not openly acknowledged, but it is definitely expected; each member family is supposed to provide assistance in proportion to its economic ability. Thus, if a nuclear family within a compound becomes economically more secure than the rest, it may find its resources taxed beyond the actual returns which it can expect from the network. As a result, the more prosperous families may stop asking for and offering services.

Case History

The network is of the compound type and includes two sisters A and B, who married two brothers. A non-kin neighbour is also included in the network. A third sister, C, brought in from the village to live with A, soon found work as a maid living in the house of their employer. On her days off, she visited A. A niece from the country has now joined the network with her husband. At first they lived with A, who obtained work for the husband and a room adjoining hers. All men in the network are currently working as tombstone-polishers. When C became pregnant, she quit her job and went again to live with A. After she had the baby, she went back to work and left the baby in A's care during the day.

Meanwhile, B's economic status had been rising steadily. Her husband did not drink and invested in home furnishings. He also found advancement at work and became a skilled worker (in the placement of tombstones). B began to refuse to lend (or ask for) favours within the network, claiming that she had no money. Her sister, niece, and neighbour gradually stopped requesting assistance, and so did their respective husbands. After a while, B found a room just beyond the limits of the shanty-town, two blocks away from her former room but with urban services. Her economic level is now rated 'B'.

Sister C found a husband and moved in with

him. The husband lived several blocks away from the sisters' compound, outside Cerrada del Cóndor. Yet C practically continued to live at A's place whenever her husband was away at work. When she works she leaves her child with A; when she needs money she borrows from A or from her niece. Her interaction with the non-kin member of the network is less intense; yet these neighbours have become double *compadres* in the meantime, and their exchange with both A and the niece and their respective husbands is very active. When a room became available, C set out to convince her husband to move in with the network. In that case, an active exchange between C and the non-kin neighbour is anticipated.

What is Exchanged?

The following items represent the most important objects of exchange in the networks, according to my observation:

1. *Information*, including directions for migration, employment, and residence; gossip; and orientation about urban life.

2. *Training and job assistance*, including the training and establishing of a relative as a competitor. Thus, a carpet-layer or a building contractor would take his newly arrived brother-in-law along as an assistant, teach him the trade, share earnings with him, and eventually yield some of his own clientele to set him up as an independent worker.

3. *Loans*, of money, food, blankets, tools, clothing, and other goods.

4. *Services*, including the lodging and care of visiting relatives, widows, orphans, old people; care and errand-running for such neighbours; and minding children for working mothers. Assistance among men includes help in home construction and in transporting materials. Children must lend a hand in carrying water and running errands.

5. *The sharing of facilities*, such as a television set or a latrine (which the men may have built jointly).

6. *Moral and emotional support* in ritual situations (weddings, baptisms, funerals) as well as in day-to-day interactions (gossip among the women, drinking among the men). It is essential to recognize that much of the socializing in the

shanty-town is based on network affiliation. This constant interaction generates an overriding preoccupation with each other's lives among the members of a network. There is little opportunity for privacy.

The ubiquity of these forms of exchange provides important evidence in support of our interpretation of shanty-town networks as economic structures which represent a specific response of marginal populations to economic insecurity in the city.

Reinforcing Mechanisms

The exchange of goods and services serves as the underpinning of a social structure: the network organization. When this exchange ceases to exist, the network disintegrates. The social structure which is erected on the basis of exchange depends on physical and social proximity of network members. Ideally, the networks are composed of neighbours related through kinship.

Actually, many networks contain non-kin members whose allegiance must be reinforced by means of fictive kinship (*compadrazgo*) and other means which will be analysed presently. Even among kin, relationships are far from secure: economic and personal differences arise frequently under conditions of extreme poverty and overcrowding. The reinforcing mechanisms to be discussed are therefore present in all networks.

Compadrazgo is widely used to reinforce existing or prospective network ties. In Cerrada del Cóndor, the *compadres* have few formal obligations towards one another. An informant says: 'When choosing a godfather for one's child one should look for a decent person and a good fiend, if it's a couple, they should be properly married. They should be poor so no one can say that you picked them out of self-interest'. Among 426 *compadres* of baptism (the most important type of *compadrazgo* in the shanty-town), 150 were relatives who lived close by, and 200 were non-kin neighbours. Another 92 were relatives who lived elsewhere in the Federal District or in the countryside, i.e. prospective network affiliates. In most cases of *compadrazgo*, the dominant factors were physical proximity and kinship. This equalitarian pattern is at variance with the frequently observed rural pattern of selecting a *compadre* above one's station in life (Forbes, 1971).

The great importance of *compadrazgo* as a reinforcing mechanism of network structure is also reflected in the variety of types of *compadrazgo* that continue to be practised in the shanty-town. These types are, by order of decreasing importance: baptism (426 cases), confirmation (291), communion (79), wedding (31), burial (16), Saint's Day (13), fifteenth birthday (10), Divine Child (8), Gospels (8), grade-school graduation (4), habit (3), sacrament (2), scapulary (1), cross (1), and St Martin's (1). All these types of *compadrazgo* mark ritual or life-cycle occasions. The formal obligations among *compadres* can be described as follows: 'They must treat each other with respect at all times, and must exchange greetings whenever they meet.' Ideally some *compadres* should fulfil economic obligations, such as taking care of a godchild if the father dies; but these obligations are no longer taken very seriously in the shanty-town.

If *compadrazgo* formalizes and legitimizes a relationship between men and women, *cuatismo*, the Mexican form of male friendship, provides the emotional content of the relationship. *Cuates* (a Nahuatl term for 'twins') are close friends who pass time together, talking, drinking, playing cards or football, watching TV, treating each other in restaurants, and having fun together; above all, they are drinking companions. Women are totally excluded from the relationship. A wife 'would never dare' to approach a *cuate* of her husband's to request a favour.

Assistance among *cuates* is ruled by social distance. Among relatives, there will be more unconditional help than among neighbours. In general, the *cuates* borrow freely from one other, help one another in looking for work, give one another a hand in fixing their homes, and stand by one another in a fight. Like *compadrazgo*, *cuatismo* is practically universal: the man who has no *cuate* and no *compadre* is lost indeed. Among 108 of the households headed by men, the circle of *cuates* of the family head was recruited as follows: 86 were groups of *cuates* who lived in the immediate vicinity, 9 were mixed groups (some *cuates* living close by and some far away), and 11 were groups whose members did not live in the immediate neighbourhood. Two heads of households had not yet made any *cuates* in the city.

It is clear that these groups of *cuates* are based primarily on the male sector of the networks described above, even though neighbours, work companions, or friends not affiliated with one's special network may be included. The existence of *cuatismo* to reinforce network affiliation is evidence that the networks are not simply built around the wives and mothers, as might be supposed from a superficial analysis. On the contrary, many networks appear to be male-dominated. If networks were based exclusively on the more visible forms of daily exchange of goods and services practised by women, the substantial overlap between networks and groups of *cuates* would be rather puzzling. Networks are constituted of nuclear families as entities; all members of each nuclear family participate actively in the relationship.

Drinking relationships among *cuates* are exceedingly important and usually take precedence over marital relationships. From a psychological point of view, drinking together is a token of absolute mutual trust which involves a baring of souls to one another (Lomnitz, 1969; Butterworth, 1972). From the economic point of view, *cuatismo* implies a mechanism of redistribution through drink which ensures that all *cuates* remain economically equal. And from a social point of view, it reinforces existing social networks and extends the influence of networks in many directions, since a drinking-circle may contain members of several networks.

The *ideology of assistance* is another important factor in network reinforcement. When questioned, most informants are reluctant to describe their own requests for assistance; yet they are unanimous in claiming to be always ready to help out their own relatives and neighbours in every possible way.

The duty of assistance is endowed with every positive moral quality; it is the ethical justification for network relations. Any direct or indirect refusal of help within a network is judged in the harshest possible terms and gives rise to disparaging gossip. People are constantly watching for signs of change in the economic status of all members of the network. Envy and gossip are the twin mechanisms used for keeping the others in line. Any show of selfishness or excessive desire for privacy will set the grapevine buzzing. There will be righteous comments, and eventually someone will find a way to set the errant person straight.

Reciprocity and *Confianza*

The types of reciprocity between shanty-town dwellers are determined by a factor which I have called '*confianza*' (Lomnitz, 1971). *Confianza* depends on cultural factors (social distance) and physical factors (closeness and intensity of exchange, as when a close friend or *cuate* enjoys greater *confianza* than a relative who lives elsewhere and is met only occasionally).

The formal categories of social distance are culturally determined. They imply a 'series of categories and plans of action' (Bock, 1969: 24) which dictate expected behaviour between individuals. These categories and plans of action can only be described ethnographically: they represent an essential part of the culture or subculture of a group or subgroup. In Mexico, within the national culture, there are subcultures of each social class, each state or region, and so on down to the level of family subculture which may imply strongly particularized sets of behaviour. Two individuals are close in the scale of *confianza* to the extent that they share the same set of behaviour expectations. These expectations include a specific type of reciprocity, extending from unconditional sharing to total lack of co-operation and distrust. The scale of *confianza* measures, among other things, the extent to which these expectations are actually fulfilled. Hence, the degree of *confianza* is not rigidly determined but may vary during the evolution of the relationship.

Social Networks in the Context of Marginality

Residents of shanty-towns such as Cerrada del Cóndor are often counted among the 'marginal' sector of the urban population in Latin America. The emergence of urban marginal populations is not, of course, exclusive to underdeveloped societies. In advanced industrial nations, such populations result from the displacement of certain social strata from the labour market through mechanization and automation of the means of production. These growing population sectors have no expectation of absorption into productive occupations, and become increasingly dependent on welfare. They represent *surplus*

population (rather than a labour reserve) and are, therefore, an unwanted by-product of the system.

According to Quijano (1970), this situation is considerably aggravated in underdeveloped countries, because the rate and pattern of industrial development are imposed from abroad. Economic dependence introduces a factor of instability, because of the hypertrophic growth of large industrial cities at the expense of the countryside. Easy access to sources of raw materials and cheap labour attract an overflow of hegemonic capital into formerly pre-industrial societies. As a result, (a) there is an increasing gap between 'modern' cities and 'traditional' rural areas on the verge of starvation; (b) new skills required by industrial growth are monopolized by a relatively small labour élite, while the great mass of unskilled peasants and artisans is displaced from their traditional sources of livelihood; and (c) superficial modernization has caused a sudden population explosion, which increases the rate of rural-to-urban migration, thus offsetting any efforts at promoting the gradual absorption of surplus populations into the industrial labour force. Thus, the process of marginalization is not transitional, but, rather, intrinsic to the system.

Quijano specifically identifies the capitalist system in general, and the dependent industrial development observed in Latin America since 1945 in particular, as responsible for the phenomenon of marginality. Adams (1970: 89–94; 1975) generalizes this analysis to apply to any large society which is subject to a process of economic development and technological change. According to Adams, any increment in social organization is achieved at the expense of disorganization among sectors of the same society, or of dependent societies. Dialectically speaking, order is the source of disorder: work creates entropy. Starting from an undifferentiated labour-force, we may build up an industrial proletariat with highly differentiated skills and a centralized form of organization; but this will generate marginalization of those populations which can no longer be assimilated or successfully utilized by the more advanced system.

In Latin America, the urban marginal strata share the following economic characteristics, which are also found among the settlers of Cerrada del Cóndor: (a) unemployment or underemployment; (b) lack of stable income; and

Larissa Lomnitz

(c) generally the lowest level of income within the urban population. Most settlers of Cerrada del Cóndor are rural migrants, or the offspring of migrant parents. Most of them are unskilled workers, such as construction workers, who are hired and fired on a daily basis; journeymen and artisans who are hired for specific jobs and have no fixed income; petty traders; and people who work in menial services. They may be described as urban hunters and gatherers, who live in the interstices of the urban economy, where they maintain an undervalued but nevertheless well-defined role. They are both a product of underdevelopment and its wards.

If this vast social group lacks any economic security and has no significant support from organized welfare, how does it survive? This question was posed by Quijano (1970: 87–96), who surmises that there must be some mechanism of reciprocity operating among marginal groups which has not been described. It is the purpose of the present section to analyse this mechanism in some detail, as a function of the socio-economic structure.

According to Polanyi (1968: 127–32) and Dalton (1968: 153), there are three forms of exchange of goods and services: (a) market exchange, in which goods and services circulate on the basis of offer and demand, without any long-term social implications attached to the exchange; (b) redistribution of goods and services, which are first concentrated in a single individual or institution from whence they flow out towards a community or society: and (c) reciprocity among social equals. Reciprocity defined in this manner is an integral part of a permanent social relationship.

In 1990 about half the population of metropolitan Mexico City was estimated to be marginal. While the dominant mode of exchange in the cities is market exchange and is afflicted by the internal contradictions described by Marx and Polanyi, no adequate systems of public or private redistribution have been created in response to the needs of a growing mass of urban marginals left to their own devices. Economic dependence aggravates the problem, because capital gains tend to be transferred abroad instead of becoming available for redistribution within the country.

Thus, marginal individuals cannot rely on the social system for the elementary needs of sur-

vival. They have nothing to offer the market exchange system: no property and no skills except for their devalues labour. Their prospects for absorption into the industrial proletariat are slim, since their numbers grow faster than the number of industrial job openings. They have nothing to fall back on: no savings and no social security of any kind. Their survival depends on the creation of a system of exchange entirely distinct from the rules of the market-place, a system based on their resources in kinship and friendship. This system follows the rules of reciprocity, a mode of exchange between equals, embedded in a fabric of social relations which is persistent in time, rather than casual and momentary as in market exchange. The three basic elements of reciprocity are: (a) *confianza*, an ethnographically defined measure of social distance; (b) equivalence of resources (or lack of resources); and (c) physical closeness of residence.

Characteristically, reciprocity generates a moral code which is distinct, and in some ways opposed, to the moral code of market exchange. In a reciprocity relation, the emphasis is less on receiving than on giving; the recipient is preoccupied with reciprocating, rather than with extracting maximum personal benefit from a transaction. Both systems of exchange may be used simultaneously in a different context: members of a reciprocity network may sell their labour as workers or servants on the urban market. Yet it is the reciprocal exchange among relatives and neighbours in the shanty-town which ensures their survival during the frequent and lengthy spells of joblessness. Market exchange represents the ultimate source of livelihood; but it is a livelihood at the subsistence level, without any element of security. Through sharing these intermittent resources with another six or ten people, the group may successfully survive whereas as individuals each of them would almost certainly fail. The networks of reciprocal exchange which we have identified in Cerrada del Cóndor are functioning economic structures which maximize security, and their success spells survival for large sectors of the population.

References

Adams, N. R. (1970), *Crucifixion by Power* (Austin: University of Texas Press).

—— (1975), 'Harnessing Technological Development', in J. H. Poggie, Jr. and R. N. Lynch (eds.), *Rethinking Modernization: Anthropological Perspectives* (Westport, Conn.: Greenwood Press), 37–68.

Barnes, J. A. (1954), 'Class Committees on a Norwegian Island Parish', *Human Relations*, 7: 39–58.

Bock, P. (1969), *Modern Cultural Anthropology* (New York: Alfred A. Knopf).

Butterworth, D. (1972), 'Two Small Groups: A Comparison of Migrants and Non-Migrants in Mexico City', *Urban Anthropology*, 1 (1): 29–50.

Dalton, G. (1968), 'The Economy as Instituted Process', in E. E. Le Clair, Jr. and L. Schneider (eds.), *Economic Anthropology* (New York: Holt, Rinehart & Winston), 1–14.

Forbes, J. (1971), 'El sistema de compadrazgo en Santa María Belén Atzitzinititlán, Tlaxcala', MA thesis (Universidad Iberoamericana, Mexico, DF).

Lomnitz, L. (1969), 'Patrones de ingestión de alcohol entre migrantes mapuches en Santiago', *América Indígena*, 29 (1): 43–71.

—— (1971), 'Reciprocity of Favours among the Urban Middle Class of Chile', in G. Dalton (ed.), *Studies in Economic Anthropology* (Washington, DC: American Anthropological Association), 93–106.

—— (1975), *Cómo sobreviven los marginados* (Mexico City: Siglo Veintiuno Editores); revised English version (1977), *Networks and Marginality: Life in a Mexican Shanty Town* (New York: Academic Press).

Mangin, W. (1967), 'Latin American Squatter Settlements: A Problem and a Solution', *Latin American Research Review*, 2 (3): 65–98.

Mayer, P. (1962), 'Migrancy and the Study of Africans in Town', *American Anthropologist*, 64: 576–92.

Polanyi, K. (1968), *The Great Transformation* (Boston: Beacon Press).

Quijano, A. (1970), 'Redefinición de la dependencia y proceso de marginalización en América Latina' (Santiago, Chile: CEPAL, mimeo).

Turner, J. C., and Mangin, W. (1968), 'The Barriada Movement', *Progressive Architecture* (May), 154–62.

14

Marital Power Dynamics: Women Providers and Working-Class Households in Istanbul

Hale Cihan Bolak

IN this essay, I address the patterns of marital negotiations in the urban working-class marriage in Turkey, where the wife is a major breadwinner. My focus is on the ways women engage male authority in mostly nuclear households. I locate these negotiations in the intersection of economic, cultural, and affective layers of marital dynamics, looking for continuities and discontinuities with the traditional patterns. Based on my research, I argue for a framework that addresses the braiding of gender and generational dynamics in the social fabric, on the one hand, and how these dynamics relate to the negotiation of new responsibilities by men and women inside and outside the household, on the other.

The primary concern of researchers who have studied the Turkish family in the last two decades has been the effects of social change and urbanization on the structure and dynamics of family relationships. Considerable attention has been focused on the enduring role of the 'functionally extended family' in buffering the impact of social change, with one example being the continuing importance of the role women play as mediators between fathers and sons. The loosening of the relationships of support and authority

among different generations of men, and the increased dependence on daughters versus sons have been identified as some of the dimensions of change (Duben, 1982; Kagitcibasi, 1984; Kandiyoti, 1984; Kiray, 1976, 1982, 1984). For example, Kagitcibasi (1984) has proposed to look at family change in terms of the changes in the system of mutual dependency relationships, noting that while intergenerational relationships may no longer involve *economic* dependency, they still involve *emotional* dependency.

The complexities of intergenerational relationships of emotional and economic dependency and their implications for marital relationships and household strategies in the urban context have not been adequately problematized by gender. Similarly, relatively little empirical research has focused on unpacking the realities of economic survival in the lives of rural migrants in the cities, especially with respect to the responsibilities assumed by men and women in the household in different socio-historical time periods. Kiray (1984) has observed astutely that while urban living has burdened the women with additional responsibilities, the men may shrug off their 'classic' responsibility—economic maintenance of the

This research was funded by grants from the Ford Foundation and the Social Science Research Council. A previous version of this essay appeared in Sirin Tekeli (ed.), *Women in Modern Turkish Society* (Atlantic Highlands, NJ: Zed Press, 1995, 173–98). I am grateful to Myra Marx Ferree for her valuable suggestions for revision, and to Josef Gugler for his editorial feedback and support.

household, concluding that women's employment *outside* the household has had little impact on the definition of male and female responsibilities *inside* the household. Similarly, Ozbay (1995) has drawn attention to the Turkish women's efforts towards raising or maintaining the status of their families that take different forms in concert with the changing opportunities for social mobility and associated values. As 'making money' became, more than ever, the prime concern of low-income families since the 1980s, resulting in men's increasing marginality to the life of the household, the valorization of domesticity among women seems to have declined while their involvement in the productive and reproductive responsibilities of their households increased.

The project on which this essay is based was carried out in 1986–7 in Istanbul and was conceived with the goal of taking a close look at the relationship dynamics in the working-class urban household in which the wife's income is indispensable to the maintenance of the family's livelihood. Given the small numbers of married women in the urban labour-force and the prevalence of the patriarchal family ideology, this particular situation represents a culturally nonsanctioned form of sexual division of labour and provides a vantage-point for studying the mechanisms for coping with possible disjunctions between the cultural ideology of the patriarchal family and the realities of economic survival. Close ethnographic studies of households and families in urban Turkey are still rare. White's (1994) research among the poor in Istanbul provides a close look at the lives of migrants, with a focus on women piece-workers who consider their economic activities not as 'work', but as 'labour'. While also focusing on mostly poor migrant households in Istanbul, my research considers a group of households in which women are self-identified 'factory-workers' with set hours of work and cash relations with strangers.

In this essay, my discussion will be informed by the following questions: how are the power dynamics constructed and negotiated in the marital relationship? through which mechanisms are these dynamics maintained, reinforced, or challenged? what are the patterns by which women engage culturally sanctioned male authority? how does the experience of these processes vary as a function of the relationship with the extended family?

Problematizing Marital Power

Without going into a lengthy discussion of the literature on power in intimate relationships, suffice it to say that research in this area is plagued by at least two major problems: first of all, informed mainly by the 'decision-making' approach, operationalization of power as the outcome of decision-making processes has been surprisingly one-dimensional. This approach has prevailed despite various theoretical and methodological criticisms (Glenn, 1987; McDonald, 1980). Secondly, how power is braided with other aspects of the marital and kinship relations—which assume a particularly critical role in Turkey—has not been a major concern.

Some recent feminist social-science research departs from convention in its insistence on relational models which not only shift the focus away from decision-making outcomes but also challenge and transcend Western culture's dichotomous construction of power and intimacy. For example, Meyer (1991: 33) has argued the need for incorporating the dilemmas involved in the 'love–power dichotomy' in a model that emphasizes the 'mutual and differential interests of parties in a dynamic relational context, including an analysis of structural aspects'. Meyer's (1991) and Komter's (1989, 1991) work has been influenced by Lukes's (1974) relational theory of power which focuses the attention on the actual content of relationships as defined by the participants. This is a three-dimensional, process-oriented framework which addresses both the ideological and structural aspects of power.

Komter (1989, 1991) conceptualizes power in terms of its exercise in a direct, observable conflict (manifest power), in terms of covert conflict and the mobilization of power or ideological values towards the prevention of issues from being raised (invisible power), and finally as the discrepancy of interests of those exercising power and those subject to this power (latent power) which can be uncovered by identifying the parties' subjective preferences under hypothetical conditions of autonomy. Applying this model to an understanding of power in different domains of marital life, she directs our attention to five elements of power processes: 'desires for (or attempts at) change; structural or psychological impediments; the partner's reaction to change; strategies to realize or prevent change; and conflicts that may arise in the process of change' (Komter, 1991: 59). Within this framework, a powerful position is one in which 'more stated preferences are realized, fewer grievances are harbored and fewer aspects of the relationship are interpreted as inevitable or unchangeable' (Meyer, 1991: 38).

I take a similar relational approach and foreground the dynamics which mediate power negotiations. Marital negotiations take place within a social-psychological landscape involving dynamics which may or may not inscribe power. While these dynamics are often acknowledged as important mediating factors, they tend to fall by the wayside in the search for causal relationships. Even research which points to the dynamic nature of marital relations and emphasizes the ongoing negotiation of the rules governing these relations often suffers from a common tendency to look for a causal relationship between an 'independent' and a 'dependent' variable. Beneria and Roldan's (1987) research on the industrial home-workers in Mexico City is an example of compromising such mediation in a final analysis which argues for the determinative effect of the relative contribution of women to the household budget on the allocation of resources. While arguing, on the one hand, that the negotiation of the terms of interaction and exchange between husbands and wives is a continuous process that is mediated by various factors, their analysis foregrounds the husband's loss of his role as breadwinner in facilitating such renegotiation.

In some empirical studies of marital power in Turkey, there has been a concern with establishing a relationship between the employment status of the wife and power outcomes. Different researchers have made varied observations about this relationship (Kuyas, 1982; Ecevit, 1986). Based on my data, I will argue the need for nuanced articulations of the mediations between breadwinner status and marital negotiations. There is no simple link between women's wage-work, their use of employment as a leverage to gain more power, and the renegotiation of gender relations in the household. The conditions for the sustenance or transformation of power relations are shaped by the interactive and mutually reinforcing effects of cultural, economic, and emotional dynamics, best captured in a multidimensional and process-oriented approach that is sensitive to the complexity and interrelatedness of the multiple layers of marital interactions. I start my discussion of the data by arguing for the relevance of the cultural construction of gender for power relationships in the households of women wage-earners. I continue with a discussion of women's expectations and how they relate to negotiations around finances, division of labour, and marital intimacy. Then, I situate the negotiations at the household level within the changes occurring in the broader cultural context and use illustrative case studies to flesh out the different patterns by which women engage male authority.

Field Research

In this project, I studied 41 households in which married women factory workers had been making a substantial contribution to the family income in the recent past. In order to determine eligibility for this study, I conducted initial screening interviews with 140 women.[1] The selection of the 41 households was partly based on my initial rapport with the women and the willingness of the husbands to co-operate. Given the conspicuous absence of data on couples in Turkey, interviewing as many husbands as possible was a major goal. Separate and often multiple tape-recorded interviews with 41 women and 27 men constituted the main source of information from the participants. The interviews were semi-structured and followed a life-history format.

Sample

The sample represents that sector of the urban low-income population in which men are not fully accountable for their most 'classic' responsibilities, namely, providing for the household. Wives worked full-time five or six days a week. Since most had shift work, their evenings were spent at work at least every other week. Wives were the major breadwinners in 24 of the households, and the only stable breadwinners in 28 of the households.[2] In 60 per cent of the households, husbands were either unemployed or had shift work/flexible hours. Families were predominantly nuclear and relatively young. Except for one recently married couple, all the couples had children.

Women respondents' ages ranged from 22 to 38, with a mean of 31. Their average residence in Istanbul was 19 years, and the mean age at arrival in Istanbul was 12. Six women were born in Istanbul, and most of the others had arrived in Istanbul before the age of 18. As such, most women's secondary socialization occurred in the city. About half of the women had an elementary school diploma. The mean age of entry into wage-work was 17, and the average time in employment was 13 years. The men were slightly older and had more education in general than their wives. At all levels of schooling, there were fewer drop outs and more graduates among the men.

The Role of Cultural Mediation in Marital Negotiations

Women's Expectations and Marital Realities

Whether or not women assume an assertive stance in marital negotiations is closely related to whether they see their employment as a resource, a source of leverage for power. This brings up the issue of the differential constructions of female employment by women and men. Elsewhere (Bolak, 1990, 1996) I have made the argument that the cultural context and particularly the construction of gender inform the internal negotiation of the dominant gender ideology by the women wage-earners and their husbands and account for their differential positioning *vis-à-vis* the 'male-provider' discourse. Women and men differ in how they construct female employment. Analysis reveals that women are not as threatened psychologically as men are by the disruption of such normative patterns as the male-provider ideal.

Women's contextualization of their experience begins with the family. They construct their employment as 'what is best for the family', and their overall evaluation of their husbands as 'responsible' or 'irresponsible' to the *family* (rather than to *them*) is an important determinant of their expectations from them. For women, more so than for men, familial concerns have a higher salience over gender conflict in marital

negotiations. And thus, for women, the strong familism characterizing the culture 'buffers', to some extent, the potential gender conflict based on men not working and women working, or women earning more than men.

Almost all the women appreciate the relative independence that comes with their monthly paychecks. However, whether their breadwinner status empowers them in marital negotiations and what stance they take *vis-à-vis* male authority varies. Limitations posed by the culturally sanctioned norms about male authority including violence, and the lack of alternatives to marriage notwithstanding, individual women's life experiences and how they make sense of them, the circumstances surrounding the marriage, the relative mesh between expectations at marriage and the current realities, and how they evaluate their husbands on the salient cultural attributes of 'responsibility' and 'respect' mediate and inform marital negotiations as well. Most women's notions of manhood centre on the man's *potential* for responsible behaviour, rather than on his current *ability* to provide for his family. So, even women who have entered marriage with the expectation that their husbands would maintain the livelihood of the household refrain from flaunting their employment even if, contrary to their expectations, they find themselves as the sole breadwinner, as long as they do not see their husbands as 'irresponsible'.

The prediction of previous research (Erkut, 1982) that role-sharing may not be a salient issue for women in Turkey is not borne out for the large majority of women in this study. Women do expect their husbands to participate in domestic tasks, especially those associated with parenting, and unmet expectations result in more conflict than suggested by previous research. It is true that having witnessed their hard-working mothers taking on both the domestic and breadwinning obligations, women feel less reliant on men whom they generally see as incompetent and peripheral to the running of the household. But they always make upward comparisons, comparing their husbands with men they hear about 'who are more helpful at home'. The conflict ensuing from these comparisons is heightened when the husbands default on both domestic and economic grounds.

The traditional women with longer rural backgrounds whose gender ideals were frustrated in their marriages, and younger, more urbanized women who are more assertive in their demands *vis-à-vis* their husbands are the most dissatisfied with the absence of male participation, because they do not see their 'sacrifice' reciprocated. The most serious confrontations occur in households in which the men are not able to ask their wives to quit working for pay and where the women are able to use their income as a source of leverage for power. These households are mostly in the category of 'Open Power Struggle' which I discuss below. Although women's economic superiority increases their expectations, and to some extent, their husband's participation, my data supports the theory and research on the division of labour in the households of employed women (Ferree, 1984, 1990) that it is only when their employment is constructed by their husbands as a 'contribution' and not a 'cost', and when the couple agree that the 'breadwinning' responsibility is a joint one that male participation becomes somewhat predictable (Bolak, 1997).

The typical discourse in the few households with the man as stable provider is an agreement that the woman works for a specific goal, whether it is to save money for a house or for retirement. Although they might both agree that they would not be able to own a home without her earnings, they hold on to the notion that she will work only as long as she can manage the 'double shift' (Hochschild, 1989), which, combined with the husband's job stability, serves to undermine any potential leverage she might have because of her breadwinner status. This ideological justification of inequality, like the use of gender-specific excuses couples construct to sustain gendered parenting and housework, are excellent examples of invisible power dynamics. Women who end up taking the major responsibility for children's discipline or grocery-shopping trace their husbands' lack of involvement to their lack of exposure to appropriate role models while growing up. For the women, the construction of the men as 'orphans' (literally or figuratively), or victims of unfortunate circumstances, provides a gender-specific excuse and justification for their husbands' inadequate involvement in family work. Again, the locus of explanation for men's irresponsible behaviour shifts quickly from gender to the family, revealing the stronger salience of the latter in the culture.

The relationship between women's breadwinner status and control over financial resources is a mediated one as well (Hood, 1986). It is true that those women whose financial autonomy is not challenged by their husbands are all primary breadwinners. Similarly, if men are unemployed, they have to consult with their wives over any major expenditure. It is also the case, however, that as long as men do bring home some money, it is very difficult for women to veto a major decision such as buying property. Finally, how women use their leverage is again intimately related to their marital expectations and the nature of marital relations. For example, in the case where the man has shrugged off all his responsibilities and the wife has assumed the primary provider role over a long period of time, she also has full control over the finances and hands him a fixed allowance. But in most other cases, the decision as to who will *manage* the finances rests with the man. For example, it is not uncommon for women to be appointed as 'financial book-keepers' by their husbands in households where resources are limited. Knowing that they can stretch the money a little further, women end up assuming what essentially is an extra burden.

In households where the traditional basis of male authority is still relatively secure, women with the financial upper hand often help maintain the status quo and corroborate their husbands' 'masculinity' by handing over their paychecks to the men even if they end up being the financial managers themselves. A woman with considerable control over the finances may nevertheless want to hand over this control to her husband and may expect him to have a say in where she goes and who she sees in her spare time. Where male authority is not so secure, women's negotiating power and the strategies they can use are a joint function of the significance of her income for the family budget, marital history, whether there is a threat of violence, and the nature of the extended family relationships. For example, while a woman who is having to maintain the household on her own despite her husband's higher income, quits her job as a way of inviting her husband to assume responsibility, another woman who is in a better financial position manages to bring herself over the years to a situation where she is able to use her whole paycheck to make monthly instalments on a flat in her name.

Although it is generally difficult for women to admit to using their employment as leverage to gain power in the household, the nature of the everyday interactions in which they are embedded inform their reference groups, which, in turn, influence their self-perceptions and the stance they take with their husbands. A traditional woman whose work exposes her to only one other woman (also from her own village) and an occasional interaction with the foreman and whose social life outside work is restricted to obligatory visits to her husbands' relatives has a quite limited repertoire to draw upon compared to a woman who works with 600 men and women on the same floor and socializes with them after work. In the first instance, the husband feels much more in control and is less threatened by the woman's employment; in the second case, the intense socializing that goes on at work has a radicalizing influence on the woman, increasing her sensitivity to inequality at home, and thus making her husband feel more threatened by the solidarity among women at work.

Men's jealousy about women-to-women relationships and their attempts to control their wives' movements have to do with the threat they feel from the solidarity women forge with other women and their feelings of being left out. Kiray (1984) identifies communication problems as a major source of incompatibility that arises after the move from the rural context where extended family relationships and sex segregation prevail, to the big city, where, on the one hand, the couples are relatively insulated, and on the other hand, they are exposed to the urban culture of 'sharing'. It is because of such 'residual isolation', that especially young couples experience this alienation as problematic. As much as women complain about their husbands' lack of understanding and appreciation, they mostly take a patient and rather nonchalant attitude towards their husbands, while urban life makes men more dependent on their wives' attention and understanding. In fact, for most men, the wife is still the confidante and the person closest to them, whereas women seem to have more options. Although women's maternalism leads them at times to overlook their husbands' expensive hobbies, it often does not make them emotionally dependent on men. It appears that female employment leaves men's needs for 'mothering' unmet at the most critical time, when they also lack extended family support.

Relationships with Extended Family and Community

In Turkey, marital power relationships are embedded in the context of extended family, kin, and other community relationships (Kagitcibasi, 1982; Erder, 1984). For example, the nature of a woman's relationship with her family of origin and particularly with her mother has a major influence on their negotiating power in marriage. Women with the most leverage in marital negotiations have mothers who, like themselves, are strong, and supportive of their decisions to keep working, while making a point of not getting involved in their marital problems. Those with the least leverage, on the other hand, are more traditional women who are isolated from their families of origin.

Even within the limits of this small-scale study, the observations of family scholars concerning the growing reliance on daughters over sons for old-age security find some support. It is in this context that the relationships with the wife's family gains significance. The situation of women who go through a period of marital separation is a case in point. Those women who emerge from such separation with an increased confidence in their abilities to support themselves and their children can more easily envision the prospect of divorce if they can count on their families, especially their brothers, to be supportive. Those women who lack this support find it harder to make this decision and are more vulnerable vis-à-vis their husbands' control.

With the erosion of the economic basis of the patriarchal relationships among different generations of men, the relationships of support and control are weakening as well. This shift in men's connection with their families of origin has implications for their relationships with their wives, contributing to a more negotiable state of affairs in the household. After their relatively sheltered lives in the village, men find themselves on their own in the city with a lot of responsibilities that they feel unprepared to deal with. For example, quite a few of the men I talked with regretted their decision to get married as having been premature. Confronted with the challenges of urban living and feeling overwhelmed by the economic responsibilities which they feel inadequate to fulfil, it is common for men to compensate by using other means to maintain their sense of manhood in the community, such as entertaining their friends, and lending money. These behaviours make them the target for their wives' accusation that 'he wants to show off even though he is poor'.

Indeed, an asymmetrical situation emerges in several households, with the wives having to be extra thrifty at home to compensate for their husbands' generosity outside the home. Spending prerogatives and priorities appear to be gendered, as well, with men more interested in ostentatious spending and women more realistic and pragmatic in their priorities. I talk elsewhere about how the cultural context that women are embedded in partly accounts for their stronger identification of their individual needs with those of their households, but rather than a focus on maternal altruism, an understanding in terms of the conflicts, struggles, and strategies in the working-class households might be more constructive (Bolak, 1996). There are several households in which men do not collaborate with women towards ensuring the welfare of their households, and instead, by competing for scarce economic and emotional resources, recapitulate, in their own relationships with their sons, their unfulfilling relationships with their fathers. Consequently, not only the wives, but the children as well are compromised in this pursuit of manhood.

Men's relationships with their agnates and fellow-villagers are uneasy ones in the urban context of economic insecurity where high expectations of financial help and assistance with finding jobs make disappointments almost inevitable. Afraid of rejection and of further injury to their sense of pride, the men may avoid going to their kin for help. When the men are broke, no longer able to maintain their ostentatious spending and hence their status in the community, they become totally dependent on their wives for support. When the woman's economic advantage coincides with the man's loss of community supports, her leverage increases and he becomes relatively defenceless. This painful process often results in a paradoxical situation with the woman burdened by a disproportionate amount of responsibility, on the one hand, and increased leverage, on the other.

With time, the woman is also able to have a greater influence on her husband's relationship with his family and friends. It is when the man is economically vulnerable and disillusioned with the lack of support from his own community that he finally starts taking his wife more seriously. Hence, especially in those areas where he is vulnerable to the influence of his kin in the earlier days of marriage, she gains an increased negotiating power, bolstered by her economic advantage. For example, she is able to assert herself more in her husband's work-related decisions, having relatively more success in discouraging him from getting into questionable business partnerships with his brothers or other kin. Likewise, she becomes less vulnerable to violence by her husband on account of a problem involving her in-laws.

A woman's increased assertiveness with her in-laws, and especially her brother-in-law is a sign of what I see as a more negotiable set of authority relations developing in the households of women providers. One concrete example of the weakening nature of patriarchal relationships is a woman's ability to avoid visiting her in-laws even when she has no financial autonomy. Resistance to socializing with her husbands' kin is often the only recourse a woman has and which she can easily legitimate by reminding him that she is a 'working woman'. The fact that two-thirds of the women preferred 'being able to do certain things without having to consult their husbands' over 'having a husband take full responsibility for maintaining his household' suggests the importance of autonomy for these breadwinners.

In what follows, I illustrate with examples how marital negotiations are not a simple function of how much money each spouse makes, and how these dimensions work in concert with each other to determine the affective tone of these negotiations.

Patterns of Negotiations at the Household Level

In her argument for deconstructing the notion of patriarchy, Kandiyoti (1988: 285) contends that 'systematic analyses of women's strategies and coping mechanisms can help to capture the nature of patriarchal systems in their class specific and temporal concreteness and reveal how men and women resist, accommodate, adapt, and conflict with each other over resources, rights and responsibilities'. How the cultural ideology about gender is negotiated at the *household* level by working women and their husbands is an empirical question which is the prime concern of this study. A close scrutiny of the sample reveals a complexity which I illustrate by presenting concrete case studies that yield different interactions of the various levels of determination. I find five basic patterns here, which are (1) open power struggle, in which each spouse minimizes the contributions that the other makes while maximizing their own control; (2) patriarchal accommodation, in which men enjoy latent (rather than manifest) power over women who do not openly challenge male power; (3) traditional defiance, in which women negotiate the situation through a combined stance of private defiance and public deference; (4) power renegotiated, in which women are able to assert some power without much challenge by their husbands; and (5) resisting conflict, in which power dynamics are hidden by an ideology of mutual interest.

Open Power Struggle

The most intense negotiations take place in the households of younger and more urbanized women with longer schooling, employment prior to marriage, apparent lack of conservatism, and child-care needs. Six households fall in this pattern. Women in these households consider their husbands to be unappreciative of them and insensitive to their needs, and engage with them in an open power struggle which often has the appearance of a 'tug of war'. They choose their female co-workers as their reference group and compare their husbands with men who 'help more', and they seem to experience the least conflict resulting from the disjuncture between the girlhood dream of being perfect caterers for their husbands and the modern ideal of 'role sharing' they are exposed to on the shop-floor. Family life is characterized by conflicts over expenditures, female access to money, and physical violence. The women are engaged in an open power struggle with their husbands and strategize to maximize their control. The men perceive their prerogatives threatened by the seeming autonomy and assertiveness of their wives and employ different

means to curb this autonomy. Their perception of their wives as having overstepped their limits as 'disrespectful women' makes them overly defensive and prone to using 'manifest power'.

Kadir and Hatice Kosan have both been employed in different factory jobs. Hatice Kosan stayed in her first job after the marriage for 2.5 years during which time the factory that her husband was working in closed and he was idle for 7 months. She quit her job to get severance pay to use towards downpayment for a place of their own. If she had stayed in her job, she would have been the primary breadwinner now. She was at home for 2 years, then worked in another job for 1.5 years until that factory closed down. She took another job, but quit after 2 years to go into shift work which she has been doing for the last 2 years. During the 2 years she worked at her previous job, he lost his factory job again and was without a job for 1.5 years. He has been in his current job for 3 years, but he wants to quit and go to back to the village to open his own grocery store.

Hatice Kosan remembers her working mother and her father as sharing household tasks, and compares her husband unfavourably with her father. While she got married with some reluctance, he rushed into it in order to get out of an arranged marriage with his cousin. Marriage severed his ties with his mother—now that he has a 'working' wife, his mother does not indulge him like she used to; she tells him: 'you two can take care of each other'. When he complains about his wife, she says, 'You found her yourself!' He feels that as the oldest of three brothers, there is nobody for him to confide in, no one to give him advice. He wants somebody older to restrain him from beating his wife, but such sanctions are not available.

Kadir Kosan had thought his wife would not be working, but found himself overwhelmed with responsibility and unable to ask her to quit. He compensates for her having to work by minimizing her contributions: 'Your money is barely enough to pay for child care and my cigarettes', and by trying to bolster his masculinity with violence. Hatice Kosan's ambivalence about being a working mother is largely mediated by her husband's put-downs and lack of co-operation. Her frustration and his sexism result in a tug or war over the allocation of family work and resources. Each refuses to help the other with their respective tasks: 'Repair work is not my responsibility if you won't go to the store!' she says. For fear that she might leave him, Kadir Kosan used to help with household tasks before their son was born; now he feels that the boy is old enough for him to take with him even if she leaves.

Knowing that her husband cannot afford to ask her to quit, Hatice Kosan resorts to all the strategies in her power. One reason she changed from a day job to shift work was so that her husband would have to take their son to day care when she was at work. Finding it demeaning to serve his son 'as if he were an old man', when she works the evening shift, he wants her to change to a day job. When her husband challenges the amount of money she spent in the market-place, she tells him 'You do it then.' Not wanting to be seen with a shopping bag and knowing that he will not be able to stretch the money like she does, Kadir Kosan feels he has no choice but to let her do the shopping, knowing at the same time that she secretly saves from the money earmarked for household expenditures. They both attempt to secure control over the resources and strive to maximize their spending money; in fact, he does not reveal the amount of his paycheck to her and retains a generous allowance before handing the rest to her for expenses. She also gets social rewards from her job and makes him even more insecure by keeping a tight network of friends. But ultimately, she lacks the support system to risk divorcing her husband. She does not have her family nearby and as a result of losing seniority when she withdrew her compensation to pay for their house, her earnings are low.

This basic pattern shows slight variations in other households with women's strategies depending on the nature of extended family relations and the relative earnings of the spouses. For example, Saray Saglam's family is supportive of her decision to divorce her husband, which, combined with her strong economic position, makes it easier to go ahead with the divorce. Another woman who practically runs the household on her income alone decides to quit working as a way to invite her husband to use his earnings more responsibly.

Patriarchal Accommodation

Under this system, there is little *visible* contest of wills, encouraging us to look more closely at 'latent' or 'invisible power'. Twelve households fall in this pattern. The wives are less challenging of the ground rules and do not seem to aspire to a renegotiation of gendered prerogatives. They are careful not to override the implicit agreement against wives' use of their employment as a leverage for gaining power. Even as the only stable providers in their families, they see the breadwinning responsibility and the rights that go along with it as belonging to the husband. The men have latent power even when they do not use violence. As in the previous group, in those few households in which the wife is the secondary provider, the discourse is that she is *allowed* to work as long as the male prerogatives in the household are not challenged. Two cases will illustrate the different mechanisms through which this system works.

Simsir and Eyup Dogar have been married for 12 years. In the first 3 years, they were both employed in the factory they are working in now. When one child became ill, the wife stayed at home for 3 years. About 3 years ago, she decided to make use of her past training to become a seamstress and found a well-paying job. Her upward mobility caused her husband's jealousy and violence quite like in the previous group and led to a marital breakdown, triggered by her husband's family who 'were envious of our lifestyle'. Toying with the idea of divorce, and after several months of separation, she finally agreed to a reconciliation. After they made up, she returned to her first job at the confectionery factory. Having lost seniority like Hatice Kosan, she started with the minimum wage. The status quo was stabilized with the husband consolidating his position as the primary breadwinner once again. The couples' interviews were similar except for the husband's denial of her ever being the primary breadwinner, and the marital problems that it caused.

Simsir Dogar's subjectivity about being a wage-earner needs to be understood in the context of her expectations and her husband's attitudes. Unlike Hatice Kosan who witnessed her parents as sharing household tasks, Simsir Dogar grew up with a mother who was at home and a father who did not do anything around the house. Hence, she does not have expectations of male participation; but even when she has occasional need for help, her husband responds 'Quit if you can't manage both! After all, you only make 50 thousand liras.' Simsir Dogar's small and thus 'humiliating' paycheck and her perception that 'a woman should not work unless she has something creative or important to do, something which requires responsibility'—as in her previous well-paying job which she had to quit to save her marriage—colludes with her husband's unaccommodating attitude and results in an unsettled stance that vacillates between modernity and traditionalism. She criticizes those women like Hatice Kosan who save money without their husbands' knowing, or others who 'order their husbands around'. Even when she was the primary breadwinner, she used to hand her paycheck over to her husband. Unlike Hatice Kosan who sees herself as not too maternal, Simsir Dogar invests all her passion in mothering her two boys.

Paternalism is one means through which patriarchal authority is maintained in the household without physical force. In the interview, Eyup Dogar presents a 'Pygmalion' story of their marriage and sees himself as her 'educator and socializer', which accords with her story. Having got married at a very young age to a man 15 years her senior, she constructs the marriage as based on 'friendship and respect, not on love.' This age difference justifies his authority, even his extreme jealousy. His treatment of her as 'daddy's girl', and their consensus on child-centredness further compensate for his authoritarianism. Furthermore, his dissatisfaction with his family's role in their separation has somewhat tempered his arbitrary control over her behaviour.

Traditionalism is another means through which authority is maintained. In the Onur household, the couple have been married for 16 years. The wife has been employed in her current job for the last 15 years; the husband had been working as a courier in a bank for 13 years when he got fired—'unjustly' they think—3 years ago. After a short period of unemployment, he started working in the neighbourhood coffee shop and his breadwinner status shifted from primary to secondary. Like Simsir Dogar, Semiha Onur observes her husband's authority on traditional grounds and sees him as basically a 'good family man' who is not altogether irresponsible. She is protective of her husband while being internally conflicted. Though she questions the inequality in the proportion of their wages that go into the household budget, she overlooks it since 'he is

not in good health and needs more pocket money to eat well at his job'; despite her fatigue, she accommodates to sexual needs and tries to bolster his ego, in a manner typically associated with working-class women in the literature (Rubin, 1976). In the Onur household, the mesh between the gender ideologies of the husband and wife provides the cement that consolidates the patriarchal order in the household even when the husband is not securely employed. However, her internal conflict surfaces in statements such as, 'If it weren't for me, he might have died somewhere, but he wouldn't appreciate what I did for him.' Women feel most conflicted psychologically if their gift of traditional deference is not appreciated by their husbands whose agenda about marriage may have changed with their adoption of the urban life-style.

Traditional Defiance

In the previous group, the men are not yet accused of being 'irresponsible', and are thus able to enjoy their wives' tolerant and protective attitude towards them. The fact that conflict does not become violent is due to the women's accommodating stance. The characteristic pattern in a third group of households is neither an accommodation to male dominance, nor an active contestation for power, but a curious combination of ritualistic adherence to traditional norms in public, on the one hand, and private mockery and manipulation of husbands, on the other, resulting from the wives' feeling betrayed and disillusioned. Six households fall into this pattern. The women can be most aptly described as 'urban villagers', and fit Kandiyoti's (1988: 282) description of female conservatism in reaction to the breakdown of classic patriarchy the best: 'Despite the obstacles that classic patriarchy puts in women's way, which may far outweigh any actual economic and emotional security, women often resist the process of transition because they see the old order slipping away from them without any empowering alternatives'.

These are women with longer rural backgrounds who experience the biggest gap between their expectations that they would not have to work after marriage, and their current situation. Although they are similar to those in the previous group in terms of their relative traditionalism, the manifestations of the men's failure to fulfil their provider responsibilities are very different in the two groups. Women's dissatisfaction with their 'irresponsible' husbands and their conviction that, contrary to their expectations, they will continue to remain solely responsible for maintaining the household leads them to challenge their husbands' authority, even if this questioning remains within traditional boundaries. This pattern resembles most closely the findings of Beneria and Roldan (1987) in Mexico City and those of el-Messiri (1978) in Cairo in that the man's provider role assumes centrality in women's depiction of manhood.

Hacer Itir is a hard-working woman who grew up witnessing her own mother in charge of the home and the farm and her father as a wanderer. Because of child-care problems, her daughter is with her husband's family in the village. They have both had different factory jobs, interrupted by plant closing or their temporary relocation to the village, none of them lasting for long. The husband has not worked for a year and a half. Hacer Itir sees herself as both the man and the woman of the house and asks: 'Of course breadwinning is man's responsibility, otherwise why call him a man?' Although she does not mind working *per se*, it is hard for her to reconcile her image of womanhood with having to support her husband while her daughter has to grow up in the village. Yet, her employment does not pose a threat to male prerogatives in public. She still lets him do the actual transaction when they make a major purchase, as it is not in her interest to humiliate her husband in public—to that extent, they collude in maintaining the status quo.

The context within which these women are embedded facilitates the maintenance of the status quo. They are mostly insulated from radicalizing influences that more urbanized women are exposed to. Hatice Itir is embedded in a tight kinship network and does not even look at men where she works let alone talk with them. Her husband has a brother who owns a coffee-house in the neighbourhood and he argues that he makes tips, enough for his own expenses and even for household expenses, by just hanging around the coffee-house. While he cannot ask his wife to quit, he protects his pride by denying his dependence on her. He has redefined the situation by seeing himself in control and her as deferential, and by projecting a self-sufficient image. The devaluation of their wives' contribution and

overstatement of their own contribution are these men's symbolic ways of maintaining the balance of power in the household.

At home, Hacer Itir talks back to her husband and even resorts to strategies like not sleeping with him and not paying attention when he is sick. In other similar households, women may mock their husbands as when one woman responds to her husband's suggestion that he will get married again: 'Who would take you?' Another may refuse to cook or do her husband's laundry unless he brings home money. The women call their husbands to task not only for failing to be adequate providers, but for failing to take care of 'male tasks' such as shopping for food. While let down, they make frequent references to not wanting to 'lose their honour' by leaving their husbands; all invoke religion in their ambivalent responses to their situation. One woman says: 'heaven is at the mercy of the husband', meaning if she wants to go to heaven, she has to be patient. In religious families, female employment causes estrangement in the marital relationship which reminds us of Mernissi's (1975) depiction of Moslem Morocco where the increased economic participation of women comes to have an 'emasculating' effect on men, who, by law, are entitled to control their women.

Power Renegotiated

There is another group of women whose primary breadwinner status and personal resources such as self-confidence, responsibility, decision-making ability, and perseverance combined with their husbands' loss of their social and economic base result in a more fluid situation with regards to power. Eight households fall in this pattern. In these households, women have gained an ascendancy which they may or may not be comfortable with, but which enables them to be assertive in the household with less challenge by their husbands than in the previous groups. Gunduz Bakir says, 'When I got married, I thought my husband was going to be authoritative and I was going to obey him, but things have not turned out that way.' She has gradually learned to enjoy her more forceful presence in the household. Her husband tries to bolster his masculinity by arguing that 'the wife fits the home just as a flower fits a vase', quite ironic since she has always worked. But he feels respected by her despite his lack of stable employment.

The extent to which these women are able to enjoy their relative autonomy depends partly on their attitudes about gender. A comparison between the Turksan and Ceviz households will reveal the difference. Necmiye Turksan's expectations were initially very similar to those of the women in the previous group, but being a strong-willed woman and seeing that her husband is not the 'responsible man' she had hoped for, she does what she had to do when she was a single woman taking care of her mother—she takes the reins herself. She started working at 10 years of age, and has been working ever since. She has been the household's primary and only steady bread-winner since marriage, and her husband has been an irresponsible provider until a year ago when they got back together after she separated from him for a year. Her decision to reconcile has partly to do with her wanting to own a flat and knowing that she cannot take care of her two daughters and pay for the flat at the same time. Now her entire paycheck goes towards the instalments for a flat, and her husband is responsible for maintaining the household.

Just as Necmiye Turksan, Remziye Ceviz sees herself 'like a man'. She arrived in Istanbul from Yugoslavia when she was 18 and declared herself as the head of her parental household. Although both are Yugoslavian immigrants, she has a more modern outlook than Necmiye Turksan. She never thought of quitting work; in fact, she told her husband, 'I might divorce you, but I would never quit my job!' She has been working in her present job without any interruption while her husband has had a bumpy work history in the last 5 years and only recently got a steady though low-paying job upon his wife's insistence: 'She told me that from now on, I had to go to work at 8 a.m. and come home at 5 p.m., and that's what I'm doing.' He now hands over his paycheck to his wife. In both households, the men have assumed more parenting responsibilities since their 'rehabilitation'.

Both women have complete jurisdiction over matters concerning the household and enjoy full independence. Their husbands appear 'childish and gullible' to them. Niyazi Turksan is alone in Istanbul and does not have any kin support. On the other hand, he is surrounded by his wife's mother and kin network. Like Niyazi Turksan,

Ozcan Ceviz has become disillusioned with kin, and especially with his mother's role in trying to break up his marriage. His loss of economic power due to his drinking brought about the loss of his support system as well, and he has taken refuge in his wife's support. Both men seem to have been humbled over the years, and have had to come to terms with their wives' strengths. Like the Turksan couple, the Ceviz couple have been socializing mainly with the wife's network of friends and family, which is a significant indicator of a wife's relative advantage. Marital compatibility is higher in the Ceviz household which makes it easier for the couple to define the wife's new role in maternal terms whereas Necmiye Turksan is not willing to mother her husband and experiences the situation as more anomalous. There are other marriages in which women benefit increasingly more from using their maternalism as a resource.

The patriarchal ideal of man as the primary provider is no longer viable for these women. If they are secretly harbouring a desire for more male authority/responsibility, and projecting their husbands as being a little more authoritative than they actually are, it is to match what they perceive to be the normative standard in society. But their lives and self-definitions appear to be quite independent of their husbands. They derive strength from their female networks, and those who live in close proximity to their mothers have a solid support system.

Resisting Conflict

Lukes's concepts of 'latent' and 'invisible' power are most relevant to understanding the marital dynamics in two groups of households. In one group identified as 'patriarchal accommodation', either paternalism or traditional attitudes deflect conflict. In yet another group of households conflict is again not the predominant reality. Six households fall in this pattern. In these households, a compromise solution has stabilized which has the semblance of domains of family life. A situation in which the interpersonal dynamics in the household cannot be adequately tapped by the concepts of domination or negotiation, it involves a relative lack of defensiveness on the part of men—which allows a definition of the situation in terms of 'Two hands are better than one', and a comfortable sense of autonomy for women.

In this group, more than any other, the women insist that their husbands' failure to maintain their households did not have an adverse effect on their relationship. Like those in the previous group, women seek more initiative and resourcefulness on the part of their husbands, but they also enjoy the appreciation they receive for being providers and for the relative autonomy that comes with being the stronger spouse. The ethic of 'shared funds' is most characteristic of this group; similarly, it is in these households and in the previous group of households that the couples co-operate most in child care. The fact that the husbands are co-operating with them for the welfare of the household makes it difficult for women to use their employment status as a resource in marital negotiations, and buffers any further claims to power they might have had.

In the Mermer household, the husband is the traditional head of household, but the wife is the manager, the central figure in the household. She makes the decisions about where she will go and what she will spend money on. There are areas in which she usually cannot go far at first when she attempts to change his mind—he likes to make his own decisions about setting up partnerships, for example. But he is not an irresponsible provider, and manages to keep bringing income, even if he is not securely employed. Nezaket Mermer is a resilient woman who holds her own and will not be dominated. She is also one of the women who like to be in charge of housework, but have enjoyed their husbands' co-operation in child care when the children were younger. Although women are again more resourceful than their husbands, the dynamics, characterized by an ideology of 'mutuality', are different from those households in which marital history in tandem with their husbands irresponsible provider behaviour and loss of social supports has empowered women.

Even when decision-making appears joint, a closer look sometimes reveals the husband's greater veto power and the ·asymmetries in the influence strategies used by the couple. An example of a less visible power differential occurs in situations where the husband and wife have been working opposite shifts to be able to take turns with child care, and have, by now, come to a point where the man avoids being together with his wife in order to stay out of conflict and to make sure he is not 'irritated'.

Conclusions

In the households of women breadwinners, the conditions for the maintenance and transformation of the traditional relations of patriarchy are shaped by the interactive effects of economic, cultural, and emotional dynamics in process. The women's dilemmas about using their employment as a source of leverage for power point to the force of cultural constructions of gender, responsibility, and male authority, on the one hand, and to the limitations on women's choices posed by the threat of violence and lack of alternatives, on the other. Despite these barriers, employment forces the limits of traditionalism '*in practice*'. Whether or not the woman's breadwinner position has a critical effect on her financial autonomy and on the allocation of household responsibilities, the traditional basis of male authority *is* ideologically challenged through a discourse that includes intra- as well as extra-household roles in the definition of 'male responsibility'. The scope of this challenge varies, depending partly on women's economic power, the extent to which their expectations are affirmed in their marriages, and on the characteristics of their cultural milieu. Nevertheless, the language of negotiation does seem to be changing, with the emotional factors like understanding and attention figuring more strongly in the equation.

Marital power gets constructed in the complex interplay between male responsibility and female respect. The women's difficulties with subscribing to the norm of 'obedience' to male authority and the men's uneasiness need to be evaluated in this light. For many women, the traditional justification of male authority as 'might makes right' retains its validity often as a *cultural script*, without necessarily being internalized, suggesting a weakening of at least the 'invisible' dimension of male authority if not of the 'manifest' and 'latent'. These dynamics are situated in the context of looser relations of support and control among different generations of men, which further contribute to the establishment of a more flexible basis for marital negotiations. Over time, women may encounter a paradoxical situation characterized by their disproportionate burden resulting from male irresponsibility, on the one hand, and their increased negotiating power, on the other. It is evident that we need relational and process-oriented frameworks which are sensitive to the interaction of multiple levels of determination of power.

Notes

1. Through a chain of contacts ranging from union leaders to company presidents, I gained access to five factories, each representing one of the five branches of the manufacturing subsector of the industry (tobacco, pharmaceutics, garment, textile, and confectionery) with the highest concentration of women workers. These factories also displayed a wide variety in terms of wages and job security, type of ownership, availability of child-care facilities, and location in the city.

2. The analysis revealed that an operationalization of women's breadwinner status based on their relative *monetary contribution* to the household budget alone was inadequate. An important component of this status resides in the stability of employment and the continuous nature of their contribution. The contribution of the stably employed women to the household made fluctuations in male income and employment possible, and their relative earnings during the last year was an imperfect measure of their relative contribution to the household. A more comprehensive assessment yielded the following: (a) in 22 of the 41 households, the wives had made the major monetary contribution to their households, while in 11 households, men had made the major contribution, and the contributions of husbands and wives were about equal in the remaining 8 households; (b) in 28 households, women had been the only stable breadwinners in the recent past (and 20 of these women had also been the major providers), while in 10 households husbands and wives had both been stable breadwinners, and in 3 households, both had been unstable breadwinners.

References

Beneria, L., and Roldan, M. (1987), *The Crossroads of Class and Gender* (Chicago: University of Chicago Press).

Bolak, H. (1990), 'Women Breadwinners and the Construction of Gender: Urban Working Class Households in Turkey', Ph.D. diss. (University of California–Santa Cruz).

—— (1996), 'He Loves to Show Off even though He Is Poor! Gender Strategies in the Working Class Household', MS in preparation for submission.

—— (1997), 'When Wives are Major Providers:

Culture, Gender and Family Work', *Gender & Society*, forthcoming.

Duben, A. (1982), 'The Significance of Family and Kinship in Urban Turkey', in C. Kagitcibasi (ed.), *Sex Roles, Family and Community in Turkey* (Bloomington: Indiana University Turkish Studies Press), 73–99.

Ecevit, Y. (1986), 'Gender and Wage Work: A Case Study of Turkish Women in Manufacturing Industry', Ph.D. diss. (University of Kent at Canterbury).

Erder, T. (1984) (ed.), *Family in Turkish Society* (Ankara: Social Science Association).

Erkut, S (1982), 'Dualism in Values toward Education of Turkish Women', in C. Kagitcibasi (ed.), *Sex Roles, Family and Community in Turkey* (Bloomington: Indiana University Turkish Studies Press), 121–32.

Ferree, M. M. (1984), 'The View from Below: Women's Employment and Gender Equality in Working Class Families', in B. B. Hess and M. B. Sussman (eds.), *Women and Family: Two Decades of Change* (New York: Haworth Press), 57–75.

—— (1990), 'Beyond Separate Spheres: Feminism and Family Research', *Journal of Marriage and the Family*, 52: 866–84.

Glenn, E. N. (1987), 'Gender and the Family', in B. B. Hess and M. M. Ferree (eds.), *Analyzing Gender: A Handbook of Social Science Research* (Newbury Park, Calif.: Sage), 348–80.

Hochschild, A. (1989), *The Second Shift: Working Parents and the Revolution at Home* (New York: Viking).

Hood, J. C. (1986), 'The Provider Role: Its Meaning and Measurement', *Journal of Marriage and Family*, 48: 349–59.

Kagitcibasi, C. (1982) (ed.), *Sex Roles, Family and Community in Turkey* (Bloomington: Indiana University Turkish Studies Press).

Kagitcibasi, D. (1984), 'Intra-Family Dynamics and Relationships: A Model of Family Change', in T. Erder (ed.), *Family in Turkish Society* (Ankara: Social Science Association), 149–66.

Kandiyoti, D. (1984), 'Changes and Continuities in the Family Structure: A Comparative Perspective', in T. Erder (ed.), *Family in Turkish Society* (Ankara: Social Science Association), 23–41.

—— (1988), 'Bargaining with Patriarchy', *Gender and Society*, II, 3: 274–91.

Kiray, M. (1976), 'Changing Roles of Mothers: Changing Intra-Family Relations in a Turkish Town', in J. Peristiany (ed.), *Mediterranean Family Structures* (London: Cambridge University Press), 261–71.

—— (1982), 'The Women of Small Town', in N. Abadan-Unat (ed.), *Women in Turkish Society* (Leiden: E. J. Brill), 341–56.

Kiray, M. (1984), 'Metropolitan City and the Changing Family', in T. Erder (ed.), *Family in Turkish Society* (Ankara: Social Science Association), 79–89.

Komter, A. (1989), 'Hidden Power in Marriage', *Gender and Society, III*, 2: 187–216.

—— (1991), 'Gender, Power and Feminist Theory', in K. Davis, M. Leijenaar, and J. Oldersma (eds.), *The Gender of Power* (New York: Sage), 42–62.

Kuyas, N. (1982), 'The Effects of Female Labor on Power Relations in the Urban Turkish Family', in C. Kagitcibasi (ed.), *Sex Roles, Family and Community in Turkey* (Bloomington: Indiana University Turkish Studies Press), 181–205.

Lukes, S. (1974), *Power: A Radical View* (New York: Macmillan).

McDonald, G. W. (1980), 'Family Power: The Assessment of a Decade of Theory and Research, 1970–1979', *Journal of Marriage and the Family*, 42: 841–54.

Mernissi, F. (1975), *Beyond the Veil* (New York: John Wiley).

el-Messiri, S. (1978), 'Self-Images of Traditional Urban Women in Cairo', in L. Beck and N. Keddie (eds.), *Women in the Muslim World* (Cambridge: Mass.: Harvard University Press), 522–40.

Meyer, J. (1991), 'Power and Love: Conflicting Conceptual Schemata', in K. Davis, M. Leijenaar, and J. Oldersma (eds.), *The Gender of Power* (New York: Sage), 21–41.

Ozbay, F. (1995), 'Changes in Women's Activities Both Inside and Outside the Home', in S. Tekeli (ed.), *Women in Modern Turkish Society* (Atlantic Highlands, NJ: Zed Press), 89–111.

Rubin, L. B. (1976), *Worlds of Pain: Life in the Working-Class Family* (New York: Basic Books).

White, J. (1994), *Money Makes Us Relatives: Women's Labor in Urban Turkey* (Austin: University of Texas Press).

15

Social Policy Transforms the Family: The Case of Singapore

Janet W. Salaff

THE intervention of the Singapore Government in moulding a new family on behalf of its development programme offers the best-known example in a market economy of a state restructuring society. Social policies shaped people's goals as well as how they reached these goals. The Singapore city-state early on offered a range of social services. We call these 'social goods', since they organized or subsidized costly services that fulfilled major family needs: these fell in the realms of jobs and retirement, housing, education, health, and childbearing. The social policies were not neutral. They were bent on 'modernizing' the family for the market-oriented development goals of the state. By tying needed 'social or public goods' to the party-state development plan, the state not only won support. These social policies extended social control by forming a common core of family values and goals. The goals took root among families from a broad range of social classes. However, more affluent families used different versions of the social services than the poor. While all were assimilated into the new social order based on the money economy, the social services conveyed variations in meaning to families of different classes. In these ways, the services reproduced social class differences in a new context.

The essay is divided into two parts. First, I give a summary of the social policies. I follow with a discussion of how families of different social groups used the policies. I studied these families during the mid-1970s and restudied them in the early 1980s.

The Economic Issues Guiding Government Policy

When Singapore gained independence in 1959, it faced a weak economic infrastructure, high unemployment, inadequate housing, and rapid population growth. Active political opposition groups mobilized the major ethnic groups for radical social change. The government, led by the People's Action Party (PAP), forged its social policies to cope with this heritage.

First, the party-state marginalized organized opposition, from political parties to labour, and on to traditional quasi-political associations, such as *landsmannschaften* (locally known as *hui guan*). The schools, health centres, and other services that they supplied were curtailed, and replaced by state organizations. Clans now mainly offered bursaries and welfare, and mounted cultural activities.[1] The new state service organizations were overarching statist organizations (statutory boards), with a lot of resources and skilful leadership, empowered to

This essay draws on and updates research more fully reported in Janet W. Salaff, *State and Family in Singapore: Structuring an Industrial Society* (Ithaca, NY: Cornell University Press, 1988).

carry forward state social policy. As sole supplier of the newly desired social goods, the PAP gained an edge over opposition groups and parties. The government also granted a monopoly to certain interest groups to provide these services. In exchange, the state demanded that these groups adhere to its developmental guidelines. A form of 'corporatism' was thus established. This corporate political structure proved central in implementing the major social policies studied here.

Singapore's Development Programmes

In the mid-1960s, Singapore's development programme centred on creating a potentially skilled but labour-intensive and low-wage industrial sector to produce for foreign corporations and absorb the many unemployed.[2] As the 1970s drew to a close, the leadership took an increasingly dim view of the country's future as a regional outpost for low-wage and low-skill manufacturing. Singapore actively solicited knowledge-based, capital-intensive industries from abroad and launched the New Economic Programme. Older heavy industries were joined by new firms that exploit advanced technology in chemical processing, machine tools, heavy engineering, computer microtechnology, technical services, and industrial research and product development. These produce goods mainly for export. The finance and banking industry, which includes offshore fund management, is also a leading growth sector.[3] These two sets of programmes, called here first- and second-stage industrialization, provide the context for the social policies reported on here.

The social services are directed towards raising family living standards—pensions and savings plans, housing, education, maternal and child health, and family-planning services—are not social-welfare or social-security programmes.[4] Indeed, the Singapore Government, like the governments of most new nations, maintained that its limited resources should not support public assistance programmes that might undermine the work ethic. Instead, state services enmesh citizens in the money economy and speed the tempo of the industrial way of life. As services gradually reached all class levels, they provided new con-

sumer goals that spurred families to work harder to pay for increased purchases. Moreover, most of the social services helped to accumulate investment capital.

Labour-Force Training

The second-stage industrial programme was accompanied by considerable investment in labour. A battery of programmes raised workers' skills to meet the many new demands of second-stage industrialization. Programmes to train workers on the job were coupled with the expansion of polytechnic and diploma-conferring education.

Pensions

A new universal pension plan, the Central Provident Fund (CPF), was introduced. That the social services have multiple functions can be seen here. Both employee and employer must deposit part of the paycheck in the employee's CPF fund. The worker can draw on these funds not only to retire, but also to buy into the public-housing sector. Once employees commit themselves to use their CPF to buy a family home, they had to keep on working to continue their payments. The CPF would also in the future free the elderly from dependence on kin.

Housing

The construction of large blocks of rental housing and the Home Ownership for the People Scheme were government programmes designed to stimulate the market in construction and finance and to promote an industrial way of life. The Housing Development Board (HDB), a statutory board, took charge of the ambitious state housing programme. Many lived in settlements of Malay-type housing, called *kampung* villages, where they neighboured, raised barnyard animals, and grew tropical fruit. The HDB demolished these along with central city slums and replaced them with high-rise buildings.

The state housing programme was one of the most drastic social policies that propelled residents into the wage labour-force and consumer economy, with wider impacts on social bonds. In the shift from *kampung* housing and small farms to HDB quarters, families lost their home garden plots and their food bills increased, as did their outlays on utilities.[5] Many suburban bungalows and *kampung* houses that were torn down housed several married couples, both related and unrelated, while few of the HDB flats did.[6] Before resettlement, these family units often shared appliances; after, they have to buy their own furnishings and other consumer durables. When they had no need to borrow these things from other people, their exchanges with kin and neighbours decreased. Deployment of women into the wage labour-force, overtime work, and moonlighting helped pay for the more costly HDB lifestyle.[7]

However, the first industrial stage, based on low-wage labour, could not guarantee a strong wage-earning position. Although many households moved into low-cost HDB units and were wage-earners, they often continued subsistence production and petty trading in their high-rise flats. Low-income families could not give up their exchanges within the community of the impoverished. Poverty was thus 'modernized' into high-rise housing.[8] Furthermore, in the early 1970s, the labour-market greatly affected a family's housing options. Families with below-average income occupied the smallest HDB flats after their relocation. Income in HDB households varied widely by housing development. Middle-class families, who applied for the more spacious and expensive Queenstown flats, earned twice as much as those in Bukit Ho Swee, mainly resettled slum dwellers.[9] Furthermore, families that purchased their own HDB flats earned more than the renters.[10]

By stage 2, however, a spurt in HDB construction, the accumulation of CPF savings by maturing workers, secure employment, and higher average wages decreased visible class differences in housing. Parts of the poorest developments were demolished. The policy of mixing flats of various sizes in new developments further loosened the link between address and class position. In 1973–4 HDB units accommodated 43 per cent of the population, and 37 per cent of them lived in flats owned by their families. By 1979–80, in contrast, 67 per cent of all Singaporeans resided in HDB quarters, and 59 per cent of these homes were their own. Engaged couples and families of any size became entitled to buy or rent an HDB flat, subject to a monthly family income ceiling of S\$2,500, in 1980. All of the families of my study met this qualification, and most of them exercised this option. In this short interval, Singapore took a long step towards becoming a society of homeowners, or more precisely, apartment-owners.[11]

Consumer financing of HDB housing through the CPF strengthened the involvement of families in the labour-force over the entire life cycle as they committed themselves to regular monthly wage checkoffs. When they saved enough for a down payment, couples could improve their housing by buying their first public-sector flat; later they might sell it and buy a larger one. In 1973 one-room, 'one-room improved', and two-room dwellings accounted for 58 per cent of all HDB households; few larger flats were built. By Phase 2, only 31 per cent of all HDB households lived in one or two rooms.[12] It appeared that the occupants of the larger flats had advanced to the later stages of their family cycle and achieved greater earning power.[13] Provisions for moving from smaller to larger apartments and from renter to ownership status demarcated the Singapore proletariat from their counterparts in many older industrial cities with inadequate housing stock and weak home-financing arrangements.

Education

For a new multi-ethnic nation like Singapore that aimed to mould a new national ideology, the schools loomed in importance. The number of all schools increased, and tuition fees were reduced and equalized. Principals, teachers, and Ministry of Education officials, all secular educated professionals, espoused the norms and values of the scientific-technological revolution, multiculturalism, and egalitarianism. The term 'meritocracy' gained currency. By this was meant that people would get jobs through their abilities and training, not personal ties. To ensure this, the school system stressed testing at early ages. A complex balloting system gave families throughout the republic greater access even to renowned schools that previously were attended solely by children with 'school ties'.

A unified admissions policy, a common core curriculum, and a bi-lingual language system were introduced. Chinese pupils chose either English- or Mandarin-medium primary school, and course work was also introduced in the second language. Increasingly, however, Chinese parents placed their children in English-medium schools. The proportion of primary school students enrolled in Chinese-medium schools dropped from 50 per cent in 1959 to 11 per cent in 1978, although not all in the English stream performed well in the English language.

Entrance to schools was competitive. During second-stage development, when the state aimed to reduce birth-rates, family-planning policies were linked to school admissions. First priority in admission to a particular primary school went to children in small families, especially if one parent was sterilized before age 40, in accordance with birth-control policies of the time. Next came children whose parents or siblings attended the same school. Finally, parental choice and neighbourhood location were considered. Secondary school admission was harder to access still. Parents ranked a secondary school by its pupils' success in the major national school-leaving examinations. Places in secondary schools were determined both by choice and by the pupil's scores on the primary school-leaving examination. Candidates who attended a primary school attached to a secondary school gained first priority. This arrangement gave an edge to middle-class parents who had the savvy and the organizational ties to stream their offspring through these feeder primary schools.

During the first industrial stage, however, it was hard for children from lower-class backgrounds to finish high school; 'educational wastage' was extensive. Out of a cohort of 1,000 first-grade pupils, only 440 graduated from secondary school ten years later. The rest repeated courses or dropped out of the academic stream without getting any of the skills required by the advanced hi-tech industries.[14] Sweeping educational reforms were introduced to accompany Singapore's second-stage economic programme. The revamped instructional system featured continuous testing, starting with a language test in grade 3. Examination-based streaming within each grade separated students of different language, technical, and academic abilities.[15] The new scheme allowed around 40 per cent of the primary school cohort to pass their O-level exams, and gain a full secondary school diploma. Following this was a pre-university college, to prepare for the crucial A-level exams that qualified students for university. The pass level for the

A-level exams was set at only 14 per cent of the primary school cohort. Only 9 per cent of that age group were enrolled in university or polytechnic institutions in 1982.[16] However, those who did not advance along the pre-university path did not have to drop out, but were eligible to attend the Vocational and Industrial Training Board's certificate- and diploma-conferring institutions.

As the educational reforms of second-stage industrialism took root, the role of class origins in educational advancement was greatly reduced. Still, the desired equality of opportunity was not fully achieved. Pupils in government-aided private schools had higher social class backgrounds and performed better on the key exams than their peers in state schools. Middle-class parents devoted much more time and money to preparing their children at home for success in school. As exams were as likely to measure pupils' familiarity with schoolroom concepts and techniques as their inherent intelligence, streaming and early testing of students could not help the poor. The battery of tests designed to identify superior pupils gave rise to widespread tutoring to improve children's performance. Again, parents with money and knowledge had an edge. The tuition fees at the local university and polytechnic institutes were low, but degree-seeking students who fail to gain admission to the sole university, the National University of Singapore, could only go abroad at great expense.

As parents devoted effort to their children's education, the nuclear family became deeply involved in the educational system. This not only channelled the economic opportunities for youth, but also gave rise to political compliance. The ideology of the Singapore meritocracy labelled those who could not continue in school as lacking qualifications, and they accepted this label. In a variety of ways, then, the educational system reproduced the class-structured society.

Health and Population Policies

The ratio of physicians and beds in government-run hospitals to population improved, while small clinics run by midwives for the poor were labelled 'below standard' and closed. By the late 1970s, maternal and child health services were delivered by 24 hospitals, 13 of them public, and an associated network of 46 clinics. Over 80 per cent of all registered births took place at public hospitals, and over 90 per cent of all new-borns were enrolled for comprehensive post-natal care at a neighbourhood clinic. This extension of state medical care enabled the public to enjoy a better quality of life and reduced occupational time lost to illness. It also laid the groundwork for what Foucault calls the medicalization of social control.[17] By this is meant that the idiom of science legitimizes the extension of state mechanisms of social control more deeply into society.

The population grew rapidly in post-war Singapore, and the government resolved early that such growth would slow capital formation. After independence, the Ministry of Health provided contraceptive services through its network of public health facilities. In the late 1960s, the state overrode the opposition of interest groups in the medical, religious, and professional communities and liberalized abortion and sterilization laws. The crude birth-rate declined in response to fertility-limitation services and first-stage industrialization. From 44.4 births per 1,000 population in 1956, the crude birth-rate fell to 21.8 in 1969 and on to 16.6 in 1977.[18]

In the early 1970s, however, the government became alarmed by evidence that the family size was still relatively high among the working poor. A 1973 survey by the Family Planning and Population Board (FPPB) revealed that fully 35 per cent of the blue-collar workers studied, but only 11 per cent of the clerical workers and 7 per cent of the professional and technical workers, considered four or more children 'ideal'.[19] The FPPB then vigorously popularized its tenet that numerous children undermine efforts to upgrade the family, because they keep the family too poor to afford their own flat, mothers cannot work, and children cannot receive a good education. Mass media presentations on population control created an atmosphere of crisis and identified large families as an imminent threat to the limited resources of the city-state. Thus the national family-limitation campaign aimed to convince people that rapid population growth endangered both their present livelihood and their future prospects.

Amid much fanfare, the economic and social 'Disincentives Against Higher Order Births' were enacted in 1973, to attain zero population growth by the year 2030, at a level of 3.6 million persons. All the services came into line to sup-port this programme with sanctions. Public hospitals' obstetrical fees increased sharply with the number of living children to whom the mother had already given birth, but the fees were waived if the mother was sterilized within six months of the birth. Though no child was denied schooling, the fourth child received lower priority in choice of elementary school than children of small families, especially those whose parents were sterilized. Though public housing was available to all, large families no longer had priority in access to HDB flats as hardship cases, nor could families with three or more children rent out one of their rooms. A female civil servant or union member obtained two months' paid maternity leave for each of her first two deliveries, but not for subsequent births. She was, however, given paid medical leave for sterilization. Finally, while all individuals whose annual income exceeded S$3,000 had to pay income tax, couples could claim no more than two children as deductions.[20]

These measures did not actually deny citizens services, for it was in the state's interest to expand social goods. They did, however, raise the price of needed social services for parents of large families, and further reinforced the view that these people are poor because they have too many children. The measures shored up the ideology of the Singapore meritocracy: only planners would succeed in raising their families above poverty. Prime Minister Lee Kuan Yew explained that a system of disincentives was needed 'so that the irresponsible, the social delinquents, do not believe that all they have to do is to produce their children and the government then owes them and their children sufficient food, medicine, housing, education, and jobs. . . . Until the less educated themselves concentrate their limited resources on one or two to give their children the maximum chance to climb up the educational ladder, their children will always be at the bottom of the economic scale.'[21] Lee's Government justified the measures as a means of helping the poor help themselves, and of reducing conflicts engendered by competition over social services. Zero population growth was attained on the eve of the new economic policy, and pressures on mothers to be sterilized abated.

In sum, the first industrial phase saw considerable social class variation in family size, owing largely to the felt need for children by the poor. As the national birth-rate declined in the 1970s, social class differences in family size narrowed. Yet, families along the spectrum of different class backgrounds still held to definitions of how children fit into their future. The poor would have more if they could only afford to feed and clothe them, while the better off, more secure families could not have more children because they felt they could not educate them properly for the new competitive society.

The social services were interlaced with the new ideology of individual merit and responsibility for one's own poverty. They extolled self-help, individualism, and competition. The educational, family planning, and housing policies moved people around. Such mobility in living arrangements, social class, and the flow of ideas atomized people and devalued reliance on kin. The extent to which the social services affected family lives can be seen in the ways the services reshaped kin ties of the families I interviewed.

The Impact of Government Policy on Poor and Secure Families: Phase 1

To understand how social services alter family lives while reproducing social class structure, I traced a panel of Chinese families with different class backgrounds from the mid-1970s to 1982.[22] They were young married adults, under age 30. I grouped them, based on scores derived from the occupational status of the parents of both husband and wife and the average educational level, occupational status, and combined earnings of the husband and wife. I divide the couples into two main socio-economic groups: the poor and the secure.

For most families, a life of poverty meant close dependence on kin and minimal use of social services, which were available only to people with money. The lives of the poor were affected by the type of work they did, the past course of their family life, their material resources, and the presence or absence of wider kin ties. In the first development stage, poverty shaped the lives of many families. With limited education, having entered the labour-force at an early age, and having contact mainly with other poor people, these men and women could not break out of their early low job status. The highest wages accrued

to men with a skill or entrepreneurial resource, especially a sum of capital. Wives who earned money helped their families eke out a minimal living, but they earned less than the men.

Poverty and their life-cycle stage placed families in close dependence on their relatives to make ends meet. They looked to their relatives for help with their early family building. Caring for and educating their numerous children was far easier when their relatives helped out. Most remained partly outside the market economy, because of close dependence on help and hand-outs from others. The closeness of poor adults to their families and to their peers of the same sex, which was fostered by their jobs, competed with the husband–wife relationship. Most husbands spent their spare time with their buddies, while wives visited kin and neighbours. Since couples had to take into account the views of kin in many spheres of their family life, they were not independent decision-makers. Their poverty and close kinship ties limited access to the state-provided housing and educational services, and they could not plan far ahead. Although they were young, they expected to bear many children. They thought this would give them more security, but many children further limited their ability to plan for the future.

Only a few poor couples had hopes of advancing in their jobs, above-average schooling, or some reason for not being deeply integrated in their kin community. These husbands and wives tried to work out together plans related to their family life. They especially tried to use the social services to a fuller extent.

In contrast, couples with above-average education, a skill, or another asset did far better under first-stage Singapore industrialism. As they held solid jobs and their futures seemed bright, I refer to them as secure, but they include couples of relatively modest means as well as the affluent.

Secure couples were optimistic about the future. Though many lived with kin, few depended on kin for daily support or finding jobs. The majority took steps to buy a home, set high standards for their children's education, and intended to have no more than three children. Most enjoyed close marriages, in which partners shared pastimes and talked things over. They were proud of their ability to plan their futures together. Only a few were unable to plan far ahead in any respect. Most took advantage of

Singapore's social services to advance their family's living standard. Secure folk expressed recognizably similar goals in the main spheres of their lives as they planned for the advancement of the small family in Singapore's industrial revolution.

The Impact of Government Policy on Poor and Secure Families: Phase 2

When I spoke with panel families of all skill levels and industrial sectors again in the early 1980s, I found them expressing increasingly similar goals as the consumer and social services society and unified market economy took hold. Living standards improved, and many Phase 1 'poor' earned enough to consider themselves secure, while the secure became even better off. Nevertheless, within the narrowed limits of lifestyle, class, and sector, differences remained. Class position still shaped diverse family styles and plans.[23]

Couples varied in their use of programmes and in the meaning of the policies to their way of life. Family relations were negotiated, and social services, such as public housing, played a central part in this process. Favoured under the terms of the state development programme, able to maintain their position in the class order, secure families made even greater use of the state services than the formerly poor. State policy, with its emphasis on training and skills for the trend towards capital-intensive industrial work, strengthened its grip on the secure families.

Measured by their use of the key institutions promoted in second-stage development, secure and less affluent families were closer together in Phase 1. Their deepening participation in wage-labour had major ramifications for other spheres of family life. The number of working men and women in my sample whose earnings lifted them above the poverty line had risen. All my interviewees became wage-workers, and their commitment to the labour-market determined how they exploited the social services. The development programme most benefited new groups of workers who found jobs in the industries of the second stage, but marginalized others who could not.

Because of economic improvements, nearly all families exploited the social services to a greater extent than before. Whereas in Phase 1 the poor either could not afford any or only the most rudimentary public goods and services, by Phase 2 all the poor bought into the public sector, and hence enjoyed a more uniform life-style. Because no poor families were excluded from basic social services, and all used these services more than before, the outward class cleavage was less visible than it was in Phase 1. Nevertheless, in Phase 1 the affluent had already received the best low-cost services and goods the public sector had to offer. In Phase 2 some could afford to top off the public sector (the highest ranked public schools) with the best that the private sector provided (privately constructed housing, private health care). A family's ability to buy a home in the public sector and the extent of its use of educational and family-planning programmes still turned on income, education, or property. Poor couples received the least, while secure families maintained a class lead. In these ways, while the social services were crucial in shaping the new and more uniform way of life throughout the class structure, the usage and meaning of these social services differed for poor and secure families.

The Singapore development strategy changed outward appearances of family life-style more than internal relations. Since families retained their distinct positions in the class structure and in wage-labour, class position shaped their style of interaction and their concerns, fears, and hopes. My interviews of panel families of different social classes in Phase 2 revealed a narrowing of differences in the many spheres of their lives specifically shaped by state policy. Their uses of social services are becoming increasingly similar. Yet variations in goals and living standards still remain.

The new second-stage programmes rooted families ever more deeply in the capitalist mode of production which opened them to control by the market and restructured them along new class lines. That the second-stage economic programmes brought workers more deeply into the market economy was seen first by the spread of wage-work and the rise in basic wages earned. The proletarianization of the men and women in my sample enabled policies that promoted certain types of industries and occupations to reach increasing numbers of workers. The policies then forced the restructuring of the labour process.

Within each of the social class groups, the earnings of workers whose skills were favoured (men and women in the industries central to Singapore's second industrial profile and in technically skilled occupations) increased the most. Men with better credentials within the favoured industrial sectors did the best. For instance, engineers with degrees did better than computer technicians with diplomas. While education tended to reproduce class position, all those who worked in jobs and industries in demand in Phase 2 fared well. These workers profited from being in the right place at the right time.

In contrast, the wages of workers in sunset industries, or those not central to the new capital-intensive, high-technology profile, did not increase so quickly, irrespective of training. Not in demand, they suffered a relative loss of income. Hawkers, for example, slid to the lower ranks of the poor. Teachers or civil servants in white-collar jobs were examples of secure employees that lost relative standing. While they kept a middle-class income, they no longer could claim pride of place. Such a realignment of opportunities occurred throughout the class spectrum.

A range of state policies aimed at improving workers' skills. But these programmes of job creation, job finding, and retraining carried forward the major class divisions and inequalities, because the state's labour investment policies affected wage-earners differentially depending on their place in the labour process and in the market economy at the time of industrial take-off. Job training programmes were highly selective. The poor, who received little formal education and possessed few skills, were least likely to upgrade their jobs through training programmes. Either such programmes were absent in their lines of work or people with low formal education lacked access to them. The secure, who already had certificates or diplomas in technical subjects, were most likely to take retraining courses. Thus those who had skills upgraded and expanded them. The secure families stayed ahead of the mainstream.

Workers changed jobs often, but did not usually change industrial sector. More than half those poor in Phase 1 found new jobs, but only a couple moved from peripheral to core-sector positions. Half the secure men changed jobs, some more than once. Here, too, few changed industrial position. Instead of widespread flux, we find stability of structural position. Job

opportunities were limited by the nature of the job search which relied on informal interaction with friends and kin, and thus it was not simply human capital that determined a worker's wage. The new programme had reshaped, not fragmented, community, and workers retained links with others in their kin, class, and occupational and industrial circles. Partly for these reasons, workers remained in the same type of jobs.

The compensation in these jobs, however, was dependent on market demands. In sum, even with little additional improvement in skills, the workers in state-promoted jobs rose. Demand, not human capital, made the biggest difference.

The new economic programme set limits to the wage-earning opportunities of women. Like the men, those women who in Phase 1 worked in the industries or skilled occupations that gained favour in the new development programme enjoyed the greatest advancement opportunities. Other women in jobs not favoured by the new economic regime have had lower pay increases and few opportunities for mobility. Worse still, those women who were not working at the time the second-stage development programme was launched, found it impossible to break into mainstream jobs, They could enter only unskilled peripheral-sector jobs or work in the undocumented, hidden economy (as waitress, babysitter, hawker's assistant).

Women had become increasingly proletarianized. Homemakers found they needed to add to the family purse, and many were compelled to work at part-time jobs. While lacking benefits and poorly paid, part-time jobs subjected women to the demands of the market economy. Whether new or old workers, through their market commitment women helped bring their families more deeply into the new structures of the second-stage economy. The state social services, such as birth-control services and housing developments in areas where factories located, enabled women to work. In turn, their wage-labour increased their families' participation in the new state social services. Women's wages helped furnish new public-sector apartments and paid for tutors for their youngsters.

Women's position in economy and society reflected their early upbringing and further contributed to the class position of the families they formed. Their educational level and social class background figured in their Phase 1 jobs, which then led to the work they did in Phase 2. Since women's paid and unpaid labour contributed to the economic status of their families, the size of their pay packet made some difference to the quality of their family life. An important element of this was the ability to use the state services fully. Thus state programmes that encouraged women to work for a wage spread the new social institutions further among families of all class levels, while women's earnings differentiated their families.

Women put even more energy into their family roles in Phase 2 than in Phase 1, efforts that limited their freedom to move ahead on the job. As women, they were also informally excluded from the key jobs of the new economy. But few in my sample struggled to overcome the gender barrier and advance in the job market, because most placed their home roles at the forefront of their commitments. The majority of women had to maintain the family domestically, and their family burden limited their alternative activities. Indeed, the work of women at home also helped the families attain the new social benefits of the new economic order, such as their children's school performance, and hence a place in a good school.

The public services that structured citizens' place in the social class order shaped their ideas about society. Access to new posts, higher wages, advancement opportunities, and other features of the job-creation schemes increased support for the new socio-economic programme of the state. As the worth of their jobs rose, some workers gained status at the expense of others. Their ability to enter these newly favoured pursuits enhanced the workers' self-image and assessment of second-stage development. The policies' true beneficiaries were men and women in the industries moving forward and especially those with technical skills, and these people gave most support to the government's development programme.

The new educational programmes supported the high-technology economy. They not only trained the next generation of workers and structured their goals for the future but also brought parents at each class level more deeply into the social order in support of the second-stage economy. Parents budgeted money to promote their children through the school system. Families also tried to encourage their children to achieve in school, and achievement was defined largely in terms of the educational system.

Because of the differential involvement of families in the educational system, a class gap remained. Under intense pressure to do well on exams and gain better school places, poor children had difficulty competing with the children of the secure. Secure families had a head start in preparing their children for the competitive exams and better connections to good schools. But the educational system added to a common conceptual framework that helped legitimate the class-divided nation. The losers redefined their loss within the dominant ideological paradigm. The ideology of the meritocracy proclaimed that children with talent faced no real financial obstacle to advancement in the school system. The cost of formal schooling was low, and children who did well in the national competitive exams were promised state support. Many poor parents pointed to these factors as evidence of an open school system; few attributed their children's failure to their disadvantaged place in the class system. Adopting the terms of discourse of the meritocracy, they explained that their children did poorly on school tests because they lacked merit or talent. Therefore, they could not expect to get good jobs and would always be poor.

The amenities that raised and equalized popular living standards—the public housing for sale and the family-planning services—also altered living standards, reshaped the popular way of life, and provided economic as well as political support for the Phase 2 economic system. Owning a home of their own became a goal for most families. Public housing was one of the most equitable of all of Singapore's social policies. The gap in ownership between poor and secure families narrowed considerably. Their homes integrated the poor into the market, which inexorably shaped a single life-style. Couples increasingly lived alike.

Still, class differences remained. Three times as many poor as secure couples lacked plans to obtain a flat of their own, and the size of a flat varied by class, as did its cost in absolute terms; as well, housing took a larger proportion of the earnings of the working class. And the sacrifices that couples had to make to fund a flat of their own, such as mortgaging their hard-gained pension funds, were greater for the poor. Come a recession, the poor would have more trouble making do on their own, without the help of their local community. Whereas in the past, the poor got help from kin who lived nearby, as well as neighbours, after resettlement, and with movement from one apartment to another, the poor lost much of this support.

Nevertheless, the Singapore public-housing programme further homogenized the social structure. By buying new flats, families became further constrained by the money economy. They became involved in long-term work goals, mortgage payments, and indebtedness to their pension plans. Continued market activity was necessary for Singapore homeowners, regardless of class position, even in the social-service society. The Singapore housing transformation created a society of homeowners, in fact or in aspiration. This transformation greatly narrowed the differences in family life-styles.

Families also came to think alike. By providing homes and other amenities, the party-state achieved social legitimacy. A home of one's own provided the new measure of family achievement in the new society. Some turned to home ownership as a standard of well-being, which compensated for dashed job hopes. The poorest couples, if they lacked even such a public good as a home, were embittered. They recognized they were socially disadvantaged. Thus public housing joined jobs and the school system in the frame of discourse for assessing the new social order.

The Phase 2 social and economic development programmes that proletarianized the populace have also increased the cost of raising children. Couples lowered their actual family size below their Phase 1 intentions, but poor parents had slightly more children than secure parents. The relative family size of poor and secure in Phase 2 did not differ much from the ratio of their intended family size in Phase 1. The reasons for reducing fertility, however, differed by class. As the groups had distinct positions in the labour-market, they also had their own views of the costs of raising children. Poor couples ceased childbearing to make ends meet from day to day, while secure couples planned fewer children to give them better educational opportunities in the future.

We saw the several ways in which the nuclear family, structurally isolated from both close and distant kin, and with independent goals and obligations, emerged as the capitalist market and the social services penetrated society. We saw the great economic and social cleavages in the first industrial stage. Poor couples then were tightly bound to their kin by poverty, uncertainty, and job links. Secure couples also depended on their kin for mutual help, but theirs was a limited dependence, linked to the early stages of their life cycle. Both groups nourished a sense of shared responsibility in the local community.

The state, through the social services it sponsored, pushed and even forced ties to the market and key social institutions, and the families I knew changed. The transformation of family- and kin-based exchanges of services and goods into a way of life in which families buy what they need was largely completed during the period of study. Families could afford to buy the everyday items they used to make by hand or exchange with kin and neighbours, as well as many other things they did not dream of having earlier. As a result, the community changed.

By altering the source as well as the distribution of resources, the state's economic and social policies restructured the fabric of community and family life. People not only became more closely integrated into a national market economy, they also entered deeply into the capitalist culture. The social services, by shaping the set of ideas that people used in daily living, played a major part in their integration. In both the economic and the ideological spheres, state services committed people to the new social order. The involvement of families in the mainstream economic and social institutions could be seen in the dependence by families at all social class levels on the money economy, public housing, the school system, and family planning organizations. But since access to these social services and institutions varied across the class structure, the meaning of the new social order to individual families also differed with social class and industrial sector.

Families in both major class groups depended on state services materially and as a source of ideas about their place in the society. But the family's position in the labour-market shaped the process. The motives and meaning of reciprocal involvement with kin varied with the economic status of the family. For some, losing their kin was less a matter of choice than an outcome of the family cycle. Few poor families were economically secure, and all had a history of uncertain income. Thus economic status, coupled with the lack of crucial family support services in the public sector (day care, homes for the elderly), shaped the structure and meaning of family life along class lines. Many poor families, unable to separate from their kin for economic reasons, made a virtue of long-term reciprocity.

In comparison with the past, however, their reciprocity narrowed. As sections of the formerly poor advanced, and with the meritocracy as the overarching ideology, the community of the poor was not cohesive. Families that bettered themselves contracted their ties to a small circle. They justified their withdrawal as an effect of their stage in the life cycle. They recognized the uncertainty of their achievements and the continued need to help out those closest to them when they were called upon to do so. But as they no longer extended such help to wider circles of kin or neighbours, very poor families lost their supportive community.

Secure couples, in contrast, keen on separating from their wider kin links from the day they married, now could attain this goal. They expressed their better living standard by living apart, while they continued to see relatives for emotional satisfaction. If they needed to support their kin materially or emotionally, they could do so while living apart. Their greater access to resources allowed husband and wife to develop a strong marriage bond. As a result of real differences in freedom from dependence on relatives, higher incomes, and better prospects for the future, the nuclear family structure and a close marital bond were more frequently found among secure couples than among the poor.

In both class groups, women's responsibility for maintaining the family increased. Homemaking tended to become even more the sole work of the wife. Now that families had purchased small apartments of their own, the wife acquired responsibilities for the upkeep of new consumer items. Child-rearing was also more time-consuming, owing to loss of help and the deepened parental concern with the rigorous state-imposed educational standards and tests that youngsters had to pass. The burden of poor women increased as the family life cycle progressed and as traditional helpers found other duties and employment, moved elsewhere, or died. For secure women, added burdens and a new and more substantial role flowed from their decision to attenuate ties with kin.

The social services that invested in labour redrew the stratification system to bring more workers deeply into the market economy and subject them to its demands. The differences in employment situation among my interviewees were large. The positions in the labour process were differentiated, but the system offered a new ideology as a standard by which to assess their place in it—the meritocracy, with equal opportunity to advance. Their behaviour varied by class, but their explanations for their positions drew on the same categories. In these ways the new development programme integrated the families while maintaining class divisions. It seems likely that their offspring will maintain the same class ranking in the next generation. As members of households became directly linked to the market economy as consumers, debtors, and pensioners, wider ties to the local community based on mutual aid weakened. At the same time, incorporating the broad populace in the market accumulated capital and stimulated the state development programme and empowered the state.

The services that invested in labour and integrated families into the commodity economy also assisted the expansion of the capitalist economy. The many changes in family lives at all class levels worked in concert to enable even hard-pressed families to have a more 'decent' life as measured by services and consumer goods. Public housing, the new school system, and small families gave the poor and people of modest means more freedom. They accepted the yoke of the market economy and viewed the social services as tools that liberated them. The market and centralized services integrated the full range of families into a single industrial way of life. But within this increasingly similar life way, their class-based options shaped their goals. The couples we met revealed the ways in which class position elaborated meaning and social ideals. The social services both promoted the capitalist system of production and legitimated the social order. Since they perpetuated Singapore's class system, the social services contributed to new forms of unity as well as to new divisions within the society.

In sum, in the first phase, social class cleavages were marked; in the second stage, industrialization narrowed some gaps. These data suggest that differences in family opportunities were still associated with social class. Nevertheless, whereas poor and secure families formed two types of Singapore families in Phase 1, they became variations on a recognizable theme in Phase 2. Families in the two class groups were constrained by similar economic and social institutions. They were still divided by their market positions, which affected their behaviour and the meanings they assigned to their actions, but the market claimed a tighter hold on all of them. They aimed for recognizably similar goals and attempted to use the same means to attain them: the public services of the second-development stage. State-sponsored public services became means to alter family lives. They provided entrance into the market and the cultural categories that became the framework within which the public evaluated the state and its socio-economic programmes.

Singapore in the 1990s

Rapid economic growth has absorbed most available labour in Singapore. In recent years, nearly all available men, and then women, have been drawn into the labour pool. By 1988, over four-fifths of women aged 20 to 24 were employed. The proportion was lower for older women, but still over half of women aged 40 to 44 were employed. Their skills are improving. In 1988, 47 per cent of the employed men and 56 per cent of the employed women had at least secondary eduction. The rapid growth of the service sector has given Singapore women more career opportunities, and they train to meet them. Since 1982 more women than men have entered the National University of Singapore. Nevertheless, women still earned only 75 per cent of what men earned in 1987.

The need for low-wage labour and young labour for industry is pressing. Singapore's labour-force is projected to grow annually by about 0.6 per cent during the 1990s, while the number of entry-level workers and those under age 24 will drop by 20 per cent between 1985 and 2000.[24] Supplementing the country's work-force with foreign labour, the government tries to stem off the social problems this entails through strict limits on their rights. It brings in low-cost domestic workers to release local women for industry.

It eases the paper work for new Chinese immigrants from Hong Kong seeking an escape from the reversion of the British colony to China in 1997. So long as they have capital or technical skills, they can become full citizens. It attempts further to accommodate to its labour shortage by shifting to higher-value-added technology-oriented industries. The National Wages Council encourages employees to hire part-time staff, draw in housewives, and raise the retirement age to 60 and over. It continues to regulate the labour-market, in particular through labour training.[25] With the absorption of unskilled labour, Singapore investors are moving their labour-intensive products to poorer nations. There is a 'regional' policy, to broaden the national economy. Again the state, for instance through the Economic Development Board, but joined by the Government Linked Companies or the Committee on Promoting Enterprises Overseas, helps small and medium enterprises to export capital and technical expertise to Indonesia, India, and other Asian countries.[26] At the same time, new government organizations support the role of multinational corporations in the economy.

To meet the needs of contemporary industry, alternative paths to academic schooling have been further broadened. The Nanyang Technological University and National University of Singapore together accept 14 per cent of the college-age group, two-thirds more than the previous decade. There are now four polytechnics, as against two a decade ago, and nearly twice as many youths take this route to enter Universities.

Singapore has become a city-state of high rises. The terms of purchase of HDB flats are easier than before, and an HDB flat has become a quasi-market commodity. This has resulted in 87 per cent of the population living in HDB apartments, with around four-fifths owned by their residents. In its external form, this way of life is more uniform than before.

In the mid-1980s, Singapore's fertility-rate dropped to 1.44 births per woman. It is estimated that in 2025, for every 100 Singaporeans of working age, 45 will be aged 60 or older. Politicians are concerned over the diminution of the so-called superior gene pool of the educated populace. They worry, too, that the nation's labour pool will dry up. And they fear that a reduced working populace will have to support a larger ageing population. Singapore is again in the forefront as the first newly industrialized economy to adopt a broad set of pro-natalist policies.

Policies were reversed for sections of the educated population. The reversal aimed to improve human capital amongst the populace and to expand the Chinese section. The new policies place the burden on women to accede to their childbearing function. In fact, married Singapore couples actually average two children, enough to replace themselves. The problem is seen to be located in those that never marry, on the assumption that educated women are unmarriageable—and want to marry. Career women are postponing marriage, and many may not marry at all. Among women in their early 30s, about one-quarter of those with a secondary education and over one-third with a tertiary education remain single. For a while, the government became 'matchmaker'. Failing this, the state tries to reduce the proportion of women in advanced studies. Women have 'flooded' the medical and law faculties in such numbers that it is rumoured that informal quotas limit their intake to one-third.

For those who want children, the various social units that had earlier fielded the Social Disincentives Against Higher Order Births have now come into line with pro-natalist policies. The Population Planning Unit replaced the Family Planning and Population Board and is responsible for developing the new policies.

Targeting the élite to bear more children, there is a wide range of benefits, including subsidies for child care, priority for public housing, and tax rebates for the better off. If the élite is encouraged to have more children, the poor are paid to have fewer. There is a sterilization cash incentive scheme for the purchase of an HDB flat, as well as adjusted hospital fees to discourage high parity births among the poor.[27]

The population transformation is a benchmark in Singapore's policies of structuring an industrial society. As in the past, however, the population profile reflects many issues that are hard to legislate. The trends in the economy and society counter the legislation for more children. First, social services have become more comprehensive than before. As this occurs, children have become even less important for parents to achieve the economic needs of security, survival, and advancement. The CPF has evolved into a comprehensive social-security savings scheme, that takes care of members' retirement, home ownership, and health-care needs. Parents can even use their CPF to finance the education of their children.[28]

Next are problems facing employed women. There is still a high level of work–family role conflicts (work as an employee, a spouse, a homemaker, mother).[29] Young educated parents put a lot of effort into raising their children. Numerous children counter the realities of daily living, so essential to the family's daily life and future. Childbearing cannot be regulated directly by policy. Thus, there is no resurgence in births.

The public has begun to take state services as their right. There are popular pressures for state subsidies of health and other social safety nets. Indeed, political conflicts emerge over the distribution of social goods. Differences in access to social services take on political significance. The opposition campaigns on the issue of equitable services, and areas that feel deprived of funds for better HDB housing did not turn out strong support for PAP candidates in recent elections. The leadership, however, does not entertain an alternative view of its social policies, and has led a frontal attack on the concept of a Singapore welfare state. It extols competitiveness, which it contrasts to 'over-spending' and the social-support systems that it contends contributed to the deep economic crisis in the West.[30] In such contradictory ways, the social services remain central to engineering the new social order. The government continues to use social policy to underpin its industrial plans, even as Singapore moves into a new industrial stage.

Within three decades, Singapore has traversed an economic and demographic path that took Western nations a century to accomplish. It now faces problems found in the mature industrial nations. Today, the party-state grapples with more complex issues of industry and society than ever before.

Notes

1. Chan Heng-chee, *The Dynamics of One-Party Dominance—The PAP at the Grass Roots* (Singapore: Singapore University Press, 1976). Lee Hsien Loon, 'Clans: Remaining Vibrant and Relevant', speech given at the Lee Clan General

Association's 86th Anniversary Dinner, Singapore, 28 Oct. 1992, reprinted in *Speeches: A Bimonthly Selection of Ministerial Speeches* (henceforth *Speeches*) (Sept.–Oct. 1992), 120–3.

2. Lee Soo Ann, *Industrialization in Singapore* (Camberwell, Victoria: Longman Australia, 1973); Kunio Yoshihara, *Foreign Investment and Domestic Response: A Study of Singapore's Industrialization* (Singapore: Eastern University Press, 1976).

3. Lim Chong Yah, *Economic Restructuring in Singapore* (Singapore: Federal Publications, 1984); D. Lim, 'Industrial Restructuring in Singapore', Asian Employment Programme Working Paper (Bangkok: ARTEP, 1984).

4. Singapore spent a relatively low amount on social-security benefits as a percentage of GDP. But if we use a broader definition of the 'social wage', which includes public housing, job retraining, and other items, Singapore would rank much higher.

5. Iain Buchanan, *Singapore in Southeast Asia* (London: G. Bell, 1972), 242. With few legal utilities, slum dwellers paid only 13% of their household budgets on utilities; poor HDB dwellers had to spend much more.

6. P. Armunainathan, *Report on the Census of Population, 1970* (Singapore: Department of Statistics, 1973), table 97, p. 208.

7. In a 1973 HDB survey, 52% of relocated families stated that they enjoyed increased employment opportunity for women in their new housing. Yeh, 'Summary and Discussion', in Stephen H. K. Yeh (ed.), *Public Housing in Singapore: A Multi-Disciplinary Study* (Singapore: Singapore University Press, 1975), 357.

8. Riaz Hassan, *Families in Flats: A Study of Low-Income Families in Public Housing* (Singapore: Singapore University Press, 1977), 85, 145. Buchanan, *Singapore*, 189–92.

9. Yeh, 'Summary and Discussion', table 2, p. 352.

10. Ibid. 333 and table 3; Chua Wee Meng and Ho Kun Ngiap, 'Financing Public Housing', in Yeh (ed.), *Public Housing in Singapore*, tables 7, 8, pp. 71–2.

11. HDB, *Annual Report, 1979/80* (Singapore: HDB, 1980), 5.

12. Yeh, 'Summary and Discussion', table 1, p. 331; HDB, *Annual Report, 1979/80*, 53, 55.

13. HDB survey figures show the average number of income earners in resettled households increase with the size of the rented flat, a finding that suggests different stages in the family cycle. Yeh, 'Summary and Discussion', tables 6–8, pp. 129–56.

14. Goh Keng Swee and Eduction Study Team, *Report on the Ministry of Education, 1978* (Singapore: Ministry of Education, 1979), 1-1.

15. *Straits Times* (Singapore), 2 Dec. 1982.

16. Goh, *Report*, 3-5, 3-6.

17. Michel Foucault, *Discipline and Punish: The Birth of the Prison* (New York: Vintage, 1973).

18. Chang Chen-tung, *Fertility Transition in Singapore* (Singapore; Singapore University Press, 1974); Ministry of Health, *Population and Trends* (Singapore: Ministry of Health, 1977), 11.

19. FPPB and National Statistical Commission, *1973 Report of the First National Survey on Family Planning* (Singapore: FPPB and National Statistical Commission, 1974), 16, 25, 54.

20. Saw Swee-Hock, *Population Control for Zero Growth in Singapore* (Singapore: Oxford University Press, 1980).

21. Cited in George G. Thomson and T. E. Smith, 'Singapore: Family Planning in an Urban Environment', in T. E. Smith (ed.), *The Politics of Family Planning in the Third World* (London: George Allen & Unwin, 1973), 249.

22. I conducted this study initially with Dr Aline Wong, National University of Singapore. We conducted it at two points in time. In the first period, we chose 100 young married Singapore Chinese couples. Over half came from the rosters of Maternal and Child Health Centres, located in the main HDB estates serving families from a range of class backgrounds. The remainder came from factories which employed married women. The period in which I first met these couples, mid-1974 to mid-1976, I designate as 'Phase 1'. During the next few years the government elaborated many of the programmes of the second development stage. In 'Phase 2', 1981–2, we met for a second set of interviews with 45 of the original families. I call these 45 families the panel sample, and it is mainly from them that we learn how the development programme affected family lives. The case studies are fully presented in Salaff, *State and Family*.

23. I have not changed the ranking of the panel families in these broad categories so as to see more clearly the relative changes to the life-styles of the class groups. Thus, in the pages to follow, I still refer to these families as they were placed in the earlier categories of 'poor' and 'secure'.

24. N. Balakrishnan, 'Save as You Spawn', *Far Eastern Economic Review*, 17 May 1990, 52–4.

25. A. Chowdhury, 'External Shocks and Structural Adjustments in East Asian Newly Industrializing Economies', *Journal of International Development*, 5: 1 (Jan.–Feb. 1993), 51–77.

26. e.g. Prime Minister Goh Chok Tong, 'Staying Competitive through Regionalisation', speech given at the Regionalisation Forum, Mandarin Hotel (Singapore) 22 May 1993. Reprinted in *Speeches*, May–June 1993, 15–22.

27. Kuldip Singh, Yoke Fai Fong, and S. S. Ratnam, 'A Reversal of Fertility Trends in Singapore', *Journal of Biosocial Science*, 23 (1991), 73–8.

28. Since my study was conducted, the CPF became a vehicle for meeting a wide range of social needs of the work-force. The CPF can be used for a wider range of investments, private-sector housing as well as public, and to buy shares and even gold. Parents can devote their CPF to pay for their children's tertiary education.

29. Samuel Aryee, 'Antecedents and Outcomes of Work–Family Conflict among Married Professional Women, Evidence from Singapore', *Human Relations*, 45: 8 (Aug. 1992), 813–37.

30. Now, competition is the most often used term, while meritocracy may be used to spur on the most disadvantaged sector, such as poor Malays. Yeo Cheow Tong, Minister for Health and Minister for Community Development, 'Social Problems among Malays', speech at the launch of AMP's Early Childhood and Family Education Programme, Singapore, Islamic Centre, 18 July 1992, in *Speeches*, July–Aug. 1992, 73–6.

16

Social Transformations of Metropolitan China since 1949

Deborah Davis

THE story of the transition to socialism in urban China has been told many times. In the first decade after 1949, the Chinese Communist Party (CCP) under the leadership of Mao Zedong implemented a blueprint for economic growth where bureaucratic plans replaced markets and urban economies were transformed from centres of commerce into centres of industry. The state became the primary employer and by 1962 private entrepreneurship only survived on the margins of the urban economy (Table 16.1).

Less frequently noted, but equally important for our understanding of the social transformations of urban society was the homogenization of city life across regions and within each city. After bureaucracies replaced markets as the nodes of economic exchange and a central plan determined the flow of goods and services, city structures and cultures became less differentiated. With the exception of Beijing which remained somewhat unique because of its singular status as the national capital, other metropolitan centres converged towards a generic urban mode. Small cities and towns similarly lost distinctive profiles as local artistic, cultural, and religious activities were submerged in the new national culture of Chinese socialism.

Parallel to the uniformity of the macro-occupational structure were more uniform life chances for individual city residents. In terms of material standards of living urban China between 1955 and 1975 became an exceptionally—and deliberately—egalitarian society. In contrast to the gaping inequalities that accompanied economic modernization after the Second World War among the developing market economies, in urban China there was growth without income polarization. In fact for urban Chinese, incomes became more evenly distributed and city residents came to expect equitable distribution of basic shelter, medical care, and education.[1] Moreover, over time CCP policies to

TABLE 16.1 Size of Private Labour-Force (using government estimates), 1949–1995

Year	Urban work-force (m)	Private employees[a] (%)
1949	15.3	67
1952	24.8	50
1962	45.7	5
1978	95.1	0.1
1988	142.7	5
1991	152.3	6
1993	159.6	10
1994 (July)	163.9	12
1995 (July)	166.3	14

[a] Includes those employed in joint ownership stock, foreign owned, and joint management firms as well as self-employed or employed by private entrepreneurs and industrialists.

Sources: *Zhongguo laodong gongzi tongji ziliao 1949–1985* (Chinese labour and wage statistics 1949–1985) (Beijing: Zhongguo tongji chubanshe, 1987), 16; *Zhongguo tongji nianjian 1994* (Chinese Statistical Yearbook 1994) (Beijing: Zhongguo tongji chubanshe, 1994), 84–5; *Jingji ribao* (Hong Kong), 5 Aug. 1995, A9.

level within-city differentials served to reduce inequities between different cities and weaken the advantages of the coastal regions. Thus, for example, in 1952 employees in Shanghai earned 36 per cent more than the national average, those in Guangzhou 21 per cent more and those in Wuhan 7 per cent more; by the end of the Cultural Revolution, average wages in all three cities had converged near the national mean.[2]

Nevertheless despite the power of the planned economy to redistribute urban wealth, all employees were not equally rewarded, and CCP policies solidified certain inequalities. For example, during the mid-1950s each city was assigned a position within a hierarchy of urban places which calibrated resident's wages and subsidies. Most advantaged were residents of the large coastal cities which had been the centres of trade and modern industry during the pre-Communist era, and which despite explicit CCP efforts to concentrate new investments in the interior were granted wage and subsidy rates consistently more generous than for employees in cities of lower 'rank'. Next in the hierarchy were residents of the provincial capitals which were the primary recipients of government transfer payments and foci of the expanded bureaucratic party-state. At the bottom were those living in the thousands of county towns and small cities in the hinterlands who enjoyed higher standards of living than nearby rural residents, but whose wealth and material security remained clearly inferior to city dwellers living further up the hierarchy of urban space.[3]

Within each city one also observed variation in income and status. Despite policies to equalize rewards between skilled and unskilled, profes-sionals earned more than routine white-collar workers, bureaucrats surpassed ordinary employees, and men maintained a noticeable advantage over women.[4] However, in comparison to differentials in other developing economies, the inequalities were small; moreover, between 1950 and 1975 they steadily narrowed.

Yet in one regard the Maoist reward structure created unique inequities which by the mid-1970s were so marked that they undercut the general trend towards greater equality. This was the variation between members of different birth cohorts. Whereas in both pre-Communist urban China and capitalist developing nations, the best educated members of the younger generations routinely surpass members of older cohorts with less education, in urban China the generation born after 1949 systematically did poorly in comparison to their parents despite their generally superior education. Thus, for example, I found that when I compared the occupational achievement at age 30 of two generations in 200 families I interviewed in 1987, 1988, and 1990, sons were less likely than fathers to hold non-routine white-collar jobs and daughters had done no better than mothers despite greater educational achievement (Table 16.2).

Three core elements of the Maoist blueprint for urban development produced these intergenerational inequalities.[5] First because after 1949 hiring and promotions followed rules of seniority and there were few alternatives to employment by the state, almost no one could move ahead faster than other members of their cohort or directly compete with members of the older cohorts. In times of slow economic growth, those who first entered the socialist work-place kept

TABLE 16.2 Comparisons of Job Statuses at Age 30 for Two Generations of Urban Residents (percentage)

Mean birth year	Fathers (N=182) 1928	Sons (N=200) 1955	Mothers (N=198) 1931	Daughters (N=214) 1955
farmer	0	1	1	2
semi-skilled manual	14	33	48	50
skilled manual	19	28	4	6
routine white-collar	8	10	20	20
non-routine white-collar	21	12	17	15
low-level cadre	16	7	7	3
middle manager	12	6	1	2
professional	8	4	3	3
high cadre	3	0	0	0

Source: interviews conducted by the author with members of 200 households in Shanghai and Wuhan (1987, 1988, and 1990).

benefits allocated early in their careers, while those who followed them paid the price of economic stagnation in terms of inferior entry-level jobs, reduced benefits, and slow rates of promotion. Secondly, hostility to commerce and individual entrepreneurship restricted the number of entry-level white-collar jobs as well as the possibility of finding a 'second chance' through success outside the state sector. Thirdly, Mao's idiosyncratic decision to close universities after 1966, and then initially re-open them on the basis of an abbreviated curriculum decisively restricted entry into professional ranks for young adults leaving secondary school between 1965 and 1976. Thus while overall socialist transformation of the urban economy and polity produced more equitable distribution of wages and wealth than had prevailed in Chinese cities before 1949, those who came of age during the 1960s and 1970s faced a more constricted opportunity structure than those ten or twenty years their senior and as a result the late Mao era was distinguished by marked generational inequality.

The Impact of Socialist Transformation on the Role and Character of Cities

In addition to altering individual life chances within cities, the CCP blueprint for economic development after 1949 redefined the relationship between cities and their rural hinterlands and thereby transformed the role and character of cities themselves. Most fundamental to this redefinition were household registration (hukou) policies which restricted rural-to-urban migration during decades of rapid industrialization and limited most new social-welfare benefits to the

minority of citizens living in towns and cities. Previous Chinese regimes had used household registration systems to control and tax the population, and the Japanese had tried to keep household registers for all residents of the occupied cities, but in the twentieth century only the Communist government so effectively used controls over migration and residence to reshape the quality of urban life.[6]

At the heart of the post-1949 household registration system was a 1955 decision to divide the entire population into a minority who were given a city hukou and thereby entitled to work in state enterprises and buy subsidized grain, and a majority who were assigned rural hukou and required to grow their own food and work in rural co-operatives or collectives. In subsequent years the administrative barriers between city and countryside established by the hukou system weakened intermittently. During the Great Leap Forward, millions of rural labourers were recruited to work in cities, and even during the Cultural Revolution decade (1966–1976) when more than 17 million urban secondary school graduates were forced to resettle in the countryside, an almost equal number of rural labourers moved into city jobs. Yet overall, the constraints operated as the planners had intended. Urban population growth was remarkably low for a rapidly industrializing country (Table 16.3) and city services were not overwhelmed by rural migrants.

However, as a result of this spatial segmentation, the majority of the population living in the less 'socialized' agricultural sector, not only received few subsidies even when their actual jobs were not in farming, but were also prohibited from moving to the city to compete for the rewards of urban employment.[7] Because of this

TABLE 16.3 Urban Population, Size (in millions) and Annual Natural Growth Rates (ANGR), 1952–1978

Year	National		Guangzhou		Shanghai		Wuhan	
	Size (m)	ANGR (%)	Size (m)	ANGR (%)	Size (m)	ANGR (%)	Size (m)	ANGR (%)
1952	35	1.9	1.3	2.6	5.1	2.9	1.4	2.3
1957	54	2.8	1.7	3.8	6.1	3.9	2.1	3.5
1962	64	2.3	1.9	3.1	6.3	1.8	2.1	2.9
1978	80	0.8	2.1	0.7	5.6	0.1	2.6	0.7

Sources: *Guangzhou sishinian* (Forty years in Guangzhou) (Guangzhou: Zhongguo tongji chubanshe, 1989), 90; *Shanghai tongji nianjian 1992* (Shanghai Statistical Yearbook 1992) (Shanghai: Zhongguo tongji chubanshe, 1992), 60, 80; *Wuhan sishinian* (Forty years in Wuhan) (Wuhan: Wuhan daxue chubanshe, 1989), 249, 251; *Zhongguo chengshi sishinian* (Forty years in Chinese cities) (Beijing: Urban Social and Economic Survey Organization of State Statistical Bureau, 1990), 40, 118.

linkage between urban residence and government entitlements, all urban places—large metropolises, cities, and towns—conferred privileges on residents in a manner without precedent in the pre-Communist era.

The second critical shift in the role and character of cities resulted from the way in which the command economy was subdivided and co-ordinated via sectors (*xitong*) defined according to end-products and industrial inputs (e.g. textile, steel, labour, energy). In pre-Communist China, cities had been nodes for regional commerce and the economy developed in terms of local and regional markets linked by the trade that flowed through riverways, sea routes, highways, and railroads. There was a hierarchy of central places but horizontal links between cities of comparable size and market-driven competition created a vast, imbricated political economy with multiple centres of power and affluence.[8] By contrast, the post-1949 socialist blueprint forced cities to disengage socially and economically from their immediate hinterland and become lower rungs on a single national hierarchy. In addition, for the Communist leaders urban areas were to be reconstructed primarily as industrial zones serving the central treasury. Development was dictated from the top down in accord with national administrative boundaries, not in response to regional comparative advantage. Profits from a textile factory in City A were remitted upward to the central ministry for light industry, and inputs for that factory were allocated downward from the centre. Cotton mills no longer dealt directly with nearby producers to negotiate price. In this way city-based industries and retail units cut off from their natural suppliers and customers, and they were separated from obvious partners and markets within city boundaries. As a result, groups of enterprises which might have benefited financially from horizontal local linkages were constrained by the sectoral divisions (*xitong*) which tied all textile factories across the country into one vertically integrated system.

Despite the inefficiencies and administrative burdens of the *xitong* structure for individual city economies, the system produced a steady stream of revenue for the central government; in fact, during the Mao years, cities were deemed to be indispensable 'cash cows' which funnelled enterprise profits directly to central coffers.[9] By 1978, 65 per cent of total industrial output was pro-duced in the 38 largest cities and nearly 60 per cent of municipal product was turned over to the national government.[10] The leadership of the central party-state may also have preferred direct extraction of profit over the more flexible approach of taxing individuals, companies, or consumers, because the vertical segmentation of economic authority constrained mobilization by local élites.

The third fundamental departure from pre-Communist city life was in the cultural realm. Throughout the 1930s and 1940s, élite and popular culture suffered due to wartime shortages, hyper-inflation, and official censorship. Yet within cities—both those occupied by the Japanese and those held by the Nationalists—there were unfettered spaces as well as long stretches of private time in which businessmen, artisans, and intellectuals pursued their varied interests. So great were these opportunities even under adversity that political scientist David Strand has described this era as a 'golden age' of public opinion because of the enormous range of ideas competing for audiences in city streets and public arenas.[11]

After 1949 the CCP moved quickly to supervise and restrict non-official uses of city space as well as the daily schedules of urban residents. They restricted political and economic activity to public venues controlled by the party-state and redirected citizen's leisure to political study or mass mobilizations. Sidewalks could no longer be commercial malls or impromptu stages. Artisans and hawkers were reorganized into co-operatives and itinerant performers and artists became state employees. Censorship over all forms of public expression and community organizations became systematic and severe.[12]

Even in the private arena of family life, the socialist blueprint left an indelible mark. Some of those consequences are self-evident.[13] Confiscation of private wealth, prohibitions against family business, mobilization of all adult women into the urban labour-force, and provision of universal and inexpensive education weakened patriarchal family relations; wives had greater equality and the young gained autonomy. However, in other ways, the socialist transformation unexpectedly strengthened traditional family ties. For example, guaranteed life-time employment and egalitarian distribution of food and medical care within the urban population, increased life expectancies and made old age the norm rather than the privilege of a few. Controls over migration kept family members of different generations in close physical proximity and fostered high levels of intergenerational interdependence. Thus, in one of the ironies of history, a Communist revolution dedicated to demolishing old traditions systematically created demographic conditions and a social environment in which a majority of urban families could realize the filial and family obligations of traditional Confucian tenets.[14]

A fifth and final consequence of the socialist transformation of Chinese cities was a distinctive institutional landscape. By this I do not mean the dreary sameness and crumbling façades of urban structures built after 1949, but rather the way in which vertical segmentation by *xitong* and dependency on employers for allocation of goods and services transformed cosmopolitan cities into clusters of urban villages. In the pre-Communist years, specialized markets created distinctive neighbourhoods and urbanites roamed freely between one location for work, another for shopping, and still others for friendship and shelter.[15] During the 1950s, the Chinese developed their cities according to a Soviet ideal of self-contained, multi-function company towns or enterprise villages (*danwei*). Urbanites found themselves increasingly dependent on their *danwei* for all basic services and less able to maintain personal networks of kin and friends outside the work-place. Not only did most urban adults work full time, six days a week, eight hours a day, many also lived in enterprise compounds which provided residents with their grocery store, medical clinic, bath house, as well as auditoriums for movies and community celebrations.

During the frequent political mobilizations, *danwei* leaders mandated daily study sessions and condemned any leisure activity which did not in some way 'serve the collective'. The outcome of these institutional arrangements were cities subdivided into inward-turning enterprise-neighbourhoods characterized by uniformity and self-sufficiency. In the oldest coastal cities where housing for employees often could not be constructed on the same premises as the work-place, enterprise-neighbourhoods were less dominant. However, even when workers lived long distances from the work-place, they still relied on the *danwei* for most daily services and consumer items

because the socialist transformation had eliminated alternative suppliers. For these reasons, it was the imposition of inward-turning *danwei* upon complex, unruly urban spaces that created the distinctive order and drabness of the Maoist city.

Chinese Cities and Urban Life Chances after the Death of Mao

After the death of Mao in 1976, Chinese cities were gradually restructured in accord with the principles of 'market-socialism'. Guided by this hybrid ideology the Communist leadership could, on one hand, praise—even advocate—privatization, commercialization, and expanded trade with the capitalist world, while at the same time require state ownership and maintain their political monopoly. The result was an urban society ambiguously balanced between a socialist past and a market future. Among new entrants to the urban work-force, there was a steady drift towards employment in the private sector, yet among those already employed the norm continued to be jobs in the state and collective sectors (compare Tables 16.4 and 16.1). Similarly, although increased commodification of goods and services reduced the dependency of urban residents on their places of employment, subsidies as a percentage of the wage packet actually increased from 2 per cent of yearly income in 1978 to 20 per cent in 1992.[16]

Nevertheless change not continuity defined urban society during the 1990s as Chinese cities and their inhabitants steadily moved away from the rigidities of a planned economy and towards the greater openness and uncertainties of a market economy. To illustrate the nature and scope of this shift, I will first discuss the changes in the role and character of cities and then changes in the life chances of urban residents.

Role and Character of Chinese Cities in the Early 1990s

The most obvious structural change from the Maoist era was a new relationship between urban and rural China. It began slowly in the early

TABLE 16.4   Destinations of New Entrants to Urban Work-Force, 1978–1994

Destinations	1978	1980	1982	1984	1986	1988	1990	1992	1994
Number (in millions)	5.4	9.0	6.6	7.2	7.9	8.4	7.8	7.3	7.1
% to state	72	64	62	58	68	58	61	50	41
% to collective	28	31	33	27	28	31	30	30	25
% to non-state/non-collective	—	5	5	15	16	11	10	21	33

Note: Percentages more or less than 100% due to rounding.

Sources: For 1978–85: *Zhongguo laodong gongzi tongji ziliao 1949–1985* (Chinese labour and wage statistics 1949–1985) (Beijing: Zhongguo tongji chubanshe, 1987), 110; for 1986–8: *Zhongguo tongji nianjian 1990* (Chinese Statistical Yearbook 1990) (Beijing: Zhongguo tongji chubanshe, 1990), 130; for 1990–2: *Zhongguo tongji nianjian 1993*, 119; Zhongguo tongji nianjian 1995, 106.

1980s when the central government permitted widespread recruitment of rural construction teams in order to support massive expansion of the urban housing stock. Millions of rural men took advantage of the opportunity and many stayed on in the cities after their first job was complete, or returned for new jobs after a short respite in their home village. Rural peddlers soon followed, and as restrictions on free markets were lifted, urban sidewalks and squares began to bustle with city residents eager to buy fresh produce from rural entrepreneurs. After the collapse of the commune system in 1982, the government took the first major step towards legalizing the more open flow of people and goods between rural and urban communities by allowing all villagers who invested in urban businesses to become permanent residents in small and medium-sized towns. The next year, the gate opened wider when the State Council allowed the 60,000 market towns below the county level to drop all restrictions on rural migrants. Subsequently this relaxation of migration restrictions extended to cities of all sizes[17] and by 1988 *official* figures estimated that 26 per cent of the population lived in urban areas.[18]

In addition to migrants who used official channels to change their status from rural to urban residents, millions of others simply took to the road, left their fields, and started a new life through 'temporary' employment in the cities. As the drive to privatize and commercialize accelerated after 1990, out-migration from villages surged and Chinese cities were no longer isolated from their rural hinterlands. Demographers estimated that by 1992 30 per cent of the population lived in urban areas and that by 2000 it would reach 35.7 per cent or 462 million.[19]

This massive demographic realignment transformed the character of urban life. Throughout the nation, markets resounded with a cacophony of different dialects. Local streets and buses were crowded with travellers from distant provinces wearing homemade cloth shoes and heaving large sacks of tools or produce. Itinerant peddlers and hawkers dominated the retail trade, and by the early 1990s, the state had effectively withdrawn from the sale of fresh vegetables, fruits, meat, and fish. The self-contained and relatively homogeneous cities which had been created and sustained by the Maoist embargo against free migration from the countryside had disappeared.

The increased movement of people between villages and cities also created the conditions for reconfiguration of metropolitan power. Traders moving back and forth between urban customers and rural suppliers, urban investors subcontracting and investing in suburban enterprises created webs of loyalty and common interests which undermined the vertical line of authority between Beijing and individual municipalities. Although the CCP systematically quashed localist ambitions and recruited the leaders of such key metropolitan areas as Shanghai and Tianjin into the central leadership, cities all over China steadily became more embedded in their local region. They established municipal special economic zones to lure foreign capital and developed wage and benefit programmes unique to their locale. In the process, cities became both more independent of Beijing and more competitive with other urban centres.[20] Overall the trend by the mid-1990s was clear. China was returning to a pre-Communist pattern where an urban core drawing on resources of a vast rural hinterland supports a regional urban élite with the potential to challenge the priorities of the central government.

The accelerated reliance on market exchange

and less restricted migration of rural labour also decisively reconfigured social life. During the 1970s, when self-sufficient enterprise-neighbourhoods controlled the mobility of their employees and political campaigns polarized the workplace, Chinese cities offered their residents little anonymity or personal freedom. In cities in the Manchurian tundra, on the arid North China plain, or in the semi-tropics of Yunnan a generic urban profile prevailed. The bulk of the labourforce had permanent employment in state jobs and had only modest worries about financing medical care or old age. Citizens were prohibited from following consumer trends in the capitalist West or from pursuing indigenous forms of worship or amusement.

Beneath the grey, utilitarian surface there had, of course, been counter-currents. Amateur singing groups, secret groups of Daoist mystics, and 'renegades' who took great risks to listen to foreign radio broadcasts on short-wave radio resisted the official demands for conformity. During the breakdown of public order at the height of the Cultural Revolution, many young people rediscovered the attraction of Western romance novels, as well as traditional folklore and pornography. But these multiple forms of resistance and alienation were constrained by the inability of participants to share and exchange their non-orthodox beliefs and pleasures publicly. Increasingly after 1980, those social currents which had been submerged surged to the surface.

By the 1990s, the work-place was neither as regimented nor as politicized, and an enlarged sphere of private time and non-state activity permitted individuals to carve out realms of life beyond the supervision of enterprise managers or colleagues.[21] Once again, cities functioned as primary locations of a wide range of entertainment. City streets became colourful, noisy, and highly commercial. New jobs in advertising and marketing emerged to support the new economy, and a vast new range of products and services lured customers into stores, boutiques, and night spots. The large coastal cities reclaimed their positions as gateways to the outside world and residents and visitors flocked to public venues where they could consume—or simply gaze at—luxury items and leisure activities previously condemned as bourgeois or even anti-socialist.[22]

Increased commodification and regional

autonomy also decisively altered land use within city centres and created new urban landscapes and skylines. Cities of the early 1990s concentrated investment around specialized nodes of commerce, restored historical landmarks, and new entertainment districts.[23] They began to sprawl into what had been for decades pristine agricultural lands. Squatters appeared at construction sites, around railroad stations, and in parks. Although by international standards crime levels were low, for the first time in decades Chinese cities were seen as dangerous places. Bars went up on windows and locked metal gates were routinely installed in new housing estates.

Altered Life Chances

During the first decade after Mao's death, the legacy of the socialist transformation continued to sustain a rather egalitarian distribution of income. The majority of new entrants still went to state-sector jobs (Table 16.4). Guaranteed monthly wages and subsidies remained the key determinants of household income and job turnover remained below 5 per cent per year.[24] Moreover, because early efforts attempted to upgrade wages of the worst paid, initial reforms *decreased* rather than increased inequality both between rural and urban areas, and within the urban work-place.[25] However, as the reforms accelerated the break with the past was decisive.

Movement of rural residents into cities became less restricted, labour and goods increasingly observed the 'laws' of supply and demand, and inequalities between the rich and poor increased across the nation. The urban–rural gap widened, and by 1993 rural residents were at an even greater disadvantage than they had been in 1980 when the reform began.[26] Within cities a highly visible minority adopted the trappings of wealth—cellular phones, expensive restaurant meals, elaborate jewellery. At the same time, a visible strata of newly impoverished and marginalized urbanities emerged.[27]

The pattern of employment and the 'rules' for remuneration shifted. The state sector no longer absorbed the majority of new entrants to the urban labour-market and jobs in the non-state sector paid the highest wages (see Tables 16.4 and 16.5). Not surprisingly jobs in the private

TABLE 16.5   Average Salary for Urban Employees, by Type of Ownership, 1984–1994 (in yuan)

Type of ownership	1984	1986	1988	1990	1992	1994[a]	1994[b]
With state ownership	1,034	1,414	1,853	2,284	2,878	4,797	6,671/7,534
With collective ownership	811	1,092	1,426	1,681	2,109	3,245	5,009/5,389
Non-state/non-collective	1,048	1,629	2,382	2,987	3,966	6,303	8,179/9,617

[a] National average.
[b] Beijing/Shanghai.

Source: *Zhongguo tongji nianjian 1995* (Chinese Statistical Yearbook 1995) (Beijing: Zhongguo tongji chubanshe, 1990), 106.

sector became the preferred destination among the best educated.[28] In addition, part-time employment and second jobs became routine. As the front runner in the move towards marketization, Guangzhou led this trend and as early as 1988 official and unofficial sources estimated that 30 per cent of Guangzhou residents held second jobs.[29] But the rush to find additional income by selling one's skill spread throughout China, and by the early 1990s approximately one-third of all urban adults earned income from second jobs. Among technicians and professionals, the rate was often twice as high.[30]

When viewed from a distance, or analysed in aggregate terms, the accelerated economic reforms of the late 1980s and early 1990s brought new prosperity to urban Chinese. In absolute terms, average wages rose from 615 yuan in 1978 to 4,535 yuan in 1994[31] and gains significantly outpaced inflation for employees in all sectors.[32] Nevertheless, when viewed from the perspective of individuals, marked polarization accompanied the aggregate gains.[33] Gini coefficients in urban areas which had been among the lowest in the world—0.16 between 1977 and 1980[34]—were estimated to have increased to between 0.40 and 0.50 by 1990.[35]

Not only did the degree of inequality increase over time, the causes of inequality changed.[36] During the Maoist years, employees in the state sector earned more than those in the collective, men did marginally better than women, and those with seniority outearned their juniors. Overall, however, the differentials did not produce significant differences in standard of living, and because most households included male and female wage-earners, senior and junior employees, disparities between households were modest. As the urban economy turned to markets, permitted second jobs, and employment in foreign-owned firms, the constraints on income inequality weakened. Enterprises were permitted

to tie wages to profits and create their own reward schedules. Foreign firms in competition for scarce expert talent bid up wages and the self-employed could earn as much as they could squeeze from their profits. The results were predictable. The greatest incomes went to the highly educated and those working in the most profitable factories. Professionals in public services—such as doctors, teachers, nurses—who during the Mao years had experienced minimal disparities between themselves and their peers in industry found themselves falling further and further behind. Unskilled labourers in old industrial plants faced economic hardship, while former shopmates with special skills—or connections—doubled or trebled old wages as foremen in joint-venture telephone or computer plants. The advantages of seniority faded, but those of maleness, college education, residence in a coastal province, and employment in a foreign-owned industry increased.[37]

The Fading Legacy of the Socialist Transformation

As China reaches the end of the twentieth century, the Communist leadership is jettisoning core institutions of socialism at an accelerating pace. Given the extent of ideological and organizational change it is possible that by the year 2000 fifty years of Communist rule will have left few physical and social traces on Chinese cities or individual urbanites. During the early 1990s, however, the imprint of the CCP socialist transformation, while fading, remains visible.

One sees continuities with the Maoist past in the physical dimensions of city life. Urban sprawl has begun to erase the formerly sharp boundary between urban and rural districts, and large settlements of transient labour have appeared throughout the country. But because land remains state property, there are constraints on the ability of markets to reshape the built environment, and the state still retains significant power over the pace and scope of development.

In the work-place, the socialist legacy is most obvious in policies towards unemployment and retirement. Officially unemployment had not yet reached 5 per cent when the government moved aggressively to mandate minimum wages, unemployment compensation, and a larger bureaucracy to handle retraining and placement of the urban unemployed.[38] Pensions for women at 50 and men at 60 remain the norm, and although they no longer replace 99 per cent of the average per capita income, they guarantee more than subsistence. The pension system itself is generally well run and reforms which will maintain the benefit even as the work-force ages and becomes more mobile have been implemented at local and national levels.[39]

One also sees the imprint of the socialist past when probing the attitudes of urban residents. Competitive labour-markets have emerged for the rural immigrants, and a shadow economy of second jobs has grown exponentially. But as of the early 1990s, a majority of urban residents remain in the state sector and among those with less than a college education a preference for steady life-time jobs in the state sector persists.[40] Moreover amidst the headlong rush for profits and material incentives, there is broad support for a basic welfare net of subsidized medicine, education, and housing, and clear dissatisfaction with the new concentration of wealth.[41]

After 1978 division of authority between the central government and Chinese cities shifted and cities became more autonomous economic and political actors. Overall the quality of every-day life improved, and residents gained unprecedented access to information about the world outside China. For the first time in four decades, international capitalists are not villains, and there is even a growing tolerance for a domestic, new middle class, if not yet a full embrace of an indigenous bourgeoisie. The language of cost-benefit analysis informs government pronouncements as well as directing decisions at the level of the factory floor. Yet at the same time, many city residents—not just CCP ideologues—remain wary of full commercialization. There is significant popular support for the socialist component

of 'market-socialism' and for restraints on private capital.

Under the Maoist blueprint for economic development, cities received the bulk of state investment and urban residents were the primary beneficiaries of socialist redistribution. Thus some of the popular support for the command economy and egalitarian redistribution simply reflects a fear that privileges will be lost. As independent entrepreneurs and enterprise managers use their new freedoms to conduct business outside the previously regimented urban enclaves or according to the rough and tumble rules of market competition, those who had been financially secure feel threatened. And throughout the nation, awareness of the social disorder in the former Soviet Union creates support for a gradual retreat from the past. The altered economic and political functions of Chinese cities and the growing financial inequalities document fundamental changes, but the continuities for the majority still employed by the state and pervasive support for redistributive income policies testify to enduring links with the Maoist revolution.

Notes

1. Comparing a broad range of developing market economies to urban China in the 1970s, sociological research found that whereas in the market economies incomes earned by the poorest 40% of households averaged 15% of total income, in urban China government policy guaranteed that the lowest four deciles garnered at least 25%. Martin Whyte and William Parish, *Urban Life in Contemporary China* (Chicago: University of Chicago Press, 1984), 44.

2. *Guangzhou sishinian* (Forty years in Guangzhou) (Guangzhou: Zhongguo tongji chubanshe, 1989), 158; *Shanghai tongji nianjian 1986* (Shanghai Statistical Yearbook 1986) (Shanghai: Zhongguo tongji chubanshe, 1986), 434; *Wuhan sishinian* (Forty years in Wuhan) (Wuhan: Wuhan daxue chubanshe, 1989), 276; *Zhongguo tongji nianjian 1992* (Chinese Statistical Yearbook) (Beijing: Zhongguo tongji chubanshe, 1992), 130.

3. For excellent overview, see Xiangming Chen and William P. Parish, 'Urbanization in China: Reassessing an Evolving Model', in Josef Gugler (ed.), *The Urban Transformation of the Developing World* (Oxford: Oxford University Press, 1996), 61–90.

4. Whyte and Parish, *Urban Life in Contemporary China*, 45, 53, 202, 207, 209.

5. Deborah Davis, 'Intergenerational Inequality and the Chinese Revolution', *Modern China*, Apr. 1985, 177–201. Deborah Davis, 'Skidding: Downward Mobility among Children of the Maoist Middle Class', *Modern China*, Oct. 1992, 410–37.

6. For an excellent analysis of this process, see Tiejun Cheng and Mark Selden, 'The Origins and Social Consequences of China's *Hukou* System', *China Quarterly*, 139 (Sept. 1994), 644–68.

7. For excellent overview, see Martin K. Whyte, 'City versus Countryside in China's Development', 56th George Morrison Lecture, Australian National University, reprinted by Australian National University, 1995.

8. G. W. Skinner, 'Cities and the Hierarchy of Local Systems', in G. W. Skinner (ed.), *The City in Late Imperial China* (Stanford: Stanford University Press, 1977), 275–351.

9. Barry Naughton, 'Cities in the Chinese Economic System', in D. Davis, R. Kraus, E. Perry, and B. Naughton (eds.), *Urban Spaces in Contemporary China* (London: Cambridge University Press, 1995), ch. 3.

10. Ibid.

11. David Strand, 'Historical Perspectives', in Davis *et al.* (eds.), *Urban Spaces in Contemporary China*, ch. 15.

12. But as Richard Kraus has argued, state takeover of the arts also produced explosive growth in the job prospects of urban artists: 'Between 1949 and 1962, performing art groups increased from 100 to 3,320, county and municipal cultural centres from 896 to 2,514, public libraries from 55 to 541, museums from 21 to 230, and film projection units grew from 646 to 18,583' (Richard Kraus, 'The Re-construction of Urban Cultural Systems', paper presented at the annual meeting of the Association of Asian Scholars, Los Angeles, 27 Mar. 1993).

13. This argument is more fully developed in Deborah Davis and Stevan Harrell (eds.), *Chinese Families in the Post-Mao Era* (Berkeley: University of California Press, 1993), 1–22.

14. Deborah Davis-Friedmann, *Long Lives*, revised and expanded 2nd edn. (Stanford: Stanford University Press, 1991).

15. Piper Gaubatz, 'The Urban Transformation in Post-Mao China', in Davis *et al.* (eds.), *Urban Spaces in Contemporary China*, ch. 2; Shamus C. Mok, 'Changchun City: The Social Production of a Dual Motor City under Chinese State Socialism', paper presented at Roundtable of the Northeast China Studies Association of the Association of Asian Studies, Los Angeles, 26 Mar. 1993.

16. *Zhongguo laodong gongzi tongji nianjian 1990* (1990 Yearbook of Chinese Labour and Wages)

(Beijing: Zhongguo tongji chubanshe, 1990), 22, 227; *Zhongguo tongji nianjian 1993*, 127, 132.

17. D. Solinger, 'Temporary Resident's Certificates', *China Quarterly*, 101 (1985), 98–101; *Guofa*, No. 141, *Guowuyuan gongbao* (Bulletin of the State Council) (1984), 919–20; Foreign Broadcast Information Service, 23 Oct. 1984, K18–19.

18. *Zhongguo tonji nianjian 1994*, 59.

19. Roger Chan, 'Challenges to Urban Areas', in *China Review 1992* (Hong Kong: Chinese University of Hong Kong Press, 1992), 12.12; *Ming Bao* (Hong Kong), 18 Feb. 1993, 60.

20. These trends are well illustrated by a case study of Shanghai by Barry Naughton, 'Cities in the Chinese Economic System', in Davis *et al.* (eds.), *Urban Spaces in Contemporary China*; and of Xinji, a Hubei county seat, by Vivienne Shue, 'State Sprawl: The Regulatory State and Social Life in a Small Chinese City', in Davis *et al.* (eds.), *Urban Spaces*, ch. 5.

21. Shaoguang Wang, 'The Politics of Private Time', in Davis *et al.* (eds.), *Urban Spaces*, 149–72.

22. The rapid growth of Japanese-style department stores in Beijing, Shanghai, and Guangzhou which in addition to offering every range of commodity also offer fashion shows, snack shops, and cosmetic boutiques are an especially good example of this transformation.

23. Gaubatz, 'Urban Transformation in Post-Mao China', in Davis *et al.* (eds.), *Urban Spaces*.

24. *Zhongguo laodong gongzi tongji nianjian 1990*, 22, 227; *Zhongguo tongji nianjian 1993*, 127, 132; Deborah Davis, 'Job Mobility in Post-Mao Cities', *China Quarterly*, 123 (1992), 410–37.

25. *Zhongguo tongji nianjian 1994*, 256; Andrew Walder, 'Economic Reform and Income Distribution in Tianjin 1976–1986', in Deborah Davis and Ezra Vogel (eds.), *Chinese Society on the Eve of Tiananmen* (Cambridge: Harvard University Press, 1990), 135–56.

26. In 1980 agricultural residents per capita consumption was 37% of that of urban residents. In 1985 it rose to 45%, but by 1993 it had fallen to only 31%. *Zhongguo tongji nianjian 1994*, 256.

27. In 1994, when average urban income was 262 yuan per capita, 20 million urban residents had less than 150 (*China News Digest*, 29 Mar. 1995).

28. *Ming Bao*, 7 Aug. 1992, 65; *Renmin ribao*, 15 July 1992, 2.

29. *Renmin ribao*, 11 June 1988, 2; Guilhem Fabre, 'Origines et dynamique de l'inflation en Chine', *Themes economiques*, 2.133 (12 July 1989).

30. *Ming Bao*, 2 Apr. 1990, 6; *Xinmin Wenbao*, 27 July 1992, 2.

31. *Zhongguo tonji nianjian 1995*, 23.

32. In four years (1977, 1981, 1988, and 1989) gains for state workers were wiped out by inflation; for

those in collective enterprises wage gains only disappeared in 1988 and 1989 (ibid.).

33. *Renmin ribao*, 4 Sept. 1991, 2; 14 July 1992, 2; 12 Nov. 1992, 2; *Far Eastern Economic Review*, 11 Feb. 1993, 45; 20 Jan. 1994, 59.

34. Zhang Dagen, Zhongguo jingji fazhan yu shouru fenpei' (Distribution of income and Chinese economic development), *Xueshu jikan*, 1 (1991), 30–6; Zhou Renwei, 'Income Distribution', in Peter Nolan and Dong Fureng (eds.), *The Chinese Economy and its Future* (Cambridge: Polity Press, 1990), 187–99.

35. Zhou Dengke, 'Lun shichang jingji xiade chengshi fupin wenti' (Discussion of problems of rich and poor in a market economy), *Shehui kexue*, 7 (1993), 46–9.

36. Deborah S. Davis, 'Inequality and Stratification in the Nineties', in Lo Chi Kin, Suzanne Pepper, and Tsui Kai Yuen (eds.), *China Review* (Hong Kong: Chinese University Press, 1995), ch. 11.

37. Yanjie Bian, John Logan, Pan Yukang, and Guan Ying, 'Income Inequality in Tianjin, 1978–1993', paper for conference on Civil Society in Tianjin, held at the University of California at San Diego, 5–9 June 1994; Feng Tongqing, 'Zou xiang shichang jingji de zhongguo qiye zhigong' (As Chinese industrial workers and staff go towards a market economy), *Zhongguo shehui kexue*, 3 (1993), 101–20; Feng Wang and Alvin So, 'Economic Reform and Restratification in Urban Guangdong', paper for conference on social stratification in Chinese Societies, held at Chinese University of Hong Kong, 10–11 December 1993.

38. Decree No. 111, State Council of PRC, *Guowuyuan gongbao*, 16 June 1993, 298–300; *Ming Bao*, 25 Jan. 1994, 12; *Far Eastern Economic Review*, 20 Jan. 1994, 59.

39. Deborah Davis, 'Financial Security of Urban Retirees', *Journal of Cross-Cultural Gerontology*, 8 (1993), 179–85.

40. A 1992 survey of urban youth found that 64% of high-school graduates still most wanted a state job. First choice for another 22% were jobs in Joint Ventures, then 10% in big collectives, and only 4% in self-employment. *Ming Bao*, 27 Apr. 1992, 9.

41. Cecilia Chan and Nelson W. S. Chow, *More Welfare after Economic Reform* (Hong Kong: Hong Kong University Press, 1992).

V Housing and the Environment

household bud
national budget
long term consequence
state move

development &
° allocation of
° forced
° construction &
° relief

Introduction

SHANTY-TOWNS and run-down tenements dramatically demonstrate the poverty of much of the urban population in Asia, Africa, and Latin America. Many urban dwellers lack adequate protection from rain and flooding, from heat and cold. Their health, and indeed their life, is threatened by contaminated water and inadequate sanitation. Still, many experience malnutrition and hunger as even more immediate problems. And limited access to medical services is reflected in high rates of increased morbidity and mortality. Also, of course, it may be argued that the underlying issue is income: given adequate earning opportunities, people will be able to secure shelter and services, as well as food and health care, that meet their needs. Nevertheless, housing and the environment command special attention in their own right.[1]

Housing, urban infrastructure, and some forms of pollution attract public attention because they are highly visible to people who make themselves heard. But there are several reasons which indeed warrant that we focus attention on housing and the urban environment. Shelter absorbs a major part of the budget of most urban dwellers. The construction of housing and urban infrastructure takes up a large share of national investment. Urban services and the maintenance of infrastructure and public housing place heavy demands on public resources. Urban planning, the absence of it, or its very partial implementation, has consequences that reach far into the future. And the neglect of the environmental consequences of rapidly growing urban agglomerations threatens global disaster. Not surprisingly then, it is in this dimension of urbanization, more than in any other, that governments are deeply involved and that political change has immediate repercussions in terms of spatial planning, the development and allocation of land, construction and rehabilitation, and the provision of services.

Jorge Hardoy and David Satterthwaite, in Chapter 17, estimate that in 1990 at least 600 million urban dwellers lived in homes and neighbourhoods that threatened their health and their lives. They trace the dismal living conditions of a large part of the urban population in the South to the failings of government at all levels: even with the limited resources available in poor countries, governments could do much better by the urban masses. But outdated institutional structures, inappropriate foreign models, and weak, ineffective, and unrepresentative governments are not equal to the task. Most housing construction proceeds informally, if it is not illegal altogether. And environment-related diseases and accidents are among the major causes of illness, injury, and premature death.

Hardoy and Satterthwaite advocate more open and participatory forms of government, urban planning with, rather than against, the urban masses, and recovery of the cost of public investments, services, and subsidies from the prime beneficiaries: landowners, businesses, and the more affluent neighbourhoods. They call for a redistribution of power and resources within the city—and across the globe.

Mohan (1996) provides a detailed account of these issues—ranging from land supply to shelter for the poor to finance for housing, services, and infrastructure to local governance—for one major country, India. And in Chapter 9 we encountered a telling example of how governments fail time and again to devise policies that take full account of the circumstances of the poor. Sumira and her neighbours were resettled on unusually generous terms, but most of them were unable to hold on to their new apartments.

The cost of urban housing and infrastructure is staggering. Regimes that could exert a measure of control over migration reduced the demand for urban housing by tailoring the urban population to the needs of the urban labour-market, as we have seen in Chapters 6, 7, and 16. Only workers needed for the urban economy were allowed into the cities, and many of them were recruited on contract; they would have to leave the city when no longer needed, and they could not bring their dependants to the city. Furthermore, attempts were made to force dependants and excess labour out of the cities. In spite of such restrictions on urban growth, socialist countries invariably had severe housing shortages. The state limited investment in the housing construction it controlled, and neglected maintenance, because such expenditures were seen as unproductive.

'Self-help housing' holds out the promise of a low-cost solution to the challenge of housing the urban poor. Kosta Mathéy, in Chapter 18, reviews three decades of academic debate and praxis. The initial impact of a few innovative planners was momentous: they influenced public opinion to the point that governments abandoned the common practice of indiscriminately eradicating squatter settlements, unauthorized structures, and slum areas. In distinct contrast, the subsequent debate remained rather inconclusive and had little practical impact.

In three decades of praxis, some important lessons have been learned: the need to define the target population realistically, to recognize the gender dimension, to take into account the interests of tenants, to avoid gentrification, and to speed up the construction process. Overall, however, 'self-help housing' programmes have rarely met the expectations of either their academic protagonists or their sponsors.

In spite of the disappointments, 'self-help housing' programmes remain popular—because of the opportunities they provide for bargaining: the poor gain access to public resources in exchange for shoring up the legitimacy of the state. Susan Eckstein, in Chapter 22, provides a detailed account of the variables affecting such a bargaining process in the political context of Mexico.

A shortage of adequate housing appears to be near universal. One of the very few exceptions in the South is Singapore, as we have seen in Chapter 15. Hardoy and Satterthwaite spell out the very special circumstances of a city-state shielded from the usual huge influx of rural–urban migrants by an international border, of exceptionally rapid economic growth providing ample resources for construction, and of public ownership of around half the island's land.

The cost of urban land, escalating rapidly as urban agglomerations spread ever further, con-

stitutes a major element in the cost of urban housing. Proximity to central locations constitutes a value that is revealed in the costs of transport and time that more distant locations entail. But more than proximity to central locations is involved in land values. As Alan Gilbert and Peter Ward (1985: 62–7) show for Bogotá, Mexico City, and Valencia, land values reflect not only the inherent but also the acquired characteristics of land. The rich may choose to live in an attractive location at considerable distance from their place of work, but an expressway is likely to shorten the distance.

How crucially the cost of land affects the provision of urban shelter for the urban masses is demonstrated by the extent of squatting, the illegal occupation of land, in developing countries in the face of often brutal repression. Many squatters chose quite unattractive locations— land unsuitable for building, or land on the distant outskirts of cities—so as to minimize the risk of being evicted. They are prepared to endure the hardships of squatting, and they take the risks it entails, because even such unattractive land is beyond their means to purchase. Organized squatter movements succeeded in improving the housing conditions of large numbers of city dwellers in a number of countries, especially in Latin America. Their fortunes varied in different political contexts, as we shall see in Part VI.

Fernando Kusnetzoff, in Chapter 19, traces the changes in housing policy in one country over more than three decades. During that period Chile experienced major political upheavals. The ideological commitment of various regimes ranged all the way from the socialism of Salvador Allende's Popular Unity to the conservative authoritarianism of Augusto Pinochet's military. With different ideological commitments came different housing policies. At the same time, each of five very different regimes found it necessary to be responsive to popular demands for housing so as not to jeopardize its electoral prospects. This was the case even of the military regime once it was constrained to hold a plebiscite and then presidential elections. Chile has thus tried out a wide range of approaches to improve the urban environment and to provide adequate housing. Still, while every regime defined a large housing deficit, none succeeded in making a noticeable impact on it—and this in a country more affluent than most in the South. Indeed, housing production in the early 1990s, higher than ever before, barely met the demands from newly established households and the need to replace obsolete dwellings.

The literature on shelter tends to focus on home ownership. And indeed, home ownership saves income for other expenditures and can provide a measure of security in old age: part of the home may be rented out to tenants, or it may be transformed into a shop or a workshop. The status of property owner may confer advantages beyond the economic. Barnes (1986: 47-69) details the power landlords in Mushin, a recently established suburb of Lagos, enjoy at several levels: their property gives them social power, especially in kinship matters, and serves as a permanent estate around which corporate kin groups can develop; it attests to their permanence and hence signifies their local citizenship in contrast to unpropertied immigrants; and it provides the basis for access to positions of local leadership.

Granted the benefits of owning shelter, the focus on ownership and the implicit assumption that it is superior to tenancy, is problematic. Many permanent residents prefer rental arrangements at certain stages of their family cycle and their career (Gilbert *et al.*, 1993: 156–60). More importantly, as we have seen when we discussed migration strategies, large numbers of migrants maintain a rural base, and many prefer to invest in what they continue to consider their rural home rather than the city. About two-thirds of urban households in Latin America, but less than half in Africa, owned their home in the 1970s, according to such data as are available (Edwards, 1990). The difference may be explained in terms of higher income levels in Latin America, but it may also be seen as a consequence of the fact that permanent urban residence is the rule in Latin America, while many urban residents in Africa maintain a rural base, as we have seen in Chapter 4.

A study of Mexico City, Santiago, and Caracas reports little conflict between landlords and tenants. Landlords and their tenants often came from similar social backgrounds, had about the same level of income, and lived in the same settlement. Rental tenure usually was long term, and landlord and tenant typically got to know each other quite well over the years (Gilbert *et al.*, 1993: 151–2). There are, of course, also situations where the relationship is quite exploitative. We have seen in Chapter 10 how affluent outsiders control much of the rental property in Mathare Valley and demand rents that are exorbitant relative to their investment in the shantytown.

The most desperately urban poor are, of course, not tenants. They cannot afford to pay rent and are left with but three options: to persuade relatives to share their shelter, to squat, or to make do without shelter (Gilbert *et al.*, 1993: 153–5). We have seen, in Chapter 19, how increasing numbers of Chileans came to share their homes when the economic crisis depressed their incomes and the military regime effectively inhibited any attempt at squatting.

The 'housing crisis' continues. The stark reality is that very few societies are prepared to assure all their members of what by the standards of these very societies is considered adequate housing. Housing, of course, is only a particularly visible element of the sub-standard conditions—in terms of nutrition, health care, education, and work—most societies take for granted for some of their members. The proportion of the population thus affected varies across societies, and it changes with economic cycles and political conjunctures, but such an 'underclass' is found nearly everywhere.

Note

1. For reviews of the rapidly expanding literature on housing and the urban environment in developing countries, see Gilbert (1992, 1994).

References

Barnes, Sandra T. (1986), *Patrons and Power: Creating a Political Community in Metropolitan Lagos* (Manchester: Manchester University Press; Bloomington: Indiana University Press).

Edwards, Michael (1990), 'Rental Housing and the Urban Poor: Africa and Latin America Compared', in Philip Amis and Peter Lloyd (eds.), *Housing Africa's Urban Poor* (Manchester: Manchester University Press), 253–72.

Gilbert, Alan (1992), 'Third World Cities: Housing, Infrastructure and Servicing', *Urban Studies*, 29: 435–60.

—— (1994), 'Third World Cities: Poverty, Employment, Gender Roles and the Environment

During a Time of Restructuring', *Urban Studies*, 31: 605–33.

Gilbert, Alan and Ward, Peter M. (1985), *Housing, the State and the Poor: Policy and Practice in Three Latin American Cities* (Cambridge: Cambridge University Press).

—— in association with Oscar Olinto Camacho, René Coulomb, and Andrés Necochea (1993), *In Search of a Home: Rental and Shared Housing in Latin America* (London: UCL Press; Tucson: University of Arizona Press).

Mohan, Rakesh (1996), 'Urbanization in India: Patterns and Emerging Policy Issues', in Josef Gugler (ed.), *The Urban Transformation of the Developing World* (Oxford: Oxford University Press), 92–131.

17

Building the Future City

Jorge E. Hardoy and David Satterthwaite

THE last four decades have brought remarkably little progress in improving housing and living conditions in most urban centres of Africa, Asia,[1] and Latin America. Overall, the number of urban dwellers living in very poor conditions has grown rapidly, even though most nations are much wealthier than they were in 1950. An estimate for 1990 suggests that at least 600 million urban dwellers live in life-threatening or health-threatening homes and neighbourhoods because of poor quality shelters, dangerous sites, and inadequate provision for safe water supplies, sanitation, drainage, or health care.[2] This is a damning indictment of the role of governments and international agencies. It becomes all the more damning when note is taken of the many innovative schemes that show how housing and health conditions can be much improved for poorer groups at low cost.

When judged against their official responsibilities in urban areas, Third World governments have failed in at least four of their most fundamental tasks: (1) to provide the legislative and regulatory system to protect citizens from exploitation by landlords and employers; (2) to ensure all citizens can find adequate accommodation and access to basic services; (3) to protect the urban environment from contamination by life-threatening and health-threatening pathogens and pollutants; and (4) to allocate the costs of implementing these three tasks among those who benefit from urban locations, urban labourforces, and government-provided infrastructure and services. Clearly, these tasks are interlinked. For instance, one critical part of protecting citizens from exploitation by employers is preventing their exposure in the work-place to life- and health-threatening pollutants. Ensuring that citizens can find adequate accommodation and access to basic services also does much to protect them from life- and health-threatening pathogens.

Most governments in the Third World—national, provincial/state, and local—are failing in these four tasks. Case studies of specific cities show that it is common for between 30 and 60 per cent of the population to live in illegal settlements, tenements, or cheap boarding-houses where infrastructure and service levels range from the inadequate to the almost non-existent.[3] Most urban centres in Africa and Asia—including many cities with a million or more inhabitants—have no sewerage system at all. Even in cities where investment has been made in sewers,

This essay is a revised, expanded, and updated version of the 'Epilogue', the final chapter in Jorge E. Hardoy and David Satterthwaite, *Squatter Citizen: Life in the Urban Third World* (London: Earthscan Publications, 1989). It also incorporates some of *Squatter Citizen*'s chapter 1 and some of the findings of the research in which the two authors were engaged between 1989 and 1993 on health and environmental issues in Third World cities. Jorge Hardoy died before this essay was completed, although he and David Satterthwaite had discussed its development for *Cities*.

connection to the system is usually restricted to the richer areas. Garbage collection services are inadequate or non-existent in most residential areas; an estimated 30–50 per cent of the solid wastes generated within urban centres are left uncollected;[4] in urban centres in Africa, less than 20 per cent is collected and properly disposed of.[5] Such waste accumulates on open land and streets, attracting disease vectors, blocking drainage ditches, and posing both a fire hazard and a hazard to children playing. The proportion of the population living with serious deficiencies in infrastructure and services may be smaller in the major cities of more prosperous nations— although in metropolitan areas such as Mexico City and Buenos Aires there are still millions suffering in this way.

The most tangible evidence of the failure of government to meet its responsibilities is the enormous burden of ill health, disablement, and premature death suffered by the urban population. This is most clearly seen in the very high infant and child mortality rates among the population living in tenements or informal or illegal settlements and in the extent to which many of the most common causes of death among this population are diseases or accidents which are easily prevented.[6] It is also clear that this failure of government impacts most on poorer groups. This can be seen in the scale of the differentials between wealthy and poor areas in environmental hazards, in access to public services, and in health indicators.[7] Infant or child mortality rates in poorer areas of cities are often four or more times that in richer areas, with much larger differentials apparent if the poorest district is compared to the richest district.[8] Causes of death in poor districts are often dominated by diseases associated with inadequate water and sanitation, overcrowding (for instance diarrhoeal diseases) or inadequate environmental management to control disease vectors within the home or neighbourhood (for instance malaria, dengue, and yellow fever) or a combination of overcrowded, poor quality (often smoky) homes and inadequate health care (acute respiratory infections, especially pneumonia). The lack of public health services and inadequate incomes ensure that illnesses and injuries often go untreated. A child who contracts bronchitis or pneumonia in the Third World is 50 times more likely to die than a child in Europe or North America.[9]

Comparable differentials are likely to exist between richer and poorer districts within most Third World cities. The incidence of tuberculosis (the single largest cause of adult death worldwide)[10] is linked to overcrowding and poor ventilation.[11] A combination of overcrowded conditions and a lack of health-care services often means that diseases that are easily prevented by vaccines, such as diphtheria, measles, tetanus, and pertussis (whooping cough), remain major causes of infant and child death.[12]

The failure of government to provide the basic investments in infrastructure and services themselves or the framework to encourage other sources of investment (individual, community-based, private sector) to do so and to ensure adequate supplies of land for housing means that Third World cities have become centres for competition for the most basic elements of life: for a room within reach of employment with an affordable rent or vacant land on which a shelter can be erected without fear of eviction; for places in schools; for medical treatment for health problems or injuries, or a bed in a hospital; for access to clean drinking water; for a place in a bus or train; for a corner on a sidewalk or square to sell some goods—quite apart from the enormous competition for jobs. In most instances, governments have the power and resources to increase the supply and reduce the cost of most of these elements.

This failure of government to ensure adequate provision of services has many other costs which cannot easily be measured. One is simply exhaustion, arising from the extra work burdens that these failures impose on poorer groups. These include the extra time needed going to and from work because of inefficient public transport and the fact that many low-income households can only find accommodation on the urban periphery. They include the time and effort needed by household members (usually women) to overcome the deficiencies in infrastructure and service provision—for instance the multiple journeys needed each day to and from the water standpipe, river, or well with heavy water containers and the time spent caring for sick or injured family members. There have been few studies on the economic impact of ill-health on poor urban households. One such study in a low-income settlement in Khulna, Bangladesh, revealed a very high economic burden because of labour days lost to ill-health or injury among principal income earners.[13] The poorest households lost most work days and much the highest proportion of monthly income; most were heavily in debt. The cost of medical treatment would have been less than the wages lost through illness or injury. An effective health-care system, perhaps combined with a community-based health-insurance scheme, can do much to lessen these kinds of burdens.

This combination of excessive work burdens, frequent ill-health, poor quality living environments, and economic insecurity contributes much to stress that in turn has major consequences for people's health. To these are added the insecurity or threat of eviction that tenants or squatters often face. There are obvious connections between these kinds of conditions and the high rates of domestic violence, suicide, drug and alcohol abuse, depression, and other psycho-social health problems now being documented in many Third World cities, even if the precise linkages between them are not well understood. But it is clear that good-quality housing and living environments with sufficient space; adequate provision for water, sanitation, and drainage; security of tenure; easy access to schools, health services, and other community and cultural facilities bring enormous and varied health benefits, including a diminution in many psycho-social health problems.[14]

Most of the diseases and injuries that are the immediate cause of ill-health and premature death among poorer groups could be eliminated or much reduced at low cost. Safe and sufficient supplies of water can be piped to the homes of poorer groups often with full cost recovery and with the beneficiaries being charged less than they previously paid to water vendors.[15] There are many examples of community-based credit schemes that have provided poorer groups with loans to improve the quality of their housing with repayment schedules which are both affordable and which generate full or close to full cost recovery.[16] Even relatively poor households can make a major financial contribution to the costs of improving water, sanitation, drainage, and garbage collection, if the most cost-effective means are used and the users can pay for the capital costs over a number of years.[17] A health-care system serving poorer groups that focuses on prevention entails relatively low costs, can usually achieve some cost recovery, and can often be financed by reallocating some of the funds that

governments currently spend on costly and ineffective centralized, curative-focused health systems.[18] The stress-related health problems faced by most poor households could also be much reduced with such measures and greater security. The financial cost of more effective government action is rarely the constraint—except in some of the poorest countries. A far more serious constraint is the institutional capacity at city and municipal level and the inappropriate nature of current responses to these problems.

Government's Failure in Housing

The gap between reality and conventional housing construction can be seen in the official statistics for the number of conventional houses constructed annually. In Third World nations, the number of conventional dwellings constructed annually is usually between 2 and 4 per 1,000 inhabitants while the population is expanding at between 20 and 35 and the urban population at between 25 and 60 persons per 1,000 inhabitants a year. The 'missing' housing units are those that are developed illegally or informally.[19] It is now common for 30 per cent of an entire city's population to live in houses and neighbourhoods that have been developed illegally and for most additions to the housing stock to also be developed illegally. The unnamed millions of people who build, organize, and plan illegally are in effect the most important organizers, builders, and planners of cities. In most cities, they have demonstrated remarkable ingenuity in developing their own homes and new residential areas. Despite limited resources and adverse conditions, low-income individuals, households, and communities have been responsible for a high proportion of all new housing units and for a large proportion of all investments in the housing stock and in housing-related infrastructure and services. The scale and nature of their investments is rarely recorded and is not reflected in official statistics. But in most cities or at least in the poorer areas of cities, the annual investment made by low-income households who are *de facto* owners of their homes is many times the average investment per household made by city and municipal authorities. This is especially so if a realistic monetary value is given to the labour time they put into improving or extending their homes.

The homes and settlements that poorer groups have developed often contain the seeds of a model for residential development and service delivery that is more appropriate to local climate, culture, and resource availability than the official model of urban development with most of its norms, codes, and preferences copied from the North.

But governments do not recognize these people as city builders. They usually refuse to recognize that they are citizens with legitimate rights and needs for public infrastructure and services. It is very rare for infrastructure and services essential to health and well-being—piped water, sewers (or other forms of sanitation), drains, all-weather roads, public transport, electricity, health care—to be supplied by government to the illegal settlements. When they are, the provision takes place years or even decades after the settlement first developed and usually only after the inhabitants have mounted a long and well-organized campaign for such provision. One hopeful development during the 1970s and 1980s was the increasing number of governments and international agencies that funded upgrading projects—for squatter settlements, tenement districts, or other settlements. However, these were usually a one-off capital investment where costs per household were minimized and where improvements in water supplies and sanitation were often insufficient to greatly improve health. In most instances, they were done with little consultation with the inhabitants and no provision was made to complement them with needed services (for instance health care) and to ensure that the new infrastructure installed was maintained. In most 'upgraded' areas, there are serious problems with maintenance within a few years.[20]

The illegal processes, by which most new city houses and neighbourhoods are developed, have been evident for decades and widely written about and discussed for over 30 years.[21] Yet few governments are prepared to acknowledge them. Most governments mix indifference with repression; some illegal settlements are tolerated as long as they do not take over prime sites, while those on land needed for other purposes are bulldozed, usually with little or no compensation for those living there. A few governments have become more tolerant and have tried to provide basic infrastructure and services. Very few governments

have taken action to ensure that poorer households can find legal alternatives to these settlements and have acknowledged that these people and their community organizations are the most potent force that Third World nations have for a more efficient, healthy, resource conserving, and democratic process of city construction.

Very few governments now have large public housing programmes, although many tried such programmes during the 1960s, 1970s, and early 1980s. Three problems dogged these programmes.[22] The first was that unit costs were high, so few units were built relative to need; indeed, in many instances, less than 20 per cent of the units originally planned were built. The second was that middle- and upper-income groups were often the main beneficiaries. The third was that even where low-income households were allocated public housing, they often informally rented the units to others because the size, design, and location ill-matched their needs.

One country, Singapore, managed to construct sufficient public housing to greatly improve housing and living conditions for most of its inhabitants. Leaving aside the social costs in the housing estates and the evictions that accompanied their construction, there are three characteristics of Singapore that make this example of little relevance to other nations in the South. The first is that Singapore has sustained one of the world's most rapid economic growth rates and yet has had very little in-migration: Singapore Island had virtually no rural population and immigration was strictly controlled. In any nation with a substantial rural population, a city that experienced comparable rates of economic growth would have been the centre for very rapid in-migration that would have swamped public housing programmes. The second reason is the fact that with rapid economic growth but relatively slow growth in the economically active population, most households experienced sustained increases in their real incomes and thus in the amount they could afford to spend on housing. The fact that they could afford to pay higher rents or, in many instances, purchase their unit made public housing less expensive for the government. The third reason is that in 1960, when the public housing programme greatly increased in scale, the government already owned large tracts of land amounting to around half the island's total area, including land close to the city centre. Thus, public housing could be built on well-located sites without incurring high land costs and without lengthy land expropriation procedures. The public housing programme was also supported by stronger land use control and public expropriation powers than in other Third World nations that have tried similar approaches.

Government's Failure in Environmental Policies

Government's failure can also be seen in the environmental problems evident in most urban centres.[23] Environment-related diseases or accidents remain among the major causes of illness, injury, and premature death; in many urban centres or in poorer districts within urban centres, they are the leading causes of death and illness. Most are caused by biological pathogens—in water, food, air, or soil. Some of the diseases they cause have been mentioned already: diarrhoeal diseases, acute respiratory infections, tuberculosis, malaria, dengue fever, and measles. Some, such as intestinal worms and filariasis, are not major causes of death, but tens of millions of urban dwellers suffer severe pain from them. A combination of safe, sufficient water supplies; provision for sanitation, drainage, and garbage collection; a minimum standard of space, security, and quality for housing; and a well-functioning education and preventive-focused health-care system brings dramatic reductions in their health impact.[24] Accidents are also a major cause of injury and premature death in most Third World cities, especially in illegal or informal settlements. Burns, scalds, and accidental fires are common in overcrowded shelters, especially where five or more persons live in each room and where there is little chance of providing occupants with protection from open fires, stoves or kerosene heaters or cookers. The risk of accidental fires is much increased in the many urban dwellings that are constructed with flammable materials (wood, cardboard, plastic, canvas, straw). The incidence and severity of accidents is also increased in the many large clusters of illegal or informal settlements which developed on dangerous sites—for instance hillsides subject to landslides, flood-plains, or sites polluted with industrial wastes.

The health impact of chemical pollutants is usually on a much smaller scale than biological pathogens and accidents. Still, in larger cities and industrial centres, rapid growth in industrial production and in motor-vehicle use, combined with government inaction or indifference to enforcing environmental legislation, often means a sizeable health burden for urban populations. It could be much reduced at relatively low cost. One of the principal health impacts is in the work-place as a result of occupational exposure to chemicals; few governments have an effective system for ensuring health and safety in work-places. In many large cities and industrial centres, air pollution levels for many pollutants such as oxides of nitrogen and sulphur, and for suspended particulates, greatly exceed the maximum level recommended by the World Health Organization. There are also serious problems of water pollution in virtually all Third World cities and serious problems with toxic and hazardous wastes in most: as governments have failed to enforce environmental legislation and to ensure that adequate provision is made to treat and safely dispose of dangerous wastes. Another major health impact is from smoke inhalation and its health consequences that affect tens of millions of urban dwellers (most of them low-income) who use coal, wood, or other biomass fuels for cooking and/or heating; the health impacts are greatest when these fuels are used in open fires or inefficient, poorly vented stoves.[25]

Most of these and other environmental problems in urban areas are more accurately characterized as political problems in that they do not arise from a shortage of an environmental resource such as land or fresh water but from the failure of government to fulfil its responsibilities. For instance, in most urban centres, it is not a shortage of land or fresh water which denies poorer groups access to land for housing and adequate provision for water, but economic and political factors. Most chemical pollution arises from the failure of government to implement regulations for pollution control and occupational health and safety.

Some Underlying Causes

Governments' failures or inadequacies usually go much deeper than this. In most urban centres, there is the failure to create the pre-conditions for democratic representation and to ensure all citizens have legal rights. Government rules and regulations in regard to building, planning, environmental protection, and employment ensure that even the most basic aspects of the lives of most low-income citizens—obtaining shelter, earning an income, obtaining food, water, and health services—are illegal. Because so many aspects of their lives are illegal, poorer groups are open to exploitation from landowners, businesses, and staff from public agencies, including the police and military personnel. As one lawyer has pointed out, the vast majority of city dwellers see the law as a tool that the wealthy and well-connected can use against them.[26] In addition, most governments' legal and regulatory systems repress the efforts of these same citizens to meet their own basic needs with their own resources and organizations. While the poorest groups are strongly penalized in their search for shelter and a source of income, major companies and institutions (both public and private) ignore or circumvent legislation on health and safety at work, on social security, on air pollution, and on the disposal of liquid and solid wastes.

This failure to develop the institutions of city governance may be the best explanation as to why cities in such diverse cultures come to present increasingly similar images. All nations are characterized by values, life-styles, customs, traditions, and institutions, in short, their culture or cultures. These are specific and unique to themselves. Culture implies knowledge, experiences accumulated and refined over long periods. People's culture has always been reflected in the houses, neighbourhoods, and settlements that they develop for themselves; one can learn much about a culture and its economy from its house designs, the materials used, and the ways that settlements are planned.[27] But among the illegal settlements, neither climate nor building materials, nor cultural and ecological differences are sufficient to allow the observer to distinguish one settlement from another. Only the historical centres and the old districts retain the characteristics that distinguished the Islamic city from the Hispano-American city, a Portuguese-American city from an Eastern city. Today, the fact that most inhabitants have inadequate incomes to meet their needs is the common denominator. It is seen in the urban landscape of recent neigh-bourhoods where much of the housing is illegally or informally developed and in the districts of low-income workers. In some illegal settlements—perhaps more so in the older, more consolidated ones—there are elements of design, site lay-outs, and use of materials that are more appropriate reflections of the nation's culture. But poverty combined with the indifference or repression of government does not allow this to develop and does not permit the full use of these people's knowledge and skills. Most governments have long ignored both history and culture as essential inputs into 'planning for development'. They see no legitimacy in the actions of their citizens who are building large areas of their cities. They cannot grasp that the house designs, the materials used, and the plans for these settlements are more realistic and often more appropriate than their own (often unfulfilled) plans for 'low-cost' housing. Meanwhile, in the residential districts of the rich, in the commercial and financial centres, in the architecture of government buildings and government housing projects, in the shopping centres and malls built to serve the middle- and upper-income areas, technological uniformity is the norm. International fashions for building design, whether in homes, shopping complexes, or offices, cross international borders rapidly, and can be seen almost as a denial of the culture and history of a city.

Governments rarely see the problems of building and managing cities in the same way as most of their citizens. The priorities of city and national governments are often almost entirely unrelated to their citizens' most pressing needs. City mayors and the vested interests who support them may want highways, metro systems, improved parking facilities, civic buildings, and water supply systems that start from the more central districts. But when the people living in the poorer neighbourhoods are questioned, their priorities are usually cheap and regular public transport, garbage collection, health centres and schools, protection against floods and, of course, water supply, drains, jobs, and the possibility of obtaining small credits. Big projects have little appeal for them. The city in which they live and work is unrelated to the city that those with power want to build.

Two parallel urban histories, closely interconnected but visually very different, are emerging. One is the official history, represented by the explicit concerns of government and major construction firms about the construction and management of the city. This official history is reflected in official statistics about the number of housing starts, the number of new factories or offices, or lengths of paved roads or water mains laid. The second, that of the low-income urban groups, has rarely been written. It is a fragmented and ill-recorded history, altogether different from the official history. It is the daily experience of millions of anonymous protagonists who must find immediate solutions to ensure their survival, with little scope for long-term perspectives. But this unofficial history is the more accurate history of city development in most of Africa, Asia, and Latin America in the last few years of the twentieth century.[28]

Cities grow and deteriorate while governments do little to control or respond to the social and environmental costs which are part of urban change. The historical moment in which Third World cities are now forming themselves is very critical. The transformation and 'globalization' of the world economy underlie an increasingly urbanized world population. But very few societies have managed to match the speed of economic and urban change with the needed institutional change. The approach of most governments to the problems that accompany this transformation is rooted in outdated institutional structures combined with inappropriate and often irrelevant foreign models and precedents. Third World city problems in the last decade of the twentieth century cannot be tackled with weak, ineffective, and unrepresentative forms of urban government. They cannot be tackled with a legal system and an institutional structure little changed from late-nineteenth or early-twentieth century precedents, most of which were imported from alien cultures and originally applied under very different economic circumstances and political structures. They cannot be tackled with models imported from the North. In the North, funding for housing construction and improvement and for the infrastructure and services associated with housing and residential areas is dependent on conditions that are not present in most Third World nations: well-developed markets for housing finance within prosperous and often growing economies, low inflation, well-established local governments with large local tax and revenue bases, and many city dwellers able to make—and sustain—large and regular payments for housing and basic services.[29]

The Future City

If there are no changes in the conventional responses of governments to urbanization, we can expect the urban future to bring tens of millions more households living in squatter settlements or in over-crowded rental accommodation of very poor quality. We can expect tens of millions more households who will be forcibly evicted from their homes as cities grow and as the urban authorities see the land occupied by illegal or informal settlements as the most 'convenient' source of land for public works and new industrial or commercial developments.

Prospects may be better for those nations which manage to sustain economic growth; a wealthier economy obviously brings greater possibilities for improving housing conditions and, in general, housing conditions are better, the higher a nation's per capita GDP.[30] But the prospects for poorer groups may not improve. The example of Seoul is an illustration of what can happen within a system of government that gives little consideration to the needs and rights of poorer groups. Despite the economic growth achieved in recent decades, housing and living conditions for low-income groups in Seoul remain poor and some five million people were forcibly evicted over the last three decades to accommodate the redevelopments which were part of South Korea's remarkable economic transformation.[31]

As a consequence of evictions and high land prices, increasing numbers of city dwellers are likely to live on sites ill-suited to housing. Cities will increasingly be made up of settlements built on dangerous sites, continuing the long-established trend of poorer groups illegally occupying or purchasing land and developing housing on steep hillsides, or around solid-waste dumps, beside open drains, or in areas subject to flooding, or in industrial areas subject to high levels of pollution.

In many urban centres, current levels of service provision are likely to deteriorate, especially where economies are stagnant. There is likely to be a rise in the number of diseases related to contaminated water, inadequate sanitation, solid-waste collection, and drainage, including those spread by animal and insect disease vectors which thrive in such conditions. Most agencies involved in running or maintaining infrastructure and services lack the resources to maintain them. Constrained national budgets imply a postponement of new investments in city infrastructure and no expansion in the provision of basic services. They allow for little or no investment in maintenance—and this is reflected in the deterioration of bus and train services, in roads and public buildings. A high proportion of Third World countries have also had structural adjustment programmes that have brought cuts to public services in urban areas and often considerably increased the number of urban dwellers living in poverty.[32]

If present trends continue, most cities are likely to become even more segregated: low-income individuals or households in the most overcrowded, worst served and often worst located sites, richer groups in the best located, highest quality, well-serviced sites. Cities' physical growth will be much influenced by where poorer groups can invade or negotiate for land. This will produce a pattern of urban development that makes the provision of basic services increasingly costly. One of the great advantages of a planned urban expansion is the large savings in the costs of providing infrastructure and services. New housing developments or commercial sites located to take advantage of existing roads, drains, water mains, and sewers ensure major cost savings. The uncontrolled process of urban expansion creates the opposite of this: large areas of land left undeveloped because the owners benefit from rising values (partly created by public investments) and a patchwork of illegal settlements wherever poorer groups find it possible to settle.

All these trends have impacts far into the future. A failure today to guarantee each city neighbourhood sufficient public space is likely to ensure that not only current but also future inhabitants of that neighbourhood lack space for recreation and leisure. Once an area is built up, it is very difficult to remedy deficiencies in public space. A failure to ensure that each neighbourhood has provision for the play, informal development, and learning needs of children of different ages, means not only much increased levels of accidents while they are children but

also a serious limitation in their development: safe and stimulating play environments have a central role in virtually all aspects of child development.[33] A failure to protect natural landscapes and historic buildings or sites means a permanent loss.

There are also issues concerning ecological sustainability. Without changes in direction, the larger cities in the Third World will have increasingly serious environmental impacts on their wider region and develop levels of resource use and waste generation that will prove increasingly unsustainable. A failure of the market to respond to increasing levels of waste and of governments to adopt or implement pollution controls and to install and operate adequate systems for sanitation, storm water drainage, and solid-waste collection will mean ever increasing pollution loads in local water bodies. A failure by public authorities to encourage water conservation and to instigate a realistic pricing structure will put increasing strains on water authorities' capacity to meet rising demands and to extend piped water to those inadequately served. Many large cities are already facing serious constraints on increasing water supplies and have long drawn on aquifers beyond their rate of natural recharge.

Among the richer cities, a failure to restrict the use of private automobiles and to provide alternative means of transport imply serious environmental impacts at several levels: from increasing air pollution and traffic congestion within the city to increased oil import bills for the national economy to increased contributions to global warming. City authorities with responsibilities for solid-waste collection and management will face increasingly severe cost problems, especially if they serve the poorer neighbourhoods scattered around the urban periphery. The larger the city, the more difficult it is to find a site for dumping solid wastes, and the more distant the site from the city, the greater the cost.

Cities ruled from above where the poor majority are segregated in unserviced squatter settlements or deteriorated and overcrowded rental accommodation, and where many of the actions that are part of their daily life are 'illegal', provide no basis for a stable society today or in the future. In many of the Latin American nations that over the last 15 years have returned to democratic forms of government, citizen expectations exceeded government capacity to deliver improvements, especially where the return to democracy coincided with economic stagnation. In addition, the habits of democracy and participation have to be relearned. Repairing and rebuilding the damage done by military dictatorships to democratic institutions and practices takes time. But the potential for change is enormous, if governments can build on and work with low-income groups and their organizations.

New Directions

If current trends in the deterioration of conditions are to be slowed, halted, or reversed, some basic questions have to be addressed. Perhaps the first is what measures should governments adopt to establish the pre-conditions for a more equitable, efficient, and citizen-directed urban development that also responds to the shortage of capital, the institutional limitations, and (for many nations) economic stagnation.

Reviewing the past 30 years, the evidence does not suggest a diminution in the role of government which for more than a decade has become the standard response. It suggests the need for a different role—more activist, more developmental, more decentralized, more representative, and more supportive of citizen efforts. One of the first steps for governments is to inform their citizens about their plans and the real possibilities for their implementation. This might appear politically dangerous, given the scarce public resources invested in cities, and the criticisms that such announcements might provoke. But more open and participatory forms of government are an essential part of addressing the fundamental problems of Third World cities. These require a frankness and honesty which have hardly characterized the actions of most governments. To elude the honest dissemination of such information is not only dangerous but it also reinforces governments' present isolation.

Third World cities have to be built with the resources available to each nation and its people. This means a recognition of the resources to hand and their best possible use. Inevitably, one turns to the knowledge, resources, and capacities of the tens of millions of people who are already the most active city builders—individual citizens and the organizations they form. At present, the

sum of their efforts is the major influence on how cities develop. But governments' failure to support and help co-ordinate such efforts represents an enormous and unnecessary loss both to themselves and to these citizens. Co-ordination on the location of new infrastructure and services, on their design, and on timing for their installation can save both governments and citizens large sums of money.

Perhaps the most poorly tapped potential source of revenue to finance such operations is the people and businesses who are the main beneficiaries of government investments, services, and subsidies. Those enterprises and households fortunate enough to have piped water supplies and services to remove household and human wastes should pay the real cost of providing and maintaining such services—and help provide the capital needed to extend such services to other households. There must be mechanisms to recapture the increments in land value that landowners currently receive as a result of government-funded infrastructure for reinvestment in public services. There must be mechanisms to recapture from businesses located in large cities the costs of the roads, transport services, water supplies, and waste services from which they benefit. The businesses that benefit from a labour force whose health and education is largely the result of government investments and services should contribute to the costs.

The work undertaken by informal community or neighbourhood organizations in providing basic services and site improvements for themselves is a rich though poorly documented source of examples from which governments can learn much. There are also dozens of positive examples of non-government organizations or groups of professionals who work with low-income groups or their community organizations—as in the case of FUNDASAL in El Salvador,[34] FUPROVI in Costa Rica,[35] SPARC in Bombay and their work with women pavement dwellers,[36] and the Orangi Pilot Project and their community-organized sewer construction and maintenance programme and other community-based programmes in Karachi.[37]

There are also the experiences of various governments—some positive, some negative—that have tried new approaches to city management, construction, or service provision—for instance the Million Houses Programme and other programmes associated with it in Sri Lanka,[38] the Presidential Commission on Urban Poverty and the Community Mortgage Programme in the Philippines,[39] the Community Development Department in Hyderabad (India),[40] the Mexican Government's national fund for popular housing (FONHAPO),[41] and the land-sharing projects and the Urban Community Development Office in Thailand.[42] There are also some examples of new coalitions in which different departments of government combine their limited resources with those of community-based organizations, NGOs, and international agencies to work together in improving housing, health, and environment in illegal settlements.[43] Certain city authorities have also pioneered new approaches as in Surabaya's *kampung* improvement programme,[44] and in a variety of environmental and poverty alleviation projects in Curitiba[45] and Cali.[46] While it would be a mistake to assume that project designs or technologies are transferable from one city to another, successful projects or programmes often contain within them principles and precedents that have relevance in other social and cultural settings.

The more innovative actions implemented in the last 20 years have had little impact in relation to the scale of the problem. Those who have benefited from such initiatives represent a tiny proportion of those in need—perhaps the equivalent of the population of a few mid-sized metropolitan areas. But these innovative actions do point to the need for new government approaches to bring out of illegality countless settlements, removing a major obstacle to co-operation between public authorities and civil societies. They give substance to the idea that there are more effective ways in which governments can work with low-income groups and their community organizations. The minimum conditions which governments must guarantee to citizens are coherent action and a connection between what they promise and what they carry out. At the same time, grass roots or neighbourhood groups want more open, wide-ranging participation in the development of their own housing and neighbourhoods without hindrance from government. This will allow governments to learn from the most active builders and designers of Third World cities.

The Role of Aid Agencies

An approach that makes best use of the knowledge, resources, and organizational capacities of citizens and their organizations implies a recognition that aid cannot provide the solution. It limits the role of national agencies as well. The solution is for governments within each city and city district to consolidate a process by which, year by year, the initiatives of citizens and community organizations are encouraged, supported, and built upon. City and municipal governments also have to remove the constraints on their actions—most often access to land in a suitable location and access to credit—and ensure that investments in basic infrastructure and services are made.

Outside agencies—whether national ministries or international agencies—so often misunderstand the nature of the problem and the range and quality of local resources on which solutions should draw. Their actions and decisions, and the forces which shape them, are generally far removed from those to whom they should be accountable—the low-income groups. This is especially so for multilateral development banks and bilateral government agencies where internal accounting procedures, staff constraints, and accountability to the governments who supervise them are a far greater influence on the form their development assistance takes than the needs and priorities of low-income groups.

The acquisition by city and municipal authorities of the capacity and power to meet their responsibilities, and to do so in ways that are accountable to their citizens, involves complex political changes. Powerful and well-organized vested interests oppose such changes—as was the case in Europe and North America as cities there began to grow very rapidly in the nineteenth or early twentieth century, also with very inadequate institutional structures. The political forces that push for such changes—which include a complex mix of citizen groups, community organizations (and their federations), NGOs, and other voluntary groups—are often opaque to outside agencies. Interventions by aid agencies and development banks who seek the efficient implementation of 'their' project may even inhibit this process. And international agencies can never stay in a city to guarantee the maintenance and expansion of the new water or sanitation or garbage collection system they helped to fund. Expanding the provision of infrastructure and services in ways that can be sustained requires effective local institutions. Outside agencies may bring support, but they cannot solve local problems unless they help develop effective local institutions. This cannot be done 'project by project'—but this is how many governments and most aid agencies operate. Successful projects usually owe their success to the fact that they removed the constraints on the actions and initiatives of citizens and their community organizations—but these constraints are only removed 'within the project'. The project did nothing outside project boundaries to remove the constraints on the vast majority of people who did not take part in the 'successful' project.

Aid is never likely to provide more than a small proportion of the capital needed for investment in delivering basic services. It cannot and should not take the lead role in making the institutional changes; indeed, some aid agencies have promoted new institutional 'solutions' which fail because they are so inappropriate to local culture and institutional arrangements. Even if aid agencies multiplied several-fold the volume of aid going to basic service provision, it would still only provide a small proportion of the needed investments. At best, aid can provide some help in funding basic needs projects. It may help develop new models to support partnerships between low-income groups and their organizations and local NGOs and municipal authorities, several private voluntary agencies such as MISEREOR already do so. But development banks and official bilateral agencies have a more important role in encouraging and supporting national governments in their task of building the capacity and competence of urban government, and of national agencies which work with urban government, to make basic investments, increase locally generated revenue, and manage and maintain urban areas. Aid agencies have some role in helping to start and consolidate the process of building the capacity of local governments to respond to citizens' needs and work with them to ensure these are met. A handful of governments and aid agencies have recognized the necessity for such changes, but these are the exceptions rather than the rule.

The search for new approaches encouraged by the economic crisis of the 1980s and early 1990s

might be turned to positive use. For example, governments' recognition of their impotence in dealing with the causes and effects of rapid urban growth suggests the possibility of an end to 'big' government, at least in relation to the construction and management of cities. This could be replaced instead by the permanent participation of neighbourhood associations in local government, thus respecting the role the people want and can take in development. It can also mean less emphasis on big urban projects and more resources for more widely distributed, smaller kinds of projects and programmes. Of course, there are large projects that are needed and which involve major government intervention, such as the control of floods and other measures to improve the environmental quality of already occupied sites, or the preparation of sites for new settlements. But multilateral and bilateral aid agencies who insist on 'big' projects, simply because their progress is easier to oversee and the loan or grant application is easier and quicker per dollar to process, are misallocating scarce resources. More representative and decentralized forms of government provide the appropriate context for decisions about the use of scarce resources. In this way, the discussion whether to favour big projects or community-based projects, whether it is convenient to reduce density or increase density in metropolitan areas, and many other aspects related to the construction and management of cities, will acquire a measure of reality and connection to everyday urban life.

New Interpretations

It constantly strikes us how narrowly the problems of Third World cities are presented. Far more material about Third World cities and their problems is written and published by people in Europe and North America than by people in the Third World. Much of this generalizes on what cannot be generalized. It reduces all Third World cities and the cultures of which they are part to 'rapidly growing cities of slums'. Researchers in the Third World are offering fresh interpretations of the city and its problems, aided by the experience of the individuals and community organizations with whom they work. These help locate city problems within a real understanding of local economies and power structures.

Many governments claim to be ignorant about the magnitude of the crisis in their citizens' habitat and are reluctant to admit that there are alternative ways to approach and to mitigate the social impact of this crisis. Degraded human environments will exist as long as there is poverty. And poverty is rooted in the way that societies are organized and wealth is distributed, both within nations and internationally. Poverty cannot be eliminated without a redistribution of power and resources. It cannot be eliminated through international aid in its current form since much of it is still oriented to supporting the survival of friendly governments in countries of strategic interest to the donor or in promoting the interests of enterprises from the donor country. Even a substantial increase in such aid will not address the problems of poverty because it does not address the redistribution of power and resources. Like many world problems that are the result of an uneven distribution of power and wealth, the solution to poverty has to be accepted as a collective responsibility. Just as cheap or free health care and a minimum income for the unemployed is accepted as a collective responsibility in many nations in the North, so too must the world community accept a collective responsibility for the world's poor.

In discussions about Third World cities, too little attention is given to the potential contributions of free people and democratic institutions. The improvement of human habitats requires the involvement of the 'users' of those habitats. One difficulty is that even while community movements are spreading in so many squatter settlements and illegal urban developments, their organization requires time. Decades of government repression or indifference have hindered their development in many countries. And new illegal settlements are established at a pace beyond the capacity of existing groups to train themselves and to develop their capacity to receive and work with professional assistance. Much as we dislike throwing on the shoulders of the low-income groups, the unskilled, and the poorly fed, the responsibility for building their own habitats, there is little alternative unless there is a major change towards a less unequal income distribution, both nationally and globally.

The catastrophes predicted in some world models developed in the 1970s, and the threats posed by climatic change noted more recently, may not impinge much, as yet, on the daily lives of those in the North. But another catastrophe is a daily reality for a large part of the world's population in the thousands of easily preventable deaths that occur each day in urban centres in the Third World and the tens of thousands of easily preventable illnesses and injuries. If city construction and the question of who receives adequate income, housing, and access to basic services is to become oriented to social need, a sharp reversal of present trends needs to be accomplished.

Within Third World nations, much can be achieved in terms of improved urban living conditions if governments no longer chain and repress but rather support individuals, households, and communities building or extending their homes and creating a living for themselves. These activities of people who work with small amounts of capital, individually and collectively, have to be supported and co-ordinated to provide certain services, to mobilize production, and to improve human habitats. Of course, this will not solve more fundamental problems. Of course, this demands complementary safeguards to protect individuals from exploitation by employers, landlords, and landowners. But supporting the efforts of the true builders of their cities, adapted to each culture and society, will help to initiate more effective and appropriate responses to contemporary problems in Third World cities.

Notes

1. Only the Third World is considered in this essay—so no consideration is given to Japan.
2. We arrived at this estimate, working with Sandy Cairncross of the London School of Hygiene and Tropical Medicine, editing the book *The Poor Die Young: Housing and Health in Third World Cities* (London: Earthscan Publications, 1990). The estimate drew on a large number of city and national studies which consider the proportion of the population living in inadequate quality accommodation and lacking basic services. It is no more than a rough estimate, but we made this in response to other agencies' estimates for the number of people living in poverty in urban areas of the Third

World or lacking basic services which suggest that a relatively small proportion of the Third World's urban population live in poverty. For instance, official WHO figures for the proportion of the Third World's urban population lacking potable water are clearly far too low; this is a result of faulty criteria applied to 'what is an adequate supply' and exaggerated government figures given to WHO (chs. 1 and 5 of *The Poor Die Young* go into more detail on these issues).

3. Ch. 3 of Jorge E. Hardoy and David Satterthwaite, *Squatter Citizen: Life in the Urban Third World* (London: Earthscan Publications, 1989) presents summaries of many such case studies.

4. Sandra Cointreau, *Environmental Management of Urban Solid Waste in Developing Countries*, Urban Development Technical Paper, No. 5 (Washington, DC: World Bank, 1982).

5. WHO, 'Environmental Sanitation Trend Analysis in the African Region', *Environmental Health*, 20 (Oct. 1993), 7–9.

6. For reviews of the many case studies, see David Bradley, Carolyn Stephens, Sandy Cairncross, and Trudy Harpham, *A Review of Environmental Health Impacts in Developing Country Cities*, Urban Management Program Discussion Paper, No. 6 (Washington, DC: World Bank, UNDP, and UNCHS (Habitat), 1991). See also Jorge E. Hardoy, Diana Mitlin, and David Satterthwaite, *Environmental Problems in Third World Cities* (London: Earthscan Publications, 1992).

7. Trudy Harpham, Tim Lusty, and Patrick Vaughan (eds.), *In the Shadow of the City: Community Health and the Urban Poor* (Oxford: Oxford University Press, 1988). See also Bradley *et al.*, *Review of Environmental Health Impacts*, and Hardoy, Mitlin, and Satterthwaite, *Environmental Problems*. For recent case studies of intra-urban differentials, see Jacob Songsore and Gordon McGranahan, 'Environment, Wealth and Health: Towards an Analysis of Intra-Urban Differentials within Greater Accra Metropolitan Area, Ghana', *Environment and Urbanization*, 5: 2 (Oct. 1993), 10–24; and Charles Surjadi, 'Respiratory Diseases of Mothers and Children and Environmental Factors among Households in Jakarta', *Environment and Urbanization*, 5: 2 (Oct. 1993), 78–86.

8. David Satterthwaite, 'The Impact on Health of Urban Environments', *Environment and Urbanization*, 5: 2 (Oct. 1993), 87–111. See also Samir S. Basta, 'Nutrition and Health in Low Income Urban Areas of the Third World', *Ecology of Food and Nutrition*, 6 (1977), 113–24.

9. A. Pio, 'Acute Respiratory Infections in Children in Developing Countries: An International Point of View', *Pediatric Infectious Disease Journal*, 5: 2 (1986), 179–83.

10. WHO, *Our Planet, Our Health*, Report of the WHO Commission on Health and Environment (Geneva: World Health Organization, 1992).

11. G. M. Cauthen, A. Pio, and H. G. ten Dam, *Annual Risk of Tuberculosis Infection* (Geneva: World Health Organization, 1988).

12. At least there has been considerable progress in recent years in increasing the proportion of children immunized against these diseases, with immunization programmes promoted and supported by UNICEF and by the World Health Organization.

13. Jane Pryer, 'The Impact of Adult Ill-Health on Household Income and Nutrition in Khulna, Bangladesh', *Environment and Urbanization*, 5: 2 (Oct. 1993), 35–49.

14. Solvig Ekblad, 'Stressful Environments and their Effects on Quality of Life in Third World Cities', *Environment and Urbanization*, 5: 2 (Oct. 1993), 125–34. See also two other articles in this issue of *Environment and Urbanization*: Larry Cohen and Susan Swift, 'A Public Health Approach to the Violence Epidemic in the United States', 50–66; and Leonard J. Duhl, 'Diversity, Game Boards and Social Entrepreneurs', 112–24.

15. Sandy Cairncross, 'Water Supply and the Urban Poor', in Hardoy, Cairncross, and Satterthwaite (eds.), *The Poor Die Young*, 109–26; World Bank, *World Development Report 1988* (Oxford: Oxford University Press, 1988).

16. Silvina Arrossi, Felix Bombarolo, Jorge E. Hardoy, Diana Mitlin, Luis Pérez Coscio, and David Satterthwaite, *Funding Community Initiatives* (London: Earthscan Publications, 1994).

17. See Cairncross, 'Water Supply and the Urban Poor', in Hardoy, Cairncross, and Satterthwaite (eds.), *The Poor Die Young*, for a discussion of how to keep monthly payments low while also fully recapturing costs in regard to water supply.

18. World Bank, *World Development Report 1993* (Oxford: Oxford University Press, 1993).

19. We deliberately use this imprecise term 'informal and illegal settlements' in that there is such diversity in the legal status of land occupation and its development for housing in low-income settlements, both within and between cities. At one end of the spectrum, there are squatter settlements where the land occupation is illegal, so any structure on it and any infrastructure or services are also illegal. At the other end of the spectrum are sites on which housing is slowly being developed— much of it of poor quality—where both the occupation of the land and its use for housing developments are legal. Illegal subdivisions fall

between these two extremes—where the land occupation is legal in that it is with the permission of the owner to whom payment has been made, but its use for housing has not been approved—and building, subdivision, and infrastructure and service standards are also usually at odds with official norms.

20. One exception to this is described in Johan Silas, 'Environmental Management in Surabaya's Kampungs', *Environment and Urbanization*, 4: 2 (Oct. 1992), 33–41, where the inhabitants in the areas to be 'upgraded' were involved in decisions about the upgrading and contributed their resources and ensured much better maintenance after the upgrading was completed.

21. See Kosta Mathéy's contribution in this volume.

22. The quantitative and qualitative failures of public housing programmes are discussed in more detail in ch. 4 of Hardoy and Satterthwaite, *Squatter Citizen*.

23. This section draws from Hardoy, Mitlin, and Satterthwaite, *Environmental Problems*, which reviewed environmental problems in Third World cities.

24. WHO, *Our Planet, Our Health*.

25. Ibid.

26. Patrick McAuslan, 'Legislation, Regulation and Shelter', *Cities*, 4: 1 (1987), 23–7.

27. Paul Oliver, *Dwellings: The House Across the World* (Austin: University of Texas Press, 1987).

28. Examples of papers that include some of this informal history of low-income settlements: Beatriz Cuenya, Hector Almada, Diego Armus, Julia Castells, Maria di Loreto, and Susana Penalva, 'Housing and Health Problems in Buenos Aires—the Case of Barrio San Martin', in Cairncross, Hardoy, and Satterthwaite (eds.), *The Poor Die Young*, 25–55; Lisa Peattie, 'Participation: A Case Study of How Invaders Organize, Negotiate and Interact with Government in Lima, Peru', *Environment and Urbanization*, 2: 1 (Apr. 1990), 19–30; Alexandrina Sobreira de Moura, 'Brasilia Teimosa—the Organization of a Low Income Settlement in Recife, Brazil', *Development Dialogue*, 1987, 152–69.

29. Arrossi *et al.*, *Funding Community Initiatives*.

30. Stephen K. Mayo and Shlomo Angel, *Housing: Enabling Markets to Work*, A World Bank Policy Paper (Washington, DC: World Bank, 1993).

31. Asian Coalition for Housing Rights, 'Evictions in Seoul, South Korea', *Environment and Urbanization*, 1: 1 (Apr. 1989), 89–94.

32. See, for instance, Augustín Escobar Latapí and Mercedes González de la Rocha, 'Crisis, Restructuring and Urban Poverty in Mexico', and Nazneen Kanji, 'Gender, Poverty and Structural Adjustment in Harare, Zimbabwe', *Environment and Urbanization*, 7: 1 (Apr. 1995). In some instances, the negative impact on poorer groups has been lessened by well-targeted compensatory measures, often implemented through emergency or social funds; see Caroline O. N. Moser, Alicia J. Herbert, and Roza E. Makonnen, *Urban Poverty in the Context of Structural Adjustment: Recent Evidence and Policy Responses*, TWU Discussion Paper, DP No. 4 (Washington DC: Urban Development Division, World Bank, May 1993). In addition, a considerable proportion of the urban poor in most countries benefited little from public services prior to structural adjustment and hence were also less affected by their cuts. If structural adjustment does succeed in reviving the economy, it may lead to greater economic opportunity for a proportion of poorer groups— although we have grave doubts whether structural adjustment will succeed in doing so for most of the poorer Third World nations.

33. Bob Hughes, 'Children's Play—a Forgotten Right', *Environment and Urbanization*, 2: 2 (Oct. 1990), 58–64; and Robert Myers, *The Twelve Who Survive: Strengthening Programmes of Early Child Development in the Third World* (London and New York: Routledge, 1991).

34. FUNDASAL, 'Proyecto Santa Teresa: Una experiencia de dotacion de viviendas a sectores de bajos ingresos economicos', mimeograph, Ciudad Delgado, El Salvador; Alfredo Stein, 'Critical Issues in Community Participation in Self-Help Housing Programmes: The Experience of FUNDASAL', *Community Development Journal*, 25: 1 (1989).

35. Manuel Sevilla, 'New Approaches for Aid Agencies: FUPROVI's Community Based Shelter Programme', *Environment and Urbanization*, 5: 1 (Apr. 1993), 111–21.

36. SPARC, 'NGO Profile: SPARC—Developing New NGO lines', *Environment and Urbanization*, 2: 1 (Apr. 1990), 91–104; E. J. Anzorena, 'SPARC—Society for Promotion of Area Resource Centres', *SELAVIP Newsletter* (Latin American and Asian Low Income, Housing Service), Mar. 1988.

37. Arif Hasan, 'Orangi Pilot Project: A Low Cost Sewer System by Low Income Pakistanis', in Bertha Turner (ed.), *Building Community: A Third World Case Book* (London: Habitat International Coalition, 1988, 81–8); and Arif Hasan, *Scaling-up of the Orangi Pilot Project's Low Cost Sanitation Programme* (Karachi: Orangi Pilot Project-Research and Training Institute, 1993).

38. Sunimal Fernando, Willie Gamage, and Dharmawansa Peiris, *Navagampura and Aramaya Place—Two Urban Case Studies on Support Based*

Housing (Colombo: National Housing Development Authority, 1987); Susil Sirivardana, 'Reflections on the Implementation of the Million Houses Programme', *Habitat International*, 10: 3 (1986), 91–108.

39. Jorge Anzorena, assorted reports in *SELAVIP Newsletter 1990–92*; and Albert Ramon, 'The Community Mortgage Programme', in Aurelio Menendez (ed.), *Access to Basic Infrastructure by the Urban Poor*, EDI Seminar Report, No. 28 (Washington, DC: World Bank, 1991), 37–8.

40. E. J. Anzorena, 'The Incremental Development Scheme of Hyderabad', *SELAVIP Newsletter* (Latin American and Asian Low Income, Housing Service), Japan, Mar.1988.

41. Enriqué Ortiz, 'FONHAPO: Mexico's National Trust Fund for Popular Housing', in Menendez (ed.), *Access to Basic Infrastructure*, 32–3.

42. ACHR, 'NGO Profile: The Asian Coalition for Housing Rights', *Environment and Urbanization*, 5: 2 (Oct. 1993), 153–65; and Shlomo Angel and Somsook Boonyabancha, 'Land Sharing as an Alternative to Eviction: The Bangkok Experience', *Third World Planning Review*, 10: 2 (May 1988), 107–27.

43. Lair Espinosa and Oscar A. López Rivera, 'UNICEF's Urban Basic Services Programme in Illegal Settlements in Guatemala City', *Environment and Urbanization*, 6: 2 (Oct. 1994), 9–29.

44. Silas, 'Environmental Management in Surabaya's Kampungs'.

45. Jonas Rabinovitch, 'Curitiba: Towards Sustainable Urban Development', *Environment and Urbanization*, 4: 2 (Oct. 1992), 62–73; and Jonas Rabinovitch, with Josef Leitmann, *Environmental Innovation and Management in Curitiba*, Brazil, UMP Working Paper, Series 1 (Washington, DC: Urban Management Programme, UNDP/UNCHS (Habitat)/World Bank, 1993).

46. Luis Fernando Cruz, 'NGO Profile: The Carvajal Foundation', *Environment and Urbanization*, 6: 2 (Oct. 1994), 175–82.

18

Self-Help Approaches to the Provision of Housing: The Long Debate and a Few Lessons

Kosta Mathéy

'SELF-HELP policies are very appealing to a wide variety of people. . . . The only problem is: self-help policies aren't likely to work', writes Peter Marcuse (1990: 10), and Tony Schuman (1986: 470) expands, 'It is precisely because self-help does not "work" as a solution that it has potential'. The assessments of the value of this ever more popular housing strategy stretch from one extreme to the other. Here I will highlight some important positions related to the academic debate on self-help housing—and to experience in the field.[1]

'Non-Conventional' Housing Production

'Much ado about nothing' could be a cynical comment on the World Bank's and other donors' recommendations to turn to self-help housing strategies in an attempt to counter the ever-worsening housing problem. The advice sounds like reinventing the wheel, while the implementation of such policies has not stopped the housing shortage from becoming more severe. The criticism against self-help policies is thus understandable. But the theoretical debate around these policies has not improved the situation either, although it kept experts busy talking and writing

for over three decades by now. But if the concept was useless anyway, nobody would care after all this time: surely self-help housing must offer benefits to some people.

One of the problems of assessing the potentials of self-help housing lies in the different meaning that people put to it—and to related concepts such as self-built, self-provision, self-promotion, illegal, and informal housing (Duncan and Rowe, 1993: 1332). I will distinguish between *autonomous* solutions and *assisted* self-help. Autonomous solutions are also often referred to as spontaneous self-help, which means that the development was not planned by official planners—although it usually is quite well prepared by the users themselves. Historically, spontaneous self-help represents the oldest form of housing provision of all, and it has worked for better or worse for thousands of years. With such a long experience there is little reason for sudden change and innovation. The only aspect that has changed over the last two or three decades is the increase in absolute poverty in the Third World, and this is reflected in less resources available for housing as well as for other basic needs of the poor.[2] Because of its typically static nature, I will not further dwell on the autonomous mode.

Assisted self-help housing, on the other hand, is characterized by the intervention of the state, or other agencies, with the aim of overcoming certain recurrent shortcomings in autonomous building activity. It has been advocated as preferable to other previously introduced, hence often named 'conventional', housing strategies. These former policies include imitations of traditional European systems of public housing. Apart from accommodation provided for public employees, housing was allocated on the basis of various criteria of 'need'. Occasionally squatters from informal settlements demolished by the authorities were rehoused. However, given increasing demand and the typically restricted budgets of public authorities in the Third World, finance could not be assured for long, particularly if the intention was to reach people who would otherwise resort to self-help solutions. Possibilities for reducing the rather high cost of this type of housing, i.e. turning to so-called 'low-cost schemes' with much lower standards, were limited. In addition, their low densities precluded the use of centrally located land and resulted in distant locations on the urban periphery. This in turn implied serious transport problems, with commuting times reaching several hours, and often resulted in a rejection of the settlements by the target population.

In order to reach a larger number of beneficiaries with limited budgets, 'non-conventional' housing policies were introduced. There are many variations in the interpretation of the concept, but the most typical solutions include sites-and-services schemes, core housing, and slum- or squatter-upgrading projects. They all have in common the participation of the future residents in the production process. Traditionally, participation was understood as a purely physical labour contribution, but the importance of administrative tasks has been recognized increasingly. The roots of assisted self-help housing can be traced to Europe and North America, where it was present particularly in periods of economic slump (Harms, 1982; Steinberg, 1993).

The (Conventional) Self-Help Debate

Assisted self-help housing strategies, as an alternative to conventional mass housing schemes, were the leading concept behind the resolutions passed in the Vancouver Habitat Conference the United Nations organized in 1976. Their point of reference was a number of progressive publications on Third World housing at the time, which suggested that slum dwellers and squatters be accepted as valued members of the community and as potent partners in the development of the city (Abrams, 1966). Particularly influential in outlining the new policies were planners like Otto Koenigsberger (1970) and William Mangin (1967), who rejected the common practice of pulling down substandard and illegal settlements (following the European 'slum clearance' model) with the unrealistic idea of providing regular housing to the affected elsewhere. Instead they emphasized the need to guide and support self-build activities by the poor as a first step towards incrementally improving housing conditions.

The proposition was revolutionary at the time and was backed by a theoretical framework generally associated with John Turner who had elaborated it in a series of publications (1963, 1972, 1976, 1978, 1988). In line with these, the main arguments in favour of self-help housing policies can be summarized as:

- Self-help housing production is assumed to be much cheaper than state- or market-provided housing because the residents are able to make the best use of local resources, can save the profits otherwise pocketed by builders and suppliers, and are much more flexible than state agencies.
- Officially defined standards—often inherited from the European colonizers—have little meaning in self-provided housing (*popular housing*) in the South. Uniform standards (*official housing*) can never match the many individual needs of millions of users. The users can assess their changing requirements for space much better than any bureaucrat. Moreover, self-builders are familiar with the site and can better adapt the building to the specific topographic and climatic conditions of the location.
- The architectural quality of a self-built house is considerably better (no monotony, personalized design) than that of official housing because the construction and design is determined by the aspired use value, and not by the exchange value as in the case of commercial builders who only have an interest in maximizing profits. Apart from appearing aesthetically more pleasant to by-passers and neighbours than monotonous mass housing to others, individual solutions can be a means to self-fulfillment and personal pride (Mühlich, 1984; Hall, 1989).

• Marginal population groups are integrated through self-help housing activity. A typical personal history pattern would be a migrant job-seeker arriving from the countryside, identified as *bridgeheader*, lodging with relatives or renting a room in a centrally located tenement for a while. Once he has obtained a secure income and perhaps married, he would become a *consolidator*. Then he will build a shack of his own, extend and improve it over time according to the growth of his income and his family. Learning to build a house can be a qualification on its own and provide a stable source of income.

These propositions not always met full agreement by architects, planners, and researchers from the Third World, particularly from Latin America. Apart from criticizing the references to unrealistic stereotypes like stable marriages or the search for self-fulfillment when building a shack, some experts involved in the implementation of self-help housing projects, such as Alfredo Rodriguez (1972) and Diego Robles Rívas (1972) repeatedly pointed to the risk of this kind of project being exploited by repressive governments as an instrument of political pacification and co-optation of grass-roots opposition. Such strategies had been all too familiar in the region since the U.S.-promoted Alliance for Progress of the 1960s which had already included self-built housing projects, and from similar but more recent projects in Mexico (Eckstein, 1977). On the other hand, the assisted self-help approach to housing also offered the potential for political mobilization resulting from collective experiences and from group discussions (Castells, 1973). Overall, the empirical evidence suggests, however, that pacification is the more likely outcome (Vaessen, 1989). Other experts criticized Turner's recommendations on the grounds that they provided a pretext for neo-liberal governments to deny or to reduce their political responsibility for shelter provision—listed as a basic right in many countries' constitution.

While these criticisms were mostly concerned with isolated aspects of the promotion of self-help policies, a coherent Marxist theory about such programmes was offered by Emilio Pradilla Cobos (1983). He explained how these programmes principally served the ruling classes' interests in capital accumulation and appropria-

tion, namely through the effects of double exploitation (prolongation of the working day through self-building activities in the evenings and during the weekends), disciplining of the labour-force (through credit and work-time commitments), and, above all, through the general process of commodification. The latter hypothesis was further elaborated by Jorge Fiori and Ronaldo Ramirez (1992), who pointed at the paradox of the promotion of self-help housing, being a subsistence form of production, by the capitalist state. After all, only the historically advanced division of labour and the manufactured form of production, with its possible economies of scale, eventually allow the accumulation of surplus value; and wage-labour and centralization, i.e. the industrial form of production, allow the appropriation of this surplus value by capitalists. Reversing this process in the development of productive forces by turning back to the subsistence form of production runs counter to the historical process.

The logic behind such an apparently contradictory move can be found in the constant need of capitalist economies to expand. The poor with their limited purchasing capacity are still largely excluded from the commodity market. But 'unconventional' self-help housing projects offer a chance for the market to expand. This happens in various ways.

• In spite of the name, most assisted self-help schemes include a considerable element of wage-labour and replace unwaged family labour or mutual aid in autonomous development.

• Land informally or illegally occupied will be made marketable through legalization schemes (the issue of land titles is a principal element in slum- and squatter-upgrading schemes, where it is expected to mobilize untapped private investment).

• Informal and traditional savings arrangements will be replaced through commercial bank credits involving a profit element.

• Free local or informally exchanged building materials (i.e. mud, garbage, cardboard, empty buildings) are replaced by industrially produced materials which are the only ones to comply with official building norms and loan requirements (and provide a safeguard for marketability in case of default on the loan).

• Legal security, higher standards, and better

services increase the market value of the affected property and fuel speculation.

- Additional, previously absent monetary expenses in the form of professional fees, administration costs, charges for infrastructure provision, permits, etc. will become mandatory. Part of the value is being reinvested by the state or semi-public enterprises for 'collective consumption' (Castells, 1977) in a manner particularly convenient for capitalist enterprises (on which the state is dependent)—but not necessarily for the local community.

It was the merit of Rod Burgess (1977, 1978) to systematically contrast John Turner's propositions with Marxist criticism of self-help housing strategies. He specifically pointed at two shortcomings in Turner's writings. First, Turner's individualistic view of self-help processes ignores the socio-political context in which they take place. The innocently praised 'freedom to build' (Fichter and Turner, 1972) is in fact very restricted in practice, and conditioned by the conflicting interests of owners and tenants, builders, financing agencies, politicians, and other social groups with their own vested interests. Self-help housing, whether assisted or not, is never really autonomous but must accommodate itself in the spaces left by others.

Secondly, the universally dominant social and economic relations tend to reproduce themselves with all their ideologies and limitations, in the self-help sector. Capitalist relationships, including exploitation, are not restricted to the formal sector, but also affect the relationships between neighbours, informal builders, private bankers, shopkeepers, etc. The proposed vision of a liberated Third Sector that is distinct from state and capital is but a myth (see also Schiffler, 1991).

Clearly, Turner's (1976: 28–9) polarization between 'popular' and 'official' housing sectors cannot be maintained. There exist, of course, qualitative shortcomings in the—mostly state-provided—mass housing schemes. But the reason for this phenomenon is not primarily institutional: the same institutions do produce high-quality housing for other clients, i.e. civil servants. In addition, Turner's prognosis of a residential trajectory from *bridgeheader* to *consolidator* is too simplistic and appears to be based on non-representative case studies: it does not reflect the social heterogeneity to be found in most self-built settlements (van Lindert, 1992).

While the discussion about self-help housing policies became polarized around his writings, John Turner has readjusted his position over time. In recent years he has emphasized decentralized decision-making and local control over resources (Turner, 1988)—rather than auto-construction. This newer interpretation of self-help became also known under the term of 'devolution' (Wakely, 1989). Non-governmental organizations (NGOs) and citizen-based organizations (CBOs), possibly supported by architect-'enablers', were to play a major role in this approach, and both NGOs and CBOs were seen as trend-setters in international co-operation. This position can also be attacked for being uncritical of the inherent weaknesses and potential abuse of such approaches. The meaning of Turner's concept of 'enabling approach' has in fact already been twisted by some World Bank experts who define it as 'policies that enable markets to work' (World Bank, 1993). But irrespective of the stringency of his propositions, Turner's authority significantly contributed to stopping the previous practice of demolitions of informal settlements around the globe.

Field Implementation and Experience: Reaching the Target Group

Apart from the initial impacts, the subsequent academic discussion on self-help housing, which often appeared rather ideological, had little practical influence on the formulation and implementation of housing projects sponsored by governments and international aid. The abstract but popular ideas of John Turner were, though unproven, taken up by political decision-makers to justify all sorts of programmes and projects. The declared principles were hardly adhered to on the project level anyway. Nobody objected to the social integration and the more beautiful built environment promised by Turner, but the politicians' basic interest was to reach a maximum number of beneficiaries, i.e. voters—and to avoid being held responsible for the miserable housing conditions of those left behind, i.e. the side-effect of devolution.

As in the case of the previous conventional housing programmes, assisted self-help made little contribution to solving the housing problem of the poor. The available public funds could only finance a rather small number of pilot projects, and replicability was never achieved. Massive and nation-wide programmes designed to address the immense and increasing demand effectively would require increasing subsidies to levels governments were unable or unwilling to meet. The alternative was to distribute the available subsidies among a greater number of people or, as the World Bank (1993) finally suggested, to do without subsidies altogether.

If public expenditures, i.e. subsidies, in unconventional housing programmes are to be reduced, someone else must pay the bill—the focus shifts to (full) cost recovery. This simple conclusion clashes with the fundamental problem of the target group's limited ability to pay. Even in the days of substantial subsidies, so-called low-income housing projects tended to be too expensive for the poor. The problem was usually resolved by simply redefining the target group. The housing was supposed to reach the poorer eventually through a filtering-down effect. In practice, such an effect was rarely observed because the housing demand of the middle classes was also increasing rapidly. If the urban poor are to be reached, the gap between housing cost and affordability must be narrowed.

Affordability and Standards

The principal reason for the housing problem under capitalism is that large sectors of the population cannot afford adequate housing. To solve the problem, either the incomes of those in need must increase, or the housing must become cheaper. Within the sectorally defined terms of reference of a housing policy there is only limited scope for the first option. In respect to cost-saving, self-help projects promised potential savings in at least three different instances:[3] the labour cost element, incremental completion, and avoiding speculation and development gains by third parties.

The most obvious possibility to reduce financial outlays in non-conventional housing schemes is self-help labour input—sometimes referred to as sweat equity. Provided that the basic subsistence expenses of the family are secured, and that the potential self-builder has no other reasonably

well-renumerated work, house ownership may thus just become affordable for many people. Indeed, 'unpaid' self-help housing can be interpreted as a job creation programme where the wage is paid in kind—a proportion of housing costs. Possible savings were found to range up to 50 per cent (Fichter and Turner, 1972: 158; Turner, 1990: 32; Duncan and Rowe, 1993: 1337). However, in most self-built schemes the participants prefer to pay someone more experienced (or poorer) to do the job, and the supervising agencies also prefer professional builders to avoid long and costly delays in completion (Abrams, 1966: 170 ff.).

The second possibility to reduce subsidies lies in further reducing the standards, possibly imbedded in the concept of progressivity: to match his or her ability to pay, the user will only obtain an unfinished house initially, neutrally labelled 'housing solution'. As circumstances allow, he or she can complete the 'core house' over time. The greater number of applicants simultaneously served in this manner provide impressive statistics, but the physical output may be less convincing. In extreme, but not unusual, cases the starting solution may consist of a sanitary unit plus a kitchen only, or even less (sanitary unit, or 'roof plan'). In upgrading schemes, which typically concentrate on improving infrastructure, the affordable improvement may be extremely modest: all that could be afforded in a particular Kampung Improvement Project in Indonesia were the provision of four toilets and nine water tabs for 12,000 residents (Ludwig and Cheema, 1987: 207).

A third opportunity to cut costs lies in the elimination of speculation and development gains of intermediaries. This option is particularly important in countries with an advanced and centralized construction industry (Duncan and Rowe, 1993)—but is less relevant in low-income housing in the South. Land speculation can also be avoided through a national land policy of land banking[4] which need not be linked to self-help approaches.

Thus the attempt to achieve replicability called for: cost recovery, reaching the target group, affordability, reduction of standards, progressivity; yet prestigious self-help housing programmes have managed to cover but a small proportion of the increase in the housing deficit, not to speak of the backlog. The other targets were not reached either. Even the biggest and most experienced donor, the World Bank, admits that most of its 'unconventional' projects did not reach the lowest 40 percentile of the population, and that cost recovery could not be assured (Keare and Parris, 1982).

Space for Negotiation

Assisted self-help housing programmes have satisfied neither the expectations of the agencies promoting them nor those of their academic protagonists. Nevertheless, in the Third World, they are still being promoted and implemented more than any other type of low-income housing programme. Even the intended target groups seem to be happy with this kind of development. What hidden virtues outweigh their apparent limitations?

Commodification provides one possible explanation for the popularity of self-help programmes with certain agents of the state who represent the interests of capitalist entities. However, the process of commodification advances in rather subtle ways and does not particularly benefit one single out of several usually competing fractions of capital. A conscious strategy to promote non-conventional housing programmes on these grounds is therefore unlikely. Commodification appears to be a self-propelled rather than a planned phenomenon. At best, a generalization of private real estate ownership may be considered a support against the infiltration of socialist ideology.

Recent research in Venezuela has pointed once again to a more convincing explanation for the popularity of non-conventional housing programmes: the potential for negotiation between the urban poor and the state (Fiori and Ramírez, 1992; Ramírez et al., 1992). According to this perspective, the state is not seen as a monolithic entity but as a rather heterogeneous body representing contradictory interests in society. By offering minimal support to housing initiatives, the state enters a bargaining process from which the target population can benefit by obtaining a greater share of the centrally administered resources or, to use Castells's (1977) term, 'collective consumption'. The outcome will depend on the community's bargaining capacity and

conjunctural circumstances. Rather than the state simply co-opting the poor, both sides negotiate, and the benefits are shared by all residents and not just their leaders. The redistribution of resources takes place in the form of infrastructure improvement[5] which is almost never charged for in practice—even while the principle of cost recovery is maintained officially. By entering the negotiation process the state safeguards its legitimacy. There are benefits for both sides.

If the popularity of self-help housing programmes in capitalist societies has its roots in the political process, questions arise as to the logic and possible implications of self-help housing strategies in non-capitalist societies.

Non-Capitalist Countries

Almost all states that consider themselves socialist have introduced non-conventional housing programmes. If the theoretical potential of self-help strategies argued by Turner and others cannot be realized in practice because of the capitalist context in which they take place, are the results under socialism much different or indeed better?

Research in Cuba (Mathéy 1993a) has shown that some of the effects of self-help strategies observed in capitalist countries do not occur, e.g. the various instances of commodification. In social terms, the collective forms of self-help construction in particular tend to be more integrative and help to avoid the exclusion of the poor. The extension of the working day is notable—whether it can be called double exploitation in these circumstances remains a matter of debate. In certain other aspects, there is surprisingly little difference compared to the experiences made in capitalist countries: the architecture is not better than in other sectors, the cost of housing is not reduced, political mobilization is not fostered, and personal living patterns—gender roles in the family, furniture and decoration within the dwelling, collective amenities—remain conservative.

Self-help programmes appear to play a much more important role in Cuban macro-economic planning. In particular, the labour-force can be redirected and usefully employed in slump periods—one of the purposes of the microbrigade movement (Mathéy, 1989; Hamberg, 1990: 60–3). The number of produced units is significant (20 per cent of the housing stock in the two major cities) and demonstrate the replicability of the scheme, a target beyond the reach of most other countries.

Managing Self-Help Schemes

Self-help housing schemes represent a major element in the housing policy of most Third World nations today. Experience over three decades has contributed to our understanding of the potential and real limitations of the strategy, and taught some basic lessons how best to proceed within the scope of such programmes and what to avoid.

First of all it is important to remember that self-help programmes may be a good solution for some people, but they certainly will not provide shelter for all. Younger families with at least two adult people seem to be able to make the best use of the option—while single-parent families and the elderly may find it very difficult to participate. Only where a well-organized community already exists may neighbours be prepared to support one or two particularly disadvantaged people. Such support is most likely in upgrading schemes where the people involved have known each other for many years. In this sense the upgrading of run-down mass housing projects through self-help initiatives has great potential, as the studies of self-built extensions in Egypt's new towns by Wilkinson et al. (1991) suggest.

Tenants conceptually have no place in self-help schemes—except as a potential source of income for the owners. Yet in some cities like in Bogotá, tenants represent more than half the population in the informal settlements (Gilbert, 1991). How can they be included to benefit from a self-help scheme along with owners? One possible approach is to divide the property among all families living on the plot, as was done in Chile to assist former *allegados*[6] under the recent densification schemes.[7] Another unconventional alternative, practised in Cuba, is the sale of roof space for the construction of additional units. Even if all residents can be served by an improvement scheme, there remains the problem that different sections of the population will benefit differently.

Gentrification represents a recurrent issue. Physical improvements in terms of better infrastructure and services provision, and legalization of land occupation, tend to make neighbourhoods attractive to more affluent income groups who displace the initial settlers. It can be argued that the poor benefit by gaining an income when selling out. But if a cash subsidy to the target group were the ultimate objective of self-help programmes, this effect could be achieved much more simply through a direct subsidy rather than through a housing programme. In order to avoid gentrification, improvements may be evenly spread over all slum areas of a city, so that no one area is singled out as a potential middle-income neighbourhood. Gentrification thus was avoided in a city-wide upgrading programme in Lusaka over long periods, as the unique time-series analysis by Schlyter (1991) has shown. Such a course of action is unlikely to appeal to politicians because improvements spread wide and far are unlikely to provide attractive photo opportunities for their campaigns.

Gentrification may even occur in newly built projects, including the standard sites-and-services approach. Where these schemes are located in the periphery, distant from the city centre and from income opportunities, the intended beneficiaries may find it extremely difficult to arrange for transport. Those affluent enough to buy a car or motor-cycle may be the only ones left who can manage. The situation becomes even more difficult in cases where the projects do not provide a nuclear unit to live in, because then participants need to commute between their old accommodations and the building site while construction is under way. This particular problem can be solved by providing a bus service from the very beginning, or by allowing people to live in a provisional shelter initially. Both facilities were provided in a relatively successful scheme in Hyderabad, India, during the 1980s. The resale of plots before building, i.e. anticipated gentrification, can be impeded by delaying the right to a plot until the intended beneficiary has actually lived on the plot over an extended period of time. This was successfully tested in Pakistan (van der Linden, 1992).

There is a major gender dimension to assisted self-help. Women tend to be excluded from schemes because of habitual prejudices about the work they are able to perform, and because they are already overburdened by their triple role in productive, reproductive, and community-managing tasks (Moser, 1993). In many countries they are also excluded for legal reasons: heads of household are assumed to be men; women do not qualify for credit. Finally, many women are not reached by the promotion for such schemes: they are disproportionately illiterate, and they have less access to newspapers. Left isolated in upgrading schemes, they are likely to be the first to move out.

Realizing that typically from 30 to 50 per cent of households are headed by women, their access to self-help housing schemes needs to be facilitated expressly. The examples of the Women's Construction Collective in Jamaica and the micro-brigades in Cuba demonstrate that women generally can do the same construction jobs as men. However, in promoting self-help participation by women, their additional responsibilities in the family and community must be taken into account. This issue can best be addressed on a collective basis, i.e. by the provision of kindergartens, communal kitchens, laundrettes, and multi-generation plots. Where women and men live together, the obvious answer to women's participation in self-building would be a redistribution of household chores. But even in a country such as Cuba which explicitly fosters women's rights, there is little evidence that such a redistribution takes place (Mathéy, 1993a).

Self-help housing construction can last over several years and thus generate substantial expenses for project supervision and administration. Also, living on a building site and sacrificing all leisure time for building tasks appears to affect the participants' family life negatively. To reduce construction periods significantly would seem highly desirable. The use of prefabrication is one approach towards such a goal (Schütz, 1992). So far, this possibility has not yet been fully exploited, although some experiences are quite promising. Successful examples include the small panel Sandino system used in Cuba and other countries in the region, or the Servivienda dwellings in Colombia, which are produced and delivered to the site by a non-profit company, and assembled by the family within a few hours with the assistance of one or two foremen of the supplier. The limitations of the latter system are its poor architecture and limited adaptability to climatic conditions (Cabannes, 1992). In Nicaragua, the Sandinista Government offered low-priced prefabricated timber units primarily to the rural population. Even apartment blocks can be prefabricated and assembled by self-help labour, as the examples of Las Vegas and Ruperto Lugo in Caracas demonstrate (Brumlick, 1992).

The construction period in self-help schemes can also be shortened by segregating various work tasks, similar to neo-Fordist production management, and improving productivity through selective reorganization. In Kenya, for example, Erkelens (1991) identified 45 different factors which influence work performance in self-build projects. Even the use of computers has been recommended to facilitate user participation in housing design and construction, and to reduce costs, though this experience seems more relevant to more industrialized countries (Uhl and Tavolato, 1992).

The World is Changing

In the Third World, self-help housing policies were introduced in a period marked by the explosive growth of major cities where the demand for shelter far exceeded the capacity of conventional models of housing provision. At the same time, purely repressive measures to counter the 'hunger for shelter' (Salas, 1992) through slum clearance or deportation was neither advisable in economic terms (because the cheap labour-force living in the slums was needed) nor politically feasible given the risk of popular resistance (there had been too many violent protests). In any case, their effect tended to be short term; the dispossessed tended to squat elsewhere, and most of the deported soon returned to the city. At the same time, foreign and international donors tended to honour a softer political profile with easier access to credits and grants.

In recent years, however, the political and economic landscape has drastically changed from what it was in the mid-1970s when the unconventional housing programmes were first initiated. Absolute poverty has increased drastically in the Third World; the so-called socialist systems in Eastern Europe have disappeared; and foreign investments of First World countries are being redirected from the South to the East.

These profound changes in external conditions require a new assessment of housing policy, too. When governments become poorer, will they turn to even more radical self-help programmes, or will they opt for a complete policy reversal? The neo-liberalism that came into vogue in the 1980s has translated into a tendency of the state to withdraw from all responsibility for housing the poor, and the emphasis on the notion of full cost-recovery fits this tendency (World Bank, 1993).

At the same time we may see the transfer of the concept of massive self-help housing programmes from the South to the East. As a significant proportion of the population in Eastern Europe is impoverished and social segregation proceeds, massive homelessness is appearing. In this context, self-help schemes involving the jobless make political sense: they do help to maintain social peace and do not compete with real-estate investment (Mathéy, 1993c). The (re-) introduction of such policies will be socially acceptable, since most Eastern European countries have a long tradition of self-help, albeit principally for the middle-income groups.

Community-based housing finance can overcome many limitations of individualistic self-help approaches (Cabannes and Mathéy, 1995). In the South, further extension of self-help projects into other sectors is conceivable. Where basic needs can be less and less satisfied through the market, direct production and bartering become more important. Thus urban food production offers possibilities for additional self-help projects. Improved techniques, such as permaculture (Mollison and Holmgren, 1978), can help to make the best use of small plots. Such developments may well reinforce the economic and spatial restructuring of Third World cities into two largely disconnected social systems.

Notes

1. More detailed reviews of the relevant literature are provided by Pamuk (1991) and Mathéy (1993b).
2. According to World Bank figures of 1992, 36 developing countries have a lower GNP per capita than in 1965. In 1990 the average population lived below the poverty line in 41 countries, whereas this was the case in only 31 countries in 1981.
3. There are, of course, more effective means of reducing housing costs than self-help. These include rent

pooling, control over land speculation (i.e. land banking), and direct subsidies. The Singapore case demonstrates that this can be done in a capitalist context.
4. Purchasing of land reserves for future development by the authorities at a period when prices are still relatively low, and releasing it to the market when demand grows.
5. Whereas according to the concept of urban social movements the people's gain obtained through the struggle goes beyond the housing issue.
6. In Chile the term *allegados* refers to relatives or friends permanently sharing the plot or the dwelling without a formal rental or co-ownership arrangement. It is a consequence of extreme housing shortage among low-income families.
7. The original idea of densification is to provide additional housing units in low-density neighbourhoods. Thus more economic use is made of the existing infrastructure and, above all, land. However, this option presupposes low densities. In the Chilean case, where densities are already very high, the programme implies just the legal division of the existing plots, which entitles the *allegados* to receive grants and subsidies for housing construction or improvement.

References

Abrams, Charles (1966), *Squatter Settlements, the Problem and the Opportunity* (Washington, DC: Departments of Housing and Urban Development).

Brumlick, Ana (1992), 'Ruperto Lugo, a High-Rise Self-Help Institutional Programme in Caracas', in Mathéy (1992), 303–10.

Burgess, Rod (1977), 'Self-Help Housing: A New Imperialist Strategy? A Critique of the Turner School', *Antipode*, 2 (9): 50–60.

—— (1978), 'Petty Commodity Housing or Dweller Control? A Critique of John Turner's View on Housing Policy', *World Development*, 6 (9/10): 1105–33.

Cabannes, Yves (1992), 'Potential of Prefabrication for Self-Help and Mutual Aid Housing', in Mathéy (1992), 241–68.

—— and Mathéy, Kosta (1995) (eds.), 'Community-Based Housing Finance: A Trialog Contribution towards Habitat II', *Trialog*, 47, special issue.

Castells, Manuel (1973), 'Movimiento de pobladores y lucha de clases en Chile', *EURE* (Santiago de Chile), 3 (7): 9–35.

—— (1977), *The Urban Question* (London: Edward Arnold).

Duncan, S. S., and Rowe, A. (1993), 'Self-Provided Housing The First World's Hidden Housing Arm', *Urban Studies*, 30 (8): 1331–45.

Eckstein, Susan (1977), *The Poverty of Revolution: The State and the Urban Poor in Mexico*, augmented paperback edn. 1988 (Princeton: Princeton University Press).

Erkelens, Peter A. (1991), *Self-Help Building Productivity, a Method for Improving House Building by Low-Income Groups Applied to Kenya* (Eindhoven: University of Technology).

Fichter, R., and Turner, John F. C. (1972), *Freedom to Build* (New York: Collier and MacMillan).

Fiori, Jorge, and Ramírez, Ronaldo (1992), 'Notes on the Self-Help Housing Critique: Towards a Conceptual Framework for Analysis and Policy Making', in Mathéy (1992), 23–32.

Gilbert, Alan (1991), 'Comparative Analysis: Studying Housing Processes in Latin American Cities', in Tipple and Willis (1991), 81–95.

Hall, Peter (1989), 'Arcadia for Some: The Strange Story of Autonomous Housing', *Housing Studies*, special issue, 4 (3).

Hamberg, Jill (1990), 'Cuba', in Kosta Mathéy (ed.), *Housing Policies in the Socialist Third World* (London: Mansell; Munich: Profil), 35–70.

Harms, Hans (1982), Historical Perspectives on the Practice and Policies of Self-Help Housing', in Peter Ward (ed.), *Self-Help Housing: A Critique* (London: Mansell/Alexandrine), 17–55.

Keare, Douglas, and Parris, S. (1982), *Evaluation of Shelter Programmes for the Urban Poor: Principal Findings*, IBRD Staff Working Papers 547 (Washington, DC: World Bank).

Koenigsberger, Otto (1970), 'Housing in the National Development Plan: An Example from Nigeria', *Ekistics*, 30: 393–7.

Linden, Jan van der (1992), 'Back to the Roots: Keys to Successful Implementation of Sites-and-Services', in Mathéy (1992), 341–52.

Lindert, Paul van (1992), 'Social Mobility as a Vehicle for Housing Advancement? Some Evidence from La Paz, Bolivia', in Mathéy (1992), 157–80.

Ludwig, Richard, and Cheema, Shabbir (1987); 'Evaluating the Impact of Policies and Projects: Experience in Urban Shelter and Basic Urban Services', *Regional Development Dialogue*, 4 (8): 190–229.

Mangin, William (1967), 'Latin American Squatter Settlements: A Problem and a Solution', *Latin American Research Review*, 23: 65–98.

Marcuse, Peter (1990), 'Why Self-Help Won't Work', *Trialog*, 23: 10–12. Reprinted in Mathéy (1992), 15–22.

Mathéy, Kosta (1989), 'Microbrigades in Cuba: A Collective Form of Self-Help Housing', *Netherlands Journal of Housing and Environmental Research*, 4 (1): 24–30.

—— (1992) (ed.), *Beyond Self-Help Housing* (London: Mansell; Munich: Profil).

—— (1993a), *Kann Selbsthilfe—Wohnungsbau sozial sein? Erfahrungen aus Cuba und anderen Ländern Lateinamerikas* (Hamburg/Münster: LIT).

—— (1993b), *Sellbsthilfestrategien als Element der Wohnungspolitik in Entwicklungsländern. Bibliographie mit Anmerkungen zum Stand der Forschung* (Kassel: Gesamthochschule).

—— (1993c), 'Wildes Siedeln? Die Dritte Welt kommt nach Europa—mögliche Konsequenzen für die Wohnungsversorgung', in Joachim Brech (ed.), *Neue Wege der Planungskultur* (Frankfurt: Wohnbund; Darmstadt: Verlag für wissenschaftliche Publikationen), 371–5.

Mollison, Bill, and Holmgren, Dave (1978), *Permakultur—Landwirtschaft in Harmonie mit der Kultur* (Shaafheim: Pala Verlag).

Moser, Caroline (1993), *Gender, Planning and Development* (London: Routledge).

Mühlich, Eberhard (1984), 'Selbsthilfe als Selbstaufklärung', *Bauwelt*, 82: 166–8.

Pamuk, Ayse (1991), *Housing in Developing Countries: A Select Bibliography and Field Statement* (Chicago: American Planning Association).

Pradilla Cobos, Emilio (1983), 'Autoconstrucción y explotación de la fuerza de trabaja y políticas del estado en América Latina', in Emilio Pradilla (ed.), *El problema de la vivienda* (Quito: CIUDAD), 77–121.

Ramírez, Ronaldo, Fiori, Jorge, Harms, Hans, and Mathéy, Kosta (1992), 'The Commodification of Self-Help Housing and State Intervention: Household Experiences in the Barrios of Caracas', in Mathéy (1992), 95–144.

Robles, Rívas, Diego (1972), 'Development Alternatives for the Peruvian *Barriada*', in Francine F. Rabinovitz and F. M. Trueblood (eds.), *Latin American Research, 2: Regional and Urban Development Policies: A Latin American Perspective* (Beverly Hills, Calif./London: Sage Publications), 229–37. Reprinted (1979) in Janet Abu-Lughod and Richard Hay, Jr. (eds.), *Third World Urbanization* (New York: Methuen), 321–9.

Rodriguez, Alfredo (1972), 'De invasores a invadidos', *Revista Latinoamericana de Estudios Urbano-Regionales* (Santiago de Chile), 4 (2): 101–42.

Salas, Julian (1992), *Contra el hambre de la vivienda: Soluciones technologicas latinoamericanas* (Bogatá: Escala).

Schiffler, Manuel (1991), 'In "Viertel der Nacht". Makler und Schulden in einem Armenviertel von Tunis', *Trialog*, 29: 25–7.

Schlyter, Ann (1991), 'Time Series Analysis: A Longitudal Study of Housing Quality in Lusaka', in Tipple and Willis (1991), 62–80.

Schuman, Tony (1986), 'The Agony and the Equity: A Critique of Self-Help Housing', in Rachel Bratt *et al.* (eds.), *Critical Perspectives on*

Housing (Philadelphia: Temple University Press), 463–74.

Schütz, Eike Jakob (1992), 'A Case for Housing Prefabrication in Squatter Settlements', in Mathéy (1992), 235–40.

Steinberg, Florian (1993), 'The Hutments of Berlin: Informal Housing in Periods of Crisis', *Trialog*, 36: 20–6.

Tipple, Graham, and Willis, Kenneth (1991) (eds.), *Housing the Poor in the Developing World* (London: Routledge).

Turner, John F. C. (1963), 'Dwelling Resources in South America', *Architectural Design*, 37: 360–93.

—— (1972), 'Housing, Problems and Policies', in G. Breese (ed.), *The City in Newly Developing Countries* (London: Prentice Hall), 507–31.

—— (1976), *Housing by People: Towards Autonomy in Building Environments* (London: Marion Boyars).

—— (1978), 'Housing in Three Dimensions: Terms of Reference for the Housing Question Redefined', in *World Development*, 6 (9/10): 1135–45.

—— (1988), 'Introduction' and 'Conclusions', in Berta Turner (ed.), *Building Community: A Third World Case Book* (London: Building Community Books).

—— (1990), 'Releasing the Potentials of Self-Help Housing', in Dennis Ingemann (ed.), *Collection of Papers from Metropolis '9, Metropolis in Ascendancy* (Melbourne: Dept. Housing and Construction), 31–3.

Uhl, Otokar, and Tavolato, Paul (1992), 'Computers, Participation and Self-Help Housing', in Mathéy (1992), 269–80.

Vaessen, Thieu (1989), 'The World Bank's Perspective on Self-Help Housing: The Case of India', in Michael Dewit and Hans Schenk (eds.), *Shelter for the Poor in India* (New Delhi: Manohar), 9–26.

Wakely, Patrick (1989), 'The Devolution of Housing Production', *Trialog*, 23/24: 19–23.

Wilkinson, Nick, Khattab, O., de Majo, G., and Kardash, H. (1991), 'Development within Development: Helwan New Community, Egypt', *Open House International*, 16 (3): 3–15.

World Bank (1993), *Housing: Enabling Markets to Work*, a World Bank Policy Paper (Washington, DC: World Bank).

19

The State and Housing Policies in Chile: Five Regime Types and Strategies

Fernando Kusnetzoff

As in most metropolitan areas in Latin America, an explosive growth of popular settlements has taken place in Chile's main urban areas, especially in the capital of Santiago, which reached a population of five million by the early 1990s. This growth occurred mostly along the periphery of those cities, by means of the rapid construction of illegal or public-financed settlements with poor living conditions in terms of the quality of the dwellings, infrastructure, and access to workplaces. This essay will explore the state's contribution to this process since the late 1950s, when the population growth in urban areas began to exert an intense pressure on the political system. The housing policies and their results will be reviewed for five successive regimes which cover a whole gamut of ideologies—conservative democratic, reformist, socialist, conservative authoritarian, and liberal democratic—through more than three conflictive decades in Chile's recent history.

In pursing this analysis, attention will be paid to two closely related factors: first, the development strategies that were pursued by each regime and in whose context the housing policies and programmes unfolded; and secondly, the political circumstances that in each case 'sensitized' the provision of housing as an important factor

contributing to the need to legitimize each particular regime *vis-à-vis* a demanding population.[1] The first three governments—Alessandri, Frei, and Allende—will be considered together, because they encompass the steady expansion of state involvement in housing under democratic rule over a period of fifteen years. Thereafter, Pinochet's housing policies will be discussed, as they represent a complete reversal of the previous trend in favour of the privatization of the housing sector until the late 1980s, when the prospect of elections—a plebiscite and later the presidential campaign—brought back the intervention of the state in housing in a way very reminiscent of the elections of the past. The last regime—Aylwin—signals a cautious return to democratic rule and an effort for more comprehensive urban and housing policies that benefit from the previous experiences.

Industrialization, the Ascent of the Working Class, and Housing

By the time of the Second World War, Chile was entering into a period of rapid industrialization, partly as a consequence of the restrictions

This is an updated and expanded version of an essay that appeared in Gil Shidlo (ed.), *Housing Policies in Developing Countries* (London: Routledge, 1990), 48–66.

imposed on the import of many goods, manufactures, and machinery by the war participants. This fact, combined with a migratory trend towards urban areas from a stagnant and obsolete rural society, explains the growing urbanization of the country. Santiago grew from less than 800,000 inhabitants in 1930 to over 1,600,000 in 1952. By 1946, a new phenomenon, the *callampas* (literally, mushrooms)—groups of shacks illegally erected on vacant public or private land—was but another symptom of the failure to absorb and employ the recent migratory waves looking for jobs in the main Chilean cities. It was clear, once the war ended, that the 'easy' import-substitution process was over, because of the competition from the recovered industrial powers of Europe, the United States, and later Japan. The Chilean industrial class understood that a more sophisticated industrial base was needed, as well as an expansion of the import of consumer goods demanded by the middle and upper classes. For that to happen, foreign investments and loans were required. The opportunity came with the presidential campaign of 1958, when Jorge Alessandri, a representative of the Chilean right, narrowly won over the candidate of the leftist coalition, Salvador Allende.

The Ambitious Housing Policies of Jorge Alessandri: 1958–1964

The presence of a strong opposition, and the fact that Alessandri inherited a state with a long tradition of social expenditures, explains his pragmatism in some areas of the economy. While he supported an indiscriminate import policy, and at the beginning took artificial means to contain inflation, it appears that he saw a possibility of expanding his political base by increasing the construction of houses for low-income families. In particular, low-income housing legislation enacted in 1959 offered the private entrepreneurial sector tax exemptions up to 20 years, depending on the square footage of the new houses. A second key law, the National Savings and Loan System, was passed in 1960. Both measures favoured the middle- and upper-income sectors. An extreme example was the building of luxury apartments in exclusive resort areas by the end of Alessandri's term. Yet housing construction increased considerably in these years—over 30,000 units were built annually. But, after two

years of growth, the economic situation of the country worsened, with a sequence of inflationary and depressive cycles. In the building sector this unleashed a speculative process generated by the demand for housing from sectors anxious to invest in real estate as a defence against inflation. Alessandri's exercise in orthodox market mechanisms proved inadequate to meet the needs of the population at large. In 1964 the candidate of Partido Demócrata Cristiano (Christian Democratic Party), Eduardo Frei, defeated the Allendista forces.

The Populist Housing Policies of Eduardo Frei: 1964–1970

The electoral programme of Frei and his party was in essence an ambitious reformist and populist effort to reach an agreement between some groups of the entrepreneurial sector—disillusioned with the failure of Alessandri's policies— and the demands of the urban and rural masses, so as to neutralize the thrust of the just-defeated left. This effort was based on a new ideology expressed in the slogan 'Revolution in Liberty'. Politically, it attempted to displace the power of the conservative agrarian sector through a mild agrarian reform, while looking for a flow of foreign capital linked to a programme of industrial expansion. Frei's first two years brought a reactivation of the economy, less inflation, and some income redistribution (Mistral, 1974: 4). But, much as in the case of Alessandri, the process lost its dynamism by mid-1966, with a return to more conservative policies that remained until the end of the Frei administration.

In the urban and housing area, Frei took bold decisions to modernize the institutional apparatus by creating a Ministry of Housing and Urban Affairs (MINVU) in December 1965. Two main objectives were defined: first, to diminish the serious housing deficit which had reached some 420,000 dwellings within the country; and second, to link the production of housing with income distribution and popular participatory effects, while at the same time stimulating the internal market. The goal was to produce 360,000 dwellings in six years, of which 60 per cent would be for low-income families. Lower standards for size and quality were adopted, and in 1967 a new concept, *soluciones habitacionales*, was introduced. This was to be one of the first tests of the controversial 'sites-and-services' approach. Instead of definitive dwellings, small modules of 330 square feet were built on partly urbanized lots. This was followed by the Operación Sitio which provided only urbanized plots, and finally by Operación Tiza (operation chalk), as the beneficiaries called it, which simply demarcated lots. The *soluciones habitacionales* were, in a sense, a realistic answer for the large number of very low-income-level Chilean families: 45 per cent within the range of 0.4 to 1.5 of the minimum wage in 1966.

The Frei Government renounced attacking the structural determinants of the housing problem. Instead, it expanded a paternalistic approach towards urban low-income groups to ensure their political loyalty. Popular savings plans, newly created Juntas de Vecinos (neighbourhood councils), and *soluciones habitacionales* were means to 'generate a specific alternative in terms of implantation and popular mobilization, against the hegemonic leftist workers unions' (Alvarado, 1973: 53). Eventually, the same mechanisms and expectations brought forth by these policies, which initially mobilized large sectors of the population, generated a reaction because of the delay in fulfilling such hopes. Thus, the drastic drop of Frei's housing programmes in 1969 was confronted by active demands, especially the *tomas de terrenos* (land seizures) which were usually answered by heavy repressive violence on the part of the government.[2]

Yet, by the end of his government, Frei had been able to achieve about two-thirds of his housing goals, with an average of 40,000 dwellings built per year. The relevance of the housing issue in those years is emphasized by Rodríguez and Marshal (1980: 71) who point out three paradoxical aspects of Frei's 'Revolution in Liberty' project:

1. On the one hand, it had not resolved the contradictions that it faced, characteristic of a reformist project: economic exclusion and an attempt at political incorporation. '. . . the social welfare modality of domination succumbed because its sustaining economic base could not bear the political weight of participative consumption without changing the nature of the material and ideological bases of bourgeois domination' (Duque and Pastrana, 1972: 273).

2. On the other hand, it had created the conditions for its own loss of control, by creating a wide network of organizations incorporating sectors which never before had made demands—the rural and urban subproletariats.

3. And above all, it established a new locale for political struggle, i.e. the new housing areas, which became a new arena for electoral and political competition in the 1970 presidential elections.

In fact, the division of forces between the conservative right that presented a strong challenge in the forthcoming elections with the former President Alessandri as its candidate, and the rather dispirited sectors loyal to the Christian Democratic Party, opened the way for the socialist option, again led by Salvador Allende. In a very heated political climate, the coalition of Allende won with a relative majority in the balloting (36 per cent) which permitted him, after dramatic negotiations with Frei's forces, to be proclaimed President by the Legislative Congress.

The 'Housing as a Right' Policies of Salvador Allende: 1970–1973

A realistic assessment of both the international hostility to his triumph and the fact that the majority of the Congress opposed him conditioned the initial policies of Allende. These policies were designed to attain fast results while respecting the existing constitutional limitations. Important changes in the forms of property, like the nationalization of copper mining, a massive agrarian reform, and the acquisition or expropriation of most banks and strategic industries, were combined with programmes to revitalize the economy and to produce a rapid change in the income distribution in favour of the country's majority. Again, the same pattern of two initial successful years followed by a period of stagnation which characterized the two previous governments of Chile, was repeated in the Allende administration. And again, urban and housing policies were at the very centre of the political conflict. By 1973, opposition activity was raised to a level that produced a virtual paralysis of the country's economic life, with the tragic outcome of the military coup of September 1973.

Allende initiated at the very beginning of his government an Emergency Housing Plan based on a diagnosis of the problems accumulated in the previous years. Four objectives were stated:

- generation of a high level of employment,
- full utilization of the installed industrial capacity,
- dynamization of housing-related activities, and
- widening of popular support bases.

These objectives, obviously, were closely related to the fact that the new government had to face the early municipal elections of March 1971, against a united opposition.[3] Thus, Allende's housing policy is distinguished from those of his predecessors by its openly political character. While both Alessandri and Frei had also well-calculated political goals, these were dissimulated behind a technocratic language in the first case, and a vague 'participatory' style in the second. In the case of Allende, the Emergency Housing Plan promised 80,000 dwellings by the end of 1971—the highest figure ever projected by the public sector in the history of Chilean housing, a figure that was almost attained. Allende gave voice to the concept of *housing as a right*, which he synthesized in his presidential message:

The basic definition of the housing policy establishes that housing is the right of all Chilean families, independent of their income level and political or religious position, it being prioritized as a function of their necessity for housing. . . . Housing stops being a commodity in order to become the right of each worker and his family. (Rodríguez and Marshal, 1980: 77)

This definition was a clear answer to the increasing popular mobilization linked to housing that evolved during the final years of the Frei Government, when in Santiago alone no less than 103 land seizures occurred, as compared to 35 in 1969 and 13 in 1968. These land seizures were backed by the leftist parties, including the radical Movimiento de Izquierda Revolucionario (Movement of the Revolutionary Left) (MIR), and were converted into settlements known as *campamentos* (encampments). In these *campamentos*, where no less than 300,000 people came to live by the end of 1970, very intense political activity took place, resulting in complex political problems for the new government. The emergency programmes were directed to face this mounting problem, but the government was aware that such a formidable effort could not be sustained over the following years. Obstacles like the limited nature of the available productive

apparatus, financial constraints, and, not surprisingly, the early boycott by the private entrepreneurs, forced the government to change its housing plans for 1972. The global problems faced by the Allende Government were exacerbated by an increasing scarcity of foreign resources resulting from a drop in the copper price and the induced international credit asphyxia. The new strategy called for expanding the supportive base of the government by means of an approach to the middle classes. It is relevant in this respect to look at the following official statement by the Ministry of Housing:

Thus, the housing sector must be considered from the point of view of the available supply capability and not from that of *demand pressures*. This implies, at the same time, producing an exhaustive analysis of the role played by the different agents in the housing process (enterprises, professionals, businessmen) and evaluating the impact of their actions on the eventual completion of the overall strategic goals. If the overall strategy requires the assistance of the middle sectors in order to widen the base of popular support and thus to fulfil the anti-monopolistic and anti-feudalistic goals of this stage, *the housing sector must not be reluctant to work with the support of professional, entrepreneurial, and industrial sectors.* (MINVU, 1972: 87–8; my emphasis)

The new strategy of the government was a desperate effort to face the severe economic crisis by measures to maintain the level of production—called 'the production battle'—demanded by the bourgeoisie, together with the restriction of popular mobilization, so as to get enough time to revitalize the economy. Actually, this strategy was not only confronted by those looking for the failure of the government, but it also created problems within the leftist organizations supporting it. This was dramatically obvious in the case of the *campamentos*, where the more radicalized groups continued pressing MINVU and the building enterprises, coming into open conflict with the authorities. It can be said that the working class reached the highest levels of both class-consciousness and confusion in that period. While some privileged workers pressed for higher salaries altogether impossible to provide, other workers participated in an untiring struggle for consolidation and progress in the 'Chilean Path to Socialism'. The final months of the Allende Government were marked by the seditious conduct of the opposition forces. The recourse by the government to bring military officers into the cabinet at the ministerial level to control the crisis, proved to be wrong and suicidal with the tragic outcome of the September *coup d'état*.

The preceding review of the Chilean experience in housing from 1958 until 1973 has shown, in a context of formal democracy, first the increasingly strategic and political nature of the housing issue as a consequence of the rapid growth of urban and metropolitan centres in a context of slow economic development. Secondly, growing state participation in formulating housing policies, with financing mechanisms to stimulate the participation of private capital, because of the weak and unfit state of the productive structure in the housing sector, as well as the inability of the market to provide for the needs of the majority of the population. Thirdly, the emergence of a greater political consciousness among the 'marginal' settlers, in the sense that they realized that even if their immediate housing needs were met, this would not resolve the basic contradictions that were perpetuating their peripheral and deprived existence. In other words, in those years the state gave increasing priority to the urban housing problem, but did not challenge the domination of the market or abandon a sectoral vision that restricted real and substantive participation of the population.

On the positive side, in those fifteen years nearly 600,000 dwellings were produced, mostly by the public sector, and thousands of lots were urbanized, revealing the gradual maturity of means to overcome accumulated deficiencies in the urban areas of Chile.

Back to the Market: Cities and Housing Under a Military Regime

From September 1973 until 1989, Chile lived under a harsh, personalistic, military dictatorship. For more than sixteen years the regime of General Augusto Pinochet had powers that no other government in the history of Chile could have dreamt of. The traditional division between the executive, legislative, and judicial branches of government was, for every practical purpose, abolished, all three branches consolidated under the control of the Military Junta, and principally, under Pinochet. With the advice of a group of staunch free-market Chilean economists—which included for a while the presence and blessings of Nobel-prize laureate Milton Friedman—the new government embarked on a radical experiment of capitalist consolidation, attempting at the same time to remodel Chilean society in terms not only antagonistic to the overthrown regime of Salvador Allende, but in fact to a long tradition of liberal democracy. In this process, a central strategy consisted of the systematic dismantling of those structures by which the Chilean state had been able for more than fifty years to advance economic and social development policies. The implementation of the new strategy required the implacable repression of the political organizations and social sectors dependent on the benefits of the previous state actions.

During this long period, high and low points, tactical changes and contradictions, were frequent, especially during the later years. For the purposes of this analysis of urban and housing policies, it is possible to distinguish two main periods, linked to important changes in the economic policies which have received ample academic attention.[4] The first period was one of orthodox neo-liberal measures until 1981, when the collapse of the main speculative and financial groups of the country—at a time when the foreign debt had increased fivefold to $20 billion—forced the government to intervene. The second period, which consisted of efforts to rescue the financial apparatus and gradually involved the state in social practices rejected by the neo-liberal ideology, was to last until the end of this regime and was conditioned by the 1988 plebiscite which was called to decide whether Pinochet was to stay in power for another eight years and, after his defeat, by the approach of the 1990 presidential elections.

The Free-Market Housing Policies of General Pinochet: 1974–1989

The urban and housing policies of the regime, in contrast to the political decisions taken in the initial years of the three previous administrations, not having to respond for a long and undefined period to an electoral constituency, permitted delaying investments in social expenditures in favour of economic goals linked with the new central strategy towards a strong insertion in the international markets. With regard to urban conditions, and in particular to housing, the new policies were directed to eliminate most forms of state control, leaving urban and housing processes to the free play of market forces. This was clearly stated in a publication of MINVU:

At the level of the urban system, planning will be aimed at making the process of urban development compatible with the global model of the country's development, creating the conditions most convenient for facilitating the operation of the urban market. . . . The state will foment and support the creation of an open market in housing. Responsibility of the production belongs to the private sector. (MINVU, 1979: 7–10)

Privatization and Liberalization of Urban Land[5]

Easy access to urban land, for those with capital to enter into real estate speculation, was soon on the agenda of MINVU. Several measures were taken to facilitate the access of the private sector to land: abolition and/or lowering of taxes related to land transactions; liquidation of state reserves of urban land; and eradication of illegal *campamentos* from the most valuable land. Residential zones of the middle- and upper-income sectors located in the municipal communes of the Oriente (East) area of metropolitan Santiago were granted special privileges. According to a study by Sabatini (1981: 223), the combined public and private housing construction in the Oriente area, which had only 12 per cent of the population of metropolitan Santiago, represented 20 per cent of the total investment in the metropolitan area between 1965 and 1973 (the Frei and Allende years); but between 1974 and 1979 it reached 36 per cent and towards the end of that year about 58 per cent of the total metropolitan expenditure. In striking contrast, in those years, particularly during the 1974–5 and 1978–9 periods, public-housing production was drastically reduced for the whole country in accordance with the strict market policies of the regime. Yet, the availability of urban land for speculative capital was limited by the existing controls over urban development and planning. Eventually, a new urban land concept abolished the urban limits of metropolitan Santiago in 1979. As the MINVU document defined it, urban land was no longer considered to be a scarce resource:

Time and again it has been insisted upon that urban land is a scarce and irreplaceable resource, which has contributed to its price suffering frequent distortions in the marketplace by artificially restricting the supply. Present policy, on the contrary, is based on the principle that land is not a scarce resource, but that its apparent scarcity is provoked in the majority of cases by the inefficiency and rigidity of the rules and legal procedures applied up until now in order to regulate the growth of cities. (MINVU, 1979: 18–19)

Some 64,000 urbanizable hectares (160,000 acres) were added to the 36,000 hectares (90,000 acres) occupied by the metropolitan area of Santiago then, instantly tripling the potential land market. The creation of such an ample supply of land for the real-estate sector was supposed to widen the access to housing by low-income sectors marginalized by the high land prices prevalent up until then. Yet, the real effects of the abolition of the urban limits were quite different. Prices in the city's preferred central areas, as well as in the new urban periphery, increased rapidly within months. In the Oriente area, the already inflated prices rose on the order of 400 to 800 per cent within two years. Thus, the policy resulted in reinforcing the speculative operations in the land market.

Metropolitan, Social, and Ecological Segregation

The urban policies of the military Junta show a combination of spatial and economic decisions that coincided in accentuating the existing rigid segregation of the population of Santiago, according to social class and income. Environmental conditions deteriorated in vast low-income areas in terms of hygiene, overcrowding, distance to work-place, and services. For several decades the Oriente zone has been the residential bastion of the well-to-do, and later, of the more successful sectors of the middle class. In the 1960s, some nuclei of low-income communities came to the Oriente zone, filling service job needs, a trend reinforced by the integrationist ideologies of the Frei and Allende administrations. The Pinochet regime's market doctrine rapidly succeeded in reversing such a social mix in prime residential areas. This happened by simple exclusion from the market, or by the drastic programmes of eradication enforced since 1979.

Subsidies strictly excluded the purchase of residential land; their explicit purpose was to support private builders in the construction of new dwellings in designated areas of lower quality. As extra-legal *tomas* were effectively precluded by the army and police, the lower classes were limited, in the first ten years of the Pinochet regime, largely to two options. First, thousands of 'homeless' newcomers to Santiago concentrated in the dense settlements in the central areas of the city, or in the poor communities around its periphery. They rented rooms in old tenements or, in the case of many young families, 'dropped' into the houses or the back patios of houses and lots of relatives and friends. In Santiago alone, by the end of the 1970s, over 135,000 families, 600,000 people, were living as *allegados* (drop-ins) in conditions of severe overcrowding.

Secondly, the low-income settlers who lived in the *campamentos* as a result of previous land invasions in predominantly middle- or upper-income areas, were subjected to displacement, dislocation, and resettlement—the *erradicaciones*. These programmes, with relocation to the most impoverished urban periphery in the south-west of Santiago, were implemented until 1989 and affected more than 40,000 families. Most of these families were moved to minimal plots and dwellings where they lived under depressing conditions of uprooting, dispersion, and isolation. To this must be added the lack of close and stable work in their vicinity, which forced many of them to enter into occasional and low-paid 'informal' jobs, until some of the former support systems created after years of struggle in the *campamentos* eventually became re-established. In some municipalities in the southern area of Santiago, the situation created by the eradication programmes was critical. According to *Hechos Urbanos* (1985, no. 46), in the municipality of La Pintana the population grew from 78,000 to 140,000 inhabitants within two years. By 1984 there were only two doctors' offices, a third of the population had no sewage service, half of the children had no schools for basic education, and unemployment was estimated at about 60 per cent.

With the displacement of nearly 200,000 people by *erradicaciones*, the segregationist logic of the capital city has shown in the 1980s the consolidation of a privileged and exclusive sector in the eastern zone of metropolitan Santiago and, bordering on it, impoverished middle-income sectors which serve as a buffer to the vast sectors where the poor live. Thus, the reshaping of the city according to distinctions of class and income

was almost completed, at least in appearance, after sixteen years of the systematic application of a free-market model of development to the urban areas.

Social Mobilization and Survival Mechanisms of the Urban Poor

The urban and housing strategies of the military regime have had two main purposes: on the one hand, to support market mechanisms in an area of previously mixed participation of private capital and public resources and, on the other hand, to dismantle the capacity for organized action by a civil society long used to benefit from such action *vis-à-vis* state institutions. Yet, from a review of the government involvement beginning in the mid-1980s, it is clear that in spite of a sustained rhetoric of free-market practices, some of the state's traditional roles were increasingly revived, as in the case of housing subsidies.

This new state involvement is not explained by a sudden awareness of the inability of the mass of the population to participate in the housing market effectively. Rather, popular opposition to the regime had become increasingly combative and organized. Massive monthly demonstrations that began in 1983 are illustrative of this confrontational style. In the housing sector itself, the exasperation of the 'homeless' and 'drop-in' families led them again to organize land seizures in the southern zone of Santiago. In spite of repression, the invaders succeeded in establishing more than 8,000 families in two new *campamentos* named after the Catholic Cardinals Raul Silva Henríquez and Juan Francisco Fresno. This surprising triumph of the popular movement was a shock to the government. Decisions were rapidly taken to avoid the recurrence of such invasions. In these two new *campamentos*, the government followed a careful plan of replacing their leaders and then atomizing the incipient organization by the forced relocation of families to other semi-urbanized lots in diverse areas of Santiago, a process completed by 1987. In this way, combining repression and concessions, the regime was able until its end to confront the popular mobilization, hoping to improve its image for the 1988 plebiscite, and subsequently, for the 1989 presidential elections.

Yet, while the mass protests were sporadic, and usually repressed, daily life had to continue for hundreds of thousands of families making up the urban poor sector of Santiago and other Chilean cities. Multiple forms of social and economic organization were developed, first to deal with the primary necessities of subsistence, and then to develop solidarity and identity within each residential unit. One study points to the variety of organizations mushrooming in these units, from those related to neighbourhood needs like communal cooking pots and family gardens, to more general ones, like committees of mortgage debtors and worker-owned workshops (Downs, 1985: 136). A study of *Hechos Urbanos* reported more than 700 popular economic organizations with over 120,000 members in Santiago in August 1984. Not only did these organizations help to mobilize previously inactive human resources, but they also generated profits, provided training, and sponsored cultural activities which were not feasible in the traditional enterprises. Workshops and permanent, part-time, and temporary work provided experience and survival skills in areas such as handicrafts, cotton-printing, and repair work, while engendering practices of social integration and solidarity that were to be important for the future democratization of the country (Downs, 1985: 19–20).

In 1989, after a period of sustained economic growth, the government expected one of its supporters to win the presidential elections that followed its defeat in the 1988 plebiscite. Yet, the Pinochetistas failed to understand that even sectors of the political right were tired of the long and personalistic dictatorship and wanted change. So, while the pro-government forces had two candidates, a consensus of pro-democracy political forces from the right, centre, and left created a movement, La Concertación Democrática, with an appealing global programme for 'the transition' and a single candidate, Patricio Aylwin from the Christian Democratic Party, which was to win the December 1989 elections with 54.5 per cent of the vote, thus calling an end to the long Chilean nightmare.

Housing and Urban Policies in the Transition to Democracy

Chile, like other Latin American countries recovering from a long dictatorship, experienced diverse and severe difficulties in re-establishing and consolidating a democratic society during the government of president Aylwin, in part as a consequence of the serious dislocations of the country's social structures by Pinochet's policies, and in part because of the presence of a united leadership within the armed forces still identified with the past regime. Yet, the Aylwin Government was able to advance the so-called 'transition to democracy' substantially. This advance was facilitated by three factors: the strength of the economy which followed the opening to the international markets of the previous government while keeping the basic macroeconomic equilibrium—GNP continued to grow at about 6 per cent annually; the purposeful continuity of the political base supporting the government; and the gradual expansion of social expenditures within the public sector and an increase in the minimum wage.

The Social Housing and Urban Policies of Patricio Aylwin: 1990–1993

The new government showed a pragmatic approach in its treatment of human settlement conditions: it kept a substantial part of the military regime's latest urban and housing policies and allowed the real-estate sector—that had benefited from years of capital accumulation—to continue a speculative process that was environmentally unsound. This approach of continuity departed from the experience of the four previous governments, where in each case the need for strong changes were presented as critical for new policies (Haramoto, 1992: 2). In a good synthesis of the views of the new regime, Greene (1993: 2–3) points to three deficits that had to be faced: a housing deficit, an urban deficit, and a participatory deficit. The remainder of this article reviews this triple problem and the policies adopted to address it.[6]

During the electoral campaign, the Aylwin forces had set out such a threefold purpose in terms of expanding the production of housing, implementing a policy of urban improvement and development, and stimulating a growing role for local and community actors in shaping the new policies and programmes. Once in power, the government defined the general guidelines to be followed (MINVU, 1990a), and a number of changes in the normative instruments were made, e.g. to secure land for recreational areas, circulation, and social equipment; to provide pavement and plant trees in low-income settlements; to permit workshops and commercial activities in low-income houses—a reality in most popular settlements; and to renovate old housing.

The 'Rather New' Housing Policies

For a long time the housing deficit has been an important political concept or 'barometer' in Chile to evaluate the success of a particular government—and it has been manipulated frequently.[7] Thus, the military regime's final estimate of a deficit of only 423,000 units in 1989 was not accepted by the experts of the new government. A new and refined survey provided quantitative and qualitative information (MacDonald, 1992: 17; Greene, 1993: 20–2) that leads to the conclusion that the housing deficit was of the order of one million units, i.e. more than 30 per cent of the population remained housed under unacceptable conditions, a situation quite similar to that three decades earlier (Kusnetzoff, 1964: 14–18).

The key element in these new estimates is the importance of the *allegados*, the households unable to afford the previous housing programmes, which in a study by CASEN, a research unit under MIDEPLAN (Ministry of Planning), reached 486,000 'drop-in' households and 408,000 'drop-in' nuclei (inside the principal household), to which must be added 181,000 units whose material or sanitary conditions were deemed irrecoverable. According to this approach, the deficit came to a total of 1,075,000 units in 1990. To this figure must be added a demand for 80,000 units per year from the constitution of new households and the obsolescence of 20,000 units per year.

The new government thus had to make an exceptional effort to face the increasing housing deficit. Table 19.1 shows an average increase of 76,000 new units per year between 1990 and 1993. This is a substantial gain over the 34,000 dwellings per year constructed under the military regime, but was still insufficient and fell short of the more than 100,000 units per year proposed in the plans of the Concertación Democrática. Still, in the last couple of years of the Aylwin administration, housing production reached levels never before seen in Chile.

The various housing programmes are presented in Table 19.2. All but one were initiated by the military government. Yet important changes were introduced in order to focus their effects on the population, by means of normative changes and a better differentiation of the programmes.

The Basic Housing Programme covers low-income families. The government buys private-sector finished houses and assigns them to the beneficiaries. Regulations were simplified to permit families with lower incomes, in particular those suffering under a critical *allegados* condition, to apply for these houses.

The Special Workers Housing Programme offers an option for groups of organized workers sponsored by public or other institutions. The houses are contracted for by the organizing entity. The value of houses is double that of those under the Basic Housing Programme.

The Unified General Subsidy Programme provides for middle-income families. The subsidy goes directly to the beneficiaries who, individually or collectively, select a house in the market. The average value of units is slightly less than double that under the Special Workers Housing Programme.

The Rural Subsidy Programme is similar to the Unified General Subsidy Programme, but is geared towards lower income families in rural areas. They can choose among three alternatives: buying a house, contracting its construction, or employing self-help procedures.

The Progressive Housing Programme was a new programme initiated by the Aylwin administration. Its main purpose was to solve the critical condition of *allegados* families. The intent was to offer an incremental housing alternative, starting with a minimal house, so as to cover a broader number of families. However, this programme has been less than successful for a number of reasons: lack of interest from the building enterprises, scarcity of land, complexity of organizational and legal services, and the need for immediate occupancy (De la Lastra, 1992). Beginning in 1993, a second phase was proposed, and new financial and technical tools are being tested to face the wide diversity of situations. The potential demand for such minimal housing is enormous, but it remains to be seen if its production can be substantially increased.

In addition to these five programmes, MINVU initiated several new programmes to improve

TABLE 19.1   Public and Private Dwellings Initiated per Year, 1959–1993

Year	Public	Private	Total	President
1959	22,731	9,109	31,840	Jorge Alessandri
1960	22,080	7,490	29,570	
1961	25,060	11,129	36,189	
1962	17,615	20,309	37,924	
1963	11,988	15,440	27,428	
1964	6,938	12,902	19,840	
TOTAL 6 YEARS	106,412	76,379	182,791	
AVERAGE/YEAR	17,735	12,730	30,465	
1965	37,514	15,054	52,568	Eduardo Frei
1966	13,433	14,328	27,761	
1967	28,285	16,653	44,938	
1968	32,730	19,683	52,413	
1969	14,460	23,310	37,770	
1970	5,914	17,792	23,706	
TOTAL 6 YEARS	132,336	106,820	239,156	
AVERAGE/YEAR	22,056	17,803	39,859	
1971	76,079	10,893	86,972	Salvador Allende
1972	20,312	13,752	34,064	
1973	20,877	14,484	35,361	
TOTAL 3 YEARS	117,268	39,129	156,397	
AVERAGE/YEAR	39,089	13,043	52,132	
1974			29,564	Augusto Pinochet
1975			37,087	
1976			15,351	
1977			17,823	
1978			22,177	
1979	15,651	3,509	19,160	
1980	16,226	7,675	23,901	
1981	15,778	11,221	26,999	
1982	12,599	7,925	20,524	
1983	13,489	8,510	21,999	
1984	20,418	13,522	33,940	
1985	25,089	15,885	40,974	
1986	27,051	22,599	49,650	
1987	25,711	22,166	47,877	
1988	26,285	34,423	60,708	
1989	44,020	33,149	77,169	
TOTAL 16 YEARS	364,358	180,545	544,903	
AVERAGE/YEAR	22,772	11,284	34,056	
1990			61,714	Patricio Aylwin
1991			71,945	
1992			81,244	
1993			91,000	
TOTAL 4 YEARS			305,903	
AVERAGE/YEAR			76,475	

Sources: SUR, *Hechos Urbanos*, no. 25 (September 1983); MINVU, *Memoria Anual* 1973–89/90/91.

TABLE 19.2   Finished Dwellings and Subsidies Paid per Year, 1990–1992

Programmes	1990	1991	1992
Progressive housing, 1st stage	0	5,081	10,500
Basic housing	16,029	24,805	22,500
Special workers housing	14,550	13,601	17,044
Unified general subsidy	24,827	22,729	22,600
Rural subsidy	6,308	5,009	8,600
TOTAL NEW DWELLINGS	61,714	71,945	81,244
MINVU sanitation	3,217	0	0
Progressive housing, 2nd stage	0	327	2,000
Sites and services	11,617	18,850	7,408
TOTAL IMPROVED DWELLINGS	14,834	19,177	9,408

Source: Greene (1993).

existing housing, to provide title deeds for families that illegally occupied urban land, and to refinance the debts of more than 280,000 families unable to service their mortgages. Another new programme was designed for rental housing, using leasing mechanisms, to induce the private sector to build units for low- and middle-income families, so as to relieve pressures on the rental market.

Comprehensive Urban Settlements Policies

MINVU put major emphasis on broadly addressing the issues implicit in the four urban guidelines of 1990. To begin with, densification of the urban space is a difficult goal. Santiago has a low density overall (less than 40 inhabitants per acre), but most new low-rise residential settlements in peripheral areas reach close to 200 persons per acre in the case of one- to two-storey buildings and up to 320 persons per acre in three-storey buildings. Their immediate environment is characterized by limited open spaces, and cultural resources to counter such limitations are severely limited. The alternative, high-rise buildings like those constructed during the 1950s and 1960s, presents difficult problems of cost, maintenance, and social acceptance. New alternatives are needed, like the development of medium-size urban centres by focusing investments that expand the employment potential of such cities. The need for urban renewal, in particular in the older central sections of Santiago and other major cities such as Valparaíso and Concepción, is recognized, but so far the government has assumed only the role of promoter and facilita-

tor of rehabilitation and remodelling projects, leaving execution to the private sector. In fact, very few projects, some of them with international support, have been implemented. Until now, private capital appears to prefer the proven expediency of new projects to the complexities of mass rehabilitation or remodelling. It may be necessary to rethink the level of public involvement. The Urban Improvement Corporation which completed important renewal projects in central areas in the 1960s offers an important precedent.

Urban segregation and land-use policies present the most pervasive problem in the principal cities of Chile. The new policies have been ineffective in addressing the socio-economic roots of the problem. With the Common Municipal Fund, an effort has been made to improve the environmental quality of low-income areas, but the existent segregation is taken as a given. As concerns the future expansion of the cities, the real value of a new urban land policy will depend on the ability of the state to influence its use. The existing legislation permits expropriations only by reason of public domain, and the proposal of a land bank has not been adopted. All in all, the pattern of urban segregation remains unchanged, and the spectacular setting of exclusive high-income residential and commercial developments towards the east of Santiago provides a clear testimony to the workings of a free-enterprise real-estate market.

The air of Santiago has been one of the most contaminated in the world since the 1970s. The Aylwin Government has been quite active in this respect by means of various measures, e.g. new normative instruments, accelerated pavement of streets, construction of new parks, and the withdrawal of thousands of older buses from the streets. At the same time, pilot projects were initiated with regard to alternative technologies and combustibles, environmental education, and traffic restrictions for private transportation. Yet, until the regional governments and the municipalities acquire more influence upon basic factors such as industrial location and the growth of the principal cities, the pollution problem will continue to severely affect the daily life of the urban population.

Community Participation in Housing and Urban Development

The repressive environment of the Pinochet years paradoxically forced people to try all sorts of means, other than overtly political ones, to organize survival, to address their mounting economic and social problems, and to exert pressure on the government. Integration and solidarity within poor urban communities thus reached a high level. The new democratic authorities had reason to expect an expansion of these attempts now that open participation was an explicit strategy of the government. For this to happen, new decentralization strategies that gave more responsibilities to the municipalities were devised, so as to establish new instances of local involvement and interaction.

Municipal participatory practice was limited by the existing legislation in several ways, e.g. while municipalities are required to inform the population on urban issues like a new Master Plan, there is no assurance that community inputs will be considered. At the same time, the classical problem remained: many local officers resisted a participatory exercise which they considered a waste of time or indeed a risk for bureaucratic incompetence to be exposed. Finally, there was a lack of a sustained participatory—rather than confrontational—approach by many communities, perhaps the consequence of more than sixteen years of severe denial of genuine community participation.

Decentralization of human settlements was the proclaimed policy of the government, but in practice little progress was made. A notable exception is the management of regional resources by the new regional authorities. Recently, a new

Regional Government and Administration Law was promulgated. It calls for substantial decentralization to regional governments endowed with ample decision-making powers for investments, territorial planning, and support of regional development. These new functions will facilitate the organized participation of community groups and institutions, but only and if the key articulations between regional and local authorities are efficiently managed.[8] On balance, the Aylwin policies for the housing and urban settlements sector appear to have been satisfactory, achieving major progress in the extent of housing production and establishing legal and normative tools for more satisfactory urban development. Still, the most ambitious goals, like an intense social participation in the planning and implementation of such policies, remain to be accomplished. Perhaps four years of transition to democracy were not enough, in a context of unresolved tensions with the military, to fully develop a national commitment to work together in the difficult field of human settlement growth and development.

Conclusion

This critical review of the varied experiences of the housing and urban policies of five Chilean governments—from 1958 to 1993—has shown several specific traits that can now be summarized. To begin with, some common patterns directly linked to the design and implementation of these policies may be singled out. First, housing and urban issues have been important public social expenditures in Chile as a direct consequence of the rapid pace of urban growth, while responding to very different social, economic, and ideological discourses. Secondly, the erratic behaviour of the national economy under each government until the late 1980s has conspired against the necessary continuity of such housing and urban policies, in a pattern of contrasting up and downs in the yearly production of urban housing and infrastructure. Thirdly, once again, the need to lower lot and dwelling standards has emerged as a prerequisite to expand the coverage of the programmes, while locational decisions have increased the social costs of many housing programmes. Fourthly, especially in the case of the Santiago metropolitan area, a pattern of ecological and social seg-regation has prevailed over short periods of efforts at social integration.

The Chilean experience shows a concern with successful housing and urban policies as an important electoral factor—with the exception of the first fourteen years of the military regime—and a powerful instrument to establish the legitimacy of government. While this was evident for the governments that preceded the dictatorship, in Allende's case this was openly stated.

For most Chileans, and especially for the urban masses, over three decades of promises, failures, and repression in their quest for a proper roof in a decent environment, has meant a hard but rich social and political practice. The brief and dramatic years of the socialist experiment of Salvador Allende and the long and painful years of the Pinochet dictatorship represent extreme contrasts in the political space allowed to the urban masses. The recent transition to democracy offers a new opportunity for the gradual consolidation of a more mature and consistent articulation of the main actors—government, the private sector, and popular organizations—towards a just and democratically planned and shared implementation of housing and urban strategies as the best, and perhaps only, way to confront the increasingly complex and persistent environmental, urban, and housing problems that affect the existence and coexistence of the Chilean people.

Notes

1. For an overview of the strategies of development applied in Chile and Latin America, see Foxley (1983). The relations between economic development and metropolitan growth are explored by Geisse (1983). Violich and Daughters (1987) provide a thorough review of urban and housing planning and policies in Chile. See also Segre (1981).
2. The political role of the actors in these *tomas de terrenos* was widely discussed by the academic community in the 1970s. See Handelman (1975), Kusnetzoff (1975), and Castells (1983).
3. In these elections, Allende's Popular Unity gained precisely 50% of the vote, which marks the high point of Allende's support. The conservative opposition was alarmed enough to begin thinking about a *coup d'état*. For some political analysts, this was the time for Allende to follow a more ambitious policy based on such wide electoral support.

Actually, his government kept to the more cautious initial strategies. The 1973 *coup* precluded the testing of the results of either option. It is the author's contention that the Popular Unity Government had fewer alternatives than commonly assumed for two reasons: first, the magnitude of the accumulated problems at the point of departure; and secondly, the strong contradictions within the alliance, as a result of the class compromise upon which it was built.

4. A provocative analysis of the bases of the Junta's model for dominance is provided by Moulián and Vergara (1981). O'Brien (1984) and Fortin (1984) expose the weaknesses of monetarist policies in Chile. Remmer (1980) analyses the demobilizing mechanisms applied by the Junta. Arellano (1985) offers elements on the Junta's social policies.
5. This section draws heavily on work by Donoso and Sabatini (1980), Trivelli (1981), Sabatini (1981), and Arellano (1985).
6. I am deeply indebted to the papers by Haramoto (1992) and Greene (1993).
7. Three decades ago, in one of the first discussions of this issue in Chile, I proposed this concept regarding the case of several Chilean governments that anteceded the ones discussed in this essay (Kusnetzoff, 1964). Polemics as to the quantitative dimensions of the housing problem continue to this day. A detailed discussion of housing production and financing during the military years is included in Kusnetzoff (1987). See also Arellano (1985). In an interview (*El Mercurio*, 15 Jan. 1988), a former Minister of Housing of the Frei Government flatly rejected the military government's claims (which at that time curiously fluctuated from 750,000 to 941,000 units dwellings built in fourteen years) and, using Central Bank data, concluded that from 1974 to 1986 only 480,000 were completed. This means a yearly average of 34,745 units, less than were built during the Frei and Allende administrations. Subsequently, the official *Memoria Anual* of MINVU (1990*b*) for the period 1973–1989 confirmed such an average, even if the Pinochet Government improved its housing production during the last two years. See Table 19.1.
8. I have explored the difficulties of a combined decentralization–local participation approach elsewhere, see UNCHS (1989) and Kusnetzoff (1991).

References

Alvarado, Luis *et al.* (1973), 'Movilización social en torno al problema de la vivienda', *Revista EURE* (Santiago), 7: 37–70.

Arellano, José Pablo (1985), *Políticas Sociales y Desarrollo: Chile 1924–1984* (Santiago: CIEPLAN).

Castells, Manuel (1983), *The City and the Grassroots: A Cross-Cultural Theory of Urban Social Movements* (Berkeley: University of California Press).

De la Lastra, C. (1992), 'Programa de Vivienda Progresiva. Documento de Trabajo 42/92, Corporación de Promoción Universitaria, Santiago.

Donoso, Francisco, and Sabatini, Francisco (1980), 'Santiago: Empresa inmobiliaria compra terrenos', *Revista EURE* (Santiago), 20: 25–54.

Downs, Charles (1985), 'Alternative Social Policies for the Poor: Lessons from the Grassroots in Contemporary Chile', manuscript, Division of Urban Planning, Columbia University, New York.

Duque, Joaquín, and Pastrana, Ernesto (1972), 'La Movilización reivindicativa urbana de los sectores populares en Chile: 1964–1972', *Revista Uruguaya de Ciencias Sociales* (Montevideo), 1: 259–93.

Espinoza, Vicente (1985), *Los Pobladores en la política*, SUR, Documento de Trabajo, No. 27 (Santiago).

Fortin, Carlos (1984), 'The Failure of Repressive Monetarism: Chile 1973–1983', *Third World Quarterly*, 6: 310–26.

Foxley, Alejandro (1983), *Latin American Experiments in Neoconservative Economics* (Berkeley: University of California Press).

Geisse, Guillermo (1983), *Economía y Política de la Concentración Urbana en Chile* (Mexico City: El Colegio de México).

Greene, Margarita (1993), 'La Gestión gubernamental en el sector vivienda y urbanismo: 1991–1992', Corporación de Promoción Universitaria, Santiago.

Handelman, Howard (1975), 'The Political Mobilization of Urban Squatters Settlements', *Latin American Research Review*, 10 (2): 35–72.

Haramoto, Edwin (1992), *La Vivienda social en Chile*, Documento Facultad Arquitectura y Urbanismo (Santiago: Universidad de Chile).

Kusnetzoff, Fernando (1964), 'Dimensiones de una política habitacional para el sexenio 1964–1970', *Revista de Planificación* (Facultad de Arquitectura y Urbanismo, Universidad de Chile), 1: 13–50.

—— (1975), 'Housing Policies or Housing Politics: An Evaluation of the Chilean Experience', *Journal of Interamerican Studies and World Affairs*, 17 (3): 281–310.

—— (1987), 'Urban and Housing Policies under Chile's Military Dictatorship', *Latin American Perspectives*, 53: 157–86.

—— (1990), 'The State and Housing in Chile: Regime Types and Policy Choices', in Gil Shidlo (ed.), *Housing Policy in Developing Countries* (London: Routledge), 48–66.

—— (1991), 'La Descentralización del estado y el estado de la descentralización', in Miguel Morales (ed.), *La Descentralización: Mito o Potencial* (San José, Costa Rica: Fundación Friedrich Ebert), 21–30.

MacDonald, Joan (1992), 'A propósito del déficit habitacional', *Revista CA—Colegio de Arquitectos* (Santiago), 70: 17.

MINVU (Ministerio de Vivienda y Urbanismo) (1972), 'Política Habitacional del Gobierno de la Unidad Popular', *Editorial Universitaria*, Santiago, Chile.

—— (1979), *Política Nacional de Desarrollo Urbano.*

—— (1990a), *Memoria Anual 1990.*

—— (1990b), *Memoria 1973–1989.*

—— (1991), *Memoria Anual 1991.*

Mistral, P. (1974), 'Reflections on the Chilean Experience', *Pacific Research*, World Empire Telegram 5 (2): 1–10.

Moulián, Tomás, and Vergara, Pilar (1981), 'Estado, ideología y políticas económicas en Chile: 1973–1978', *Revista Mexicana de Sociología*, 43 (2): 845–903.

O'Brien, Philip (1984), 'Authoritarianism and Monetarism in Chile, 1973–1983', *Socialist Review*, 77: 45–79.

Remmer, Karen L. (1980), 'Political Demobilization in Chile: 1973–1978', *Comparative Politics*, 12 (Apr.): 275–301.

Rodríguez, Alfredo, and Marshal, María Teresa (1980), 'Chile 1960–1980: Dos décadas de políticas de vivienda', in Manuel Manrique and A. Maquina (eds.), *Problema Urbano y Trabajo Social* (Celats, Peru), 59–93.

Sabatini, Francisco (1981), 'Santiago: Sistemas de producción de viviendas, renta de la tierra y segregación urbana', Instituto de Estudios Urbanos, Documento de Trabajo, No. 128 (Santiago).

Segre, Roberto (1981), 'The Territorial and Urban Conditioning of Latin America Architecture', in Roberto Segre and Fernando Kusnetzoff (eds.), *Latin America in Its Architecture* (New York: Holmes & Meier).

SUR (1982–1990), *Hechos Urbanos*, 12–95 (Santiago).

Trivelli, Pablo (1981), 'Elementos teóricos para el análisis de una nueva política de desarrollo urbano: Santiago', *Revista Interamericana de Planificación* (Mexico), 60: 44–69.

UNCHS (United Nations Centre for Human Settlements) (1989), *Decentralization Policies and Human Settlements Development* (Nairobi, Kenya).

Violich, Francis, and Daughters, Robert (1987), *Urban Planning for Latin America: The Challenge of Metropolitan Growth* (Boston, Mass.: Lincoln Institute of Land Policy/Oelgeschlager, Gunn & Hain).

VI Patterns of Political Integration and Conflict

Introduction

A WIDE gap in income and wealth, power, and status separates the élite from the mass of the population in most developing countries, and the middle class is usually quite small. The majority of the urban population have a standard of living so low that it is inconceivable to the average citizen of an industrialized country. While most urbanites are better off than the rural masses, there are some who have no shelter, who can barely clothe themselves. Malnutrition is common. For many, the quest for food for themselves and for their children is a daily struggle for survival.

The urban masses cope with their condition in a variety of ways. In Chapter 13 we saw how urban poor secure the help of kin, neighbours, and friends in the crises which regularly threaten those living close to subsistence. But government officials loom large as a threat or a resource. Police officers and tax collectors have to be evaded; trade licences and building permits are required; some secure a job with the government or a public hand-out; many are desperate for adequate public services.

Most government leaders are less than responsive to the needs of the masses. Some are preoccupied with ensuring the loyalty of the armed forces, most are concerned with promoting the investment of indigenous and foreign capital. They are constrained by multinational corporations and foreign governments which provide investments, credits, and grants, buy exports, and may threaten subversion. The mass of the population typically has little impact on the decisions of government and cannot count on effective legal protection in dealing with its agents. The leverage of citizens has increased as elections have become more common and more competitive in recent years, but the prime beneficiary is usually the middle class.[1]

Rural populations rarely participate significantly in the political process. Where they do take part in elections, local élites commonly control their vote through patronage or outright coercion. Powell (1980) concludes from a comprehensive review of rural voting patterns that almost everywhere, just below the surface, there is a 'rural mafia' system. The rural masses are thus usually without a voice and left with but three options. They may rebel when their land is taken away from them, taxes are raised, or the prices paid for their crops are lowered, but such upheavals tend to be isolated. Alternatively, peasants can withdraw from the market and revert to subsistence farming, jeopardizing food supplies for the cities and exports. But the most common response of the rural masses to the neglect, or even outright exploitation they face, is to vote with their feet. They form the large stream of rural–urban migrants who cannot be absorbed productively into the urban economy. They put severe pressure on the limited urban resources. And they swell the ranks of the urban masses strategically poised at the centres of local, regional, and national decision-making. The politics of developing countries are by and large played out in these urban arenas.

The following chapters focus on ethnic identities that usually divide the masses, but at other times serve to focus opposition to a regime; on systems of clientelism that muffle the voice of the masses, even while dispensing some resources; on social movements that articulate and press popular demands; and on the role of organized labour. We will conclude with an account of the various forms of mobilization that initiated the process of democratization in Chile and eventually forced the military from power.

Major political alignments in developing countries are often based on distinctions of origin, religion, or caste. In much of Africa, and in major parts of Asia, most migrants in the cities and their children categorize themselves and others in terms of region of origin. In parts of Africa and Asia, adherents of two or even three world religions live within the same city boundaries. At times they have clashed with each other, most dramatically Hindu and Muslim in South

Asia. But conflict can also be articulated in terms of divisions within a religion, such as the Muslim sects in Syria. Caste is of central importance in the Hindu religion, and for the majority of Indians it remains the primary referent. Determined by birth, and supposedly immutable, it can also change its guise according to context.

Alignments of shared origin, religion, or caste commonly crystallize around political and economic grievances. An ethnic group may move ahead of other groups because its land offers exceptional opportunities, a religious group may have the edge in education, or a caste may enjoy advantageous occupational opportunities. Once some members of such a group have established themselves in a privileged position, they wield considerable influence over the opportunities open to others. When their patronage goes to fellow-villagers, co-religionists, or members of the same caste, an entire group can be seen to enjoy privilege, others to be excluded. Furthermore, the continuous modifications of identities of origin, religion, and even caste typically increase the congruence between such identities and political and economic interests. At the same time, such divisions usually obscure major aspects of stratification and defuse class conflict among those who affirm a common identity.

Political conflict and competition over economic opportunities often have ethnic connotations.[2] Today, nearly every African country is deeply divided along ethnic lines (Gugler, 1996). These divisions are exacerbated to the extent that regimes take on ethnic identities and major economic opportunities are seen to be monopolized by members of particular ethnic groups. To refer to such a pattern as 'tribalism' is unfortunate because of the pejorative connotations the term 'tribe' has acquired. It is also misleading. Such identities in the urban setting, and their cultural content, usually bear a rather tenuous relationship to traditional societies and cultural differences. The past provides the raw materials, but ethnic identities are fashioned in the confrontations of the urban arena.

Heribert Adam and Kogila Moodley, in Chapter 20, analyse the ethnic conflict that has come to the fore now that South Africa has achieved majority rule. The premier political party, the African National Congress, pursued a multi-ethnic and indeed multi-racial policy from its very establishment in 1912. It always had a fair succession of leaders from among the Zulu as well as the Xhosa and had managed to overcome ethnic consciousness to a large extent, at least as far as the political activists were concerned. The ANC initially encouraged Mangosuthu Buthelezi, a former member of the ANC Youth League, to assume the leadership of the KwaZulu 'homeland' the South African minority regime had created. The Inkatha movement he established was envisaged as an internal wing of the ANC under the protective umbrella of the 'homeland'. Buthelezi, however, pursued an independent line. While he consistently maintained an anti-apartheid stance, his rejection of the armed struggle and international sanctions drew Inkatha closer to the South African regime and away from the ANC exiles abroad. The acknowledgement by the South African regime in 1989 that the majority could no longer be excluded from the legitimate political process, gave urgency to the conflict over the respective roles of the ANC and Inkatha in post-apartheid South Africa. The conflict has escalated in cities where competition over jobs and housing has heightened—and where confrontations between more- and less-skilled workers, between long-time urban residents, squatters, and migrants in single-men hostels, are readily cast in ethnic terms: a majority of the Zulu are migrants with limited skills and they predominate in the hostels. At the same time, a Zulu cultural revival has taken place. Hostel dwellers and unemployed migrants, in particular, find solace from their material and symbolic deprivation by taking pride in their King and the reconstructed memory of Zulu resistance against colonial conquest.

Adam and Moodley observe that relationships of patronage complement the ideological identification of Zulu. Patrons and clients often share a common identity. It provides an affective dimension to relationships between patrons and clients that are instrumental and profoundly unequal. Such relationships are found throughout the world in the most diverse settings, from national politics to the politics of academia, but their significance varies. In many developing countries they play a central role in fashioning political integration and delineating lines of conflict.[3]

Michael Johnson and Susan Eckstein, in Chapters 21 and 22, describe and analyse two systems of patronage that present a sharp contrast. In Lebanon, the origin, and some of the peculiarities of patronage, can be traced to a feudal past. Johnson presents a case study of Sunni Beirut where until the 1970s political bosses operated machines that usurped much of the powers of the independent state established in 1943. They distributed economic and other services to their clients in exchange for consistent political loyalty. Alliances with strong-arm neighbourhood leaders served to recruit clients into these machines and to control them, by force if necessary. The political bosses maintained control because they were in a position to play arbiter: to shield their followers, and in particular their strong-arm retainers, from the authorities, or to call in the police to punish the recalcitrant. The structure of the machines, based on local neighbourhoods and clans, prevented the emergence of self-conscious, horizontally linked social categories such as interest groups or classes. The electorate was further fragmented because most favours remained in the gift of the political bosses and went to individuals.

If the political bosses largely controlled both the allocation of state resources and the state's coercive potential, their position was predicated on the state providing resources and holding the means of coercion in the police and the army. The collapse of the state in the civil war in 1975 deprived the political bosses of their base. They lost control over their strong-arm retainers, some of whom emerged as prominent militia leaders over the course of successive rounds of fighting. Initially a progressive coalition was established, but it broke down, not least because of foreign interventions, and anarchy resulted. Solidarities of neighbourhood and clan remained as the only defence in an age of violence.

Eventually the imposition of peace by the Syrian army brought a stunning reversal. State authority was backed by the force of Syrian arms, and public opinion had turned against the militias. A new generation of political bosses emerged that was closely linked with the bosses of the pre-war era of stability and security. Unlike their fathers and uncles, they were no longer dependent on strong-arm retainers. But new forms of allegiance had emerged as well. During the civil war, the trade unions had boosted their recruitment and played an active role opposing the war: they had begun to act independently of the competing political leaders.

In Mexico, after the devastations of a long civil war, political power was effectively concentrated at the apex of the state, in the hands of the president and his close advisers. Co-optation throughout the political and administrative system effectively channelled and scaled down demands and kept the opposition in check. Elections were held regularly, but the monopoly of power of the Institutionalized Revolutionary Party (PRI), ruling under a variety of 'revolutionary' names since 1928, was not threatened until recently. A party that legitimizes itself on the basis of a revolution, yet consistently flouts the officially proclaimed revolutionary goals, has managed to stay in power for more than two generations.

Susan Eckstein, in Chapter 22, reports on her research in three low-income neighbourhoods in Mexico City over two decades. In the late 1960s, a large number of organizations were active in these neighbourhoods. However, the local leadership was invariably co-opted and integrated into the power structure, either through explicit institutional ties or through covert arrangements with functionaries outside. When Eckstein revisited the three neighbourhoods in 1987, organizational life had atrophied in two of them: residents no longer saw any reason to organize, and the ruling party had lost the organizational structures to bring out the vote.

In the inner-city neighbourhood, in sharp contrast, the victims of the 1985 earthquake had established local groups that pressured the state. Their disruptive tactics, and the support they received nationally and internationally, eventually persuaded the regime to respond to their demands. However, it required the movement to work with, not against, the state. It thus succeeded in co-opting opposition through reform until order was restored. Old forms of co-optation through organizations no longer served, but the regime was still able to contain popular pressures. Over the following years the regime developed the art of centrally allocating goods and services to serve political ends: its judicious use of patronage assured PRI of electoral victory in 1994 once again.

Bryan Roberts, in Chapter 23, focuses on the notion of 'citizenship' to assess the changing outcomes of the contest between political élites and social movements. Comparisons across Latin America allow him to suggest major differences. Roberts distinguishes three types of countries in the region. The countries of the Southern Cone had reached high levels of economic development and urbanization by 1940: they have the most extensive sets of political and civil as well as social rights, and of supporting institutions such as trade unions, political parties, and civil associations. Countries such as Brazil, Mexico, and Colombia had high rates of economic growth and urbanization only after mid-century: the extension of citizenship came later and was less complete. Finally, in the countries that lagged behind, rather poor and largely rural, citizenship remained quite limited.

Roberts suggests that political and civil rights were in jeopardy throughout Latin America for much of the period from 1940 to 1980 because the priorities of the urban masses revolved around establishing themselves in the city and improving their position. They did indeed obtain major improvements in their social rights during this period: health and educational standards rose rapidly throughout the region. This pattern began to change in the mid-1970s. Social rights became less salient, while civil and political rights were strengthened. Roberts explains the change in terms of two trends: an increasingly large proportion of the urban population was established in the city and social mobility declined; and the economic crisis and pressures from foreign creditors left the state with little scope to expand social rights. As the state abdicated responsibility for social and economic welfare, a wide variety of social movements organized—and demanded civil and political rights. They derived strength from the increasingly sophisticated media and the growing presence of non-governmental organizations. In the process, new groups found an effective voice: Indian communities, the young and the old, and women. A multitude of social movements appeared, disappeared, and emerged afresh in new forms—and increased the awareness and participation of the citizenry.

The austerity measures imposed by international agencies since the mid-1970s provoked unprecedented protests in debtor countries, ranging from mass demonstrations to organized strikes and riots. Walton and Ragin (1990) show in a cross-national analysis of austerity protests that one of the principal conditions for their occurrence and severity was overurbanization—defined as the level of urbanization relative to GNP per capita. Many governments felt com-

pelled to soften the impact of austerity measures or to rescind them altogether. Some regimes collapsed. Eventually the protests played a major role in persuading external actors—foreign governments, the International Monetary Fund, private bankers—to retreat from austerity policies.

Social movements, rather than elections, have been the most common means by which the masses in developing countries have been able to extract concessions from the political system or indeed to transform it. The most conspicuous action taken by urban poor have often been squatter movements. In many countries, especially in Latin America, squatters have frequently organized on a scale so large as to persuade the authorities to condone the illegal occupation of land, especially so when the squatters succeeded in mobilizing outside support. The political role of squatter movements, however, remained invariably narrowly circumscribed by political élites. Established squatters are particularly prone to be drawn into patron–client relationships: to avoid eviction, to secure infrastructure and public services, and eventually to obtain legal title to the land.

The success of squatter movements has varied with the scope for organizing, the extent of outside support, and the reaction of public authorities. Different political contexts have led to diverse outcomes across countries, in a given country over time, and even across locales within a country at a particular historical juncture (Gilbert and Gugler, 1992 [1982]: 193–9). Thus, Gay (1994) shows how the strategies pursued by two *favelas* in Rio de Janeiro vary from the exchange of resources for electoral votes to a combative stance that secures assistance without committing votes.

Urban workers, unlike squatters, have significant leverage. Any country's economy is dependent on the sustained labour of its work-force. The leverage different sectors of the labour-force can exert varies greatly though. Unskilled workers can be easily replaced by eager recruits from among the unemployed and underemployed, but many skills are in short supply. At the same time, underdeveloped economies tend to be heavily dependent on the effective operation of a few key sectors. A strike by railroad or dock workers brings the entire economy to a halt in many a country; a work stoppage by miners can threaten a foreign-exchange crisis. Those who operate such crucial economic resources, and whose skills are not easily replaced, potentially enjoy considerable political power.

By now nearly all developing countries have a history of worker protest.[4] We have already seen, in Chapter 21, that trade unions became prominent in opposing the civil war in Lebanon. In the democratization movements that swept many countries in recent years urban workers often played a key role. Drake (1988) demonstrated the resilience of the labour movements under repressive military regimes in Chile, Argentina, Uruguay, and Brazil. Seidman (1994) speaks of 'social-movement unionism' as a response to authoritarian industrialization strategies in Brazil and South Africa. In both countries, semi-skilled workers in heavy industry, organized first in factories and increasingly in impoverished urban communities, took advantage of conflict in élite circles to press demands for more far-reaching changes.

Maria Helena Moreira Alves, in Chapter 24, provides a detailed account of the labour movement that emerged in Brazil in the late 1970s. The military regime had fostered a culture of fear. In May 1978 the metal workers in the Greater São Paulo region responded with innovative tactics: they checked in for work and sat in front of their machines with their arms crossed; they had no visible leaders; and they organized no pickets. Community support ensured that there were virtually no scabs. The strike spread to other cities and to other industries: by the end of the year more than half a million workers had been on strike. A vast and peacefully organized social movement of civil disobedience developed a 'culture of confidence'. From the labour movement the powerful Partido dos Trabalhadores emerged. Eventually civilian rule came to Brazil. Luis Inácio da Silva, the president of the Metal Workers' Union popularly known as Lula, had gained such national prominence as to come close to being elected president of Brazil in 1989 and again in 1994.

Manuel Garretón, in Chapter 25, offers an account of the transition to democracy in Chile where a military regime was particularly well entrenched. The transition was achieved by mass mobilizations in the urban arena. After a decade of ruthless repression, opposition reaffirmed itself, from 1983 onward, in mass protests in the cities, in particular in the capital Santiago:

strikes, absenteeism, slowdowns, and demonstrations at work; assemblies and demonstrations at the universities; absenteeism at schools; street demonstrations and honking of horns; boycotts of stores; residents in middle- and lower-class neighbourhoods turning out the lights and banging pots and pans; shanty-towns erecting barricades. Garretón describes the 'invisible transition to democracy' as civil society thus reasserted itself and various interests confronted the military with their demands. Many groups won concessions, but the military remained firmly in control, repression of dissent the rule. The impact of these mobilizations remained limited because ideological and expressive differences fragmented groups similarly committed to democracy.

Garretón casts this as a paradox: popular mobilizations are not able by themselves to end military regimes and effect a transition to democracy, but effective political mobilization for a democratic transition cannot emerge without roots in popular mobilizations—even as it abandons the heroic, symbolic, and autonomous dimensions of collective action. In Chile, the divisions of the opposition allowed the military to reassert itself and set the agenda: a plebiscite to be held in 1988 to confirm Pinochet in power rather than competitive elections. Nevertheless, nearly the entire opposition united to reject the proposal decisively, unleashing a process of transition that culminated in elections in 1989 that re-established democratic rule.

In 1980, about two-thirds of Latin Americans lived in states under military rule or dominated by the military. In some countries the military had come to power in response to guerrilla movements which, inspired by the Cuban example, fought for a revolutionary transformation. The military succeeded in eliminating each of these movements and all subsequent attempts to confront them by force of arms. They tended to brutally repress any form of dissent and invariably clung to power. Eventually, they were dislodged by mass mobilizations rather than armed insurgents. The Pinochet regime was the last to go. In 1996, not a single military regime is to be found anywhere in the Western hemisphere, even if the military continue to hold considerable power in some countries where they brutally repress dissent, Amerindian peasants the prime targets.

In Asia, a number of authoritarian regimes

have been toppled over the last ten years. Elsewhere in the region, political change is barely perceptible even as dramatic economic growth sweeps through an increasing number of countries. The stark contrast between the world's two most populous countries remains the same as it has been for close to four decades: an authoritarian regime continues to rule China with an iron fist, while Indians live democracy with all its discontents.

In Africa, the military were in power in a majority of countries in the 1980s, and most civilian regimes were similarly authoritarian. The example of Eastern Europe, rather than Latin America, inspired dramatic change. Quite a number of repressive regimes were dislodged, others forced to make concessions, in a chain reaction across the continent. A large majority of African countries had presidential and/or parliamentary multi-party elections in the early 1990s. Still, some of the most repressive and exploitative regimes remain entrenched in power.

The wave of political liberalization that swept across the globe over the last ten years brought a degree of democratization to the majority of developing countries. These dramatic changes have nearly always been triggered by mass movements: street demonstrations, strikes, boycotts. They took place in urban areas, and their impact was most strongly felt in capital cities. In many cases they enfranchised the peasantry along with everybody else, but their very success reminds governments everywhere of the revolutionary potential of the urban citizenry and bodes ill for rural interests.

External pressures played an important role in bringing about political liberalization in some cases. Western governments began to show a measure of reluctance to support repressive regimes now that they could no longer be used as pawns in Cold War games. And creditors—Western governments, banks, the International Monetary Fund—imposed austerity measures that eroded support for the regimes forced to implement them. Structural adjustment programmes tended to benefit the rural sector, as we have seen in Chapter 2 for Africa, and hit hard much of the urban population which saw its real income drastically reduced—and revolted. The peasantry now has economic argument on its side, but the impact of urban unrest will continue to be remembered.

Rural-based guerrilla movements brought about the collapse of military regimes in Liberia, Somalia, and Uganda within the last ten years. There is little to indicate that they represented rural interests, even less to suggest that in their wake the peasantry will wield political power commensurate with its numbers. These guerrilla movements constitute notable exceptions to Garretón's contention that contemporary military and authoritarian regimes can only be terminated by negotiations and institutional mediations—and to my argument that contemporary revolutions are urban in character (Gugler, 1988 [1982]).

Notes

1. For a detailed analysis of urban stratification in Latin America, which emphasizes major changes over the last six decades, see Oliveira and Roberts (1996).
2. For a wide-ranging and penetrating discussion of ethnicity, ethnic conflict, and strategies of conflict reduction, see Horowitz (1985).
3. For a recent discussion of political clientelism, see Roniger (1994).
4. Bergquist (1986) provides accounts of close to a century of labour struggles in Chile, Argentina, Venezuela, and Colombia.

References

Bergquist, Charles (1986), *Labor in Latin America: Comparative Essays on Chile, Argentina, Venezuela, and Colombia* (Stanford, Calif.: Stanford University Press).

Drake, Paul (1988), 'Urban Labour Movements under Authoritarian Capitalism in the Southern Cone and Brazil, 1964–83', in Josef Gugler (ed.), *The Urbanization of the Third World* (Oxford: Oxford University Press), 367–98.

Gay, Robert (1994), *Popular Organization and Democracy in Rio de Janeiro: A Tale of Two Favelas* (Philadelphia: Temple University Press).

Gilbert, Alan, and Gugler, Josef (1992 [1982]) *Cities, Poverty, and Development: Urbanization in the Third World*, 2nd edn. (Oxford: Oxford University Press).

Gugler, Josef (1988 [1982]), 'The Urban Character of Contemporary Revolutions', in Josef Gugler (ed.), *The Urbanization of the Third World* (Oxford: Oxford University Press), 399–412. Earlier version in *Studies in Comparative International Development*, 17 (2): 60–73.

—— (1996), 'Urbanization in Africa South of the Sahara: New Identities in Conflict', in Josef Gugler (ed.), *The Urban Transformation of the Developing World* (Oxford: Oxford University Press), 210–51.

Horowitz, Donald L. (1985), *Ethnic Groups in Conflict* (Berkeley: University of California Press).

Oliveira, Orlandina de, and Roberts, Bryan (1996), 'Urban Development and Social Inequality in Latin America', in Josef Gugler (ed.), *The Urban Transformation of the Developing World* (Oxford: Oxford University Press), 252–314.

Powell, John Duncan (1980), 'Electoral Behavior among Peasants', in Ivan Volgyes, Richard E. Lonsdale, and William P. Avery (eds.), *The Process of Rural Transformation: Eastern Europe, Latin America and Australia* (New York: Pergamon Press), 193–241.

Roniger, Luis (1994), 'The Comparative Study of Clientelism and the Changing Nature of Civil Society in the Contemporary World', in Luis Roniger and Ayşe Güneş-Ayata (eds.), *Democracy, Clientelism, and Civil Society* (Boulder, Colo.: Lynne Rienner), 1–18.

Seidman, Gay W. (1994), *Manufacturing Militance: Workers' Movements in Brazil and South Africa, 1970–1985* (Berkeley: University of California Press).

Walton, John, and Ragin, Charles (1990), 'Global and National Sources of Political Protest: Third World Responses to the Debt Crisis', *American Sociological Review*, 55: 876–90.

20

'Tribalism' and Political Violence in South Africa

Heribert Adam and Kogila Moodley

The Shell House Affair

IN March 1994, among the high-rise office tow-ers of Johannesburg, a single incident epitomized the unresolved clash of traditionalism and modernity, of countryside and city. It showed insecure resentful migrant workers being taught a lesson by self-confident new political masters, ethnic loyalties being manipulated by self-serving élites, and new urban ethnic legends being con-structed for future battles. The ongoing violence between the African National Congress (ANC) and the Zulu-based Inkatha Freedom Party (IFP) usually occurs out of sight in the Natal countryside with the ANC being portrayed as the main victim. Suddenly the conflict had culmi-nated in an embarrassing reversal of victimhood for the forces of libration. The new rulers showed everyone that they could behave as ruthlessly towards their opponents as their own oppressor had done for decades. Moreover, the moral high ground that the ANC and Mandela justifiably held in the eyes of the world was momentarily shattered by the incident, better to be forgotten and reinterpreted than thoroughly analysed. The event also revealed why it is impossible to arrive at an agreed upon truth in an ongoing ideolo-gical battle even if dozens of 'objective' foreign observers witness the occasion.

On 28 March 1994, a month before the first democratic election in South Africa, about 50,000 Zulu migrant workers converged on cen-tral Johannesburg from outlying hostels and squatter camps for an anti-election mass rally in support of the Zulu king. At the end of the day, eight of the Zulu marchers were killed by ANC guards around the ANC headquarters at Shell House, more were shot randomly by unknown gunmen from surrounding buildings at Library Gardens and scores of injured, both protesters and bystanders, lay in their blood in different city streets. Despite hundreds of eyewitnesses, jour-nalists and film crews, statements in Parliament, and a 172-page submission by the police to the Goldstone 'Commission of Inquiry Regarding the Prevention of Public Violence and Intimidation', totally contradictory versions of the events surrounding the Shell House shooting are believed in by the partisans. The Acting Attorney-General for the Witwatersrand, Kevin Attwell, says South Africa may never get the answers it awaits, while a researcher for the lib-eral Institute of Race Relations writes in an insti-tute publication of a 'Chappaquiddick-like sore that will fester politically for years to come' (Paul Pereira, *Fast Facts*, Sept. 1995, 2). Inkatha com-pares the Shell House shooting to the massacre of fleeing protesters by the apartheid police at Sharpeville 34 years earlier.

The ANC version speaks of disorderly fanatic Zulu marchers bent on attacking the ANC head-quarters, defended by guards reluctantly forced to shoot in legitimate self-defence. Indeed Mandela himself in a surprise statement in Parliament several months later revealed that he had given explicit orders to shoot to kill, if necessary. Before the event, he had phoned the then state president de Klerk, and the Commissioner of Police, requesting special security precautions around ANC headquarters as party leaders were at risk. The police report details the extra forces and soldiers deployed around the city and Inkatha-friendly hostels in response to these well-founded fears. Nobel-prize laureate Nadine Gordimer (1995) however states: 'No roadblocks were set up around the city; the police outside Shell House that day ran away.' Yet film footage shows nervous policemen with guns guarding against snipers kneeling over bodies of shabbily clad peasants. Gordimer, normally an outspoken champion of poor people's causes, calls these Zulu traditionalists 'undisciplined hordes coming to the city' with 'the threat of violence and violence itself'. The police, on the other hand, argue 'that at no stage were marchers seen attacking Shell House', and that the shooting had, in any case, taken place away from the entrance to that building, when ANC guards opened fire on marchers 50 to 70 metres away 'for no reason whatsoever'. The police assert that the trigger-happy ANC guards constantly pointed firearms at passing marchers by standing outside and on the balcony of the building and being hostile and uncooperative when asked to withdraw by the police. Another plausible version asserts that right-wing police generals deliberately re-routed the marchers past Shell House in order to provoke inter-black pre-election clashes. This suspicion is backed by the absence of roadblocks around the ANC headquarters.

The events mark an important moment in the history of South Africa's political transition. Competing political forces laid claim to city areas and exclusive sovereignty; the Zulu migrants were mobilized in the name of a threatened ethnicity by expedient leaders who shortly afterwards switched sides: the Zulu king in whose defence against alleged ANC assaults the protesters rallied, joined the ANC camp after securing his economic fiefdom; Buthelezi's Inkatha decided to participate in the election at the last minute after the promise of international mediation of outstanding issues. They now work together in a government of national unity while their followers continue to kill each other. Each Christmas or Easter holiday when the male migrant workers return to their rural homesteads, the killing rate soars as old and new scores are settled among people who all live on the poverty line. Since Zulus fight mainly Zulus, the clashes cannot be labelled tribal or ethnic; as hardly any whites or Indians are ever affected, the conflict is clearly not a racial one either. Nor can it be described as entirely a political competition, in which unscrupulous leaders mobilize ethnic sentiment. Unlike Bosnia, the political élites of South Africa constantly preach peace and sit in the same cabinet. Nonetheless, they deal with each other as semi-sovereign entities akin to nation-states. Since the political heads come from different ethnic groups, their relationship is inevitably perceived as ethnic or tribal animosity.

Explaining Political Violence

Probably no other aspect of the South African conflict has elicited more divergent explanations and misinterpretations than the ongoing political violence. It is variously attributed to: (1) the old regime's double agenda and unreformed police; (2) a 'third force' of right-wing elements in the security establishment; (3) the Inkatha/ANC rivalry, engineered by an ambitious Buthelezi who fears being sidelined rather than treated as an equal third party; (4) the ANC's campaign of armed struggle, ungovernability, and revolutionary intolerance; (5) ingrained tribalism, unleashed by the lessening of white repression that resulted in 'black-on-black' violence; (6) the legacy of apartheid in general, migrancy, hostel conditions, and high unemployment among a generation of 'lost youth'. Helen Suzman, for example, singled out sanctions for at least 'part of the blame' in her 1991 presidential address to the Institute of Race Relations.[1]

Our analysis refutes any single-cause explanation. Rather than focus on the policies of various leaders, we find it more useful to examine predisposing social conditions, such as the rural–urban divide, the intergenerational cleavages, and the differential living conditions, social status, and heightened competition of long-time urban residents, shack dwellers, and migrants in single-men hostels. Regardless of peace accords signed at the top, antagonistic groups at the bottom often act violently, independent of leadership control. Such behaviour has, in particular, been undertaken by elements of the security establishment and the KwaZulu police, linked to agendas of destabilizing democratization and preserving chiefly rule.

Obviously, youthful activists are challenging traditional African patrimonialism, the role of petty bourgeois powerbrokers in the townships and chiefs in the countryside. Traditional authority clashed with the newly autonomous, better-educated segments of the urbanized working class. Analysts of political violence are aware of this larger context, but they list, label, and categorize a variety of contributory factors without explaining their origin or relationship. Here is a particularly comprehensive example by Cape Town political scientist Peter Collings (1990: 46): 'Gangsterism, vendettas, banditry, protection rackets, individual and group psychosis, competition for turf and treasure, a spreading mood of anarchy in which everyone thinks they possess a license to kill, the resurgence of antique hatreds, desperation born of unendurable poverty or fear—all these factors have variously contributed to outbreaks of violence.' At the least, one would want to weigh these causes against each other and denote the historical conditions under which they express themselves simultaneously.

Other South African academics readily point fingers at the system of apartheid as the cause of the township conflict. In the words of Rupert Taylor (1991), 'Apartheid has succeeded in engineering group divisions among the oppressed.' While one can agree with Taylor that the conflict 'is not of some essential ethnic forces', to blame it only on the manipulations of apartheid is an oversimplification. Ethnic antagonism exists in societies that do not have apartheid. Above all, in demystifying ethnicity, the analyst needs to show why the manipulators are so successful in constructing and exploiting ethnic cleavages. Apartheid did not invent all ethnic divisions; it skilfully utilized collective memories and distinct histories. Analysts who deny the historical reality invoke magic formulas to wish away deep-rooted perceptions that can be mobilized for progressive as well as retrogressive ends.

A much more promising and sociological approach to township conflict has been adopted by Lawrence Schlemmer (1991), who distinguishes between general background conditions, predisposing factors, and triggering events. Background conditions include the dislocation brought about by urbanization, the high levels of unemployment and dependency, the breakdown of the traditional family structure, and the erosion of normative restraints on murder. Short-term measures cannot easily address these conditions. On the other hand, predisposing factors and, in particular, triggering effects, Schlemmer points out, can be reduced or directly counteracted. Among these predisposing conditions are the social alienation of rural migrants in an urban youth subculture and the heightened competition for limited opportunities in a social climate of politicized mass action. Once set in motion, the violence becomes self-perpetuating. Distrust increases and is reinforced by partisan reporting. Clearly, interventions aimed at apportioning blame or taking sides are doomed to failure. Instead, bringing the feuding groups together, supporting independent monitoring committees, ensuring impartial policing, taking grievances seriously, and balancing the press reporting through a better understanding of both sides are more likely to reduce violence.

Schlemmer has rightly pointed out that the South African violence is not of the type of an internal war with 'massive revolutionary motivation at both the élite and popular levels'. Most striking is the constant talk of peace by the leadership but ongoing clashes on the ground, despite an overwhelming popular sentiment for reconciliation. Therefore, the violence is better classified as conspirational and turmoil-producing in nature, originating from politically active networks of people in several camps (the former National Party government, Inkatha, and the ANC) who act according to their own agendas. They view the leaderships' overtures cynically. In the words of the late ANC Natal Midlands leader Harry Gwala, 'No amount of talk between them will bring an end to the violence' (*Cape Times*, 29 Oct. 1992). Similar sentiments can be heard from extremists on the right-wing and in Inkatha. Localized acts of aggression against leaders and supporters of the various factions provoke immediate retaliation or else the many grievances are exploited to teach the other side a lesson. Even the long-standing factional fights between local communities in certain areas of

Natal are now carried out under the banner of political labels, though these disputes do not necessarily have an ideological content. The mediators and peace commissioners originate mostly from outside these semi-rural communities and, as ethnic outsiders, carry little weight. When local strongmen are literally dragged into these peace-making sessions, they consider the occasion at best another forum to assail the enemy or achieve a propaganda victory.

The State, Inkatha, and the 'Third Force'

At the height of the sanctions campaign and the civil war with the ANC, the South African Government viewed the anti-sanctions Inkatha movement as a valuable ally. It courted the free-enterprise advocate Buthelezi as a useful counterforce against the 'socialist' alliance between the ANC and the South African Communist Party (SACP). Initially, Pretoria was concerned that Inkatha might be eclipsed by the ANC. To prop up the only credible black moderate who was assumed to have a large following, the government provided propagandistic educational and military training assistance, delivered by front organizations of military intelligence, which a besieged Inkatha had invited, and readily accepted without considering the consequences of colluding with racist supporters.

The instigators and financiers of the violence can be found in the same circles of semi-independent military intelligence operatives who, disagreeing with de Klerk's policy change, wanted to see negotiation fail and the right-wing agenda succeed. P. W. Botha had accorded these securocrats unprecedented autonomy. De Klerk had cautiously begun to dismantle their institutions and allowed their illegal activities to be publicly exposed by the Goldstone Commission. Yet once part of a state-sponsored destabilization campaign to make the surrounding region economically dependent, the hard-line ideologues had acquired a life of their own, long after the state ideology had fundamentally changed. The securocrats' dealings with blacks merely reflected the much more vital struggle among whites. In the conflict over strategies for Afrikaner survival, even the faction in control of the state could not

afford to suppress its opponents directly. There is no question that the South African police, especially its black members, and particularly in KwaZulu, do not behave in an impartial manner; police have frequently colluded with Inkatha at the expense of ANC supporters. However, the right wing's obvious sympathy for Zulu warriors does not mean that Buthelezi is a mere paid stooge. However, such an image explains why various surveys report that among urban blacks Inkatha is more unpopular than the neo-Fascist 'Afrikaner Weerstandsbeweging' (AWB). Inkatha is blamed for instigating the violence by more than 50 per cent of supporters of all parties other than the Inkatha Freedom Party.

Widespread acts of omission, rather than rarer acts of commissioned violence, characterize the South African conflict. With the deeply ingrained stereotype that blacks are by nature violent people, police fail to protect ordinary residents from the fallouts of battles for political turf. Until August 1992 the state allowed 'traditional weapons' to be carried for likely confrontation and generally failed to secure prosecutions once clashes had taken place. The resulting distrust of the police among black activists does not facilitate collecting evidence for convictions. Witnesses are often intimidated. Police work for prosecutions is carried out so casually that only determined attorneys are able to pass the stringent hurdles to obtain conviction in independent courts. Since the police themselves have become frequent victims of counterattacks by political activists and criminal gangsters alike, frightened middle-class members of all groups sympathize with the few remaining 'agents of law and order' rather than demand their drastic reform.

To attribute all violence to a state-directed 'third force' does not explain the attacks against Inkatha officeholders and the police. Among the fatalities are equal numbers of ANC and Inkatha members or sympathizers. There is also evidence that both movements have their own 'third forces' that are not under the direct control of the national leadership. When former members of the ANC's armed wing (MK) 'take out' specific Inkatha targets in Natal and Inkatha warlords, policemen, and apartheid-trained special black units organize attacks against ANC leaders and sympathizers, it is difficult to ascertain which side has started the violence and which exercises revenge. Generally it can be said that the MK

318 H. Adam and K. Moodley

violence is carried out more professionally and with the use of sophisticated firearms. In comparison, the more primitive ('traditional') weaponry of Inkatha members lends itself more to random violence by excited mobs against anyone who is not part of the crowd. For example, unable to find those who have attacked the hostels, some hostel dwellers take revenge on those whom they believe are sheltering their enemies. In turn, this incenses the surrounding communities against all hostel dwellers—whether or not they were involved in the violence or are Inkatha supporters—and their children often cannot attend schools in the townships for fear of attack. Bus drivers are sometimes told by commuters 'not to stop for that hostel dog', and the mere identification as a Zulu-speaker carries the risk of death. Hostel dwellers compared the once proposed fencing of the hostels to the caging of animals in a zoo. The carrying of shields, spears, axes, and knobkieries ('traditional weapons') has become elevated to a question of asserting Zulu identity after a sensible government decision to ban the display of all weapons in public demonstrations. The two camps have become so polarized that in some areas nurses in hospitals refuse to treat patients who belong to the other side. While Inkatha accuses Mandela of using 'inflammatory language' at the United Nations when he suggested that Inkatha was a 'surrogate' of the transition government, Buthelezi responds in kind by urging his youth brigade to 'bugger up' the ANC or they and their future would be 'buggered' by a reckless ANC strategy (*Cape Times*, 7 Sept. 1992, 5). On the other side, the call by the late SACP Chief Christ Hani 'To clear townships of puppets', means exactly the same for IFP-aligned black councillors and policemen.

John Argyle has argued that much of the Natal conflict is now motivated principally by the desire for revenge and therefore resembles century-old blood feuds. In the past, such feuds originated not only over land but also over insults to honour or violations of women. Both sides cite provocation as a defence. As Kentridge (1990: 19) notes, 'In a war there are no aggressors; ostensibly no side ever initiates an attack . . . if an attack is made, it is always retaliatory'. Yet, while it is wise 'to believe neither side' until independent conclusive evidence is available, as Argyle cautions, one cannot simply blame the conflict on long-standing cultural traditions of

habitual feuding. The waxing and waning of the feuds can be traced to the changing conditions that precipitate or repress intergroup and interpersonal violence.

The continued violence by current and former state agents was nurtured by the failure of the de Klerk Government to make a moral break with apartheid. By not offering any apology, let alone compensation to the victims, the government reinforced the impression that previously a just war had been fought, though now, for tactical reasons, it had come to an end. Pik Botha, for example, in defending the Inkatha funding, reiterated defiantly that he would repeat it under the then prevailing circumstances. During the undeclared war, the state's killers had been celebrated as heroes. Their bravery in a war the de Klerk Government never refuted could not now suddenly be redefined as an atrocity, particularly since the perpetrators disagreed with the government's new tactical policy of reconciliation with 'terrorists'. By treating apartheid as a mere mistake and costly error, not a crime or a moral aberration, the de Klerk Government forfeited its opportunity to pressure unreformed state agents to now conform to the rules, rules that everyone knew had been ignored in the unofficial war.

Of course, de Klerk's cabinet could not announce a moral rebuff of apartheid because they had been involved in its conception and execution. They broke with the system not because it was wrong but because it did not work. Just as former Stalinists in the new Soviet republics now pose as democrats, so the former apartheid rulers now behave as reformers. In the absence of any programme for the moral rehabilitation of a society, the depth of social transformation remains unsure. When unrepenting incumbents remain in office, the past literally haunts the future and reveals itself in the continuation of the same practices, regardless of the new era.

'Tribalism' in Perspective

In a truly divided society it is almost impossible for an individual to assume any identity other than one of those prescribed by the communal division. In Northern Ireland, people see only Protestants and Catholics; in Israel, only Jews and Arabs. Within such frameworks, as many social scientists have suggested, people define and interpret social existence. In divided societies the everyday social reasoning based on familiar labels and expectations has little to do with doctrinal issues of religion. Rather, it signifies likely power and status differentials, based on a long history of communal conflict that encompasses every member of the community.

South Africa differs from those communal conflicts in that apartheid labels have engendered so much opposition and comprise such a variety of cross-racial common characteristics that many whites and blacks can afford to act as if they live outside their communally imposed category. In fact, the democratic non-racialism espoused by most major actors in South Africa is quite distinct from the entrenched, exclusivist communalism elsewhere. The state construction of official ethnicity in South Africa has no precedent in any other divided society, although the post-apartheid ethnic strife may in the end come to resemble communal divisions elsewhere.

One of the great surprises of 1990, therefore, was the sudden emergence of the 'tribal' factor. Hitherto, it had been taboo for the disenfranchised to talk publicly about Zulu and Xhosa forces. Only apartheid's ideologues used such labels. Why then has simmering ideological conflict suddenly been recast by participants as a tribal clash? While both leaderships proudly display their non-tribal stance, some of their followers, nevertheless, kill each other as Zulus and Xhosas.

Despite historical competitions and conflicts between the two Nguni-speaking people, the ANC, the premier organization to combat tribalism politically, has always had a fair succession of leaders from both ethnic groups and had managed to overcome tribal consciousness to a large extent—at least as far as the political activists were concerned. The ANC's last president before its banning, the Nobel Peace Prize winner Albert Luthuli, was a respected Zulu chief, and the ANC's National Executive Committee (NEC) in exile always included some, though disproportionately few, Zulus. NEC members assert that the ethnic background of candidates was never an issue; in some cases, they say, they were not even aware of the origin of comrades in exile, although this is flatly contradicted by other insider accounts (Ellis and Sechaba, 1992).

Similarly, the SACP's ideology had always embraced cosmopolitan internationalism, at least in theory. Attitudes gradually changed, however, after 1979, when Buthelezi's Inkatha movement began to pursue an independent policy. That the ANC initially approved of Inkatha is nowadays often unmentioned. Oliver Tambo in particular encouraged Buthelezi to assume his Bantustan role. The ANC served as the midwife of Inkatha, which was envisaged as an ANC internal wing, a Trojan horse, under the protective umbrella of a Bantustan, led by Buthelezi. A shrewd former member of the ANC youth league, Buthelezi knew his aristocratic descent from a family of the king's advisers would aid his popularity.

The ANC offspring quickly outgrew its parent, and Inkatha began to pursue an independent line. Buthelezi refused to recruit for the MK camps; to have complied would have been to risk direct clashes with Pretoria. Among the ANC's cadres Zulus became underrepresented. The much higher support for the ANC in the Xhosa heartland of the Eastern Cape and the ANC's invisibility in rural Natal are also reflected in the composition of the organization's leadership. Tribal separateness was reinforced not so much by Buthelezi's style or cultural symbolism as by his divergent policies. After all, the Xhosa Transkei was the first Bantustan to accept tribal independence, while the allegedly tribalist Buthelezi harboured national ambitions and refused to steer a secessionist course. But his rejection of the armed struggle and support of foreign investment deepened the rift between Inkatha and the ANC, and drew Inkatha closer to the South African establishment than to the anti-apartheid exiles abroad—despite Buthelezi's consistent anti-apartheid stance.

The turn of events traumatized the ANC, whose former ally had now become a potential partner in a dialogue that could further marginalize the exiles. The ensuring war of words culminated in the South African Youth Congress (Sayco) declaring Buthelezi 'an enemy of the people' in 1990. For the ANC, Buthelezi's name had become anathema. During the various meetings between ANC executives and South African academics that preceded the legalization of the resistance, one could talk rationally about any controversial subject, but not about Buthelezi. Similarly, many overseas South Africa watchers have adopted the ANC line that Buthelezi is 'a pathological case' and that Inkatha consists of 'a bunch of murderers' akin to the Khmer Rouge.

The violent struggle for territorial control of townships, squatter camps, and hostels has also entered the even more vulnerable factory floor. Owing mainly to the explicit political stance adopted by the Congress of South African Trade Unions (Cosatu), the formerly integrated union movement became fragmented and labour relations heavily politicized. Conflicts on the shop-floor emerged as the consequence of Inkatha's founding of the new union Uwusa, which was applauded and endorsed by shortsighted employers in retaliation for Cosatu's backing of sanctions. Elijah Barayi, the president of Cosatu, promised at a mass rally in Durban in November 1985 'to bury' Buthelezi. In retrospect, even Cosatu activists consider this declaration of war a serious political error. Jeremy Baskin (1991), a former national co-ordinator of Cosatu, now deplores Barayi's speech: 'It gave the impression that Cosatu's major aim was to oppose Buthelezi and the homeland system. His speech ignored the lesson learnt by Natal unionists over the years: winning workers in the region to progressive positions was achieved by hard organizational work and not by attacks on Buthelezi.'

In July 1990 Inkatha's perception of being under siege was heightened by the ANC–Cosatu decision to elevate the Natal regional violence into a national issue. National marches and strikes were supposed to demonstrate that Inkatha was not a national force, that it could be sidelined in the forthcoming negotiations and Buthelezi could be buried politically. Inkatha now felt compelled to demonstrate its clout on the Rand as well. The migrants in the hostels became the obvious force to be mobilized. If the ANC was going to demonstrate the irrelevance of a rival, Inkatha was going to prove its relevance through ferocious *impis*, as angry battle-ready groups are called. Isolated hostel dwellers, who were looked down upon by the ANC youths in the townships, were ready to teach them a lesson, although both segments of the urban proletariat had coexisted side by side for decades. With the exception of clashes between residents and migrants during the Soweto upheaval in 1976—disputes largely instigated by the police—tribal cleavages had never played a role in the multi-ethnic townships. In fact, many township residents, including the majority in Soweto, are

of Zulu origin. Migrants and permanent residents of Zulu, Xhosa, and mixed origin lived side by side in shabby hostels and backyards. Aggressors would be unable to distinguish township residents according to ethnicity. But with the hostels labelled 'Zulu' and the townships 'Xhosa and ANC', an ideological conflict and socio-economic cleavage became transformed into a tribal war. The transformation was triggered by the ill-advised ANC strategy to isolate Inkatha rather than include it in the broad anti-apartheid alliance. Former Inkatha General Secretary Oscar Dhlomo rightly concluded from his independent insider's perspective: 'Buthelezi has skillfully utilized ANC blunders to his advantage. He is now able to claim, thanks to the ANC, that anyone who demands the dismantling of the KwaZulu government is challenging not only the Zulu nation but also the Zulu King' (*Sunday Tribune*, 26 Aug. 1990). Although the violence in Natal is between Zulus, any conflict acquires a tribal connotation as soon as the chiefs, *indunas* (headmein), *sangomas* (witchdoctors), and shacklords present it to their large followings as a matter of defending the traditional order against 'outsiders'. And by labelling hostilities in tribal terms, the leaders reinforce this perception and broaden their own constituencies. 'I want to make it quite clear that ANC attacks are not only attacks against Inkatha', said Buthelezi at a rally in Bakkersdal, 'they are attacks against Zulu people just because they are Zulu' (*Guardian Weekly*, 12 May 1991). Similarly, KwaZulu minister Ben Ngubane charges, 'We are in a critical situation. The police are stretched to the limit and cannot protect us. Our options are stark: either we run for cover or we defend ourselves. Obviously we are opting for the latter' (interview, 25 Oct. 1992). Thus, as in most ethnic conflicts, each faction views itself as the victim of the other's aggressions, against which self-defence is only natural.

Intergenerational and Urban–Rural Cleavages

Traditional Zulus who have a reconstructed memory of a pre-colonial independent kingdom are deeply offended by the young comrades' rejection and denial of this identity. Inkatha supporters represent the poorer, more rural, more illiterate, older and more traditional section of the 8 million Zulu group. Historically, Zulus were conquered last and were therefore considered better organized and more resistant to colonization. They developed, like the Serbs in their battle with Muslims during the Middle Ages, a 'heroic warrior' mythology of bravery and invincibility. Only Zulus cherish the memory of powerful kings. Only Zulus developed a state that pre-dates colonialism. This ethnic memory and cultural heritage are bound up with Zulu identity, invoked, manufactured, and mobilized by traditional élites who are threatened by democratic modernity. They deeply resent the uprooted, urbanized youth that reject the tribal authorities and embrace the ANC.

In the view of traditional Zulu leaders, the younger generation's attitudes threaten the established cultural hierarchy. Interviews of Inkatha members in shack settlements around Durban encountered frequent resentment about generational disrespect: 'You get these young comrades in the township who come to the older men in the hostels and tell them they must stay away. They show no respect.' A midday Saturday march of 40,000 ANC supporters through central Durban two days before the 1994 election was comprised almost exclusively of young men and women, who made no bones about their mocking attitude towards IFP-inspired Zulu traditionalism. Despite an IFP youth league with support among religiously oriented, rural youth, particularly women from conservative families, it is obvious that the IFP has lost the support of most urban, secularized, and alienated young voters. Since this constitutes the fastest growing segment and the urban influx is likely to accelerate, the long-term prospects of Inkatha electoral gains are dismal for mere demographic reasons.

While ethnic revival in the context of self-help associations among newly urbanizing peasants occurred frequently elsewhere in Africa, rural Zulu ethnicity has become so stigmatized in the polyglot South African townships outside the hostels, that this historical legacy works against ethnicity maintenance. The hegemonic sentiment has stereotyped Zulus as backward, violent, unsophisticated newcomers to the extent that only the very insecure or ignorant would want to seek shelter in the proud display of heritage. In short, as an organizing principle outside the hos-

tels, Zulu identity is a liability rather than an asset in gaining access to scarce resources controlled by political opponents. Employers now shy away from hiring 'Inkatha trouble-makers'. Such attitudes in a climate of silent disdain further reinforce the marginal position of industrial latecomers.

The call for cultural revival is heeded by the most deprived among the Zulu people in search of responses to their humiliation. The invocation of a mythical past and images of pride and success in battle offers a source of dignity and identity to the rural poor, the hostel dwellers, and unemployed migrants. In the predicament, tribal identification is a badge of honour. Similar frustrations exist on the other side, and the conflicts between youthful ANC supporters and traditionalists among Xhosa squatters follow a similar pattern. In Cape Town, for example, a vicious war between two black taxi organizations reflected the conflict between two patronage groups, each claiming allegiance to the ANC. The heightened competition for city routes and an outdated permit system that favoured newcomers over old-timers represented a local variation of clientelism among a deprived Xhosa group, a situation that had been played out among Zulu speakers in Natal many times before. In Cape Town it is neither tribal animosity nor political ideology that has caused such deep rifts and violence among the deprived.

Strikes and mass protests enforced through intimidation by overzealous youth constitute another trigger for violence. In December 1990, at the Mandelaville squatter camp near Bekkersdal, youth belonging to the ANC-aligned Bekkersdal Youth Congress wanted to stop pupils in the area from writing end-of-year examinations. The rival Azanian Students' Movement and the Azanian Youth Organisation rejected the move, and the ensuing fighting left a trail of death and destruction (*Star*, 20 Dec. 1990).

Urban–rural tensions in the townships are marked by generational, cultural, and political differences. The predominant urban black identity emerged from a mixture of traditional elements and rural customs, the street wisdom of survival in the townships and in the modern work-place, and the consumerist aspirations of secular Western society. This politicized, individualistic urban culture defines itself in sharp contrast to the ethos of the rural inhabitants and migrants, who are considered illiterate, unsophisticated country bumpkins. In the status hierarchy of the townships, the people with rural ties are often scorned as ignorant ancestor-worshippers who do not speak English and practice a social life of tribalism and witchcraft. The denial of ethnicity and rejection of most cultural traditions by urban blacks reflect not only the apartheid attempts to manipulate ethnic differences, but also an arrogant predilection to associate rural customs with false consciousness.

The ANC leadership embodies the urban views of those who have left tribalism behind and now wear suits and ties. At most they may stage tribal traditions as ceremonial events, which they attend with amused smugness in much the same way as some urbane Westerners enjoy folk dances. The ANC's internationalism and cosmopolitan universalism jars with the attitudes of the traditional African rural population. For many of them, the ANC appears as an élitist urban group whose leadership speaks English and looks down upon the ethnic customs of the peasants.

Many people in the rural communities and the migrant hostels deeply resent the political activism of the urban-based youth as a subversion of the traditional order in which children obey and politics is left to the elders. For the older generation, youthful activism is an ungrateful waste of the educational opportunities for which the parents sacrificed so much. On the other hand, youth accuse their parents of having compromised themselves with the system. This generational conflict has torn apart many families and pitted communities against each other, particularly in the semi-urban settlements surrounding Pietermaritzburg, where rural and urban values clash directly under conditions of dire poverty.

In some parts of Natal, in the Durban townships of Lamontville and Chesterville for example, youthful activism also has greater space because the area has traditionally been one of free-hold settlements not under the jurisdiction of the KwaZulu Bantustan authorities. Their potential incorporation into KwaZulu was particularly resented by the residents in the 1980s. In Durban, in contrast to black life in Johannesburg or Cape Town, KwaZulu reaches right into the suburbs, and the rural and urban exist side by side throughout much of Natal. Since KwaZulu never applied influx-control measures, unplanned squatting on the outskirts of cities was common, while the rural newcomers found a much more regulated and planned environment in Cape Town or Johannesburg.

The late conquest of Natal compared with other parts of South Africa meant that the traditional economy remains more intact there than elsewhere. The failure of employers to hire Africans as supervisors and middle managers also impeded African upward mobility in Natal. Although the majority of homesteaders in rural Zululand are dependent on remittances from migrants in Johannesburg, fewer families have moved out permanently, and the majority of migrants consider the rural area their home, to which they periodically return and plan to retire. Together with the reconstructed and revitalized memory of more successfully organized resistance against colonial conquest under powerful kings, a more traditional way of life has survived among segments of Zulu speakers. Both their self-definition as proud warriors as well as their objective differentiation in attitudes and geographical movements form the background to the clashes in the cities.

Isolated Migrants in Single-Men Hostels

The single-men migrant hostels on the Rand, the majority of whose inhabitants are Zulus, provided a flashpoint of resentment for both hostel dwellers and local residents. In Alexandra in March 1991 close to a hundred people died in riots that erupted after an unprecedented accord was signed between the Transvaal Provincial Administration and the ANC-controlled Alexandra Civic Organisation. The agreement amounted to a local model of what had yet to be negotiated at the national level: phasing out of the black council, which was widely considered to consist of corrupt collaborators, and the placing of township land on the Far East Rand under the joint control of ANC-aligned civic organizations and the surrounding white areas. To all interests and purposes, the accord represented an ANC victory over the traditionalists who, under 'Mayor' Prince Mokoena, had controlled Alexandra. Mokoena's office had been occupied for several weeks by community activists who put up posters declaring 'Away with Mokoena—out of our hostels'. Many councillors had been killed or driven from office during the ANC campaign to render the townships ungovernable. Many remaining councillors therefore aligned themselves with Inkatha for protection. Like the councillors, the hostel dwellers expected they too would be driven out of Alexandra, although the civic organization denied such intentions. However, the accord provided for the 'upgrading and possible conversion' of the hostels, and the perception easily spread that the hostels were to be demolished. The political power struggle acquired an ethnic dimension when the besieged Mokoena mobilized the hostel dwellers by appealing to Zulu pride with the slogan that the other residents were 'undermining the Zulus'.

The migrants, at the bottom of the social hierarchy, had always felt humiliated by the better-off township residents, and sexual rivalries were not uncommon. Now the single men felt that their homes and very existence were threatened. Although the civic organization claimed it had consulted the hostels in all negotiations, such consultations clearly did not forge any bond or political loyalty. On the contrary, the promise of better housing for families was perceived as abolishing the last foothold of the illiterate migrants in the city, many of whom did not want their rural wives and children to stay with them in the ramshackle hostels. Yet they had attached meaning to their own deprivation. The *Weekly Mail* (22–27 Mar. 1991) observed: 'The stench-ridden "single-men's" hostel, built decades ago for migrants laboring in the factories and homes of Johannesburg but with no legal right to bring their families to the city, resembles a prison. Yet it is home to these men and their loyalty to such a place is surprising.'

Inkatha's hardline attitudes towards illegal immigrant workers also proved more popular than the ANC's exhortations for empathy and solidarity. Gangs in Alexandra claiming to be ANC and SANCO (South African National Civic Organisation) officials, for example, have evicted illegal immigrants from their homes to move in themselves. The local ANC, despite pleas for 'humane treatment for people who have come to South Africa in search of peace and work', nonetheless had to back Home Affairs Minister Buthelezi's call for deportations and a crackdown on employers of illegal foreigners. 'Trade union's hard-won settlements are being undermined by bosses who use immigrant workers' declared Gauteng ANC leader Obed Boysela (*Cape Times*, 24 Jan. 1995). The xenophobia rose following rumours that the community would benefit from reconstruction and development finances for housing if all illegals were removed. The ethnic tensions in Alexandra clearly point to the material impoverishment and scarce elementary resources as the underlying cause of xenophobic frictions. As in California or Western Europe, poorer foreigners are blamed by the underprivileged for their own misery.

The sense of social isolation and marginalization felt by the migrants was reinforced by trends in union politics. As Eddie Webster, professor of sociology at the University of the Witwatersrand, has pointed out, union leaders are increasingly drawn from the better educated and more skilled urban-based stratum. At the same time, he notes, union leaders have also become involved more in national politics, and the sectional interests of rank-and-file members have been sacrificed to the overall demands of a national agenda. Among those most hurt by these trends are the shop-floor representatives of the migrant underclass. Webster concludes: 'Indeed, the failure of unions to address hostel-dwellers' grievances has contributed to the feeling of alienation among many and made them an ethnic constituency more easily mobilized by Inkatha's labor wing, Uwusa' (*Business Day*, 2 Aug. 1991).

As the political violence in the Transvaal continued, the Inkatha leadership also felt besieged by a prospective ANC–National Party deal. In response to the labelling of Inkatha as a 'minor party'—while both the government and the ANC referred to themselves as 'senior players'—Buthelezi bluntly warned de Klerk at the opening of the KwaZulu Legislative Assembly that Inkatha would 'tear down piece by piece and trample on' any plan that the NP and the ANC designed in a private arrangement. In this respect, the political violence in the Transvaal benefited Inkatha by demonstrating the national scope of the party. The defensive aggression need not have been orchestrated from above, as the ANC asserts, given the enmity on the ground. However, the clashes weakened the ANC and gave Inkatha the profile it could not expect from elections.

The fury and irrationality with which gangs of hostel dwellers have lashed out indiscriminately against township residents can only be understood in this context of isolation and anxiety. The magic rituals among the combatants in Natal, the fact that many could not even identify the cause or the name of the leader for who they were fighting, point to their search for symbolic compensations to counter depths of powerlessness. Such despair can readily be exploited by the strongmen on each side. During the battles in Natal, both sides 'press-ganged' youth into the fighting. In some areas, each household was required to pay 10 rand for 'equipment', a euphemism for weapons and *muti* (traditional medicine). 'People's courts' implement the 'call-ups' and war taxes with a hundred lashes for offenders. On both sides elements of traditional superstition motivate the combatants: 'Before we go into the fighting, some people at the houses near the battlefield stand outside with buckets of water and *muti*. They dip a broom into the mixture and sprinkle it over us as we run past. If you want extra protection, you can also go to an *inyanga*, but that costs more. Comrades believe that the *muti* will stop the bullets from hitting them and will give them courage' (Carmel Rickard, 'When You See the Enemy's Shacks Blaze, You can't Help Feeling Good', *Weekly Mail*, 23 Feb. 1990, 3). The same 'comrade' describes an enemy who was shot and tried to run away but fell and was stabbed. Then someone cut off his genitals and took them away.

Mutilations are reported in many communal conflicts. People are not just killed in Yugoslavia or Azerbaijan but in addition are often grossly disfigured. This unexplained practice points perhaps to deep-seated feelings of emasculation. It has yet to be satisfactorily explained by psychological insights, which are usually neglected in favour of the focus on rational competition. Such mutilation robs the enemy not only of his life but of valued qualities that the victor symbolically appropriates: potency or eyesight or brains. To possess the vital organs of the enemy is to possess power, invincibility, and immortality. The more powerless people are, the more they become obsessed with the symbols of power. The rituals of protest and the preoccupation with an imagined armed struggle reveal other dimensions of the same phenomenon. Like the foot-stamping *toyi-toyi* (war dances) and strident war

songs ('Kill the Boers') once ritually performed at rallies and the repetitive shouts of 'Amandla!' (power), militant attitudes and militaristic gestures express something deeper than mere militancy. For the powerless, images and symbols of power must substitute for real clout.

Deep humiliation results in fantasies of power. The more the ill-fated armed struggle fades into the background, the more some of the township youth want to resurrect it. At rallies they sport a new folk art: imaginatively designed and carefully assembled imitations of homemade guns, MK47s, and bazookas, often grotesquely oversized. The grim-looking bearers shout martial slogans and brandish their war toys, hoping from them to borrow the strength needed to conquer their own anxieties, like the children who pose on the tanks at war memorials in other societies.

Conclusion

There are two common objections to a policy of reconciliation with regard to Inkatha. First, support for Inkatha is so weak that the movement can be ignored, pre-empted, or even eliminated. Second, the price demanded for incorporation by Buthelezi is too high. Against these arguments Schlemmer (1991) has pointed out that conflict resolution has to take into account not only the size and scope but also the intensity of interests: 'The intensity of the IFP's interaction in the political process has clearly signalled the potential costs of excluding it, or reducing its leverage in negotiations.' From a moral point, this position can be interpreted as yielding to violence. From a pragmatic perspective, however, there is little choice if greater damage is to be avoided. Weighing the costs of continued confrontation against the potential benefit of peaceful competition through compromise amounts to a political calculation that separates ideologues from pragmatists.

Among the majority of Zulu supporters of Inkatha, relationships between leaders and followers are based not only on ideological identifications and ethnic symbolic gratifications, but also on reciprocal instrumental advantages. Inkatha's poor and illiterate constituency depends on patronage, handed out by strong leaders and local powerbrokers in return for loyalty, regardless of a leader's ideological outlook or ethical behaviour. Political powerlessness reinforces the importance of African auxiliaries to whom the impoverished can turn for protection and favours. When the apartheid state decentralized control by letting trusted African clients police themselves and administer their own poverty, the leaders' status and importance were further strengthened. Thus emerged a classical system of clientelism and patrimony. Clientelism flourishes in conditions of inequality, where marginalized groups depend on patronage networks for survival, or at least for small improvements. It is the exclusive control of scarce goods (permits, houses, civil-service positions) that give patrons their power. Clientelism thrives with rightlessness.

Once equal citizenship, however, gives formal access to basic goods to all, and all are entitled to equal treatment, the monopoly of patrons is undermined. If the police, for example, act impartially, there is no need to be protected by a warlord. If people acquire confidence in the law, they need not rely on vigilantes. If justice is administered by impartial courts, it need not be sought through private vendettas. Thus, a democratic equality that allows claims to be made through formal channels pre-empts local dependence on informal patronage.

It remains to be seen whether equality before the law and new life chances for the formerly disenfranchised will pre-empt clientelism and the quest for ethnic separateness. Observers remain sceptical. Former *Sunday Times* editor Ken Owen notes that a relatively autonomous Natal in a federal structure might be a co-operative partner in a greater South Africa, 'while a KwaZulu forcibly incorporated in a structure controlled by its bitterest enemies, might become as indigestible as the IRA in Britain, or the Turks in Cyprus, or the Basques' (*Sunday Times*, 8 Sept. 1991). Buthelezi too has threatened that the civil wars in Angola and Mozambique could pale in comparison with the future destructive upheavals in South Africa. The ANC, on the other hand, is not inclined to heed such predictions and would rather risk a repetition of Biafra than compromise on the relative centralization of political power or bend towards recognition of Zulu claims. This approach sets the new South Africa on a collision course.

The ANC has treated the IFP more as a security problem than as a political opposition with

which the ruling party has to co-exist. Calling the IFP a 'bandit organisation' (Steve Tshewete of the ANC) or dismissing Buthelezi as a 'mean-spirited, gutter-mouthed politician' (Roger Burrows of the DP, *Business Day*, 12 Apr. 1996) may escalate the verbal warfare and delegitimize the party in the eyes of its detractors. However, the fact remains that the party received 50 per cent of the 1994 votes in South Africa's most populous province, even if the result continues to be suspect as a negotiated election (Johnson and Schlemmer, 1996). Instead of hoping that the IFP would cease to exist 'after Buthelezi', the new powerholders would act more wisely if they were to strengthen the moderate elements in the IFP who share much of the ANC's concerns about avoiding the ongoing strife in a province that desperately needs economic growth. With a declining support base due to continued urbanization, the IFP-oriented chiefs could well be driven into more regressive stances unless the stronger ANC accommodates the feudal remnants in a similar prudent compromise as the dangerous white right-wing has been neutralized. Concessions towards an imaginative federalism and sincere efforts to include the boycotting IFP in policy-making promise to be more effective long-term solutions to the ongoing political violence than deploying more soldiers and policemen.

Note

1. A vast literature now exists on political violence in South Africa, though mostly of the propagandistic kind. Among the noteworthy academic works are Brian McKendrick and Wilma Hoffman (eds.), *People and Violence in South Africa* (Cape Town: Oxford University Press, 1990); and N. Chabani Manganyi and André du Toit (eds.), *Political Violence and the Struggle in South Africa* (Johannesburg: Southern Books, 1990). The Natal conflict is perceptively analysed by Herbert Vilakazi, 'Isolating Inkatha—A Strategic Error?', *Work in Progress 75*, June 1991, 21–3; and by Mike Morris and Doug Hindson, 'South Africa: Political Violence, Reform and Reconstruction', *Review of African Political Economy*, 53 (1992), 43–59.

On Inkatha and Buthelezi, the published accounts are almost unanimously hostile. See e.g. Gerhard Maré and Georgina Hamilton, *An Appetite for Power: Buthelezi's Inkatha and South Africa* (Bloomington: Indiana University Press,

1987); and Mzala (pseudonym of a deceased ANC activist), *Gatsha Buthelezi: Chief with a Double Agenda* (London: Zed Books, 1988). Empirical work is rare but see e.g. John D. Brewer, 'The Membership of Inkatha in KwaMashu', *African Affairs*, 84 (Jan. 1985), 111–35.

On the malleable nature of ethnicity as a mental construct, juxtaposed to its endurability and primordiality, see Leroy Vail (ed.), *The Creation of Tribalism in Southern Africa* (Berkeley: University of California Press, 1989), particularly the contribution by Shula Marks, 'Patriotism, Patriarchy and Purity: Natal and the Politics of Zulu Ethnic Consciousness'. Gerhard Maré, *Brothers Born of Warrior Blood* (Johannesburg: Ravan Press, 1992) writes with a similar perspective, as do, from a more general viewpoint, E. Hobsbawm and T. Ranger, *The Invention of Tradition* (Cambridge: Cambridge University Press, 1983). Against this literature stands an equally adamant body of writing on the primordial and socio-biological nature of ethnicity, perhaps best exemplified by Pierre van den Berghe's *The Ethnic Phenomenon* (New York: Elsevier Press, 1981).

Matthew Kentridge's *An Unofficial War* (1990), like most other journalistic efforts, describes the impact of the war on people's lives but does not analyse the forces that have caused and sustained the conflict.

For an account of the 1994 election battle in KwaZulu–Natal, see Alexander Johnson, 'The Political World of KwaZulu', in R. W. Johnson and Lawrence Schlemmer (eds.), *Launching Democracy in South Africa* (New Haven and London: Yale University Press, 1996). See also G. Hamilton and G. Mare, 'The Inkatha Freedom Party', in A. Reynolds (ed.), *Election 94* (Cape Town: David Philip, 1994), 73–88; and A. Johnston, 'The IFP Enigma', *Democracy in Action*, 31 May 1995, 5–6.

References

Baskin, Jeremy (1991), *Striking Back: A History of Cosatu* (Johannesburg: Ravan Press).

Collings, Peter (1990), 'Pigs, Farmers and Other Animals', in *The Watershed Years* (Cape Town: Leadership Publication), 43–54.

Ellis, Stephen, and Tsepo, Sechaba (1992), *Comrades against Apartheid: The ANC and the South African Communist Party in Exile* (Bloomington: Indiana University Press).

Gordimer, Nadine (1995), 'The Shell House Affair', *New Left Review*, 213: 125–9.

Johnson, R. W., and Schlemmer, Lawrence (1996) (eds.), *Launching Democracy in South Africa* (New Haven and London: Yale University Press).

Kentridge, Matthew (1990), *An Unofficial War: Inside the Conflict in Pietermaritzburg* (Cape Town: David Philip).

Schlemmer, Lawrence (1991), 'Negotiation Dilemmas after the Sound and Fury', *Indicator South Africa*, 8 (3): 7–10.

Taylor, Rupert (1991), 'The Myth of Ethnic Division: Township Conflict on the Reef', *Race and Class*, 33 (2): 1–14.

21

Political Bosses and Strong-Arm Retainers in the Sunni Muslim Quarters of Beirut, 1943–1992

Michael Johnson

In the mid-twentieth century Lebanon's service-based economy developed such that Beirut became the banking and trade centre for the Arab Middle East. The relative decline of Lebanese agriculture, coupled with the emergence of capitalist farming in what were previously neo-feudal estates, led to an increasing migration to the city. A small and dependent industrial sector remained subordinate to the interests of commerce and finance, and did not provide work for significant numbers of immigrants who simply swelled the ranks of the semi-employed sub-proletariat.[1] The overwhelming majority of the labour-force was employed in the service sector. Just as there was peasant individualism in Lebanese villages, so Beirut's society was characterized by sub-proletarian and petty-bourgeois individualism. There was little or no class consciousness, and political parties (with one exception, the Christian Phalangists) were weak to the point of insignificance, as were trade unions and other interest groups. Religion divided Beirut, as it did the rest of Lebanon, into two roughly equal confessional groups, the Christians and the Muslims, which were further subdivided into a plethora of sects.

Partly as a response to the fragmented and individualistic electorate, Lebanon's democratic political system became dominated by locally powerful leaders called 'za'ims' (Arabic: *za'im*, pl. *zu'ama*) who in the cities developed sophisticated machines to recruit a clientele. This essay starts by describing how these machines operated in the Sunni Muslim quarters of Beirut before the outbreak of civil war in 1975, and emphasizes the role of neighbourhood strong-arm men who recruited and policed the clienteles of the za'im. The strong-arm retainers were called 'qabadays' (Arabic: *qabaday*, pl. *qabadayat*), a word which implies the use of physical force to defend and promote honour and leadership. In the latter part of the essay it is shown how the Sunni qabadays of Beirut began to assert themselves as leaders in their own right during the early 1970s, and how they became completely independent of the za'ims in the Lebanese civil war.

Although a few published sources on Lebanon's political economy are cited in the notes, most of the information was collected by the author in periods of field-work in the Sunni quarters of Beirut. Thus the 'clientelist system' and its breakdown, which are analysed here, are

This is a revised and updated version of an essay that appeared in Josef Gugler (ed.), *The Urbanization of the Third World* (Oxford: Oxford University Press, 1988), 308–27. It is based on field-work conducted by the author in West Beirut during 1972 and 1973 as part of a Manchester University project, funded by the British Social Science Research Council, on Lebanese politics and society, and on interviews with Beirutis over the years.

specific to Sunni Beirut. There is, however, some evidence to suggest that similar changes took place in similar political systems in other parts of urban Lebanon.

Political Bosses

At the end of the First World War, to the delight of Lebanese Christians, the League of Nations awarded the Mandate for Syria to France. At first the Sunni notables of Beirut, accustomed to being the agents of Ottoman rule, refused to have anything to do with what they regarded as a foreign and Christian imposition on a Muslim majority. But Sunni za'ims gradually began to participate in the new democratic institutions, and they eventually co-operated with the Christians in founding the independent state of Lebanon in 1943. Parliamentary democracy provided access to governmental patronage and, in order to maintain a large clientele, it became essential for za'ims to be regularly elected as deputies and appointed as ministers. As part of the elaborate system of confessional checks and balances in Lebanon, the office of president was reserved for a Maronite Christian. But the Sunnis were given the premiership, and conflict between Sunni za'ims in Beirut was mainly concerned with competition for this powerful office.

Most of these Sunni za'ims were descended from the notable families of the early twentieth-century Ottoman period. They inherited the wealth and clientele of their fathers, forming the basis of their electoral support in independent Lebanon. The clientele was bound to the za'im by a network of transactional ties, where economic and other services were distributed to the clients in exchange for consistent political loyalty. This political support usually took the form of voting for the za'im and his allies in parliamentary elections, but the clientele could be required to support the za'im in other political conflicts, and could even be expected to take up arms in disputes with other za'ims. A considerable amount of ritual support was involved in the patron–client relationship, and clients had to demonstrate their loyalty in a variety of ways: on feast days, clients visited their za'im to wish him the compliments of the season; and when a za'im returned from a journey, his supporters usually turned out to welcome him home. Such occasions often involved the supporters slaughtering sheep and holding a great reception, during which the za'im's armed retainers would fire their automatic rifles into the air—a popular method of expressing loyalty, political strength, and jubilation.

The za'im maintained his support in two important ways: first by being regularly returned to office, so that he could influence the administration and continuously provide his clients with governmental services; and secondly, by being a successful businessman, so that he could use his commercial and financial contacts to give his clients employment, contracts, and capital. Depending on the wealth and influence of his clients, the za'im provided public-works contracts, governmental concessions, employment in the government and private sectors, promotion within the professions and civil service, free or cheap education and medical treatment in government or charitable institutions, and even protection from the law. In order to survive electoral defeats and periods in opposition, the za'im had to be rich, or have access to other people's wealth, so as to buy the support of the electorate as well as the acquiescence of ministers and officials responsible for particular governmental services. Although election to the Assembly was of considerable advantage, it was not always essential, and a za'im could survive temporary periods of opposition and political weakness by using the credit he had built up in the past. Thus, for example, a judge, who owed his original appointment to a particular za'im, would probably continue to be lenient to the za'im's criminal clients even if the za'im were no longer in office; and a businessman, who had been enriched by the za'im granting him a contract or concession, could be expected to donate some money to the campaign fund when the za'im was in opposition. In both cases, the grateful clients might eventually be bought off by rival za'ims. Such changing allegiances were not uncommon, and lower down the social hierarchy large-scale defections from the clientele took place when a za'im failed to deliver the goods.

Za'ims were not elected on the basis of a programme, but on their ability to provide their clientele with services. In this sense, the clients' support was a transactional obligation, rather than a form of moral, ideological loyalty. But although national and programmatic appeals were not usually made by za'ims, they did make some moral appeals, and sought to woo their electors by posing as local champions and confessional representatives. Thus a Beiruti Sunni za'im appealed to Beirutis *qua* Beirutis, and to Muslims *qua* Muslims. Za'ims did win the support of constituents who were not of their religion. Indeed, the electoral system compelled them to do this.[2] But the major part of their support usually came from their own sect. During the 1950s, for example, Sunni za'ims in Beirut put forward demands for increased Muslim representation in the administration, and increased influence in what they described as a Christian-dominated state. Similarly, by providing governmental services for his constituency, the za'im was able to demonstrate that he was an active local representative. During the 1960s, when the reformist 'Shihabist' regimes of Presidents Shihab (1958–64) and Hilu (1964–70) diverted governmental resources to the poorer, outlying regions of the country, Beirutis complained that contracts and jobs were going to non-Beirutis at their expense. Sa'ib Salam,[3] consistently excluded from the premiership during this period, was thus able to capitalize on this feeling, portray his competitors as traitors to their city, and maintain the support of a large part of his clientele even though he was politically weak for such a long period of time.

Strong-Arm Retainers

A defining characteristic of the za'im was his close coercive control of the clientele. All Lebanese politicians entered into transactional relationships with their constituents, but what distinguished the za'im was his sophisticated machine and his willingness to use force, not only in attaining political objectives, but also in maintaining the loyalty of his clientele. A za'im did not necessarily set out to bully his clients into accepting his leadership. But in building his apparatus to recruit and control the machine, he made alliances with strong-arm neighbourhood leaders or qabadays, whose support was based, in part at least, on coercion.

The word *qabaday* is generally supposed to be derived from the Arabic verb 'to grasp or hold'

(*qabada*). But the same word is found in Turkish (*kabadayi*) where it means 'swashbuckler, bully, tough'. In Lebanon *qabaday* is an ambiguous word which can have positive or negative connotations, although it is usually a sign of approval and is used to describe someone who is quick-witted, physically strong, heroic, or possessed of some other supposedly masculine attribute. Negative connotations include bullying, throwing one's weight around, and other forms of unruly and obstreperous behaviour. In its most specific sense, *qabaday* is a title given to a street or quarter boss who combines both positive and negative attributes into a leadership role. Here the qabaday recruits a following on the basis of his reputation as a man of the people, as a helper of the weak and the poor, as a protector of the quarter and its inhabitants, and, most important, as a man who is prepared to defend his claims to leadership by the use of force. All these characteristics might apply to the za'im as well as the qabaday, but prior to the outbreak of civil war in 1975 the two leaders could be distinguished by their social origins and levels of leadership. Whereas the za'im was born of a rich notable family and was recognized as the leader of a large following, the qabaday's parentage was of low socio-economic status and his political influence was limited to his immediate neighbourhood. Typically, the qabaday was a criminal involved in protection rackets, gun-running, hashish smuggling, or other similar activities. The za'im provided him with protection from the police and the courts in return for his political loyalty and services. These services included recruiting and controlling the za'im's clientele, organizing mass demonstrations of support, and, if necessary, fighting for the za'im in battles with his rivals.

The qabaday was particularly well suited for ensuring the loyalty of the za'im's clients because although he could use his relative wealth and his underworld connections to help the inhabitants of his quarter, he could also use physical force to control the clientele. One informant described in the following way how a qabaday established himself:

A qabaday starts by throwing his weight around, picking quarrels and beating up those who don't treat him respectfully. For example, if he is sitting on the street corner playing *tawila* [a type of backgammon], and someone from the quarter walks by without wishing him a good day and enquiring after his health, he might beat him. Later he'll probably shoot someone. Perhaps he'll kill another criminal, perhaps someone in a quarrel. The important thing is not so much who he kills, but the way he does it. Anyone can shoot someone. A qabaday has to do it openly and be willing to accept responsibility, and possibly go to prison for about four years.

In other words, the importance of the killing was the symbolic nature of the act. A qabaday was someone prepared to promote his leadership claims by an open murder in complete disregard for the law. He accepted the risk of imprisonment and a protracted vendetta, but in committing the act openly he won the respect of his fellows. Because physical strength and notions of honour were (and are) highly prized values, the qabaday attracted a local following and could therefore expect the protection of a za'im and a reduced prison sentence for his crimes. He acted outside the law, but was governed by another code of norms accepted by the society in which he lived. Some Beirutis, particularly the highly educated, did not accept this code, and for them the qabaday was az'ar, a criminal, thug, or murderer. But in the poorer quarters, where the qabadays were particularly powerful, az'ar was a term reserved for a man who killed dishonourably, for a robber of the poor and, most significantly, for a qabaday who worked for an opposing za'im. One man's qabaday was thus another's az'ar, but most people had ambiguous feelings about their local strong-arm men. They supported them partly out of respect for their honourable qabaday qualities, and partly out of fear of their willingness to use force.

One way Muslim Beirutis expressed and attempted to resolve this ambiguity was to indulge in myth-making, particularly about those qabadays who operated in the early part of this century. In these stories the qabaday aided the weak and the poor, fought against Christian attacks on Muslims, and protected the Sunni quarters against the excesses of the Turkish and French governments. Although the myths admitted the criminal activities of many of these leaders, they always emphasized the qabaday's basically honourable character and the services he performed for his neighbours. Of course, the stories changed according to who was the narrator and who the listener, and it is interesting to note that, as a 'Christian foreigner', I was often told the qabaday was a repository of all Arab

virtues, that he was hospitable to strangers, and that Muslim and Christian qabadays were friends who regularly visited one another's houses. So according to the myth, the paradigmatic qabaday was the strong, honourable quarter leader of the Ottoman period. A coffee shop that I used to visit in the early 1970s was owned by just such a leader. As an old man of some 90 years, he was able to lay claim to the 'true qabaday status'. He never smoked cigarettes, nor drank alcohol; he wore traditional dress, and much regretted the decline of traditional moral standards. His predominantly youthful customers accorded him great respect, greeting him with elaborate courtesy and ritual, including kissing his hand. Although they broke most of the old man's rules of morality, they always compared him favourably with what they called the 'new qabadays':

The real qabaday has disappeared. Nowadays the new qabadays drink in nightclubs, smoke hashish and deal in cocaine and heroin. In comparison with the 'Hajji' [the coffee-shop owner had made the pilgrimage to Mecca] they are all weaklings. Look at him; even at his age he is still physically strong. These new men are only strong because they carry a pistol.

Nevertheless, despite their opinion, which they shared with many other people, most of the young men respected the new qabadays, valued their friendship, and even had qabaday aspirations themselves. This was largely due to the qabaday's proven ability to defend and champion his local community. Periodic confrontations between the Muslim and Christian halves of the city encouraged the local population to see the qabaday as their communal protector. Most of the more powerful qabadays in the early 1970s had established themselves during the civil war of 1958 when, as young men, they had demonstrated their ability to mobilize armed bands of shabab (young bloods) to fight for the insurrectionist za'im, Sa'ib Salam, against the government of President Chamoun. Their bravery in battle, and their ultimate victory over the Chamoun regime, turned them into heroes. Often uneducated men from relatively humble backgrounds, they emerged as popular leaders of their quarters and, as such, were extremely useful to the za'ims. The za'im, born of a rich notable family, had great social and economic status, and he won support partly because of that. But by working through the qabaday, he was able to show his willingness to come down closer to his low-status clientele. While he gained increased moral support through his alliance with a popular leader, the za'im also derived considerable pragmatic advantage, because the qabaday, as local boy and local hero, was in an ideal position to recruit and maintain a loyal clientele.

The Qabaday as Part of the Za'im's Apparatus

The most important part of the za'im's apparatus was the core of qabadays. During elections, they acted as 'election keys' (*mafatih al-intikhabat*) and ensured that the za'im's clientele voted for him and his allies. Between elections they recruited supporters, channelled requests for services, and organized mass demonstrations of support. All politicians used intermediaries to recruit and maintain support, but the za'im's strength lay in his ability to give protection and assistance to the popular leaders of the street. Other politicians were either unable to offer this protection, because they were new to the game, or unwilling because of their commitment to some form of social change.

Successful za'ims recognized the necessity of keeping their organization as simple as possible. Even if the za'im could find the time to see each of his clients when they asked for services, he would have found it impossible to ensure their loyalty. In the predominantly Sunni constituency of Beirut III, a Sunni za'im could expect between 12,000 and 16,000 or even 18,000 votes. A proportion of these voters would be clients of other politicians on the za'im's electoral list (a temporary alliance of candidates in multi-member constituencies), but the majority would in some way be beholden to the za'im. In constructing his machine, the za'im sought to organize these voters into manageable groups, where he could leave the responsibility of controlling the clientele to his lieutenants.

The za'im kept the number of lieutenants to a minimum, and usually had something in the region of 15 to 20 qabadays in his core group. Typically, a member of the core was an established neighbourhood leader who controlled a network or gang of lesser qabadays. Two major

types of core qabadays can be distinguished: quarter bosses, who looked after the za'im's interests in particular localities, and family qabadays,[4] who were often also organized by quarters.

Perhaps the most important type of qabaday was the boss of a quarter. The boundaries of Beirut's administrative districts were virtually the same as those of the nineteenth-century quarters which were established as Beirut developed beyond the walls of the old city. The confessional war of 1860 prompted thousands of Christians to settle in the relative safety of Beirut, and the expansion of trade attracted members of all confessions to seek their fortune in the service sector of the economy. The different confessions tended to settle in particular districts. The Christians settled in the east, the Muslims in the west, and these two regions were subdivided into quarters dominated by particular sects. Native Beirutis, generally Sunnis, also moved to a pleasanter environment outside the walls, and while immigrant Muslims (usually Shi'ites) tended to settle in the inner and outer rings of the western city, the native Sunnis were predominant in the quarters of the middle ring. The Sunni mercantile families were the first to move, but the large residences of rich merchants were soon surrounded by the homes of poor and middle-class Sunnis. Although this meant that the Sunni quarters were not inhabited by homogeneous occupational groups or classes, they did acquire some distinguishing characteristics. Basta, with its port workers and stevedores, had a more popular character than Musaytiba where established mercantile and sheikhly families tended to settle; and the quarters of the inner and outer rings (respectively, Bashura and Tariq al-Jadida) were considerably poorer than the generally middle-class quarters of Mazra'a. The process of urbanization gradually changed the character of these quarters. The middle ring, in particular, became much more heterogeneous than before, and the boundaries between quarters became less precise. Nevertheless, there is still a sense of inhabitants 'belonging' to a quarter. There are differences between quarter dialects, and particular families or clans continue to be associated with particular districts.

In pre-civil war Beirut, the za'im could conveniently organize his electoral clients by quarters, and he established local bosses to control them.

Sometimes these bosses were associated with the traditional occupation of the quarter. In Basta, for example, one qabaday inherited a family business in the port where there was a qabaday tradition of gang fights between competing families and factions over such prizes as government concessions to manage the barges and to transport cargo from the ships. Some quarter bosses were protection racketeers; others were alleged to be political assassins or agents in the pay of foreign embassies; but most were outwardly respectable businessmen. The important thing was that they were powerful local leaders who knew their quarter and its inhabitants, were always ready to see and help the za'im's clients, and could mobilize lesser qabadays to fight for the za'im.

The other important type of core lieutenant was the family or clan qabaday. In a sense, every qabaday represented both his family and his neighbours in the quarter, but some qabadays came from such large clans that they were particularly important for ensuring the loyalty of their kin. In Sunni Beirut the 'Itanis formed perhaps the largest family, with around 4,000 voters. Traditionally, they supported the Salams, both Sa'ib and his father Salim. But because the family was so large, and spread all over the Sunni quarters of Beirut, it needed a certain amount of organization if it were to form an efficient unit in Sa'ib Salam's clientele. Other za'ims recognized the value of the family's votes, and various attempts were made to woo the 'Itanis away from their traditional za'im.

So as to maintain his hold on the family, Salam established a number of 'Itani qabadays to control their kin. This was not a novel nor a unique tactic. What usually happened was that the za'im chose particular individuals who had a certain amount of support in the family, and then built them up to a position of monopoly leadership. Thus according to some informants, Salam let it be known that any 'Itani living in Musaytiba, who wanted a service from him, should first see Hashim 'Itani, his chosen representative and the 'official' 'Itani qabaday of the quarter. In 1958 Hashim had been a minor qabaday fighting under Salam's leadership. Over the years, he increased his local influence, and it was said that when Salam recognized Hashim's strength, he gave the qabaday considerable economic assistance. It is certainly the case that

from being the owner of a small coffee shop, Hashim 'Itani extended his business interests such that he became part-owner of two of Beirut's biggest cinemas, and one of the directors of a company that owned restaurants, bars, and cafés in the fashionable quarter of Ras Bayrut. Although the amount of governmental patronage that Salam could dispense was limited in the 1960s, he was able to capitalize on his close relations with right-wing Christian politicians, including his former enemy, ex-President Chamoun, and give help to his clients in the fields of business and commerce. Thus a number of informants told me that Salam helped to find Hashim 'Itani Christian capital and expertise, and that in doing so he created an indebted ally. These same informants claimed that by refusing services to those 'Itanis not vetted by Hashim, Salam could control his 'Itani clientele in Musaytiba through one loyal and grateful lieutenant.

The example of Hashim 'Itani emphasizes the moral, as opposed to transactional, relationship between the za'im and his qabadays.[5] All za'ims had a moral core of lieutenants, who over time received so many transactional benefits from their patron that the relationship acquired a degree of permanency. Members of the core remained loyal to the za'im not simply because of their expectation of future services. They also had a debt of gratitude for past services. This debt changed the character of the za'im–qabaday dyad from a patron–client exchange to a leader–follower relationship, which was often further transformed into a condition of friendship. Perhaps Sa'ib Salam's greatest strength was his ability to make such a transformation with a few key people. Reference has already been made to his exclusion from high office during the so-called Shihabist regimes of Presidents Shihab and Hilu in the 1960s. During that period, the agents of the Deuxième Bureau (intelligence apparatus) skilfully operated the za'im–qabaday system against Salam and other za'ims opposed to Shihab. They suborned criminal qabadays who needed government protection, and they established their own powerful qabaday network to organize the clientele of Shihabist za'ims. But although Salam lost a large part of his clientele to the Shihabist qabadays and their masters, he was able to maintain his apparatus by giving commercial and financial assistance to his lieutenants, and by forming close moral relationships with them. Thus a man like Hashim 'Itani, who did not rely on criminal activities for his economic well-being, was not subverted by the Deuxième Bureau.

The core qabaday's role should be contrasted with that of the notable (*wajih*). Notables were accorded high status because of their wealth, philanthropy, and reputation for religious piety. But they were not usually significant as election keys, as they were not prepared to spend time dealing with the individual problems of the poor, and were not able to coerce the clientele during elections. If a notable did use his contacts in government to build up a large client following of his own, he was likely to become a politician in his own right and ultimately pose a threat to the za'im's dominance. For this reason, the notable could not be trusted. The qabaday, on the other hand, 'came from the masses', was protected or enriched by the za'im, and was much more easily controlled. In addition, because he usually lacked education and the statesmanlike qualities required of a politician, he was not likely to stand as a candidate in parliamentary elections.

The Qabaday's Functions in the Machine

In using the qabaday as an intermediary, the za'im cut down the risk of performing services for his rivals' supporters, left the policing of the machine to a few individuals over whom he had close control, fragmented the electorate into politically artificial entities of quarter and family groups, and further bound his individual clients into the patron–client debt relationship.

The za'im's patronage resources were not infinite, and it was necessary to supervise carefully the recruitment as well as the loyalty of his clientele. Thus he and his secretaries operated a selective procedure when a potential client requested a service. Most attention, and a personal audience with the za'im, were granted to rich clients, clients from large families, and other voters who could provide the za'im with significant monetary and electoral resources in return for patronage. A second category of clients were those poorer and politically less important voters who made up the majority of the za'im's support base, but individually could only pledge their own votes, the votes of their nuclear family, and possibly the votes of a few friends. Usually such people had relatively simple requests and were dealt with by the za'im's secretarial assistants. On behalf of the za'im, secretaries wrote a note or made a telephone call to the relevant government department or whatever institution or individual was concerned with the client's case. A third category of clients included those who voted in another constituency and those who were not enfranchised. An example of the latter group was the Palestinian community. Most Palestinians were not granted full civil rights in Lebanon, did not have the vote, and, as a result, had nothing to offer the za'im. Such people were refused services, while Lebanese citizens from other constituencies were told to approach their own za'ims.

The qabaday was particularly important with regard to the second category. As scores of people came daily to the za'im's house or office to ask for services, the za'im and his secretaries could not easily distinguish which potential clients were voters in the constituency, and which of those were consistent supporters. Elaborate records were kept, but the best way of establishing a client's credentials was to insist on his first seeing a quarter or family qabaday. The qabaday had extensive knowledge of his local domain, and was well placed to vet the client's loyalty and political reliability. He could see which clients regularly presented themselves on feast days, and other similar occasions, to demonstrate their ritual support and continuing allegiance to the za'im, and he could usually ensure that clients voted in the way they were instructed.

The use of the qabaday also served to fragment the electorate and prevent the emergence of self-conscious, horizontally linked social categories such as interest groups or classes. By forcing clients to approach him through their quarter or family qabaday, the za'im encouraged the individual client to see himself as a member of a particular quarter or family grouping, and discouraged the formation of other social categories that might have posed a real threat to the status quo. Although urbanization tended to lead to the break-up of the quarter as a homogeneous unit, and the extended family gave way to the nuclear family, the za'im capitalized on his clients' mythological conceptions of social organization, and in some cases actually created artificial quarter and family identification.

But although the za'im used family and quarter as units of organization, he did not deal with them as corporate groups with their own elected spokesmen. He did not, for example, make a contract with the president of the 'Itani Family Association to provide the family with a sum of money, or a certain fixed number of services, in return for the whole family's support. The electorate was further fragmented because the contracts were made between the za'im and *individual* members of the family or quarter. The qabaday played a crucial role in establishing and policing these contracts. Although the qabaday independently performed some services, such as mediation of disputes, protection of the quarter, and charitable distributions of small sums of money, his access to governmental patronage was limited. He therefore tended to act as a broker who did not speculate in his own right, but effected and facilitated the formation of contracts between the individual clients and the za'im. Claims that election keys in Beirut delivered vote blocs should be treated with some scepticism. When informants told me that a qabaday was worth at least 2,000 votes to a particular za'im, this did not mean that if the qabaday defected the za'im would lose all 2,000 votes. He would inevitably have lost some, but he would only lose a large number if he failed to find another qabaday to take over the defector's role. Even if a client went first to his street qabaday, then to the quarter boss and only after this went to the za'im's house, where he perhaps saw a secretary rather than the za'im himself, the contract was made between the client and the za'im, not between the client and the qabaday. The important function of the qabaday was to reinforce this za'im–client relationship. When a client received a service from the za'im or his secretary, he was making an agreement with a notable who was socially distant and only occasionally seen in a face-to-face situation. But the transaction was channelled through the qabaday whom the client saw in the quarter possibly every day, and this regular contact between client and qabaday served as a continual reminder of the client's debt to the za'im.

The Retainers become Bosses

Until the outbreak of civil war in 1975, the clientelist system of the za'ims worked remarkably

well. It is true that the lack of political parties and national political structures meant there were few arbitrating mechanisms which could control local conflicts. This resulted in costly local feuds between notables, and contributed to the outbreak of a full-scale civil war in 1958. But the system was flexible enough for enemies to become reconciled. From 1958 to 1964 the reformist President Shihab dealt with some of the socio-economic and political problems which had given rise to civil war; and his successor, President Hilu, carried through a number of Shihabist reforms. In addition, the factional squabbles between za'ims usually served the interests of the dominant social class, for a relatively weak state could not interfere in trade and finance. The most serious danger to the Lebanese commercial-financial bourgeoisie was the prospect of social unrest among the urban subproletariat. So long as the za'ims could control these people, the persistence and reproduction of the social formation was not endangered.

The success of the clientelist system was, however, predicated on the continued availability of patronage, particularly with regard to the provision of employment, promotion, and career opportunity. Furthermore, given the strong social attachments to religious primordial loyalties, the distribution of this patronage had to be fairly shared between the confessions. In the early years of independence, these conditions appeared to be being met, but as time wore on fundamental problems emerged which led ultimately to the revolutionary climate of the 1970s. The rate of unemployment rose, and it became increasingly obvious that the Muslim community—and especially the Shi'ite sect—was underprivileged as compared with the Christians. With the development of social and communal conflict, the clientelist structures of Sunni Beirut came under increasing pressure and, during the civil war, eventually collapsed into a murderous anarchy.

During the late 1960s, two groups of qabadays could be distinguished in Sunni Beirut. One was composed of strong-arm businessmen loyal to Sa'ib Salam, and these included Hashim 'Itani in the quarter of Musaytiba. The second group was Shihabist and was protected or paid by the Deuxième Bureau. These leaders of the street were more committed than the Salamists to the popular ideology of Arab nationalism. They were at least nominally Nasserist and had remained loyal to Cairo after Sa'ib Salam had changed his own allegiance to Saudi Arabia. The most important and influential qabaday in this group was Ibrahim Qulaylat from the popular quarter of Tariq al-Jadida. A somewhat mysterious 'Robin Hood' character, Qulaylat was generally considered to be an agent in the pay of Egyptian intelligence or the Deuxième Bureau, and probably both. In the late 1960s, he was unsuccessfully prosecuted for the assassination of a right-wing and pro-Saudi newspaper editor, and many observers thought that pressure had been applied on the court from higher echelons of the Shihabist regime.[6]

The election of President Frangieh in 1970 brought about a change of regime. Sa'ib Salam became prime minister and, with his president's blessing, set about purging the Deuxième Bureau. Principal officers were put on trial and others were removed from positions of influence. The new internal security agency was staffed by officers loyal to the Frangieh–Salam regime, and although these men were competent they lacked access to the intelligence networks which had been assiduously built up since 1958. As a result, the regime lost control of the political underworld.

During the Shihabist period (1958–70), the Deuxième Bureau had been able to tolerate some minor excesses on the part of the racketeers, while at the same time keeping them under a tight political control. After the purge of the Bureau, these Shihabist qabadays were able to operate more independently. While some pro-Frangieh za'ims were able to recruit certain criminal qabadays and bring them under state control by offering them continued protection from the law, other street leaders were able to build links with powerful patrons outside the Lebanese clientelist system. These patrons were principally the Palestinian organizations and the Libyan and Iraqi governments.

By the beginning of the 1970s, a number of Palestinian gangs, largely recruited from the refugee camps, had developed smuggling and protection rackets of their own and were emerging as competitors of the indigenous Lebanese gangs. Through their connections with commando organizations, the Palestinian criminal networks were well armed; and as the Cairo Agreement of 1969 allowed the Palestinians to

enforce their own laws in the refugee camps of Lebanon, the gangsters had a secure base from which they could operate. The competition between these new gangs and the Lebanese qabadays led to armed clashes. The za'im protectors of the qabadays were powerless to prevent these, and were often unable to bring the Palestinians to justice. Revenge was exacted through vendettas which exacerbated the existing state of tension.

Some of the Muslim racketeers dealt with the new situation by working in partnership with Palestinian gangs, and it appears that the commando movement itself took control of some qabaday networks. Palestinian and Muslim Lebanese gangs were also brought closer together as a result of their common opposition to some of the Christian gangs. There was a traditional rivalry between the communal champions of Muslim and Christian Beirut—a rivalry that was usually contained by the mediation of za'ims and notables from both communities. It seems that in the 1970s, one of the objections of Christian parties like the Phalangists to the armed Palestinian presence in Lebanon was that it was upsetting this delicate balance of the political underworld.

Of perhaps greater concern was the fact that some Muslim qabadays were becoming more overtly political than the normal type of racketeer. A prominent example of this trend was Ibrahim Qulaylat, who after the death of President Nasser had switched his allegiance to Arab paymasters more radical than President Sadat and had founded his own organization called the Independent Nasserist Movement. The populist ideology of Nasserism was attractive to such 'primitive rebels',[7] appealing particularly to their sense of Arabism and to their political identification with Islam. It was during the 1950s that it had first provided a populist expression of Sunni discontent, and many of the more powerful qabadays had established themselves in the 1958 civil war as the communal champions of their quarters. Qulaylat was 16 years old at the time and he, like other Sunni qabadays, fought for the 'Nasserist' insurrectionary za'im, Sa'ib Salam. The adoption of Nasserism by Salam and other za'ims reduced the already limited revolutionary potential of the ideology, incorporating it and the qabadays into the clientelist system. After Salam broke with Egypt in the 1960s, the

Nasserist qabadays were brought under the supervision of the Deuxième Bureau. Again the radical element of the ideology was controlled, possibly more effectively this time by the apparatus of a quasi-police state. Manipulated first by the za'im representatives of the state, and then by the state itself, Nasserism provided a framework for socializing the urban poor and their qabaday leaders into an acceptance of the status quo. The end of Shihabism, however, provided an opportunity for freeing the ideology from the control of the state and liberating its populist appeal.

The radicalization of popular Nasserism was a gradual process, and it became a significant political force only when state authority finally collapsed in Beirut during the fourth 'round' of fighting in September 1975. Throughout the first three rounds of the civil war, the quarters of Sunni Beirut remained relatively free from the fighting. Barricades were erected around some quarters and there were cases of kidnappings, mutilations, and murders, but there was no heavy fighting as there was in the suburbs. The Sunni establishment only lost control of the situation in September when the Phalangist militia began to bombard the western half of Beirut's commercial district in an apparent attempt to force the government to send in the Lebanese Army to restore order.[8] This brought the fighting into the centre of the city and, by the end of October, Ibrahim Qulaylat's militia, the 'Murabitun', had become involved in a strategic battle which earned Qulaylat the right to be considered as one of the foremost leaders of the National Movement (the coalition of 'progressive' forces fighting the predominantly Maronite rightist militias).

In October fighting had again broken out along the 'confrontation line' which divided West and East Beirut, and later spread westwards along the coast from the commercial district towards Ras Bayrut. The Christian rightist militias were attempting to control access to the port, and as part of their strategy they occupied the high-rise building of the Holiday Inn. Opposing them were a number of leftist and Muslim militias under the overall leadership of the Murabitun. The Muslim militias inflicted a significant defeat on the Christians, and the Murabitun demonstrated that it was a relatively well-disciplined and powerful fighting force. In the fifth round of fighting, in December, Ibrahim

Qulaylat again achieved notoriety when he led another attack on the Christian strongholds in the hotel district, and forced the Maronite militias out of the Saint George and Phoenecia hotels.[9]

Before the hotel battles of October and December, Qulaylat's Independent Nasserist Movement had been just one of many Nasserist and populist groups in Sunni Beirut. As a result of the bravery and success of its fighters at the end of 1975, however, the Murabitun militia became an influential force in Lebanon's political mosaic. This period marked the collapse of the za'ims' clientelist system in Sunni Beirut. Men like Ibrahim Qulaylat were no longer beholden to their traditional patrons. Although Sa'ib Salam, for example, made persistent communal appeals to the Muslim community, forming a new political party based on the 'teachings of Islam' and condemning what he called the 'destructive left',[10] the initiative had already passed to the new Nasserist leaders. In contrast to 1958, the war of 1975 brought about the establishment of a popular movement in Sunni Beirut completely independent of the za'ims and their clientelist system.

After March 1976, when the Murabitun and its allies finally took the Holiday Inn, the National Movement and the Palestinians extended their control to take over almost 80 per cent of Lebanese territory, including part of the Maronite stronghold in Mount Lebanon. The rightists faced imminent defeat and were rescued only by shipments of weapons from Israel and by the intervention of Syrian troops in June. The latter invasion, and the rightist counter-attack which it allowed, inflicted severe losses on the Palestinian and progressive coalition and led to a military stalemate which amounted to a *de facto* partition of the country. Whether the left could have brought about a new political order in Lebanon is a matter for speculation. All that can be said is that once the Syrian army had prevented a leftist victory, the progressive parties were not very efficient at imposing such an order on their much-reduced sphere of influence.[11]

In East Beirut the Phalangists controlled their 'street' and made it illegal to carry a gun without a permit. Criminals were arrested and sometimes summarily executed, and taxes were raised and services provided. Such an order was built on the fact that the Phalangists had either eliminated their rivals or brought them into a unified military force under the command of Bashir Gemayel. Gemayel's rise to power was destructive and bloody,[12] but many inhabitants of Christian Lebanon considered the rough discipline he imposed as preferable to the chaos which prevailed in West Beirut.

One major division amongst the progressive forces in Muslim Beirut had occurred when a number of organizations, such as the pro-Syrian Ba'th faction, formed a separate front sympathetic to Damascus. Relations between this new coalition and the Lebanese National Movement (LNM) were especially tense during the Syrian invasion, but improved a little in the latter part of 1977 when the pro-Syrian Ba'th and the Progressive Socialist Party (predominantly Druze Muslim) issued a joint proclamation as a step towards an improvement of relations between Damascus and the LNM. Events outside Lebanon also had an influence on the making and breaking of alliances. The 1978 Camp David agreements between Egypt and Israel promoted a *rapprochement* between Syria and Iraq, which was reflected in Lebanon by close relations between the two wings of the Ba'th Party. But hostilities broke out once more when Syria supported Iran in its war with Iraq. Similarly, the Shi'ite 'Amal' militia supported Iran while Arab nationalist forces tended to support Iraq. Such divided loyalties led to non-Lebanese conflicts being fought by proxy in West Beirut and the suburbs, and these conflicts became inextricably wound up with the parochial rivalries between street and quarter gangs.

The number of fighting forces in Muslim Beirut seems to have increased after the 1975–6 civil war. Syria, Iraq, Libya, Iran, and Saudi Arabia, either directly or indirectly funded and supplied militias, and there were probably other countries involved as well. Some Arab regimes such as Syria and Libya supported more than one militia, and by the late 1970s there were around 30 separate fighting units in West Beirut alone. Many were the clients of one particular regime, while others received material aid from a number of different sources; some were financed by wealthy Lebanese to protect their businesses and property, and many were connected to protection rackets and smuggling. There were disciplined political organizations amongst them, but a large proportion were little more than adoles-

cent street gangs—inheritors of a qabaday tradition which had degenerated into its most anarchic form. In a complicated process of fission and fusion, such gangs and militias fought and allied according to the nature of the dispute. Often they fought street to street over the extent of their territories and protection rackets, but faced with a common foe they would band together, as when pro-Syrian groups united to fight supporters of Iraq. During the Israeli siege of Beirut in the summer of 1982, the various groups tended to co-operate in readiness for an invasion of their quarters, but even then there were some battles over who would defend which street. The most petty disputes could give rise to gun-battles: motorists shot at each other after quarrels over a right of way or parking-space; dissatisfied customers gunned down shopkeepers; jilted lovers killed their rivals; and because many people could count on the support of their local gunmen, such incidents often widened into artillery duels between one street and another. The Syrian army, already unpopular because of its intervention on the side of the Maronite rightists, was unwilling to move too hard against the Muslim street. Instead, Syria preferred to work with a new clientelist system and seek out favoured clients through whom it could manipulate the violent and unstable politics of West Beirut.

The Bosses Return

After the Israeli invasion of 1982, it looked as though some form of political order could be re-established. Sa'ib Salam became more prominent and played an important role as a mediator between the Palestinians and the US envoy who negotiated the withdrawal of the commandos from Beirut. For a brief period a reconstituted Lebanese army patrolled the streets of West Beirut and there were hopes for a lasting peace. But disputes between the Druze Muslims and Maronites of Mount Lebanon led to more fighting which eventually resulted in Shi'ite militias taking control of large parts of Sunni Beirut in 1984 and destroying the Murabitun in 1985.[13] During ten years of civil war, ideological politics had gradually retreated before the powerful forces of confessional allegiance. It had become impossible to talk of a political left and right in

Lebanon, and what remained of the country was a patchwork of territories, in and around Beirut, under the control of the different confessional militias and their foreign backers.

The clientelist system which had maintained political order before 1975 had given way to anarchy in some areas and to enclaves of confessionally based authority elsewhere. In Sunni Beirut the relatively ordered structure of the za'ims' political machines had long been replaced by unbridled gangsterism and, in the mid-1980s, communal conflict between Sunni and Shi'ite Muslims became more prominent than ever before. In response, Sunnis increasingly turned inward on the quarter and family solidarities which had been manipulated so effectively by the pre-war za'ims. Without a central authority which could enable the za'ims to control the qabadays, such primordial solidarities provided some sense of community and security in a highly competitive and violent society.

Whereas quarter and family solidarities in the 1960s and early 1970s can be described as essentially 'mythological' and were often artificially created by za'ims to maintain control of the clientele, they had become powerful forces by the 1980s. The divisions between quarters and clans, fuelled by rivalries between local militias, had contributed to a fragmentation of the Sunni community. The pre-war za'ims had been remarkably effective at maintaining political order, but their system had relied on their access to the superior force of the Lebanese army and security apparatus which could be used to control rebellious qabadays. Criminal qabadays had been protected by their za'im, but if they stepped out of line they could be handed over to the police. The break-up of the army and police during the civil war weakened the Sunni za'ims and left their community open to parochial conflicts and divisions.

The eventual imposition of a *pax Syriana*, however, finally restored the za'ims' authority. As a reward for supporting the United States and its allies in the Gulf War, Syria was allowed to pacify most of Lebanon and, in 1992, elections were held for a new parliamentary Assembly. Looking down the list of those elected, it is not surprising that the pre-war za'ims' sons or close relatives and associates were prominent. The Syrian occupiers inevitably turned to those leaders who could help re-establish order and, in Sunni Beirut, they assisted a new generation of

politicians whose family names linked them with a, by now, almost mythical past of stability and security.

The former qabaday, Ibrahim Qulaylat, had been decisively defeated by the Shi'ite Amal militia in 1985. Even if he had been able to maintain his leadership role, it is highly likely that most voters would have associated him with anarchy and turned against him in favour of the élite of za'im families. Civil war had fashioned some qabadays into warlords, but peace invited the return of more diplomatic and benevolent leaders than the malevolent racketeers. Sa'ib Salam retired to Geneva, although his family was still active within the political élite and it would be surprising if a Salam were not elected to the Assembly in the near future. For the moment, though, the anti-Salamist faction in the Sunni élite were in the ascendant. Among those elected in Beirut were two former prime ministers, Salim al-Huss and Rashid as-Sulh, the latter belonging to a notable family which had been a rival of the Salams since the early twentieth century. Also elected was Usama al-Fakhuri, a doctor who had stood unsuccessfully against Sa'ib Salam in 1972, the last time elections were held in Lebanon prior to the civil war.[14]

But although the élite stratum of the pre-war system was back in place, some changes had occurred. In particular it seemed that the existence of a state authority backed by the force of Syrian arms, coupled with popular resentment against the militias, meant that the post-war Sunni parliamentarians in Beirut no longer had to take account of a subaltern rank of qabadays.

The conventional view of Lebanon during the civil war was that society had collapsed irremediably into an anarchic conflict between the militias. At times, of course, there was widespread demoralization,[15] but what was remarkable was the extraordinary resilience of Lebanese civilians. Attitude surveys conducted in the 1980s revealed that an overwhelming majority opposed a partition of Lebanon along confessional lines and favoured the reconstitution of a democratic political system.[16] Support for the militias fluctuated according to the extent of the fighting: at times they were seen as protectors but increasingly they were widely resented.

New forms of allegiance emerged. Very significantly, trade unions recruited more members and became active in campaigning against the war. In contrast to the pre-war situation, they began to act independently of competing political leaderships. Strikes and demonstrations in favour of peace were organized in Beirut by the non-confessional Confédération Générale des Travailleurs Libanaise (CGTL) in 1986, 1987, 1988, and 1990. These attracted the support of tens of thousands of people—Muslims and Christians—and clearly undermined the influence of the militias who responded by threatening the lives of union leaders.[17]

We should not, however, generalize too much about the attitudes of the Lebanese to the militias. In 1992, in the Baalbeck constituency in the Bekaa valley, for example, the candidates of the Shi'ite 'Hizballah' militia defeated members of a slate composed of establishment politicians; and in Christian regions most people boycotted the elections as instructed by their warlords. Indeed, such was the dissatisfaction with what was seen as Syrian gerrymandering, that the turnout was low throughout the country, and many Muslims as well as Christians refused to vote, the Salamist faction in Beirut being a prominent example.

The complexities of the Syrian-imposed peace in Lebanon are beyond the scope of this essay. What we can say is that in Sunni Beirut the old political families have re-emerged and the strong-arm leaders of the street have been neutralized. Civil war had provided the qabadays with an opportunity to operate outside the controls of clientelism, but it had ultimately destroyed them as a political force. First, the conflict between Sunni and Shi'ite militias in the 1980s had led to pitched battles between Qulaylat's Murabitun and the Shi'ite Amal that ended in complete victory for the latter in 1985. Secondly, the resultant control of the Sunni quarters of West Beirut by Shi'ite militiamen was widely resented by the local population, perhaps particularly when the rival Hizballah and Amal militias fought each other in 1990 with scant regard for civilian casualties.

When Syria finally imposed order and the militias were disarmed in January 1991, the Sunnis of Beirut breathed a sigh of relief and looked forward to a secure future. After years of warfare and social chaos, they turned again to the traditional élite. There were still divisions between those who were close to the Syrians and those, like Sa'ib Salam, who wanted to ensure that the Syrian occupation would be a short one. But

members of the élite and their respective clients and supporters were all basically agreed that their ultimate aim was to reconstitute a Lebanese democracy which would reunite the nation and turn its back on the horrors of strong-arm violence and depredation.

Notes

1. One of the best analyses of the Lebanese economy in the modern period is S. Nasr, 'Backdrop to Civil War: The Crisis of Lebanese Capitalism', *MERIP Reports*, 73 (1978). For more details, see C. Dubar and S. Nasr, *Les classes sociales au Liban* (Paris: Presses de la Fondation Nationale des Sciences Politiques, 1976).
2. The seats in the Assembly were allocated to the various confessional communities according to their supposed size in the population. For the elections of 1960, 1964, 1968, and 1972, in the constituency of Beirut III, four of the five seats were reserved for Sunni Muslims, and one for a Greek Orthodox Christian. In order to win, a Sunni candidate usually had to recruit the support of Christians as well as Muslims.
3. After 1958 Sa'ib Salam was the most powerful Sunni za'im in Beirut, and he remained influential even after the outbreak of civil war in 1975. Other Sunni za'ims, such as 'Uthman ad-Dana, lost virtually all influence during the wars of the 1970s and 1980s.
4. The word 'family' refers here to a clan of people bearing the same surname and having the same confession, who trace their genealogy to common ancestors. Often such clans have family benevolent associations, open to all members of the clan, and these give some organizational solidarity to the group.
5. Such concepts as 'moral', 'transactional', and 'core group' are derived from F. G. Bailey, *Stratagems and Spoils: A Social Anthropology of Politics* (Oxford: Blackwell, 1969), esp. 34–49.
6. M. Johnson, *Class and Client in Beirut: The Sunni Muslim Community and the Lebanese State, 1840–1985* (London: Ithaca Press, 1986), ch. 3, case study 7.
7. E. J. Hobsbawm, *Primitive Rebels: Studies in Archaic Forms of Social Movement in the 19th and 20th Centuries* (Manchester: Manchester University Press, 1959).
8. K. S. Salibi, *Crossroads to Civil War: Lebanon 1958–1976* (Delmar, NY and London: Ithaca Press, 1976), 97 ff.
9. Ibid. and *MERIP Reports*, 44 (1976), 14.
10. *MERIP Reports*, 44 (1976), p. iv.

11. For further information on the 1975–6 civil war and its aftermath, see, for example, W. Khalidi, *Conflict and Violence in Lebanon: Confrontation in the Middle East* (Cambridge, Mass.: Harvard University, Center for International Affairs, 1979).

12. J. Randal, *The Tragedy of Lebanon: Christian Warlords, Israeli Adventurers and American Bunglers* (London: Hogarth Press, Chatto and Windus, 1983), 109 ff.

13. See Johnson, *Class and Client in Beirut*, ch. 8.

14. For the election results, see J. Bahout, 'Liban: Les élections législatives de l'été 1992', *Monde arabe: Maghreb-Machrek*, 139 (1993), 53–84.

15. S. Khalaf, *Lebanon's Predicament* (New York: Columbia University Press, 1987), ch. 11.

16. T. Hanf, *Coexistence in Wartime Lebanon: Decline of a State and Rise of a Nation* (London: I.B. Tauris, 1993), 490 ff. This book is the most important recent reference on Lebanon. Others include N. Shehadi and D. Haffar Mills (eds.), *Lebanon: A History of Conflict and Consensus* (London: I.B. Tauris, 1988); and H. Barakat, *Toward a Viable Lebanon* (London: Croom Helm, 1988). For Lebanese women's perspectives on the civil wars, see M. Cooke, *War's Other Voices* (Cambridge: Cambridge University Press, 1987).

17. Hanf, *Coexistence in Wartime Lebanon*, 639–40.

22

Political Incorporation and Populist Challenges in Mexico City

Susan Eckstein

WHY does poverty persist when poor people organize in groups concerned with their welfare? How can income distribution become more in-egalitarian as national wealth increases and a mass-based party officially commands the reins of government? Poor people's organizational effectiveness, as we shall see, may be limited if the groups with which they associate, or leaders of those groups, are formally co-opted and if informal pressures impede those co-opted from using their positions to serve their constituents. These processes are illustrated with data from Mexico.

The study focuses on organizational relations in Mexico City in: (1) a centuries-old centre-city slum; (2) a now legalized squatter settlement formed by an organized land invasion in 1954; and (3) a low-cost government-built housing development, designed in part to house low-income families, that opened in 1964. The three areas were selected to represent different types of lower-class residential settlements, in order to ascertain whether and how different dwelling environments impact on their residents.

I have studied the communities over a twenty-year period, between 1967–8 and 1987. This essay is based primarily on about a hundred in-depth open-ended interviews that I had over the years with officeholders in local organizations and institutions and with 'influentials'. Some interviews took four to eight hours, and several appointments. The people interviewed were asked about the activities of their groups, the history of their

communities, and their personal backgrounds. My initial research also included a survey of a hundred residents in each area about political, economic, and social matters; detailed interviews with a smaller sample of men residents about their work experiences; and a survey of local businesses. In addition, I attended group meetings, talked informally with many community residents, and consulted other available data, including documents, newspaper articles, and other surveys.[1] The first portion of the essay focuses on dynamics through the early 1970s, the latter portion on changes in the mid-to-later 1980s.

The Making of Political Order

Formal Co-optation and Incorporation

In the late 1960s there were over two dozen economic, political, administrative, and civic (non-religious) groups in the three areas. Most of them were either formally incorporated into the party that has ruled since its formation in the late 1920s, the Party of the Institutionalized Revolution (PRI), or into government-affiliated organizations; if they were independent their leaders tended to be exposed to PRI and government influences.[2] The local groups included territorial-based divisions of political parties, party-affiliated groups, and diverse government agencies; economic associations; and athletic and other social associations.

Typically the local groups upon formation addressed local concerns. Since concerns in each area differed somewhat, divisions of the same parent organization at first did not always address the same sets of issues. In this respect the organizations adapted to their constituencies.

Independently of specific conditions that gave rise to the formation of local groups, group leaders came with time to claim a common concern about the welfare of their rank-and-file members. The first groups in the squatter settlement, for instance, began as loosely structured followings around 'natural leaders' who tended to problems residents had securing legal title to land appropriated illegally, and attainment of basic social and urban services. However, over the years they came to stress a general commitment to moral, social, and economic matters, not specific local concerns.

When originally affiliating with national PRI and government groups, members received a variety of coveted benefits, such as schools, roads, pavements, markets, and public transportation. In the squatter settlement they also secured property rights. Their ability to secure goods and services through the groups typically proved, however, to be limited and short-lived. Local residents and local organizational leaders, with time, stopped pressing for material and social improvements, even when local needs continued. For example, in the 1970s people in the squatter settlement did not even try to pressure the government to build additional schools when 2,000 children could not be accommodated in existing facilities. And the groups lacked power and budgetary discretion to address such needs on their own.

Congressmen were among the only popularly elected *politicos* with official institutionalized power. Yet even they rarely used their authority on behalf of their constituents. They did not view their main task as one of representing their electorate in the legislature, as evidenced by the way they divided their time, defined their responsibilities, and voted in Congress. Nonetheless, they did not entirely ignore people in their district, for they informally solicited government agencies for community facilities and helped some residents secure jobs. Furthermore, their district headquarters offered such services as medical care and courses in sewing, homemaking, and typing, which residents, if they so chose, could make use of.

Groups in the areas, whether they had institutionalized access to power or not, came to place constraints on members which undermined their effectiveness. After affiliating with national organizations once active members became largely inactive,[3] and they shifted from seeking benefits collectively to dealing individually with the organizational leaders about common concerns. To the extent that residents continued to partake in group-sponsored activities they typically did so in return for, or in anticipation of, favours from the group leaders. Members also learned to solicit, not demand, goods and services, and they learned what they could solicit.

Even social groups that did not formally affiliate with national institutions were subject to political-administrative constraints. Although these groups, unlike the PRI and government-affiliated groups, technically had authoritative power, because their constituents were poor they had limited financial resources. Under the circumstances their leaders sought help from *politicos*, and in so doing they were subject to political-administrative influences. The group leaders, either because they individually formally affiliated with PRI or government-linked groups, or because they anticipated or aspired political patronage, organized their groups similarly to party- and government-linked groups. They introduced hierarchically structured territorial-based organizations, with representatives on each block; they urged members to attend civic and political manifestations; and they spoke of political-administrative concerns at their group meetings. Once the head of a church-initiated social group privately affiliated with a PRI-linked association, for example, he began to give political and civic matters high priority in his own association. He began to report regularly on the activities of the PRI group, whose leader monopolized a number of rallies the social group organized. The group, in due course, was diverted from its original social and economic and deliberately non-political objectives. Similarly, the head of a large, formally autonomous soccer league in the inner-city area allowed PRI to offer social services at the group's headquarters, required members to partake in civic celebrations, and organized his group by blocks, as PRI did, so that, as he phrased it, he would be ready should the party call on him.

Residents' increased inability over the years to

secure collective benefits from the government was not the result of any 'organizational incapacity' or 'political incompetence', as some commentators on Mexico have claimed.[4] The very organizations with which they affiliated, directly and indirectly, weakened their collective strength. As local groups and leaders of local groups established 'extra'-local ties, and as relations within the groups became routinized, informal pressures operating in the society at large affected relations locally: pressures that made residents, in the main, apathetic and submissive to the powers that be.

Informal Processes Constricting the Political Power of Organized Residents

Since formal power potentially serves as a basis of effective power, co-optation and incorporation into ruling party and government groups cannot account for residents' limited political efficacy. My field-work suggests that when inter- and intra-group relations are hierarchically structured, when class biases discriminate against the poor, when the government monopolizes resources that groups themselves lack, 'have-nots' cannot readily use formal power and formal organizations for their own ends.

For one, the hierarchical way that *políticos* were appointed and removed from office and the hierarchical channels through which urban and community services were obtained constrained local leaders to conform with 'rules of the game' that they did not establish. They felt that they otherwise could neither personally advance politically nor secure benefits for their constituents. The possibility that higher-ranking authorities could remove them from office further induced them to conform with the expectations, or their perceived expectations, of higher-ranking functionaries. In the process 'appropriate' local concerns were delimited, and heads of local groups became subservient to their non-local organizational superiors. A politically ambitious head of a local division of a government agency went so far as to say, 'If you do more than your boss you're in real trouble.'

Secondly, the hierarchical system of political appointment and advancement typically kept functionaries from becoming entrenched in the community. Thus, a market administrator explained to me that the head of the Division of

Markets regularly reassigned administrators so that none of them became subservient to local interests and failed to fulfil obligations to their superiors.

Thirdly, the hierarchical promotion system drained the areas of the energies of their most effective leaders. Ambitious leaders were pressured to partake in activities in other parts of the city, to secure or maintain the support of higher-ranking functionaries. During the electoral campaigns, for example, the leading *políticos* engaged almost exclusively in political activities in other electoral districts. While higher-ranking functionaries did not necessarily deprive the local communities of their most effective 'natural leaders' deliberately, their own opportunistic concerns had such an effect.

The hierarchical system also prevented groups from organizing 'horizontally'. At times local groups were isolated from, and in competition with, other divisions of their same parent organization. In the case of unions, local divisions sometimes were incorporated into different sectors of the PRI (into the labour or 'popular' sector). Since benefits were allocated to the different sectors, and to groups within each sector, separately and unequally, local unionized workers did not identify with one another and organize collectively. And rank-and-file members, through their union affiliation, were exposed to non-local, non-economic party and government influences. In response to 'vertically' channelled commands, union members, such as locally based market vendors, participated in civic and party rallies and they were encouraged to vote for the PRI, but they were encouraged otherwise to be politically acquiescent.

Independently of the hierarchical structure of political and administrative groups, class biases that informally operated in the society at large served to constrict the political effectiveness of local groups. Access to local leadership posts, opportunities for political advancement, and the ability of local groups to secure benefits for their members—once relations in the areas were routinized—hinged more on the socio-economic status of group members and the group leaders than on formal objectives of the groups.

While local group leaders were not the local economic élite, they tended to rank among the more economically successful in the areas, and their political mobility tended to vary with their class standing. Leaders of humble social origin stood the greatest chance of promotion within the PRI and government hierarchies when members of their groups were highly mobilized, which rarely occurred once national political and administrative institutions established local affiliations and relations within the areas were routinized.

While functionaries might preoccupy themselves with concerns of resident poor even if they themselves were better off, because they were primarily interested in their own advancement they tended to be hierarchically rather than community-oriented. Local leaders rarely joined together to promote programmes for the benefit of residents. They had much more contact with persons of higher rank in their own organizations outside the local communities than they had with one another, even if affiliated with the same formal organization.

The communities, like their leaders, were rewarded inversely to their economic need, once residents ceased to be mobilized. Residents of the housing development, the most middle class of the three areas, secured the greatest range of social services, even though from the onset they had the best facilities. Moreover, within the development the sections housing the most middle-class residents, and with middle-class leaders, benefited most from government assistance. Their greater effectiveness derived not from the specific groups with which they affiliated, for local divisions of the same PRI and government groups in the centre-city area and the legalized squatter settlement were not equally rewarded. Their better fortune derived from higher-ranking non-local functionaries favouring their middle-class subordinates and from local middle-class functionaries generally having the best non-local political contacts.[5]

The personal manner in which *politicos* were appointed and removed from office, goods and services distributed, and local demands and conflicts articulated and resolved, further limited residents' ability to use local groups for their own ends. The national power structure induced such *personalismo* and made residents—particularly in the two newer areas—feel dependent on and indebted to government authorities for material benefits, including land and pavement for which they paid. The benefits generally were personally 'given' to them by high-ranking functionaries at public rallies.

The very process of securing personal 'favours' from the government had the effect of weakening residents' collective strength. In exchange for the 'favours' local leaders in the two newer areas helped higher-ranking party and government functionaries establish political order and party and government loyalty where institutional life had been previously non-existent.

Reflecting both the importance and the structural basis of *personalismo*, local leaders reported that higher-ranking functionaries instructed them to personally request, not demand, goods and services and to petition for them either individually or in small groups, not *en masse*. Consequently, leaders in the squatter settlement who initially mobilized large followings subsequently went to government offices alone or with small delegations. The government was thus under less pressure to be responsive to residents' wants.

And because politics was personalized so too was conflict. When residents' expectations were not fulfilled they blamed individuals, and competition for political and economic spoils pitted local leaders against each other. The hierarchical system of appointments induced leaders to compete for local followings and influence, including when the groups they headed were affiliated with the same parent institution and when the groups concerned themselves with the same sets of issues. Such petty jealousies undermined the potential collective strength of the communities. The tensions turned people's energies 'inward' against each other, rather than collectively 'outward' towards the institutions that could help them. Furthermore, in personalizing conflict, residents did not criticize the institutions in ways that might have contributed to a delegitimation of the inegalitarian socio-political order.

Any political power which residents might have enjoyed through collective organization was further weakened by the sheer multiplicity of groups operating locally. Higher-ranking functionaries tacitly supported this multiplicity by periodically distributing gifts to the different associations, by attending local group meetings, and by seeking collaboration from the groups in civic events.

Finally, a law of 'social dissolution' gave the administration the right to intervene in groups. The very existence of the law probably inhibited residents from forming groups and engaging in activities that might have been officially condemned. The law also was used occasionally to chastise controversial local leaders. Yet, overt and violent government repression was rare. From the state's vantage-point, co-optation and informal constraints generally sufficed to induce conformity.

Structural Conditions Inducing Leadership Co-optation and Formal Political and Governmental Affiliation

The preceding discussion highlighted how and why organizations publicly committed to poor people limited members' efficacy. But why did resident poor affiliate with organizations that had such effects? The situation within the three areas suggests that local leaders established ties with national political and administrative groups because they felt such ties to be expedient: that is, for opportunistic reasons more than shared ideological and moral commitments. Viewed from the local level, leadership co-optation and group affiliation with national organizations occurred because the people in local command positions felt they could thereby secure political and economic spoils and expand their sphere of influence. The hierarchical, centralized nature of the regime induced them to think so. They did not by choice establish ties that left them without authoritative power.

Despite such structural constraints local leaders on occasion deliberately established no formal ties with national government and party groups: when they felt their constituents or potential constituents were antagonistic to direct association with the government or the PRI. In such cases, however, leaders typically established on their own informal relations with people in the party and government. The politically ambitious head of a large soccer league in the centre-city area, for example, never formally affiliated his group with any official national organization because he sensed that his team members would feel compromised by such affiliation. Yet, as he admitted to me, he collaborated with political and government functionaries and he ran his group in ways that he thought would please 'higher-ups' because of this own political ambitions. Aware that prior leaders of the league and prominent local athletes had been awarded posts in the PRI and the government, he hoped to secure political patronage on the basis of his

organization. Similarly, the head of a church-sponsored social group in the housing development informed me that he never allowed his group to officially join a PRI or government group because he thought he would be forced to resign if he did. But he deliberately affiliated with a PRI-linked group on his own, for he felt he could thereby secure material assistance that would enable him to maintain a following.

The higher-ranking non-local functionaries with whom I spoke told me that they formally sought the collaboration of local leaders and formally incorporated local groups as branch affiliations of their organizations because they felt that their superiors, in turn, would be impressed by greater numbers under their command and by their success at establishing social and political order in the areas. They felt that their own chances of promotion would thereby improve.

The Unravelling of Political Order in the 1980s

The Atrophy of Formal Organizational Life at the Grass-Roots Level

In the late 1980s everyday involvement in local group life was less in the shanty-town and housing development than it had been on my visits in the late 1960s and early 1970s. In both areas civic groups that had functioned in the past had become defunct. One leader of a moribund group had become disillusioned with politics, another had died. The head of the civic group in the poorest section of the housing development, formed at the initiative of a foreign missionary in the 1960s, by the 1980s was very much an individual in search of a group to lead. The Mexican priest who replaced the missionary refused to allow the group to meet.

The priest who put a stop to the lay group typified the stance of all clergy in the three communities in the late 1980s towards independent civic groups. The archbishop of Mexico City at the time discouraged church-based social action groups. The one US missionary in the inner-city area who had been influenced by the Catholic left already in the late 1960s, and who had encouraged lay community-oriented groups and lay leadership, had been reassigned by the hierarchy to a provincial parish in the State of Mexico in the late 1970s. In discouraging lay groups the church helped the state restrict the 'organizational space' of groups that might challenge the status quo.[6]

No new independent or quasi-autonomous civic groups replaced the groups that had become defunct in either the shanty-town or the housing development. There were only government- and party-affiliated groups, and they had little support. Residents over the years had become less inclined to participate in civic activities, less inclined to attend group meetings, and, as detailed below, less inclined to vote. The only civic groups to hold meetings on a regular basis were new, government-linked so-called Residents' Associations, which involved some two to three dozen people. Not even all the block representatives of the Associations attended the organization meetings. In the shanty-town only original settlers were at all active, and only a small number of them. The growing tenant population took little interest in community matters. In the housing development, one Residents' Association president said that she held the office for a second three-year stint (after stepping down for one term) because 'hardly anyone wanted to work for the group'.

Government–party hegemony in the shanty-town and housing development had prevented new groups from emerging, and residents saw no reason to organize on their own. But because organizational life had atrophied, co-optation no longer served the regulatory effect of earlier years.

Electoral Disobedience

The atrophy of formal organizational life, despite and because of state monopolization of the 'organizational space', was only one sign of a change in state-societal relations. PRI's success at the polls also dropped precipitously in the 1980s. It dropped with the state's organizational hold over the populace, on the one hand, and with the plunge in living standards, on the other hand.

Residents who had unhappily seen their living standards deteriorate with the country's debt crisis and who lost all hope that their situation might soon improve, refused to support the regime to the extent that they had in years past. 'Defection' was particularly marked among the younger generation.

Yet few of the 'defectors' turned to parties of

the Left. Only in the inner-city area had the Left any presence to speak of, for reasons explained below. The disenchanted who voted instead typically turned to the main party of the Right, the National Action Party (PAN). They did so because the PAN was the strongest and most visible opposition party nationally. They typically were not party loyalists, but in voting for the PAN they sought to register their hostility to the PRI. However, self-employment has historically been a conservatizing force, and with the economic crisis increasing numbers of people had to eke out an existence on their own; the work situation therefore also undoubtedly added to PAN's improved draw at the polls.

Others who were disenchanted withdrew from politics altogether. Contributing to the rise in non-voting was a diminution of ties to groups that instil in members a sense of duty to vote: both in the communities and through work.

Abstention rates were highest in the shanty-town and inner-city area. Many residents in the housing development were formal-sector workers with ties to the state through their PRI-affiliated unions. Also, their earnings, unlike those of informal-sector workers, had been adjusted over the years (even if only partially) to inflation, and they enjoyed highly subsidized housing. These factors together help explain why people in the development remained loyal to the PRI to an extent that residents in the other areas did not.

Reflecting the changed political mood in the shanty-town, the president of the Residents' Association there told me that in 1982 PRI polled about 35 per cent of the votes, PAN polled about 25 per cent, and an equal proportion abstained.[7] The rest of the votes went to other parties (mainly to the conservative, semi-clerical Mexican Democratic Party). In contrast, PRI had won about two-thirds of the local vote in the 1970s, and about three-fourths of the vote in the 1960s.

Even long-time PRI voters who did not break with tradition, including local PRI and administrative cadre, voted less out of conviction than in the past. They felt that the party no longer did as much for them as it used to.

But why did the politically disaffected not turn to parties of the Left? The Left at the time was too small to have significant presence throughout the city and it was too divided to be politically effective. No party of the Left had concentrated its meagre resources in the shanty-town or housing development. The Left had been active in the inner city in the 1980s, but its potential electoral impact was weakened owing to clever state manipulation of a housing programme, detailed below.

Grass-Roots Mobilization for Housing in the Inner-City Area: The Limits of Co-optation

In the 1970s the government tried to implement an urban renewal plan in the inner-city area that would have driven many residents away because rents in the proposed commercial housing were going to be beyond their means. But residents did not passively acquiesce to the plan. They organized in self-defence, with the help of some 'new social movement' type groups. Catechist groups (partly at the instigation of the foreign missionary before he was reassigned to a provincial parish) played an active role; so did CONAMUP (the National Co-ordination of the Popular Urban Movement), a national 'front' comprised of 'popular' groups with no state affiliation. The groups successfully helped stop slum clearance, but the quality of housing in the area remained poor.

Then, in 1985 the area suffered damage from the earthquake that killed thousands of people and caused millions of dollars of damage in the city. Interested in clearing the area for 'development', the government initially ignored the housing needs of residents whose dwellings fell into a hopeless state of disrepair. Still unwilling to leave, residents organized on their own, with earthquake victims in other parts of the city, and with the support of parties of the Left, university students and faculty, and 'socially conscious' priests, architects, engineers, and other professionals. The movement built on the rich informal social and economic life of the area, and on the 'social infrastructure' of the tenants' movement of the 1970s. Some of them affiliated with the CUD, the Sole Co-ordinating Committee of Earthquake Victims. The CUD was another 'new social movement' type group that deliberately resisted state affiliation. It was a city-wide umbrella organization of earthquake victims that staged, as part of its strategy, various protests and other acts of civil disobedience.

The groups that mobilized locally for housing were, however, not all in accord. They differed in objectives, tactics, and political leanings. Some groups with links to the Left supported solidarity activity with Central American guerrillas, and pressed for 'proletarian justice'. They presented themselves as a cultural vanguard and sought to form a unified front. They wanted not merely new housing but also the right to build it themselves. Another set of groups, by contrast, built on the values of the *barrio*. They emphasized the cultural roots of the community, including its artisanal traditions. Their approach was more restorative: to preserve the *barrio* and its traditions. They did not advocate a unified front, and they resisted affiliation with the CUD and formal ties with the Left because they believed their interests would thereby be compromised. They saw the CUD as dominated by middle-class earthquake victims, whose concerns differed from theirs.[8] Although the PRI had initially opposed the mobilization for housing, it too organized people in the area once its opposition failed to stem the grass-roots movement for housing reconstruction.

To restore order, the government either had to repress the grass-roots movements or be responsive to their demands. Repression would have been politically costly, especially since Mexicans of all classes sympathized with the plight of the so-called *damnificados*. Consequently, the government agreed to implement a housing reconstruction programme, locally and in other sections of the city affected by the seismic eruption.

The earthquake victims succeeded in getting the government to rebuild the area for them because they organized *en masse* outside formal administrative and party channels and because they relied on defiant tactics, not the deferential petitioning encouraged by the powers that be. Precisely because the government disapproved of tumult, it responded to the grass-roots movement: to restore order. It was responsive to their wants also because inner-city residents had the support of middle-class folk, including middle-class earthquake victims elsewhere in the city. In the highly politicized context, the government could not easily respond to middle-class concerns while excluding the poorer victims'.

Before responding to local demands or rebuilding the *barrio*, the Minister of Urban Development and Ecology, who was in charge of overseeing earthquake reconstruction, made all groups active in the housing movement sign an accord: *El Convenio de Concertación Democrática para la Reconstrucción de Vivienda.* The document specified technical, social, and financial aspects of the housing programme.[9] The groups had to conform with the rules of the game before construction of the new housing began. Signatories of the *Convenio* included not only a dozen local groups and other groups of earthquake victims, but also university groups and technical and professional associations. In signing the agreement the groups agreed, in effect, to work with and not against the state. In so doing, the accord put an end to 'popular' protest.

The *Convenio de Concertación* was consistent with long-standing inclusionary corporatist state–society relations in Mexico. However, it relied on a new means of establishing social order where old mechanisms proved ineffective. The *Convenio* involved a public pact between the state, on the one hand, and groups that had resisted formal affiliation with the government and the PRI, on the other hand. The pact involved housing concessions in exchange for termination of civil disobedience.

The pact, moreover, allowed the government and party to manipulate housing allocations in a manner that strengthened their influence locally. The PRI's hold over the inner-city area had long been problematic, as reflected in the previously noted high rates of electoral abstention. By means of the *Convenio* the government masterfully regulated how local earthquake victims organized to attain housing, and how they organized upon receipt of the housing. It insisted that residents in each apartment building be organized, that the organizations have a specified structure, and that the heads of each group, and they alone, represent residents in all dealings with the government. The persons selected had to have not been previously active in tenants' groups. The government thereby sought to bypass and undermine groups that had deliberately formed independently of the state, and to deprive the leadership of the housing movement of its social base.

Gratitude for the housing which the government sold residents at cost made recipients more positively predisposed towards the PRI than in years past.[10] To fuse (confuse) gratitude to the government with gratitude to the ruling party, the PRI, in addition, offered residents a free tank of propane gas when they moved into their new quarters. In the state's paternalistic tradition recipients of the new apartments were told to solicit for the tanks—to ask, in essence, a favour of the party. The 'favour' was granted, less than a year before the presidential election. Meanwhile, the head of the ministry in charge of the housing reconstruction used the pact to build up a political following, locally and among earthquake victims elsewhere in the city. Thus, he individually, as well as the PRI institutionally, responded to the disorder in a manner that resolved the housing crisis to their political advantage. Grass-roots activism subsided.

Conclusion

The experiences of the low-income neighbourhoods demonstrates that poor people's organizations may contribute to the persistence of poverty and inequality when members are co-opted into groups that provide no institutionalized access to power and when membership serves to regulate what and how demands are made on the government. Neither the mass base nor the populist ideology of the groups in the areas studied sufficed to counter the effects of formal and informal social and political dynamics that subordinated members' concerns to those of more élite groups.

Co-optation resulted from the opportunity structure that induced leaders, largely for opportunistic reasons, to affiliate with national political and administrative groups, either individually or organizationally. In turn, the PRI's populist doctrine and grass-roots organization facilitated acceptance of the official party and the government among resident poor, and enabled both to stand as the champion of local residents. However, as co-optation with time brought little to the neighbourhoods, organizational life in effect atrophied, and residents in growing numbers either lent support to opposition parties at the polls, or abstained from voting.

Since mass-based formal political and administrative institutions served more to regulate than to empower their members, residents attained most state concessions when they organized, *en masse*, outside official channels. Mass mobilizations were especially effective in the shanty-town in the early years of the settlement's existence, and in the inner-city area after the earthquake. They were effective because the government responded to local concerns in order to restore political order and restrain additional demands on the state.

Mobilization did not, however, automatically bring material concessions. The state demanded compliance with 'political rules of the game' it established in return for goods and services. And mobilization was not always a precondition for material improvements. More privileged groups, in the housing development, obtained better benefits without recourse to defiant tactics; they did because of the government's class bias, concealed at the level of official discourse and organization.

By the late 1980s the nature of political accommodation demanded in exchange for favours shifted, as the state faced growing challenges from 'new social movement' type groups and opposition parties. In the past, local groups were pressed to affiliate with the state in exchange for coveted items. But in the 'pact' established with earthquake victims, and their sympathizers, the government 'only' required political demobilization, not state affiliation, in return for housing. The pact is a model of how the state can contain pressure 'from below' where old forms of co-optation no longer serve.

Ethnographic studies, of course, raise the question of generalizability. Electoral and other forms of 'disobedience' are most probable in the recently settled low-income areas on the periphery of Mexico City and other cities where the government has provided fewer services than in the older shanty-towns. Government success at incorporating grass-roots groups, and co-opting their leadership, tends to be contingent on allocations of goods and services. With this objective in mind, a new government ministry was established after the political crisis of 1988, when the PRI barely won the election (and possibly only with fraud). It allowed for centralized allotments of goods and services to the country's poor. *Solidaridad*, as the public-works programme came to be known, deliberately targeted communities on the basis not merely of local need but political disenchantment. The programme was premised on the potential co-optive effects of the provisioning of community services. The programme helped PRI regain mass support in the 1994 presidential election. However, the devastating economic recession that shortly after caused living standards of the poor to plunge made the state's political task formidable and the party's hegemonie hold more problematic than ever.

Notes

1. For a more detailed discussion of the methodology, see Susan Eckstein, *The Poverty of Revolution: The State and Urban Poor in Mexico* (Princeton: Princeton University Press, 1977; 2nd edn. 1988).
2. None of the other political parties had an organizational apparatus approximating that of the 'official' PRI- and government-affiliated groups.
3. In contrast, Karl Deutsch and Daniel Lerner assume that participation is an ever-increasing phenomenon. See Karl Deutsch, 'Social Mobilization and Political Development', in Jason L. Finkle and Richard W. Gable (eds.), *Political Development and Social Change* (New York: John Wiley & Sons, 1966), 205–26; and Daniel Lerner, *The Passing of Traditional Society* (Glencoe, Ill.: Free Press, 1958). Changes in political involvements over time, I argue, are explicable at the structural, not individual, level.
4. Oscar Lewis, *Five Families: Mexican Case Studies in the Culture of Poverty* (New York: New American Library, 1959); Susan Kaufman Purcell and John F. H. Purcell, 'Community Power and Benefits from the Nation: The Case of Mexico', in Francine F. Rabinovitz and Felicity M. Trueblood (eds.), *Latin American Urban Research, 3. National–Local Linkages: The Interrelationship of Urban and National Polities in Latin America* (Beverly Hills, Calif.: Sage Publications, 1973), 49–76; Gabriel Almond and Sidney Verba, *The Civic Culture* (Princeton: Princeton University Press, 1965); Roger Hansen, *The Politics of Mexican Development* (Baltimore: Johns Hopkins Press, 1971).
5. Kenneth Johnson argues that on the national level particular *camarilla* or political clique memberships are generally of greater political consequence than formal political or administrative career patterns. The limited access residents have to such cliques informally limits their political power. See Johnson, *Mexican Democracy: A Critical View* (Boston: Allyn and Bacon, 1971). Nora Hamilton, *The Limits of State Autonomy: Post-Revolutionary Mexico* (Princeton: Princeton University Press, 1982), also highlights the informal class basis of the Mexican state.
6. On informal church–state ties, despite constitutional separation of church and state, see Eckstein, *The Poverty of Revolution*, 108–24.

350   Susan Eckstein

7. If the non-registered population were taken into account, the non-voting population figures would be even higher.
8. The middle-class earthquake victims who dominated the CUD lived in high-rise condominium apartments. Many of them had property title problems, and no strong commitment to their place of residence. By contrast, the people in the inner city were tenants in rent-controlled one- or two-storey buildings with strong ties to their community.
9. The agreement specified the following guidelines: dwellings were to be constructed on the same plots and for the original tenants of the damaged buildings; safe provisional housing was to be provided for families during the reconstruction period, close to their original homes, or families were to receive economic assistance, if they found their own temporary accommodations; the new dwellings would all be forty-square metres and include a living-dining room, two bedrooms, a bathroom, kitchenette, and a space for washing; beneficiaries would only have to repay the direct building costs; and a committee, comprised of representatives of the organizations participating in the agreement, would evaluate proposed alternative projects in terms of the norms of the agreement and existing building codes.
10. In the bitterly contested presidential election of 1988, when Cuauhtémoc Cárdenas carried the Federal District, the PRI won in the old *barrio* section of the inner-city area. In the newer and more politicized section of the area, though, where ties to the Left had been strongest, the opposition won. This information comes from a telephone interview with a local leader conducted after the 1988 election. In another interview I was told by a leading architect of the *Convenio* that 85% of the housing recipients in the city are believed to have voted for the PRI in 1988.

23

The Social Context of Citizenship in Latin America

Bryan Roberts

CITIZENSHIP and the contests over its meaning have been key themes in Latin America development since the Independence movements of the nineteenth century. Latin American states have been strongly nationalistic, determined to make their members foreswear parochial loyalties, whether of community, region, or ethnicity, in favour of the central authority. Moreover, the liberal Constitutions that were adopted throughout the continent, often directly influenced by the example of the US Constitution, created the basis for a civil society independent of the state. Constitutions throughout Latin America established the inviolability of private property and the equality of all citizens before the law, ending status privileges and restrictions such as those attached to aristocratic status, slavery, or to community and church lands (Pérez and Gonzalez, 1927).

The rights attached to citizenship have, however, often been of little practical value to the inhabitants of the region. The extension of citizenship rights from the nineteenth century onwards has had the effect of classing the native Indian populations of Latin America as suspect members of the new nations, since, in practice and often in law, citizenship has required literacy in Spanish or Portuguese and a foreswearing of indigenous practices and loyalties. Violations of basic human rights, such as through arbitrary arrest, have been widespread up to the present. Voting and democratic politics have been interludes within a recurrent theme of authoritarian government.

Governments have, however, been more ready to extend the social benefits of citizenship: education, health care, and social security.

Citizenship in Latin America appears, then, to have primarily served the purposes of élites as they have struggled with each other over the making of the modern nation. Citizenship has been used by élites to rally support for projects that have varied with time and place. In the twentieth century, these élite projects have often assigned the state an authoritarian role in economic development, particularly as the Latin American economies shifted towards industrialization and tried to lessen dependence on the export of primary materials. These projects sought, however, to curtail popular participation in state-led development. They ranged from simple appeals to nationalism against an external or internal enemy, as in the many justifications of military rule, to the more complex projects of populist regimes seeking to incorporate sectors of the working class with the offer of welfare benefits or job security.

I shall argue, however, that the authoritarian strategy, though prevalent, is not a consistent limit on the evolution of citizenship in twentieth-century Latin America. Moreover, it is a strategy that is proving to be increasingly irrelevant in face of the new global context in which the region is placed. Particularly important aspects of this context are the ending of the phase of import-substituting industrialization and the move to open markets and export-oriented industrialization strategies. This change in the political economy of the region places important limits, as we will see, on the capacity of states to co-opt opposition through the partial and targeted extension of certain social rights, while at the same time these same states curtail political and civil rights. Furthermore, countering top-down strategies to limit citizenship are popular pressures that, while weakly organized, exercise a cumulative force for change against even the most consolidated authoritarian élite strategies. In practice, citizenship in Latin America has been defined by a combination of élite and popular pressures that varies historically and from country to country.

Social movements in Latin America are the most visible signs of the struggle to define and re-define citizenship. Their field of action differs from what it is in the advanced industrial countries because in Latin America the project of modernity is far from accomplished, particularly in terms of formal democracy and an adequate standard of living for the majority of the population. As Schuurman and Heer (1992: 26–8) argue, social movements in Latin America, whether 'new' or 'old', are more likely to seek political participation within the state, not autonomy from it, with the aim of making government more responsive to citizens' needs.

The liberal legacy in Latin America and the strains of a rapid urbanization and industrialization made the limited expansion of citizenship through education, health, and social security a central element in the development projects of Latin American élites. In turn, increased aspirations and economic needs made non-élites question the limitations on citizenship, particularly in the political sphere. Also, the change in the global context within which nations develop, particularly trade and market integration on a world scale, has altered the meaning of citizenship both for élites and for non-élites. An important issue is the close involvement of international agencies and non-governmental organizations in the internal debates over citizenship in Latin America. Though élites are likely to contest the further expansion of citizen rights, they are handicapped, as we will see, by the logic of their own neo-liberal development projects that have come increasingly to emphasize individual, market-based, rights and obligations and the withdrawal of the state.

The Context of Citizenship in the Post Second World War

The general trends of development in twentieth-century Latin America have been diverse. Though all countries have been dependent economically on the advanced industrial countries, particularly the United States, the history and nature of their insertion into the world economy differentiated their political and social evolution (Cardoso and Faletto, 1979). Simplifying a complex history, I argue that the more diversified the internal social and economic structure of a Latin American country, as measured by urbanization and economic growth, the more powerful were the 'new' class interests, of industrial bourgeoisie, the salaried middle classes, and the industrial worker. In this situation, class alliances became more complex as traditional classes defended their interests or sought accommodation, and the civil, political, and social attributes of citizenship became one of the major issues over which these conflicts and accommodations took place.

In the twentieth century, the Latin American countries fall into three broad groupings with respect to urbanization and economic growth. These groupings are not exhaustive, but illustrate some of the main contrasts in the development experience of the region. The first grouping consists of those countries which had high levels of urbanization and per capita income, by the 1940s. There are the three countries of the Southern Cone—Argentina, Chile, and Uruguay —which are similar to each other in their high levels of development on these two indicators and in their relatively slow rate of growth subsequently (Table 23.1).[1]

The three countries were heavily populated by European migrations at the end of the nineteenth century and the beginning of the twentieth century. Agriculture was substantially commercialized and none of them had a large peasant population.

After mid-century, a group of countries both urbanized rapidly and had high rates of economic growth. These countries are the three largest Latin American countries in population— Brazil, Mexico, and Colombia. By the 1980s, they had attained levels of per capita income approximating those of the early developers and had more than half their populations living in towns and cities of substantial size. All three

TABLE 23.1   The Pace of Development in Latin America

Countries	Beginning of period[a]		End of period[b]		Rates of growth[c]	
	% in cities of 20,000+	Per capita income constant ($[d])	% in cities of 20,000+	Per capita income constant ($[d])	Urbanization (%)	Per capita product (%)
Early developers[e]	39.0	2,193	69.9	2,789	1.5	0.7
Fast developers[f]	15.9	1,121	52.9	2,258	3.0	2.0
Slow developers[g]	11.7	796	25.8	1,008	2.0	0.7

[a] Beginning is 1940 for urban population and 1960 for per capita income.
[b] End is 1980 for urban population and 1988 for per capita income.
[c] Rates are 1940–80 for urbanization and 1960–88 for per capita income.
[d] The dollar per capita income is in constant dollars of 1980.
[e] Argentina, Uruguay, and Chile.
[f] Brazil, Colombia, and Mexico.
[g] Bolivia, El Salvador, Guatemala, and Honduras.

Sources: Economic and Social Progress in Latin America 1989 Report (Washington, DC: Inter-American Development Bank, 1989); *Statistical Yearbook for Latin America and the Caribbean, 1992* (Santiago: United Nations, 1993).

countries have considerable natural resources and a regional diversity which, from early in the century, contained areas of modern as well as traditional agriculture. They had a diversified urban system that included important industrial centres such as São Paulo in Brazil, Medellin in Colombia, and Monterrey in Mexico, as well as the national capitals and provincial administrative centres.

In contrast with this group are the slow developers, Bolivia and El Salvador, Honduras, and Guatemala in Central America. Nicaragua also fits this category, but the Sandinista revolution of 1979 and the US-imposed blockade led to a different trajectory characterized by, on the one hand, the rapid extension of citizenship and, on the other, by economic collapse with a negative per capita growth rate between 1979 and 1994 of –4.0 per cent (Inter-American Development Bank, 1995).

These countries were among the least urbanized and economically developed countries by mid-century. Their subsequent rates of change were also relatively slow so that by the 1980s, they remained mainly rural with low levels of per capita income. The slow developers had entered the world economy as exporters of primary materials, agricultural in the case of Central America and extractive in the case of Bolivia. Their specialization in this role had barely changed by the 1980s, providing a weak stimulus for industrialization and for the development of the internal market. The Bolivian case differs from that of

the others in this category in that Agrarian Reform in the 1950s stimulated the development of a network of small towns and economic diversification in the countryside.

Peru conforms to the slow development pattern in terms of low initial levels of urbanization, low initial per capita levels, and low per capita growth thereafter. It departs sharply from it, however, in having one of the highest rates of urbanization in the region between 1940 and 1980. In terms of the relation of per capita growth to urbanization, Peru is the most extreme case of 'overurbanization' in the region. By 1990, 70 per cent of Peru's population was classed as urban, compared with 51, 43, 40, and 40 per cent of those of Bolivia, El Salvador, Guatemala, and Honduras, respectively (ECLAC, 1995).

These and other country variations mean that there is no one pattern of urbanization in Latin America in the twentieth century. There is, however, one feature of the Latin American agrarian structures that shaped urbanization in most of the region, with the exception of the Southern Cone countries. Latin America, apart from the Southern Cone, was still predominantly agricultural by mid-century, contained a substantial peasant population, and had a weakly developed internal market. Indeed, by the 1980s, peasant farmers were a larger proportion of the agricultural population in Latin America than they had been in the 1950s and had increased in absolute numbers (De Janvry, Sadoulet, and Young, 1989). The peasant, semi-proletarian character of many Latin American agrarian structures influenced the development of the cities, placing, I suggest, an unusual emphasis on networks of kin and fellow-villagers in migrant adaptation to the city, as Gavin Smith (1984) shows for highland Peruvian migrants in Lima (see Roberts, 1995: ch. 4, for a review of this literature emphasizing the interdependency of rural–urban economic and social relationships in the peasant economies of Latin America, whether 'Indian' or 'Mestizo').

The distinction between early, fast, and slow developing countries provides a first approximation to understanding the evolution of citizenship in Latin America. The early developers have the most extensive set of citizenship rights and of supporting institutions, such as trade unions, political parties, and civil associations. The middle classes early became an important force in politics, followed by substantial sectors of the urban working class (Cardoso and Faletto, 1979: 74–126). It is in these countries, also, that the social rights of citizens became most extensive due to their long history of confronting urban employment and welfare problems. The extension of citizenship among fast developers will be later and less complete than among the early developers, but considerably more advanced than among the slow developers.

However, the evolution of citizenship in Latin America is not linear, nor did the extension of one set of rights, whether civil, political, or social, necessarily entail the extension of others. The evolution was complicated by the uneven pattern of economic growth. Even in the early developing countries, population concentration in the cities did not result in patterns of urban residence and occupational stratification that incrementally fostered class consciousness and organization to extend and defend rights. Though the strength of citizen organization in Argentina and Chile explains, in part, the violence of the dictatorships that sought to eliminate that organization, it is still the case that civil and political rights in these countries were effectively suppressed in the 1970s and earlier. Also, despite the destruction of political opposition by the Chilean military after 1972, social rights with respect to health, education, and family welfare continued under the dictatorship, though at lower levels. To explain these developments, we need to explore the nature of the urban culture that arose during the rapid urbanization of Latin America.

The Myth of Marginality Revisited

In the 1950s and 1960s, commentators in various countries of Latin America emphasized the social and economic marginality of a major part of the region's urban population. They emphasized the recent rural origin of many urban inhabitants, their low levels of education, their lack of familiarity with urban culture, and their failure to find stable wage-work, particularly in large and medium-size firms.

In both city and countryside, urbanization had increasingly mobilized populations in the sense of weakening old patterns of livelihood and making people seek new ways of coping, both politically and economically (Di Tella, 1990: 22–3). These new patterns of coping often were embedded in apparently traditional practices, such as the use of kinship or community of origin networks to survive in the cities. Now, as never before, networks had to be used to manage relationships with the state, not indirectly as through a rural *cacique*, but directly over the multitude of everyday ways in which state policies and practices affected urban inhabitants. In the cities, these populations were attuned to national politics, were directly affected by economic cycles, and were aware of the impact of economic policies on their lives.

Mobilization did not mean, however, that the popular classes were able to organize to exercise effective and consistent pressure on the state. The overwhelming impression derived from reports of organization among the mass of the urban populations of Latin America up to the 1980s is of their fragmentation. They were fragmented economically by urban labour markets in which employment failed to concentrate in large firms. The heterogeneous sector of the services increasingly predominated over manufacturing as a source of employment. By 1980 manual employment in manufacturing industry had declined to approximately 16 per cent of Latin American urban employment (Oliveira and Roberts, 1994: table 5.2). Even the informal economy contained a heterogeneous series of statuses and income levels that differentiated the interests and lifestyles of its members (Portes, 1985; Oliveira and Roberts, 1994: 296–7). The popular classes were fragmented socially because to survive they needed to adopt individual household strategies, including networking throughout the city.

The issue was not one of lack of interest in collective solutions on the part of the popular classes, but of the structural conditions that foster individualism. The poor were rarely in a similar life situation to their neighbours, but were differentiated by type of employment, by stage in the life cycle, and, above all, by the importance of the individual household and its labour resources as a means of survival. The main coping device used by the poor in Latin America was to increase the number of household members in the labour-market, either from within the nuclear family or by adding members to the household. Social and economic fragmentation encouraged vertical, rather than horizontal, political relationships as individuals sought patronage and protection from above as a means of securing what little they had gained in housing or as a means of obtaining more benefits for themselves and their neighbourhood.

The pressure to resolve everyday necessities did, however, generate co-operation as well as fragmentation. Jelin (1993a: 19), referring to Campero's (1987) assessment of the neighbourhood movements in Santiago, points to the tension between an individualistic differentiation among the urban poor and the shared difficulties in subsisting in the city. The first undermines collective identity, while the second generated collaboration and created a basis for collective action. Also, rural–urban migration frequently grouped kin and fellow-villagers in the same neighbourhood and united them in reciprocal favours (Lomnitz, 1977). Formal associations of migrants from the same village or region were organized in certain cities, most notably in Lima, but also in Mexico City, some of them from areas of predominantly Indian culture (Altamirano, 1988; Doughty, 1970; Hirabayashi, 1993; Kemper, 1977).

Under these conditions, collective, neighbourhood-based strategies to cope with economic difficulties had a limited development. In most countries, collective neighbourhood-based movements to seize land, subsequently defend it, and provide it with infrastructure attained a degree of success (Roberts, 1978: 152–7). These movements frequently developed a city-wide organization, as has been reported for Chile (Castells, 1983: 175–212), Peru (Van Garderen, 1989; Blondet, 1991), and Mexico (Bennett, 1992). However, the success of movements often depended on the

patronage they received as a result of intra-élite conflicts, as in the case of the Frente Popular Tierra y Libertad, a coalition of neighbourhood movements in Monterrey, Mexico (Castells, 1983: 197). When patronage was withdrawn, the movements weakened (Tamayo, 1994). Another reported alternative was the co-option of the movement leadership by governing élites, leading eventually to a distancing of the organization from the people it represented. Even where a movement's leadership was committed ideologically to increasing participation, as in the case of the leadership of the Chilean squatter settlement, Nueva Havana, a distancing occurred. This was based, in part, on the priority that the leadership gave to the interests of their political party, the MIR, over the interests of the neighbourhood, and, in part, to the exigencies of surviving under poverty. The struggle to make ends meet divided members of the neighbourhood from each other and gave them different preoccupations than had their leadership (Castells, 1983: 199–209).

Perhaps the major sources of independent organization for the population were trade unions. Industrialization from the 1940s onwards gave a certain bargaining-power to workers in key sectors of the economy. Workers in transport were early organized, as were workers in sectors such as oil and public utilities. Large-scale manufacturing industry, in branches such as textiles, steel, engineering, and automobiles, were organized into trade unions throughout the region. In most countries, trade unions exercised an important influence at the national level for most of the period, often interrupted, however, by military government. Registered trade-union membership had reached relatively high levels by the 1980s, but varied in coverage between countries. In the early developing countries, union coverage remained extensive, though interrupted by the periods of military dictatorship. Union coverage was lowest in the slow developing countries. Thus, in the early 1980s, Argentina, Chile, and Venezuela were estimated to have between 30 and 40 per cent of the economically active population affiliated with unions, Colombia and Mexico between 20 per cent and 30 per cent, while El Salvador and Guatemala had less than 10 per cent (ILO, 1985: table 1.2; Latin American Bureau, 1980).

Trade unions represented only a limited form of citizenship participation for two reasons. One is that they mainly covered workers in large firms, not reaching the informal sector nor many workers in the services. The other is the hierarchical organization of many of the unions in Latin America. The incorporation of unions in the governing structures of Brazil and Mexico often made them unresponsive to the demands of their members and vehicles for the imposition of government policies. However, despite these caveats, labour relations were an important arena in which citizenship was contested and redefined. Despite the appearance of the tight control from above of labour relations in Latin America, there is also a long history of grass-roots activism. Jelin (1975) pointed out that this activism, based on 'bread-and-butter' issues, is more common than is supposed and, at various times and places, has escalated into broader-based movements questioning labour relations and the lack of worker participation. In the 1970s, the importance of grass-roots activism became clear in the case of the independent union movement in Mexico (Roxborough, 1984) and in the automobile workers unions of São Paulo (Humphrey, 1982; Moises, 1982). In the São Paulo case, union activism was reinforced by community participation, since the automobile plants employed a large part of the population of the townships in which they were located (Alves, 1992). In a comparative study of São Paulo and South African workers' movements, Gay Seidman (1994) shows how neighbourhood and community concerns reinforced the solidarity of the workers in both situations.

The majority of the urban poor were not members of unions and rarely participated, in the 1950s and 1960s, in associations or organizations that gave them a basis of collective organization independent of the state. Political parties throughout Latin America made explicit appeals to the poor, but there were few parties with a national importance whose policies explicitly catered for all the poor. Perlman (1976) documents the support that *favela* inhabitants in Brazil gave to the military regime that overthrew the left-leaning civilian government of Goulart. That support, Perlman makes clear, was not due to a political ignorance based on marginality, but to a reaction to the economic anarchy of the previous regime and a calculation that military rule would improve the economy. The major exception to this absence of a grass-roots, community-

based politics linked to political parties was Chile from the end of the 1950s to the military coup of 1972 (Valdés and Weinstein, 1993: 48–62).

Under these conditions, despite the militancy of some sectors of the urban population, both political and civil rights were in jeopardy in Latin America for much of the period from 1940 to 1980. The priorities of most urban inhabitants were focused on 'making the city': finding shelter, incrementally improving their situation, and expecting their children to have a better education and better job prospects than they had been able to obtain. Within the urban employment structure, there were considerable opportunities for education-linked social mobility as employment in non-manual 'white collar' occupations increased from 22 per cent of total employment in 1940 to 35 per cent in 1980 (Oliveira and Roberts, 1994: table 5.2; CEPAL, 1989: cuadro 1–9).

In contrast to the partial eclipse of civil and political rights, social rights increased rapidly from 1940 to 1980. By and large, countries throughout Latin America were successful in raising both health and educational standards. Levels of education rose rapidly throughout the region, with even those countries, such as Peru, with low rates of economic development, showing relatively high levels of education by 1980 (Oliveira and Roberts, 1994: table 5.3). A broadly similar picture can be seen with respect to a health-related measure, that of life expectancy. The improvement in life expectancy was dramatic in Latin America, irrespective of the level of economic development of the country. To be sure, the earliest developing countries still had the highest life expectancies by the 1980s, but the gap between them and both the fast and slow developing countries had narrowed considerably. In Nicaragua, Honduras, Guatemala, El Salvador, and Peru, life expectancy at birth had increased from between 40 and 45 years in 1950 to averages of between 65 and 68 years by 1995 (ECLAC, 1983: table 9; 1995: table 12).

The extension of social security to the economically active population and their dependants was another area in which social rights were advanced during the period of import-substituting industrialization, with the countries of early development having the first and most complete extension of social security (Mesa Lago, 1978, 1991). The timing of this extension varied according to the sector of employment. In general, those occupational groups who received pensions and health protection earliest were effective pressure groups because of their political or economic importance.

The Changing Definition of Citizenship

From approximately the mid-1970s, macro-level forces created a new context for the development of citizenship in Latin America by reducing the saliency of social rights, while increasing that of civil and political rights. Furthermore, micro-level forces linked to the macro changes altered some of the bases for participation, particularly amongst women and in relation to changes in household and community organization.

Macro-Level Forces

By the end of the 1980s, the urban systems of most of Latin America were mature. In the 1980s, natural increase accounts for most of the growth of cities and rural–urban migration for a relatively small proportion. The rural population was, by 1990, only 23 per cent of the Latin American population and too small relative to the urban population to be able to contribute substantially to urban growth rates (Roberts, 1995: ch. 4). In Mexico, for example, my calculations based on the Census indicate that migration from smaller places contributed only about 10 per cent to the growth of cities of over 100,000; but there were considerable transfers of population between these large cities. Indeed, inter-urban migration replaced rural–urban migration as the most important component of internal migration (Lozano-Ascencio, Roberts, and Bean, 1997). Thus, in Mexico City, which had experienced high rates of rural–urban migration in the 1950s and 1960s, most of the economically active population were born there by the late 1980s, and the 1990 Mexican Census suggests that Mexico City had a net loss of population as a result of migration. The new 'cities of peasants' in Mexico were growth centres such as the cities of Mexico's border with the USA. Even in these cities, a substantial component of the migration flows were not peasants but migrants from other urban centres, notably Mexico City.

Though the generalization needs to be tempered by differences in the stage of urbanization, it is likely that by 1990 the period of 'making the city' was over in much of Latin America. Land invasions and the self-construction of houses continued, but, increasingly, the priorities of urban residents turned to consolidating and defending on an individual basis the gains that had been made. Susan Eckstein (1990a) provides an account of this process in a peripheral squatter settlement of Mexico City which changes from a 'slum of hope' to one of 'despair' in the period from the 1950s to the 1980s. She reports how the increase in renting rather than home ownership, the regularization of the situation of the shanty-town as a normal subdivision of the city, and the installation of urban services led to a decline in collective organization, noting a reliance on individual household strategies, rather than community ones, to resolve the increasing problems of poverty, underemployment, and crime which the neighbourhood experienced in the 1980s.

As citizens gained the basic right to shelter but faced the more intractable problems of urban poverty, so, too, their demands posed a more complex problem for the authorities. Patronage, the old stand-by of Latin American governments, was not enough to buy-off opposition, since it was no longer a case of buying-off particularly well-organized neighbourhood groups, but of satisfying a general urban need. Also, the chaos and pollution of the major cities posed a problem for élites also. In the new economic ideologies of the 1980s, poor urban infrastructure was seen as an obstacle to modernization likely to deter foreign investment and hold countries up to ridicule in an international community committed, at least on paper, to environmental concerns.

A new emphasis became apparent, stronger in some countries than in others, on the technical efficiency of government in delivering services. Budgets were inadequate to meet all demands, but one widely practised solution to this excess of demand-making was to base service delivery on impersonal criteria of need and community contribution (Ward, 1993). This service delivery might still have political ends, but it marked a new non-patronage stage in state community relations, as Eckstein (1990b) argues in her analysis of the Mexican Government's implementation of a social pact in 1986 (*Convenio de concertación democrática*) to resolve the housing needs of the earthquake victims of Mexico City.

The maturing of the cities coincided with the ending of the dramatic changes in the occupational structures of Latin America. The major shift from agriculture to non-agricultural occupations was over for all but the slow developing Latin American countries, and, within the cities, the shift to non-manual occupations, particularly higher non-manual occupations, had decelerated. The 1980s were, by and large, years of economic crisis in Latin America in which most of the region's economies had either negative rates of economic growth or rates lower than 1 per cent annually. This performance meant that the per capita GDP was often lower in 1990 than in 1980 and the proportions of the region's population living in poverty increased during this decade (ECLAC, 1993: table 23). It also meant that the number of jobs generated by the formal sector of the economy did not keep pace with the growth of the economically active population in either rural or urban areas. By 1990 the urban informal sector was larger in most Latin American countries than it had been in 1980. This was particularly marked in some of the countries of early development, such as Argentina and Uruguay, that previously had low levels of urban informal employment. The data suggest that most of the growth of the informal sector was in jobs that paid wages below the poverty line and which were disproportionately taken by women.

Open unemployment became an issue to an extent that it had not been previously. In the 1960s and 1970s, unemployment was often characterized as a condition most likely found among the better educated and wealthier sections of the population who could afford the 'luxury' of waiting until a suitable job appeared. In contrast, in the 1980s, data from Brazil and elsewhere suggested that open unemployment was concentrated among the poor and least skilled (Humphrey, 1994). Even the informal economy could not absorb the labour surplus to the formal economy because many of the opportunities within it, such as self-employment, required some capital and depended on formal-sector workers as customers.

As the decline in social mobility and the increasing complexity of the city began to shift the priorities of the population to civil and political issues, so, too, events moved the state in a similar direction. Throughout Latin America, with some exceptions, the state reduced its directive role in the 1980s. The external debt and the need to attract foreign investment exercised a fairly uniform effect on the countries of Latin America. It made governments unusually sensitive to the requirements of the international lending agencies. These requirements included fiscal austerity to reduce government deficits and counter-inflationary pressures. Governments were urged to privatize state industries, make administration more efficient, reduce subsidies for consumption and production, and cut back state employment. They were also urged to reduce tariff protection for local industries and to ease restrictions on foreign ownership of local resources.

Chile was one of the first to follow this road to economic growth under the Pinochet military dictatorship from the mid-1970s onwards. Civilian regimes in other Latin American countries followed suit in the 1980s, Mexico, Argentina, and Peru being some of the foremost examples. The change in state policies meant that nationalism became less important as a state ideology. Governments from Mexico to Argentina talked less in the 1980s about the need to protect national sovereignty and more about the need to open up their markets to the outside. In Mexico, even a symbol of the state's revolutionary origins, the *ejido* (a form of social property in land belonging to village-based peasant farmers and their heirs), was permitted to be sold as a means of increasing private investment in agriculture. The corporate features of many Latin American states were undermined in this situation, since neither entrepreneurs, unionized workers, nor peasants could any longer count on protection from external competition. Indeed, to attract foreign investment, governments throughout the continent sought to weaken labour protection and to give more freedom to firms to determine work organization.

The Bismarkian model of economic development, with the state taking a directive role, became less appealing to the élites that controlled the Latin American states. The extension of social rights had become burdensome for some states and promised to soon be burdensome for others. Instead of co-opting groups by extending them social protection, governments in various

Latin American countries sought ways of shifting some of the burden to the different groupings of civil society, urban residents, employers, and workers. Thus, urban neighbourhoods and rural communities were asked to contribute part of the costs if they were to receive help under the *Solidaridad* programme in Mexico. Employers were urged to pay the real cost of their workers' social security. Self-reliance and community caring were stressed as possible substitutes for state-provided welfare.

By throwing more of the responsibility for social and economic welfare onto the populace, states, both directly and indirectly, promoted the independent organization of citizens. Neighbours, for example, were encouraged to organize and take fiscal responsibility for their neighbourhood affairs. Fiscal austerity also meant that some of the state's key supporters, the multitude of government health, educational, and administrative workers, felt that their interests were in jeopardy. A wide range of groups organized in order to cope with the new rules of the game. Entrepreneurs were amongst the first to do so. Where previously they could rely on tacit agreements with the state to protect their economic interests in face of competition, they now found themselves facing fierce competition from manufactured imports. Also, though free trade brought export opportunities, these required organization on the part of entrepreneurs to generate the resources and information to compete in new markets. In Mexico, entrepreneurial organizations became more active in the 1980s, particularly to lobby government and influence politics (Tamayo, 1994). Also, small and medium-scale entrepreneurs, under severe pressure from foreign imports, organized independently of larger entrepreneurs and took a different political stance. In Brazil, Gunn (1994) describes a similar pattern occurring amongst regional entrepreneurs in the state of Ceará who entered politics through the Brazilian Social Democratic Party. They wrested control from what they saw as a corrupt and inefficient oligarchy, with the explicit aim of modernizing and democratizing the state administration.

The significance of this entrepreneurial organization is that it emphasized civil and political rights. Entrepreneurs' ability to influence government began to depend more on their political weight than on their economic power. Govern-

ments, after all, could now rely on the free market to stimulate investment and productivity and were, in words if not in practice, not committed to any one set of entrepreneurs. For entrepreneurs, political opening gave them their best chance to influence government. Thus, in Mexico, the conservative party, PAN, became reinvigorated in the 1980s as a party supporting entrepreneurial interests and opposed to the lack of civil and political rights. PAN's platforms in the late 1980s and early 1990s were basically in agreement with the government's economic policies, but opposed the governments' lack of democracy and respect for civil rights (Tamayo, 1994).

By the 1980s, citizens had access to a sophisticated media, newspapers and especially radio and television, that provided information over which states had little control. The media are often controlled by governing élites, as in the case of Televisa in Mexico, but their influence on public opinion is broader than their political propaganda. It includes images of life-styles and family practices, national and international, conveyed through the *telenovelas* (soap operas) that are avidly watched by an estimated half or more of the region's population. The presence of the national and international media covering the Zapatista uprising in Chiapas was an important factor giving it national and international visibility and a degree of protection from state repression. Indeed, the media has become the forum through which an independent public opinion is being formed. Opinion polls have became routine in Latin America in the 1980s, even for authoritarian governments. Published widely during the 1994 Mexican and Brazilian Presidential elections, they included analyses of the social bases of the different parties, their regional strength, and reported respondents' evaluations of presidential performance.

The increasing presence of non-governmental organizations (NGOs) has also been a key factor in broadening and protecting citizenship. In the 1970s and 1980s, NGOs expanded their presence in Latin America as they did in other regions of the developing world (Livernash, 1992). They became means of channelling humanitarian aid and of promoting community self-help in the face of the indifference or hostility of authoritarian governments towards citizen rights. NGOs could by-pass national governments because they had powerful international backers. The Catholic and other churches in Chile, for instance, sponsored a range of NGOs concerned with human rights issues and with collective organization to improve living conditions such as neighbourhood associations, soup kitchens, and craft workshops (Valdés and Weinstein, 1993: 130–59). International civil rights associations, with the backing of the United Nations, of the Western European and United States governments, began to make their presence felt even under dictatorships. International relief organizations, such as OXFAM, and international charitable foundations sponsored a host of community development and community self-empowerment projects in Latin America. The relations between the international sponsors, the local NGOs, and the communities they serve is often a difficult one because of differences in power and agenda, but NGOs have helped train and provide opportunities for leadership for large numbers of the Latin American poor, rural and urban, men and women (Schuurman and Heer, 1992: 46–59). By the end of the 1980s there were an estimated 700 organizations in Chile linked to international donor funds, highly concentrated in Santiago (ibid. 55).

The Formation of Identities

The increasing openness of markets and state reluctance to protect the victims of economic competition provides an impetus to ethnic identity in many parts of Latin America. For centuries, indigenous cultures had either been ignored or repressed by national governments, being allowed to survive in economically marginal areas. The spread of commercial farming, ranching, and improved road infrastructure had, by the 1980s, brought even these marginal areas of Latin America into close economic dependence on national and international trends. Indian communities in the Chiapas area of Mexico became more economically differentiated and more sensitive to economic cycles, experimenting with new cash crops, engaging in various forms of seasonal labour, and producing handicrafts (Wasserstrom, 1983: 214–15). Though differentiation produced class divides within communities, it did not override the sense of Indian identity. Instead that identity acquired new potential as a basis of ethnic alliance in face of increasingly complex and pressing external relations. Similarly, Hirabayashi (1993: 129) points out that ethnic alliances among mountain Zapotec in Oaxaca, Mexico, increased in order to defend their rights and autonomy against the intrusion of large businesses, Protestant missionary groups, and the state.

As the Mexican state withdrew its protection of the *ejido*, ending the possibility of new grants of land, so, too, through the North American Free Trade Agreement, it threatened the livelihood of the Indian farmer unable to compete with cheap imports of grain and fodder crops. Ethnicity became, in this situation, a basis of collective action through the Zapatista movement that shook Mexico in January of 1994. This movement was both a 'new' and an 'old' social movement. It represented the long struggle of Indian farmers against mestizo and 'white' landowners. It was partly based on a sense of being abandoned by a corrupt and uncaring government, an age-old grievance of Indian rebellions in Latin America. But it also had new elements, such as an emphasis on democratic reform and a degree of regional autonomy as a solution to local injustice, rather than a reliance on the beneficence of central authority.

The cumulative consequences of economic transformation in Latin America effect the formation of ethnic identity even when there has been no substantial change in the relations between state and civil society. Thus, the Guatemalan state, dominated by its military, continues to repress independent organization. Its policies of combining repression and top-down community development have, however, resulted in a more widespread sense of pan-Indian identity than existed previously (C. Smith, 1990: 271–83). In Bolivia, the declining significance of the mining sector and the abolition of the traditional hacienda as a result of the 1953 agrarian reform appear to have made class a less significant basis of collective organization and ethnicity a more significant one, though not in new areas of colonization where the peasantry has adopted mestizo culture (Albó and Barnadas, 1984). Territorially based Indian political organizations have recently acquired national political significance, as in the case of the CIDOB (the Indigenous Confederation of the East of Bolivia) founded in 1982.

Making ethnic rights a national political issue

owes much to increasing Aymara self-awareness as they migrated in large numbers to La Paz in the 1970s and 1980s. Despite the considerable pressures to adopt Spanish, Aymara ethnic consciousness has been strengthened by their common life situations, such as residence in the higher sections of the city, poverty, participation in the informal economy, and the availability of both radio and television in Aymara (Albó, 1990).

Age has become a more important political identity as the demographic structure of societies matured. In the Southern Cone countries, pensioners became increasingly strident in defence of incomes that were severely eroded by inflation. Even in a country with a younger age structure, such as Mexico, pensioners demonstrated publicly against inadequate pensions. The young were particularly hard hit by these changes. Just as in the 1950s and 1960s, they had been the first to benefit from the new occupational opportunities, so, too, in the 1980s, they were the first to be hit by the declining opportunities. Youth (15–24 years) unemployment became an important issue in the Latin America of the 1980s, with levels estimated at between 25 and 30 per cent (ILO, 1989: 30).

The frustration of the young occurred at all educational levels. Indeed, it may have been highest in the highest levels of education. This frustration was a factor in the appeal of *Sendero Luminoso* to university students in the Peruvian highlands (Gianotten, de Wit, and de Wit, 1985). University education was by the 1980s widespread in Latin America, and not simply the privilege of a small economic and social élite. However, the number of professional or managerial jobs available to University graduates was probably in decline in the 1980s. Lorey and Mostkoff's (1991) data for Mexico shows, for example, that the production of university-trained people far exceeded rate of job creation for professional and equivalent jobs. Students, always a politically active group in Latin America, had new reasons to organize politically. University budgets had not kept up with inflation nor with enrolment in the years of crisis. Moreover, government sponsorship of private higher education was seen by students in state universities to be a direct attack on public education. Throughout Latin America in the 1980s, students were to supply much of the organization for social movements in both urban and rural areas. Grievances there were in plenty, but students often supplied the organizational catalyst.

Perhaps the most significant social trend changing the nature of citizenship in Latin America was the altered status of women, in the economy and in the family. The structural changes in the occupational structure combined with economic crisis to increase women's economic participation throughout the region. Up to 1980, women in Latin America had low rates of economic participation, particularly in the cites, where some 25 per cent of women between the ages of 12 and 60 were economically active. On marriage, women rarely worked outside the home for pay and did not enter the labour-market later, even when their children were grown. However, in the 1980s, more married women with children entered the labour-market and more of those who left to marry and have children returned when their children were of an age to look after themselves. Women sought employment in greater numbers partly because there were now increasing numbers of jobs available to women, in clerical, sales, and even manufacturing occupations and partly because the lowering of male real incomes during the 1980s made it necessary for additional members of a household to seek paid employment.

This increase in women's economic participation had ambiguous results for their status. Various studies report that women felt that having their own source of income gave them more independence within the family (Benería and Roldan, 1987). Yet, working for pay also increased the burden on women, since studies also report that there was little reduction in the time they spent on household chores. Women in effect worked the *doble jornada* of paid and unpaid work. The pressures on male heads of household also appear to have increased as a result of declining real incomes. Furthermore, having working wives outside the home violated cultural norms of the appropriate roles for men and women. The combination of women's feelings of greater independence and negative male reactions was probably an important factor in the domestic violence reported by many studies of urban Latin America in the 1980s and in the increase in the numbers of female-headed single parent households (González de la Rocha, 1988, 1994).

Though most women in Latin America still pursued their traditional roles as wives and mothers, there were clearly some changes in the way women saw their role in society. This was particularly true of countries, such of those of the Southern Cone in which women's economic participation was higher and had increased earlier. There, too, non-traditional forms of the family, such as single-parent households, were more common. Throughout Latin America, however, women appeared to be taking a more independent role not only economically but in community affairs. Blondet (1991), for example, describes the importance of women to the neighbourhood movements in Peru from the early 1970s onwards. Her case study of Vila El Salvador shows the growth of a women's movement, successfully organized around the collective provision of food through soup kitchens. The active leadership in this movement was disproportionately taken by women who were single parents and who were economically independent.

Teresa Valdés (1993) provides a comprehensive and detailed study of neighbourhood movements in Chile and of women's role in them. Women gained experience of organizing during the 1970s and 1980s and this experience, along with the economic pressures on them, made women and women's organizations increasingly important in community and national politics. Women's demands tended to be somewhat different from those of men, more oriented to ensuring that the welfare of the family was secured by organizing to reduce the cost of food, to gain needed services, including education, and to improve their living environment. While men tended to dominate the more directly political organizations, women dominated those concerned with subsistence issues (ibid. 210). Jelin (1993b: 42–5), discussing the movement of the Madres de Plaza de Mayo, makes the general point that Latin American women who are active in the struggle for civil rights are not necessarily feminist, since they are responding to the specific issues that affect them and, consequently differ from country to country in their form and rationale of action. In Argentina, the Madres used maternity, their traditional role within the family, as both a defence against repression and as a means of attacking the dictatorship.

The increased political participation of women broadens the basis of political participation. To traditional male issues, usually tied to the workplace, are added a range of issues to do with the management of daily life. These issues have given salience to civil and political rights. Freedom to associate, not to be subject to arbitrary arrest, and to be able to express dissatisfaction through voting against a government become more important rights when the state loses its capacity to co-opt through the extension of social rights. In face of rising unemployment, arbitrary violence, and rising prices, people may still value the social security attached to certain types of employment. But the extension of social rights is a less immediate prospect than the regaining of civil and political rights.

Changes in community organization, particularly in the cities, but also apparent in the villages, eroded some of the safety valves that had made viable the authoritarian strategy of social rights. Though there is no conclusive evidence on this point, it is likely that there is a weakening of the networks of kin and fellow-villagers that had provided much of the mutual aid that enabled people to cope with urban life. The increase of inter-urban mobility is likely to combine with intra-urban mobility to fragment support networks. Thus, when a son or daughter leaves a city to assume work in another, it is unlikely that they will be helped in adjusting to the new place to the extent that their parents were when they moved from village to city. The common interests in property and the communal traditions of reciprocity that gave village-based networks their strength are barely present in the cities and are fast disappearing in the countryside.

Argentina, an early developing country, demonstrates some of the problems of urban social isolation that can arise from high urban residential mobility. Thus Redondo (1990) reports the difficulties faced by the elderly in La Boca, a port neighbourhood of Buenos Aires. They had settled in the neighbourhood when it was a bustling centre of port activity and remembered the cohesion of the mainly Italian-origin community. With the years, their children had moved away and the neighbourhood became increasingly filled by new immigrants from the interior of Argentina, from Bolivia, and from Paraguay. Redondo's elderly informants report their feelings of isolation both from their neighbours and from their children and other relatives.

They must make do on their own and felt that there was little community support.

Any reduction in the community's capacity to care will give rise to more demands on the state to provide substitute services. Remember that a similar process is likely to be occurring in the villages of Latin America as children leave for the cities or migrate internationally, and those who are left become increasingly dependent on state services. In the current period of fiscal austerity in Latin America, meeting increasing demands cannot readily be resolved by the state extending social rights. The issue thus becomes a political one around which various groups lobby to advance their interests. In this situation, political rights and civil rights become a more attractive focus for economic élites as well as the mass of the population. Democratic politics involves a competition for resources which remains biased in favour of the wealthy, and, for élites, this is one way of handling the problem of not having enough resources to meet demand.

Conclusion

The increase in demand-making at the bottom combines with a partial withdrawal of the state from providing rights from above. Social rights attached to employment, the Bismarkian model, becomes unviable as a means of co-opting opposition given the enormity of the fiscal burden. Social rights have thus become in the 1980s and 1990s another basis of conflict between élites and the population. As market integration on a world scale, make the nationalist, authoritarian state less attractive to economic élites, so, too, they have taken a renewed interest in the original liberal project of Latin America. Associations and political parties become more acceptable ways of advancing their interests. Élite control of crucial institutions, such as the media, gives them some assurance that politics can be directed as they wish. But the population is also more participant, not perhaps in terms of traditional movements, such as trade unions. There is the emergence of a powerful public opinion. This is fed by the difficulties that most of the population faces in making ends meet and is partly articulated by the media. More groups are ready to organize to defend their interests: women, the

young, the old, ethnic and neighbourhood groups. No single social movement covers these various interests or has successfully and consistently articulated any one of them nationally. But their constant appearance, disappearance, and emergence in new forms increases citizen awareness and participation.

It is the cumulative effect of these different trends that underpins the current 'transition to democracy' in Latin America. The transition is unlikely to result in stable two-party politics. Interests are too volatile and unorganized for that. However, it is equally unlikely that the transition will be stalled and return to the authoritarian/corporatist practices that were common in Latin America for much of the twentieth century. The economic basis for that model—the import-substituting phase of Latin America's development—has gone and, with it, its limited conception of citizenship. Instead, there are now multiple bases for citizen participation in Latin America and active contests over the nature of citizen rights and duties.

Notes

1. The data in Table 23.1 is based on averaging the figures for the countries in each category. There is, however, practically no overlap between individual countries in the different categories on either urbanization or level of per capita income. The one exception is Mexico's per capita product in 1988 which exceeded that of Chile by a small margin ($2,588 versus $2,518). By 1992, however, the World Bank reported a per capita income for Mexico that was the highest in Latin America with the exception of Venezuela.

References

Albó, Xavier (1990), 'La Paz/Cukiyawu: Las Dos Caras de Una Ciudad', mimeo, Centro de Investigacion y Promocion del Campesinado (CIPCA), La Paz, Bolivia.

—— and Barnadas, Josep (1984), *La Cara India y Campesina de Nuestra Historia* (La Paz: Unitas/Cipca).

Altamirano, Teófilo (1988), *Cultura Andina y Pobreza Urbana: Aymaras en Lima Metropolitana* (Lima: Pontifia Universidad Catolica del Peru).

Alves, Maria Helena Moreira (1992), 'Cultures of Fear, Cultures of Resistance: The New Labor

Movement in Brazil', in Juan E. Corradi, Patricia Weiss Fagen, and Manuel Antonio Garreton (eds.), *Fear at the Edge: State Terror and Resistance in Latin America*, (Berkeley: University of California Press), 184–211. Reprinted as Chapter 24 in this volume.

Benería, Lourdes, and Roldan, Marta (1987), *The Crossroads of Class and Gender* (Chicago: University of Chicago Press).

Bennett, Vivienne (1992), 'The Evolution of Urban Popular Movements in Mexico between 1968 and 1988', in Arturo Escobar and Sonia Alvarez (eds.), *The Making of Social Movements in Latin America* (Boulder, Colo.: Westview Press), 240–59.

Blondet, Cecilia (1991), *Las Mujeres y El Poder* (Lima: Instituto de Estudios Peruanos).

Campero, G. (1987), *Entre la Sobrevivencia y la Acción Política: Las Organizaciones de Pobladores en Santiago* (Santiago: Instituto Latinoamericano de Estudios Transnacionales (ILET)).

Cardoso, Fernando H., and Faletto, Enzo (1979), *Dependency and Development in Latin America* (Berkeley: University of California Press).

Castells, Manuel (1983), *The City and the Grassroots* (London: Edward Arnold).

CEPAL (Comision Economica para America Latina) (1989), *Transfomación Ocupacional y Crisis Social en América Latina* (Santiago: United Nations).

dc Janvry, Alain, Sadoulet, Elisabeth, and Young, Linda Wilcox (1989), 'Land and Labor in Latin-American Agriculture from the 1950s to the 1980s', *Journal of Peasant Studies*, 18: 3–35.

De Tella, Torcuato (1990), *Latin American Politics* (Austin: University of Texas Press).

Doughty, Paul (1970), 'Behind the Back of the City: Provincial Life in Lima, Perú', in W. Mangin (ed.), *Peasants in Cities* (Boston: Houghton Mifflin), 30–46.

Eckstein, Susan (1990a), 'Urbanization Revisited: Inner-City Slum of Hope and Squatter Settlement of Despair', *World Development*, 18 (2): 165–81.

—— (1990b), 'Poor People versus the State and Capital: Anatomy of a Successful Community Mobilization for Housing in Mexico City', *International Journal of Urban and Regional Research*, 14 (2): 274–96.

ECLAC (Economic Commission for Latin America and the Caribbean) (1993), *Statistical Yearbook for Latin America and the Caribbean, 1992* (Santiago: United Nations).

—— (1995), *Statistical Yearbook for Latin America and the Caribbean, 1994* (Santiago: United Nations).

Gianotten, Vera, de Wit, Tom, and de Wit, Hans (1985), 'The Impact of *Sendero Luminoso* on Regional and National Politics in Peru', in David Slater (ed.), *New Social Movements and the State in Latin America* (Amsterdam: Centro de Estudios y Documentacion Latinoamericana (CEDLA)), 171–202.

González de la Rocha, Mercedes (1988), 'Economic Crisis, Domestic Reorganization and Women's Work in Guadalajara, Mexico', *Bulletin of Latin American Research* 7 (2): 207–23.

—— (1994), *The Resources of Poverty: Women and Survival in a Mexican City* (Oxford: Basil Blackwell).

Gunn, Philip (1994), 'Social or Entrepreneurial Innovations in the Geography of Production: The case of Ceará in Brasil', paper presented to the XIII World Congress of Sociology, Bielefeld, Germany, 19–23 July.

Hirabayashi, Lane (1993), *Cultural Capital: Mountain Zapotec Migrant Associations in Mexico City* (Tucson: University of Arizona Press).

Humphrey, John (1982), *Capitalist Control and Workers' Struggle in the Brazilian Auto Industry* (Princeton: Princeton University Press).

—— (1994), 'Are the Unemployed Part of the Urban Poverty Problem in Latin America' *Journal of Latin American Studies*, 26 (3): 713–36.

Inter-American Development Bank (1989), *Economic and Social Progress in Latin America, 1989 Report* (Washington, DC: Inter-American Development Bank).

—— (1995), *Latin America in Graphs* (Washington, DC: Inter-American Development Bank).

ILO (International Labour Office) (1985), *World Labour Report, 2* (Geneva: International Labour Office).

—— (1989), *World Labour Report* (Geneva: International Labour Office).

Jelin, Elizabeth (1975), 'Espontaneidad y Organización en el Movimiento Obrero', *Revista Latinoamericana de Sociología*, 2: 77–118.

—— (1993a), 'Prólogo para un texto comprometido', in Teresas Valdés and Mariza Weinstein (eds.), *Mujeres que Sueñan: Las Organizaciones de Pobladoras 1973–1989* (Santiago: Facultad Latinoamericana de Ciencias Sociales (FLACSO)), 13–22.

—— (1993b), *¿Ante, De, En, Y?: Mujeres, Derechos Humanos* (Lima, Peru: Red Entre Mujeres/Diálogo Sur-Norte).

Kemper, Robert (1977), *Migration and Adaptation: Tzintzuntzan Peasants in Mexico City* (Beverly Hills, Calif.: Sage).

Latin American Bureau (1980), *Unity is Strength* (London: Latin American Bureau).

Livernash, Robert (1992), 'The Growing Influence of NGOs in the Developing World', *Environment*, 34: 12–43.

Lomnitz, Larissa (1977), *Networks and Marginality: Life in a Mexican Shantytown* (Orlando, Fla.: Academic Press).

Lorey, David, and Mostkoff, Aida (1991), 'Mexico's "Lost Decade", 1980–90: Evidence on Class Structure and Professional Employment from the 1990 Census', in J. Wilkie, C. Contreras, and C.

Weber (eds.), *Statistical Abstract of Latin America*, Vol. 30, Pt 2 (Los Angeles: UCLA Latin American Center Publications).

Lozano-Ascencio, Fernando, Roberts, Bryan R., and Bean, Frank D. (1997), 'The Interconnectedness of Internal and International Migration: The Case of the United States and Mexico', *Soziale Welt*, forthcoming.

Mesa-Lago, Carmelo (1978), *Social Security in Latin America: Pressure Groups, Stratification and Inequality* (Pittsburgh: University of Pittsburgh Press).

—— (1991), *Social Security and Prospects for Equity in Latin America*, World Bank Discussion Paper, No. 140 (Washington, DC: World Bank).

Moises, Jose A. (1982), 'What is the Strategy of the "New Sindicalism"?', *Latin American Perspectives*, 9 (4): 55–74.

Oliveira, Orlandina, and Roberts, Bryan (1994), 'Urban Growth and Urban Social Structure in Latin America, 1930–1990', in L. Bethell (ed.), *The Cambridge History of Latin America*, vi (Cambridge: Cambridge University Press), 253–324.

Pérez Serrano, N., and Gonzalez Posada, C. (1927), *Constitucciones de Europa y America, Tomo II* (Madrid: Librería General de Victoriano Suárez).

Perlman, Janice E. (1976), *The Myth of Marginality* (Berkeley: University of California Press).

Portes, Alejandro (1985), 'Latin American Class Structures', *Latin American Research Review*, 20 (3): 7–39.

Redondo, Nelida (1990), *Ancianidad y Pobreza* (Buenos Aires: Editorial Humanitas).

Roberts, Bryan (1978), *Cities of Peasants* (Beverly Hills, Calif.: Sage Publications).

—— (1995), *The Making of Citizens: Cities of Peasants Revisited* (London: Edward Arnold).

Roxborough, Ian (1984), *Unions and Politics in Mexico* (Cambridge: Cambridge University Press).

Schuurman, F., and Van Naerssen, Ton (1989) (eds.), *Urban Social Movements in the Third World* (London: Routledge, Chapman and Hall).

—— and Heer, Ellen (1992), *Social Movements and NGOs in Latin America* (Saarbrücken: Verlag Breitenbach).

Seidman, G. (1994), *Manufacturing Militance: Workers' Movements in Brazil and South Africa, 1970–1985* (Berkeley: University of California Press).

Smith, Carol (1990) (ed.), *Guatemalan Indians and the State* (Austin; University of Texas Press).

Smith, Gavin (1984), 'Confederations of Households: Extended Domestic Enterprises in City and Countryside', in N. Long and B. Roberts (eds.), *Miners, Peasants and Entrepreneurs* (Cambridge: Cambridge University Press), 217–34.

Tamayo, Sergio (1994), 'The 20 Mexican Octobers: Social Movements and Citizenship in Mexico, 1968–1988', Ph.D. diss. (University of Texas at Austin).

Valdés, Teresa, and Weinstein, Mariza (1993) (eds.), *Mujeres que Sueñan: Las Organizaciones de Pobladoras 1973–1989* (Santiago: FLACSO).

Ward, Peter (1993), 'Social Welfare Policy and Political Opening in Mexico', *Journal of Latin American Studies*, 25: 613–28.

Wasserstrom, Robert (1983), *Class and Society in Central Chiapas* (Berkeley: University of California Press).

24

The New Labour Movement in Brazil

Maria Helena Moreira Alves

The Organization of Trade Unions as Institutions for Social Control

IN analysing recent Brazilian history one is struck by the juxtaposition of old and new mechanisms of social control. The corporatist regulations that limit the freedom of labour to organize were established in the Brazilian Labour Code of 1944 (Consolidação das Leis do Trabalho, henceforth referred to simply as CLT). Based on Mussolini's Carta del Lavoro, they persisted through the democratic governments (1944 to 1964) and the subsequent authoritarian rule. Thus, the labour movement in Brazil has been shaped by regulations that both limit the representation of workers and provide the background for its current militancy. Section V of the CLT carefully regulates the organizational structure of trade unions in Brazil. It is worthwhile to devote some attention to the most pertinent mechanisms of social control embedded in it.[1]

In the CLT, trade unions are defined as organizations for the mediation of social conflict. Their primary purpose, under the law, is to facilitate the resolution of conflicts between labour and capital through tripartite negotiations that involve the active participation of the state. Their secondary purpose is to provide social benefits. The organizational framework is corporative, forming a pyramidal structure that encompasses associations of blue-collar workers, of white-collar professionals, and of employers. The pinnacle of the pyramid is the Ministry of Labour.

Workers (or employers) who wish to form a trade union must apply to the Ministry of Labour for formal recognition. All unions must be formed in accordance with a predetermined organizational charter, which places all administrative power in the board of directors. The regulations are intended to establish control by the Ministry of Labour, which not only has the power to withhold formal recognition of a trade union considered too radical but can also subdivide a specific union into several different sectors.[2] Electoral regulations concentrate power in the hands of incumbents by facilitating manipulation of votes. As a result, one of the most frequent complaints of opposition trade unionists (apart from lack of access to voters) is of open fraud and violation of the secrecy of the ballot.[3] In addition, the Ministry of Labour has the power to annul elections at all levels of trade-union organization or to cancel the right of certain candidates to run for office. Thus, even if opposition candidates win elections, another battle is often necessary in order to begin to carry out the mandate.[4]

The corporative nature of the trade unions is also ensured by voting procedures at the federation and confederation levels. Each local trade union in Brazil has the legal right to two representatives (but only one vote) at the federation level. Hence, a trade union with 100,000 mem-

This is an edited version of an essay that appeared in Juan E. Corradi, Patricia Weiss Fagen, and Manuel Antonio Garretón (eds.), *Fear at the Edge: State Terror and Resistance in Latin America* (Berkeley: University of California Press, 1992), 184–211. By permission of the University of California Press.

bers has the same representational voice as a trade union with 100 members. Furthermore, each federation has only one vote for the election of members of the boards of directors of confederations. Through such procedures, the Ministry of Labour maintains its almost absolute control of elections for the higher levels of trade-union representation. Because the Ministry of Labour has the power to recognize trade unions, it often actively seeks to create unions with few members (known in Brazil as 'ghost unions') so as to ensure a countervailing electoral weight to more active and representative unions with large numbers of members. For this reason it is difficult indeed to find representative opposition trade-union leaders on the boards of directors of federations and confederations.

Another important mechanism for the control of trade-union representation is financial ties to the Ministry of Labour. Most trade-union funds come from a compulsory tax levied on all Brazilian workers, whether or not they are members of unions. This tax is automatically deducted from paychecks and collected by the Ministry of Labour, which also has the sole power to redistribute the funds to confederations, federations, and local trade unions. All funds must be deposited in a state bank and may be frozen by the government. The CLT carefully establishes the portion of funds for welfare services. The CLT prohibits the utilization of these funds for political campaigns, for the purpose of trade-union organization in the work-place, or for financing strikes. These restrictions are legally justified because the funds come from federal taxes and are considered public money to be administered by the state.[5]

The Ministry of Labour also has the power to directly intervene in trade unions, federations, and confederations. By a simple decree, therefore, the government can cancel the electoral mandate of one official or of the entire board of directors. In such cases the Ministry of Labour appoints new administrators to run the union until new elections are organized.[6] According to the CLT, trade-union officials who have lost their post through governmental intervention become permanently ineligible and are prevented from participating in elections at any level of trade-union organization.[7]

The Imposition of New Controls

When the military took over state power in 1964, the mechanisms of control included in the CLT were strictly enforced. But the military also drafted new legislation. Only a few months after the takeover, comprehensive anti-strike legislation was passed.[8] Public employees were denied the right to strike, as were workers in 'essential services', a category that included health workers, workers in the pharmaceutical industries, transport workers, and bank workers. The law also prohibited all strikes that were defined as 'social, political, or religious in nature' or that were called in solidarity with other strikers. Strikes for better working conditions or salary raises were allowed in principle. However, the regulations for calling a legal strike were so stringent that it was nearly impossible for trade unions to fulfil all requirements. According to Kenneth Mericle, the law made illegal most strikes except those called to replace lost salary when employers failed to pay workers for more than three months.[9] Kenneth Erickson, in turn, states that the rigorous implementation of the anti-strike law, when considered in conjunction with the norms of the CLT, did succeed in eliminating most strikes for a number of years. In 1963 there were 302 legal strikes in Brazil. This total fell to 25 in 1965, 15 in 1966, 12 in 1970, and none in 1971.[10] Between 1973 and 1977, there were only 34 strikes or slowdown operations.[11] As we shall see, this law became the focus of opposition by the new trade-union movement after 1977.

Another important measure for the control of workers was introduced with the reform of job-security legislation in 1966. The previous legislation established tenure for all workers employed by a company for more than ten years. The reforms introduced in 1966 cancelled previous rights to job security and established a special fund for the payment of compensation to fired workers, the Time-in-Service Guarantee Fund.[12] The cancellation of job security allowed firms to rotate the labour-force, often firing trade-union activists and replacing them with non-unionized labourers. The regulations established a flexible system of trade-union control in the work-place, making it difficult for members of the opposition to organize in the factories.

Finally, the military governments enacted a

series of laws to control wages. Collective bargaining over wages between workers and employers was prohibited. A system of wage indexing was introduced whereby the government, once a year, decreed the percentage raise in salary for all workers (whether or not they were unionized). This policy removed trade unions from any major role in the negotiation of salaries and reinforced their position as merely social-benefit distributors. The increases in salary were always below the real rate of inflation and had the effect of decreasing purchasing power.[13] The indexing of salaries neutralized collective organizational efforts and greatly benefited corporations by regulating industrial relations for long-term planning. The major purpose of the legislation thus was both to decrease salaries and to attract foreign investment by ensuring a 'safe climate' *vis-à-vis* labour.

An Ideology for Fear-Mongering

The civil/military coalition developed a cohesive ideology that placed Brazil in a geopolitical context of superpower confrontation and clearly defined the limits of opposition. The Doctrine of National Security and Development (from 1964 to 1980) had three components: a geopolitical analysis of global warfare in the nuclear age; a theory of the 'internal enemy'; and a model of economic development combined with security.[14]

In the world-view of the Doctrine of National Security and Development, Brazil belongs within the influence area of the United States as an important country. Its mineral and human resources, coupled with its geographical position in the South Atlantic, make it a key nation for the defence of the West.[15] Within the overall geopolitical context of the doctrine, military strategists developed the theory of the internal enemy. According to this view, in a nuclear age, traditional concepts of warfare no longer applied. The Soviet Union would avoid direct confrontation by supporting groups inside each country to overthrow pro-Western governments. Thus, the main danger would come from an internal enemy rather than from direct foreign aggression. The third component of the doctrine is its model of development. To stimulate the rapid accumulation of capital, the civil/military

coalition in power designed a developmentalist model based on foreign, national, and state capital. Multinational corporations were seen as efficient and beneficial investors in the country. The model required large state infrastructural investments, a system of fiscal incentives for private investment in key sectors of the economy, the concentration of income to ensure a market for durable goods, and, finally, a disciplined labour-market to facilitate planning and assure 'safe' investment.

Because the benefits granted to multinational corporations were often at the expense of national firms, the model met with continued opposition from labour and national entrepreneurial sectors. In turn, the activities of various opposition groups were carefully followed by state intelligence services and catalogued according to their degree of organization. Surveillance followed guidelines developed at the Superior War College and the intelligence schools. According to these guidelines, it was necessary to determine the exact degree of 'nonconformity' of an activity before taking coercive action. The state sought to determine whether an activity was simply 'opposition', or 'pressure', or a 'contesting action'. The *Manual básico da Escola Superior de Guerra*, for example, provides guidelines for the assessment of the level of dissent:

(1) The [size] of nonconformist groups or nuclei. (2) The intensity of their political or social activity. (3) The quality and quantity of the people who belong to these nuclei. (4) The emotional repercussion that their activity manages to provoke among the population. (5) The . . . number of nuclei that are a direct, organized challenge in relation to the less-organized nuclei of opposition. (6) The proportion of voters who are members of the government's party and of the parties of the opposition. (7) The . . . number of votes that were obtained by the government's party and by the opposition parties. (8) The quantity, quality, and degree of actual influence of opposition ideas in public opinion.[16]

The greater the potential impact of the opposition on public opinion, the graver the threat to the state. At higher levels of organization, the opposition was no longer considered simply an 'obstacle' to national security policies but became defined as 'antagonistic', in which case it required greater coercion.

In order to carry out such policies, it became necessary to take full control of state power, to centralize this power as much as possible in the executive branch, and to place those closest to the information network and programming of internal security policy in key government positions. In practice, therefore, the theory of the internal enemy in the doctrine led to the establishment of a vast network of information and the implantation of a repressive apparatus capable of guaranteeing the elimination of the enemies hidden within the nation. The rigorous implementation of security policy, with time, tended to create a climate of suspicion, fear, and mutual denunciation among the population.

Several policies were implemented to control the population. First, military and intelligence operations were expanded. The role of the armed forces was modified to include among its primary responsibilities the maintenance of internal security. The military police, a *paramilitary force* that was traditionally responsible for crime prevention and answerable to the governors of the various states, were put under the direct command of the army and turned into an auxiliary branch of the armed forces. The military police became a major tool in the direct repression of opposition. To provide for intelligence and information gathering, the military government created the National Information Service (SNI), a sprawling intelligence agency answerable only to the executive's National Security Council, with a partly secret budget and the legal right to transfer military and civilian public employees to its service. By 1978 the repressive apparatus comprised an estimated 562,750 people (277,750 men in the army, navy, and air force; 185,000 military police; and approximately 100,000 members of the various agencies of the intelligence community), with an official budget of two billion dollars.[17]

The second policy involved the constant organization of surprise road blockades to check documentation (in Brazil it became a criminal offence to fail to carry identification documents). The frequent 'blitzes' organized by the army and the military police were patterned after the methods of political control used by German occupying forces during the Second World War. Finally, the military, particularly in the rural areas, organized programmes that transferred parts of the population to controlled areas and involved permanent utilization of large numbers of troops.[18]

One must also note the importance of two other state-encouraged policies for the development of a culture of fear: first, the widespread use of torture, beatings, and police violence against the population; second, the unofficial encouragement of paramilitary organizations such as the death squadron and the *mão branca* (white hand) that operated with impunity after 1968.[19]

In Brazil during the period of military governments, control was attained not only through physical coercion but also through psychological repression. The aim was to make individual citizens feel uninformed, separate, fragmented, and powerless. The atomization of citizens creates a feeling of helplessness and hopelessness, which sharpens the underlying fear of participation in political or social opposition activities. Uncertainty was also encouraged. The various legal Institutional Acts enacted by military governments for the purpose of political control, as well as the National Security law, provided only vague definitions of criminal acts of opposition.[20] Legislation was therefore open to different interpretations, with the result that any specific act of opposition was potentially punishable. Whether a particular activity could be subject to indictment depended more on its place in the categories of the theory of the internal enemy than on a strict definition of the law. This element of uncertainty was crucial in creating, in every citizen, a fear that his or her act of opposition could result in indictment, legal sanctions, or even imprisonment and torture.

Legal recourse was also limited through the utilization of special military tribunals to judge crimes against national security, the withdrawal for many years of the right to habeas corpus and other individual guarantees, and the limitations imposed on civilian judges so as to control their actions. During the worst years of repression lawyers who defended political prisoners were also threatened, and consequently the availability of legal counsel was limited.

The underlying fear was enhanced by the lack of information. Official censorship played an important role here. Stories of imprisonment, indictment, and torture abounded, but often it was difficult to be truly informed, with any certainty, of what was going on. Because so many citizens had personally experienced political repression or witnessed a public demonstration of physical force in the repression of demonstrations, assemblies, strikes, document searches, or blitzes, people did not hesitate to believe the stories of pain and torture. Censorship reinforced fear because the truth of such stories could not be determined, and they tended perhaps even to be exaggerated through popular communication. In addition, censorship limited the opportunities to influence public opinion, to demand justice, or to impose constraints on power. Therefore, censorship increased the citizens' feeling of being isolated from the community and rendered them powerless, helpless, and hopeless in the face of state coercion.

The development of a culture of fear was a consequence of a combination of elements: vagueness of legal definitions for criminal activity; limitations on legal and individual rights; lack of access to information and difficulty in communication; fragmentation of community and collective efforts (particularly through impediments to social organizations such as trade unions, political parties, and community grass-roots groups); and, finally, open utilization of physical coercion combined with semi-clandestine activities such as torture and illegal executions.

Winning Back Organizational Power

The debate over the best strategy for obtaining autonomous organization of trade unions in Brazil is as old as the corporative regulations of the CLT. During the period of democratic and populist governments prior to 1964, the controls of the labour code were not strictly enforced. Governments sought the support of trade unions and, increasingly, forged an alliance with labour leaders based on issues of national concern. Because, on the one hand, the mechanisms of control were not strictly enforced and, on the other hand, labour leaders believed that the union tax and the system of *sindicato unico* (which guaranteed trade unions single representation for each branch of industry) were beneficial to the organization of labour, there was little incentive to organize outside the official trade-union structure. The major attempts to break the ties with the Ministry of Labour were, therefore, limited mainly to the organization of a parallel trade-union central—the Central Geral dos Trabalhadores.[21]

Circumstances changed, however, with the military coup of 1964. Because so many labour unions became directly controlled by government-appointed officials, it was important for workers to find alternative strategies of organization. The debate revolved around two courses of action: to work within the 'official' trade unions to pressure for more representative leadership, or to concentrate on the parallel organization of independent but illegal trade unions.

Given the repressive conditions, most trade unionists supported efforts to fight within the 'official' trade unions to wrest control from government-appointed officials. In mounting resistance to the military governments, workers eventually organized a widespread grass-roots movement of union opposition. Members of opposition groups connected to various underground political parties or to the progressive sectors of the popular Catholic Church began slowly to organize pressure groups in both the urban and the rural areas. The major aim was to organize workers in factories and in agricultural areas. Once set up, these groups could collectively pressure the government for elections in particular trade unions and then challenge the incumbent boards of directors. Because the union opposition gave strong emphasis to the organization of workers in the factories, these workers gained experience in grass-roots, democratic forms of organization. It was also argued that once an opposition board of directors took office, the controls could work to the benefit of workers by ensuring strength to the new incumbency, guaranteeing *sindicato unico*, and placing in the hands of the opposition a large machinery with sufficient assets to ensure continuity. Of course, the government could combat this strategy through the use of its right to annul elections, to cancel the candidacy of certain activists, or simply to intervene in the union by decree.

In spite of these threats, most trade-union activists supported the alternative of union opposition. Illegal opposition seemed too costly. Leaders who attempted to organize underground were ruthlessly persecuted, arrested, tried, and sometimes even killed or forced into exile. The most dramatic example was the effort of workers to set up factory committees to support illegal strikes in Contagem (Minas Gerais) and Osasco (São Paulo) in 1968. Particularly in the case of Osasco, the violence of the repression used to end the strike and break up the factory committees was remembered long afterward. The memory of this experience led workers to believe that industrial action could be successful only if carried out on a massive scale and with careful preparation.[22]

The years between 1968 and 1977, the most repressive phase in the history of the Brazilian labour movement, were largely characterized by the abandonment of efforts to organize parallel structures of union organization and by the increased activity of opposition groups within the official trade unions. The movement of union opposition was more successful in gaining legal office in local urban unions during this period than in gaining control of urban federations and confederations. The movement had even greater success in the rural sector. Activists of the union opposition—particularly in those groups that organized together with the Catholic movement in the countryside—were elected to the boards of directors of many federations of agricultural labourers. This position enabled them to gain control of the Confederation of Agricultural Workers in 1968 and to acquire representative leadership in all subsequent elections. This success may be explained by the fact that rural labour unions tend to have fewer members than their urban counterparts, so that ghost unions do not have as great an impact as they do in urban areas, where large unions are common.

Legitimizing Civil Disobedience

One of the remarkable results of the process of development in Brazil since the late 1950s has been the concentration of industrial production in São Paulo. By 1976, metropolitan São Paulo had the largest industrial park in the Third World. The most important industries were located in four towns known as the ABCD.[23] Multinational corporations in the automobile industry, responding to tax incentives and cheap labour, installed major production units in the region. A few giant factories congregated a high percentage of the labour in the area. This concentration of production allowed the formation of a modern urban proletariat comprising an estimated 210,000 workers in the metal-working industries of the area.[24]

The automobile companies encouraged a rapid turnover of the work-force, regularly firing thousands of workers just before the official date for wage increases. The turnover policy was organized by the major automobile corporations so that workers fired, say, by Volkswagen, would be hired by Ford to do the same job but with a lower starting salary. Management thus utilized the Time-in-Service Guarantee Fund to increase exploitation.[25] Ironically, this turnover may, in reality, have aided in the organization of metal workers of the ABCD industrial region. The mobility of labour allowed leaders to move from factory to factory, carrying with them their organizational experience. This turnover also contributed to the development of a strong sense of collectivity and community. Although the corporations benefited from the elimination of job-security guarantees, they also cut one of the last bonds that tied a large number of workers to their employers.[26] Therefore, one of the major mechanisms of control of labour—fear of unemployment—did not work well with a labour force accustomed to high rates of turnover and frequent dismissals. In fact, though job-security guarantees would be one of the main demands of the labour movement in future years, it was not as prevalent a component of the culture of fear as the threat of physical repression.

I shall concentrate on the metal-working industry in the ABCD region for several reasons. First, the automobile and related industries are the most important and dynamic sectors of the Brazilian economy. Secondly, in this region one can best understand the process of the formation of an urban proletariat and the development of organizational skills capable of overcoming the initial paralysis of fear. Finally, although, as I have already pointed out, the rural social movements of trade unions and squatter/settler organizations are extremely well developed, the dispersion of the many groups in the vast interior of Brazil makes it difficult to achieve consistent structural strength sufficient to counter the culture of fear. Study of the ABCD region helps explain a process by which fear is dispelled, community ties are strengthened, and a new dynamic develops.

The Metal Workers Union of São Bernardo do Campo and Diadema is the only legal representative of the metal workers in both São Bernardo and Diadema, where most of the largest factories are located. The two other metal-working unions represent workers in São Caetano and Santo

Andre. For the metal workers of the ABCD, the system of *sindicato unico* would prove to be an enormous advantage, for it concentrated power in a few local unions. Furthermore, the accumulation of so many thousands of workers in a few plants facilitated the organizational efforts of trade-union militants. Metal workers of the ABCD developed a clear consciousness of the power implicit in their numbers and in their strategic position in production. If they could stop the giants of the automobile industry, most of the industrial complex of Brazil would grind to a halt.

In the union elections of 1976 a new and militant group of union leaders was elected. It was headed by Luis Inacio da Silva (known by the nickname of 'Lula'). The new board of directors began to organize groups of workers in each plant. Meetings were held to discuss trade-union legislation, the anti-strike law, regulations for salary increases, job safety, employment security, and other labour issues. Past strategies of organization in the plants were analysed carefully so as to draw appropriate lessons from the experiences of many years. In addition, the union leaders asked opposition economists, sociologists, and lawyers to make presentations and to aid in the educational process of trade-union members. It was important to study strike legislation and to understand legal controls so as to develop proper strategies of industrial action. Any action taken would have to be carefully planned in order to prevent repression or intervention by the government. It was also important to understand the production and investment plans of the transnational corporations so as to pinpoint the best moment to take action.

For two years the small groups met, discussed alternatives, and learned to understand their legal, political, and economic environment. These meetings were crucial for the formation of a new leadership who understood the overall organization of production in the automobile industry and the specific limitations of the labour legislation. By 1977 the organization of workers in the major plants were already well advanced. The next step was the holding of meetings, by factory, to discuss issues of concern to workers in each corporation and to tighten the organization inside the plants. Each group became the core of trade-union representation in the factories, the link between the union and the workers.

Through this process of discussion and analysis an overall strategy was developed. It was necessary to dispel the workers' fear and show them that if they acted together, in large numbers, the government repression would be hampered. It was also necessary to find a way to gain time for success to build so that negotiations could begin directly with management, thus overcoming the impediments to collective bargaining. One small victory would be crucial, for it would allow other workers to hope, to gain courage from collectivity, and to act in defence of their interests. With the aid of lawyers, workers determined that according to the terms of the anti-strike legislation, a strike was defined as 'not going to work'. Careful examination of production and investment plans also showed that the best moment for industrial action was when export quotas had to be met by multinational corporations and smooth production was most important. The strategy relied on tight organization inside the factories, careful utilization of legal loopholes, and non-violent action to eliminate possible justification for repression.

At 7.00 a.m. on Friday, 12 May 1978, 2,500 workers of the Saab Scania plant in São Bernardo do Campo came to work, punched in their cards, and sat in front of their machines with their arms crossed. Soon workers coming in for different shifts in twenty-three other major corporations began to act in the same way. By the second week the movement had spread to Santo Andre, São Caetano, and Diadema, with a total of 77,950 workers on strike. When asked, workers denied they were on strike, claiming that they were in fact coming in to work and punching in their cards as specified in the law. The major demand was for collective bargaining by the union and the employers. The military government was taken by surprise. Management of the large automobile corporations worried about the increasing disruption of production. If the strike continued, they would be unable to meet shipment deadlines for vehicles already sold abroad. On 31 May the Metal Workers Union signed an agreement with the auto companies to provide salary raises above the government's official index.

It was an important victory. The government's rigid legislation for the control of salaries had been overturned, and the principle of collective bargaining had been accepted by management.

In addition, workers had shown that it was possible to override anti-strike legislation through co-ordinated action. The government had been forced to accept the negotiations and had been unable to repress the strikers directly. It had been important to act within the letter of the law. The effect in this case was to cause momentary confusion and allow time for more workers to lose their fear and join the strike. It was also important to keep demands flexible and small. Management, fearful of huge losses, entered into negotiations to reach a quick agreement. Surprise, confusion as to interpretations of the law, co-ordinated action by thousands of workers, and minimal demands were elements that, when combined with the huge daily losses from disrupted production, led management to urge the government not to interfere in the negotiations.

The fear had been overcome by a new kind of strike action, one without visible leaders, without pickets, and virtually without scabs. Within ten days the strike had spread to ninety metal-working firms. From 12 May to 13 June 1978, 245,935 metal workers were on strike in São Bernardo, São Caetano, Santo Andre, Osasco, Jandira, Taboao da Serra, Cotia, and Campinas. Over 400 factories were affected, and the movement spread to eighteen different towns in the state of São Paulo. Other workers, in different trades were quick to follow the example of São Bernardo do Campo. The heart of Brazil's industrial park had come to a halt. Throughout 1978 there were mass strikes among metal workers and workers in the ports, in urban transport, and in the tobacco, glass, ceramic, textile, chemical, and pharmaceutical industries. They were joined by white-collar and professional workers: school teachers, university professors, doctors. By the end of 1978, 539,037 workers had been on strike.[27]

A vast and peacefully organized social movement of civil disobedience developed over the next few years. The military government charged that the strikes were illegal and threatened the security of the nation. Trade-union leaders, echoed by prestigious lawyers and by the hierarchy of the Catholic Church, considered the decrees of the military government illegitimate:

Therefore, we affirm that there is a *legitimate legal order* an an *illegitimate legal order*. Legislation that has been imposed by an act of force constitutes an illegit-imate juridical order. It is illegitimate because, above all, its origin is illegitimate. The only legitimate juridical order is the one composed of legislation rooted in the sovereignty of a people, born from the legitimate and elected representative system.[28]

Basically, the civil-disobedience movement challenged the decrees, Institutional Acts, and complementary legislation that had been elaborated by the executive without the approval of, or even consideration by, a freely elected national Congress. The civil-disobedience movement derived its legitimacy from the struggle to undermine an imposed legal system and to pave the way for a constitutional assembly capable of restructuring the legal order on a foundation of democratic legitimacy.

In 1979 and 1980 the civil-disobedience movement continued to be spearheaded by the metal workers of São Bernardo do Campo and Diadema. The first confrontation with the military authorities exploded the day before President João Batista Figueiredo took office in 1979. An estimated 185,000 metal workers in the ABCD industrial region went out on strike. Although the strike had been carefully prepared by the trade unions, the corporations this time were ready to resist. Factories were locked so that workers could not take refuge inside the buildings. Workers were forced on to the streets in mass picket lines. With thousands of workers assembling in front of the gates of the major plants, there were violent incidents, traffic was disrupted, and the military police broke up meetings to maintain public order. Unable to meet in the factories, too numerous to meet in the trade-union headquarters, metal workers took over the Estadio Vila Euclides, the largest football stadium in the region. There they assembled by the thousands every day to discuss the continuation of the strike. A new kind of decision-making process developed in which all matters of importance were discussed in small groups and put to the vote at the mass meetings. In 1979 the trade unions were thus able to connect grass-roots organization with mass decision by direct vote.

The corporations refused to negotiate unless the workers went back to work. A mass meeting of close to 100,000 metal workers in the Estadio Vila Euclides rejected the proposal. The next day the military government ordered intervention in all the trade unions involved in the strike. Having lost their unions, the workers appealed for community support. The Catholic Church, for the first time, moved directly to provide open support to striking workers. Not only did the *communidades de base*, the Church's grass-roots organizations in the neighbourhoods, provide material support, but all the churches of the region were opened to the workers. From then on, strike assemblies were held in the Cathedral of São Bernardo do Campo.

The metal workers now faced difficult circumstances: the union was under intervention, there was widespread and violent repression of workers in picket lines, and, in spite of generous donations to the strike fund, the economic situation of workers and their families soon became grave. Part of the preparations had involved attempts to save enough to survive a period of strike. However, inflation rates of over 90 per cent made this goal impossible for most families to meet. When the corporations offered a 63 per cent raise (the original demand had been for a 78.1 per cent raise so as to recover at least most of the purchasing power lost to inflation), Lula called another mass assembly in the Vila Euclides and urged workers to accept the agreement. For the first time Lula used his personal influence to swing the opinion of large numbers of workers. Once they had reluctantly returned to work, the military government cancelled the intervention. The move was calculated to undermine Lula's influence by spreading doubt as to his motives in deciding to end the strike. But it backfired: in order to demonstrate support for their leaders, 150,000 workers held a massive rally on 1 May 1979. The rally had the support of all opposition parties and cemented an alliance that was to prove important for the further organization of the movement.

The year 1979 is a landmark in the social and political history of Brazil. The strike movement now spread to most important industrial, service, and white-collar sectors,[29] and to almost all states of the country.[30] In all, more than 3,207,994 workers were on strike throughout 1979.[31] The civil-disobedience movement succeeded in legitimizing strike actions in such a way that, although still technically illegal, strikes became an accepted and integral part of industrial relations in Brazil.

During 1979 members of the Metal Workers Union of São Bernardo do Campo and Diadema met frequently with their leadership to discuss aspects of their organizational framework. The experiences of the past two years were carefully analysed so as to develop alternative strategies. The first conclusion was that although it was important to avoid governmental intervention in the union, this should not be a primary goal.[32]

The second conclusion was that the organization was too dependent on the top leadership. In order to limit the influence of top leaders, to increase the participation of the rank and file in decision-making and in the implementation of strategies, as well as to protect the movement from being immobilized by the arrest of its leaders, the internal structure of the union was modified. Assemblies were held by the production sector of each major corporation to elect representatives to an informal factory committee. These committees together—an estimated 20,000 to 30,000 metal workers—formed the intermediate levels of leadership. Next, general assemblies of all workers were held, inside the plants, to form a committee of 450 members. The Committee of 450, elected with representation from each factory, would be responsible for the co-ordination of the movement. Open assemblies, in the union itself, elected the members of the two higher levels of organization: the Committee of Salary and Mobilization—composed of sixteen members, and the *diretoria de base*, the rank-and-file-board of directors—which had thirteen members who became *de facto* shop stewards. The regular board of directors of the union (twenty members) was responsible mainly for overall co-ordination and for the legal representation of the metal workers in the negotiations to be carried out with the corporations and the government.

Finally, it was concluded that mass pickets were a dangerous mistake, to be avoided at all cost. Experiences in most strikes, in all trade categories, indicated that the military government used the pickets as an excuse for overt, violent repression. Hence, it was decided that the strike actions of 1980 would have to be tightly organized inside the factories so as to enable the continuity of the movement even in the face of lockouts, repression, intervention in the union, or arrest of the top leadership.

When the metal workers of the ABCD region decided to go out on strike again, in April 1980, the new organizational strategy was implemented. The strike action was organized by production sectors in each plant. The intermediary leadership at the rank-and-file level was responsible for the continuity of the movement and for the overall network of internal communication. The Committee of 450 and the Committee for Salary and Mobilization co-ordinated the strike and implemented major decisions made by vote in the huge daily assemblies held in the Estadio Vila Euclides. The *diretoria de base* acted as an alternative to the legal board of directors, who were almost immediately placed under arrest. This time, to avoid the repression of pickets, all workers stayed home. At the end of the first assembly, which voted to begin the strike, Lula simply told workers: 'OK. It has been decided. You all know what to do.' Thousands filed peacefully out of the soccer stadium either to go home or to aid in the organization of the strike fund.

The metal workers of the ABCD, who, with their families, numbered close to one million, received decisive support from community grassroots organizations throughout the country. A strike fund was organized by members of the opposition and funnelled through the Catholic Church. Every week, metal workers formed long lines around the cathedral to pick up packages containing food. Some 480,000 tons of foodstuffs were delivered to the strikers, and churches became focal points for collecting donations and organizing distribution. In addition, the strike fund was able to collect the equivalent of $3.4 million dollars throughout the country.[33]

This time the military government decided that the metal workers had to be crushed. The corporations were ordered not to negotiate. The Minister of Labour ordered immediate intervention in the trade unions, arrested the board of directors of the Metal Workers Union of São Bernardo do Campo and Diadema, and occupied the union building with troops armed with machine-guns. Nineteen days after the strike began, 10,000 soldiers occupied all the cities affected by the strike. Blitzes were launched, neighbourhoods were surrounded, and leaders were taken from their homes, from meetings, even from inside the churches, at gunpoint.

The very force of the repression, however, cemented an alliance between the metal workers and opposition political parties, Church leaders, and the Brazilian Bar Association. The evidence of state violence against a strictly non-violent movement shocked the public and delegitimized the government.

The physical repression also mobilized the women in support of the striking workers. Women's support groups mushroomed across the country. The wives, mothers, and daughters of strikers organized resistance to the violent intervention of the army. A mass meeting of 40,000 women affirmed support for the strike but added a further demand: the release of all those arrested. The rally ended with a march through the city that gathered support from other women so that by the end of the demonstration police estimated the crowd at over 100,000. The women were accompanied by their children and carried armfuls of roses. At a certain point, the demonstrators were met by army troops. After a tense period, the soldiers refused to shoot and allowed the march to proceed peacefully. For the first time since 1964, members of the military openly showed their opposition to repressive orders. The march of the women marked a turning-point in the history of the labour movement: while there was still fear, a culture of resistance to oppression gained increasing force. A new feeling of confidence developed in the people of São Bernardo do Campo—confidence in themselves, in their collective power, in their insertion into the community. Joy, hope, and positive aspirations began to replace feelings of isolation, powerlessness, and dread.

The 'Republic of São Bernardo do Campo'

The strike of 1980 ended in defeat. None of the original demands was granted. The workers lost their union, and leaders of the rank and file in the plants were summarily dismissed without compensation. The management policy of labour turnover now specifically targeted those who had been most important to the continuity of the movement. The intelligence services, including the powerful SNI, compiled long lists of leaders and worked closely with the corporation managers to eliminate them from the factories. The blacklist came to be a predominant criterion for the hiring and firing of workers in all major plants. This policy was so thorough that, to take a single example, Volkswagen fired 13,621 workers in 1981.[34] In addition, many of the national members of the opposition who had previously supported the strike movement now openly criticized the leadership as 'irresponsible'.

And yet, in São Bernardo do Campo itself there was a strong feeling of victory. Public debates were organized almost every night, in the churches or in headquarters of neighbourhood organizations. People participated in analysing the movement. The military police were still everywhere, but the fear had disappeared. There was a new sense of dignity, of self-respect, of independence in relation to the coercive power of the state. Visitors coming from other areas of the country or from abroad often expressed surprise at encountering such hope and joy. After all, they commented, the strike had been defeated and the workers not only had been forced to return to work but had actually lost many of the benefits that they had so painfully struggled to gain. What could account for this new confidence?[35]

The strikes of 1978, 1979, and 1980 had all been fundamentally organized around a specific and limited list of demands. But even though wage raises were undoubtedly significant in the mobilization of workers, it was the process of organization itself that proved to be the key element in producing the emancipatory effects of the strikes. The forms the struggle took—with the decision-making process through the mass assemblies coupled with increasing levels of immediate participation in groups organized by sectors inside the factories—allowed the development of confidence and dignity. The process of organization, which concerned the democratization of power, enabled workers to feel that each person could be both a member of the collectivity and a significant individual actor in the overall movement. Hence, the emancipatory effect of the strikes was that workers created a new way of exercising political power, a new manner of finding community solutions.

After 1981 residents of the area began to talk about the 'Republic of São Bernardo'. The concept expressed their collective awareness that something profoundly liberating had happened—a chain of fear had been broken with their resistance. To their previous keen awareness of their privileged place in the system of industrial production, from which they derived their power to act, was now added a sense of history.

In the endless discussions in the local clubs, squares, and bars, people often referred to themselves as active participants in history. After 1981 the social and political movement in the area can perhaps best be described as a developing 'culture of confidence'.

Union members were now entering a phase of construction, of deepening the organizational structures that had allowed them to live and to experience the joys of democratic participation in major decisions. Shop-floor representation was still illegal. The regulations of the CLT prevented the recognition of shop stewards, of union delegates in the work-place, and of factory commissions. Yet workers were increasingly aware of the importance of improving and broadening their organizations inside the factories while maintaining institutionalized connections to their union. Hence, the next years were devoted to industrial actions aimed at forcing a *de facto*, or even *de jure*, recognition of their representative structures. The overall strategy aimed at achieving structural transformations through the establishment, by stages, of a series of *faits accomplis*. The first step, which gained general acceptance of strikes and collective bargaining, had been accomplished with the massive strikes of 1978, 1979, and 1980. The next step was to force management to recognize freely elected representation inside the plants—both of shop stewards and of factory commissions. Eventually, efforts would be made to win back the union and to challenge the provisions of the CLT that prevented former union directors from running for office in any elections.

The management of some multinationals in the area attempted to neutralize the growing organization in the plants through the creation of 'company representative committees'. This was particularly an objective of Mercedes-Benz and Volkswagen. However, workers protested the managerial interference and openly rejected the initiatives of the corporations.[36] Instead, workers organized informal provisional commissions and enforced their recognition through the pressure of strikes.

The process by which workers in São Bernardo do Campo gained legal recognition for their factory commissions is perhaps best exemplified by events at Ford Motor Company. A careful analysis of Ford's investment plans showed that the best time to act was just before the multinational company began production of its World Car model, which was to be produced largely in São Bernardo do Campo and exported. The members of the informal provisional commission of Ford wrote, and then widely distributed, a note that stated their position:

The government says that this country is in crisis. Meanwhile, the large multinationals are making ever higher profits and program the expansion of their factories. Volkswagen says that it is in crisis, but, at the same time, it buys huge extensions of land in the state of Para, invests to gain control of the production of motorcycles and trucks, and announces the development of new automobile models. However, it fires 8,000 workers. Mercedes-Benz programmed, and is already implementing, a very large expansion of its production, opening even a whole new plant in Campinas. However, rumors have it that it will also dismiss a large number of workers. And Ford do Brasil, in 1980, had a raise in profits of 412.5 percent, and its capital investment was increased by 65.1 percent. The factories of Ford are being enlarged to be able to produce the World Car. The workers in its tractor factory are also now under a continuing program of enforced extra hours. We all know that the value of labor in the overall final cost of production is only around 5 percent. Even so, it is always the workers who pay the price of a crisis: right now, 450 are being thrown into the streets. In our country, this is the way things are. So that the corporations can have high profits the government does not care if thousands of unemployed workers and their families die of hunger. . . . This is why we are fighting. We, the workers of Ford, have begun today a decisive struggle to gain representation, so we can best organize to stop this situation. If we do not, all of us workers, react now, tomorrow will be too late. This is why, even though we are employed, we are on strike. We have stopped work, are running all risks, because we believe that the working class supports us and no longer accepts the injustice and exploitation that weigh so heavily on the poor.[37]

A strike began to protest the policy of mass firings, but it also aimed at establishing a permanent, and legally recognized, factory commission.[38] The occupation lasted forty-six hours; management finally signed an agreement that provided for the establishment of a civil association, with a legal charter and legal recognition. The members of the Ford factory commission were to be directly elected by secret ballot of all Ford employees over eighteen years of age and with more than three months in the firm. They would have a two-year mandate. The first clause

of the charter established the major objectives of the commission:

I. To establish an effective communication channel between the corporation, the employees, and their representatives as well as to improve the relationship between the corporation and the union. II. To ensure a just and impartial treatment of employees based on strict application of labor rights contracted in a negotiated settlement. III. To improve the relationship between employees and their supervisors, based on mutual cooperation and respect. IV. To aid the development of a harmonious relationship of work in the factory and to be a channel to deal with tensions, misunderstandings, and confrontations. V. To solve conflicts with management through the establishment of direct negotiations.[39]

Finally, Ford also agreed to rehire the fired workers and not to dismiss any workers without prior consultation with both the union and the factory commission.

The first elected board of directors of the Ford factory commission took office on 1 March 1982. Although later analysis of the specifics of the legal statutes led workers to point out many flaws, none the less they became the basis for all further negotiations with corporations and for the establishment of other factory commissions.

The experience of democratic organization in the plants enabled workers of São Bernardo do Campo to gain increasing confidence. The establishment of factory commissions led to the formation of inter-factory councils, coordinating different levels of organization. Their increased confidence, in turn, allowed them to challenge other institutionalized forms of state control, such as its continuing intervention in the union. In 1981 pressure by workers became so severe that the Ministry of Labour acquiesced in holding new elections in the Metal Workers Union of São Bernardo do Campo and Diadema. Although Lula and other members of his board of directors were prevented from being candidates, nothing could keep them from actively campaigning in support of other leaders who had been important organizers of the strikes of 1978, 1979, and 1980.[40] As a result, the slate supported by Lula received 97 per cent of the vote. The new board of directors, headed by Jair Meneguelli, took office in July 1981. The intervention by the Ministry of Labour had lasted only a year.

In July 1983 Meneguelli and the Metal Workers Union of São Bernardo do Campo spearheaded a general strike called to oppose the federal government's agreements with the International Monetary Fund. In retaliation, the Ministry of Labour once more ordered intervention in the union. This was the third direct military intervention within four years. These recurrent interventions, however, increased the sense of unity, solidarity, and determination among workers in São Bernardo do Campo. They clearly demonstrated, furthermore, the need to challenge the entire edifice of control embedded in the CLT. As they had done before, workers began to meet to discuss alternatives capable of undermining the legality of the CLT.

Study groups analysed different aspects of the CLT to decide which elements were beneficial to the organization of workers and which should be eliminated. They proposed to draft a legal document to be introduced in Congress by the Partido dos Trabalhadores (Workers' Party). In addition, they devised a strategy to challenge in court the constitutionality of some electoral impediments included in the CLT. A new wave of strikes, public rallies, marches, and assemblies mobilized the population of São Bernardo do Campo to demand new elections in the union. The Ministry of Labour came under pressure from the multinational corporations in the area, who were beginning to see some advantages in having a representative board of directors in the union.[41] Finally, a new date for elections in the union was set for July 1984.

The metal workers sprang a further surprise: the only slate to register for the elections included five former union officers, among them both former presidents, Lula and Meneguelli. None of them, according to the CLT, could participate in union elections. The Ministry of Labour acted swiftly to annul their registration. This annulment allowed the metal workers of São Bernardo do Campo and Diadema to enter a suit in the civil courts arguing for the maintenance of the five candidacies and questioning the constitutionality of the cancellation by the Ministry of Labour. The case went to the Supreme Court, which, in a historic decision, allowed these leaders to run for office. Because this time the government could not even find sufficient candidates to fill its own union slate, the opposition received 100 per cent of the votes and took office, legally, in September 1984.

The steps taken to force *de jure* recognition of representative rights corroded the structural mechanisms of control built by the national security state. The metal workers successfully carried out a strategy of change through pressure from the bottom. To use their own imagery, they were like thousands of little termites who go largely unnoticed until they have eaten away at the very foundations of a house. And, they hasten to add, in the 'Republic of São Bernardo do Campo' they are not even waiting for the house to fall down, they are already busily constructing the new building.

Conclusion

Solidarity overcomes fear; in unity there is strength. The experience of the new Brazilian labour movement, particularly as exemplified by the metal workers of the ABCD region of São Paulo, offers confirmation of these rather banal truths. Yet it also shows that such concepts need to be sharpened and qualified so we may grasp how it is that powerless, exploited, intimidated people can escape from a culture of fear and begin to build a new culture of confident resistance.

Central to the Brazilian experience, first, was the recognition that solidarity breeds strength only if it is a solidarity of participation. The resourcefulness, the resilience, and the political acuity of workers of the ABCD were rooted, in large part, in the strength of their commitment to democratic forms of decision-making. Organization was always constructed from the base up; leaders were expected to consult continually with and be directed by those whom they led. Leadership itself was systematically underplayed. This feature of the movement was misunderstood not only by the military government but by the 'élite opposition' as well. Moreover, the general assumption that workers could not possibly act save at the direction of a vanguard cadre of troublemakers provided an enormous advantage to the movement. Thus, preparations for the 1978 strike were made virtually under the noses of management and the repressive apparatus of the state. Yet, because these preparations were carried out quietly, openly, and in small community and factory groups at the base of society, they went largely unnoticed. On the morning when 2,500 workers crossed their arms and refused to work, virtually the whole of the 'élite'—the government, the security agencies, the press, the academic community—was astounded. 'Where did this movement come from?' 'How did they ever get so organized?'

Secondly, we have seen that under conditions of superexploitation, the circumstances that might otherwise obstruct the creation of solidarity (fear of unemployment, fear of arrest, fear of being branded part of the internal enemy and therefore brutalized by torture) may actually diminish. We noted the ironic consequence of the policy of job rotation in the automobile firms of the ABCD, and we observed how once community organizations (such as those sponsored by the Church) began to allow people to communicate with one another about their experiences and their fears, some of the effects of intimidation began to wear off. Even direct repression and torture have limits. The force of the repression pressed thousands of women to organize effective support and resistance. And as workers would readily point out, the pain of watching one's children go hungry every day far outweighs whatever pain results even from torture.

This does not mean that those at the bottom of the heap, those who have, so to speak, nothing to lose, are in the best position to turn the logic of a culture of fear around. The most forceful movement did not come from the constantly unemployed and underemployed, from the millions who, in Brazilian society, have to fight to survive from day to day. The automobile workers formed, to a considerable degree, an aristocracy of labour, and they were acutely aware of the privileges this position afforded them. But they were also aware, and surely it was the experience of base-level democratic politics that helped to make them so aware, of the responsibilities that accompanied that privilege. By virtue of their importance to the industry, by virtue of their numbers and their skills, by virtue of their concentration in a relatively small geographical location, they could challenge structures of labour control and labour repression that affected not just themselves but the entire workforce of the country. They were acutely conscious of this historical mission, and from it they derived both confidence and strength.

Notes

1. For analysis of the legislation and its various mechanisms for state control over trade unions, see Maria Helena Moreira Alves, *Estado e oposição no Brasil (1964–1984)* (Rio de Janeiro: Vozes, 1984); Jose Rodrigues Albertino, *Sindicato e desenvolvimento no Brasil* (São Paulo: Simbolo S. A. Industrias Graficas, 1979); H. J. Fuchtner, *Os sindicatos brasileiros de trabalhadores: Organização e função política* (Rio de Janeiro: Edições Graal, 1980); R. A. Silva, *Estrutura sindical e perspectivas de mudança*, Document 3 (São Paulo: CEDEC, 1983). For sources in English, see Maria Helena Moreira Alves, *State and Opposition in Military Brazil* (Austin: University of Texas Press, 1985); K. P. Erickson, *The Brazilian Corporative State and Working Class Politics* (Los Angeles: University of California Press, 1977); Kenneth S. Mericle, 'Conflict Regulation and the Brazilian Industrial Relations System', Ph.D. diss. (University of Wisconsin, 1974); and R. Munck, 'State and Capital in Dependent Social Formations: The Case of Brazil', *Capital and Class*, 10 (1980), 125–54.

2. In the state of São Paulo alone, from 1980 to 1985 there were examples of both forms of control. The association of subway workers had to wait more than five years for recognition by the Ministry of Labour before it was allowed to register as an official trade union. And the Ministry of Labour repeatedly attempted to break the organizational power of the Metal Workers Union of São Bernardo do Campo and Diadema by subdividing it into separate unions of automobile workers, auto-part workers, and electrical workers in the car industry. However, because of the level of militancy that workers in these sectors had reached, the government's immediate strategy failed. Nonetheless, the experiences demonstrate the nature of the controls embedded in the CLT.

3. In the elections of 1981 and 1984 in the powerful Metal Workers Union of São Paulo, which represented 400,000 metal workers, the opposition accused the board of directors of manufacturing voting boxes with false bottoms and counting more votes than there were registered voters. Such complaints rarely lead to legal victories because of the difficulty entailed in providing evidence; the opposition lacks access to both lists of voters and the polling stations themselves.

4. Here again, an example can shed light on the workings of the electoral controls. The Bank Workers Union of Rio de Janeiro held three different elections in 1978 and 1979 before the opposition trade-union leaders, who were elected by a large majority, were finally allowed to take office.

This is not an uncommon occurrence; often the Ministry of Labour has given in to pressure only after a number of electoral victories by opposition candidates.

5. It is interesting to note that although a large percentage of trade-union officials recognize that the trade-union tax is a significant mechanism of control by the state, they are divided in their opinions about whether the tax should be abolished. An important survey of trade-union leaders found that most officials believe that unions could not survive financially if the tax were abolished. Most, however, would like to see reform of the CLT so that administration of the tax funds is transferred to trade unions. For a detailed report of the findings of this research on a state-by-state basis, see *Sindicatos em uma epoca de crise* (Rio de Janeiro: CEDEC/Vozes, 1984).

6. The elections may be postponed, sometimes for a number of years, and the union remains under official intervention. The Bank Workers Union of Rio de Janeiro, for example, remained under intervention for eight years before the government finally gave in to pressure and allowed elections.

7. The military governments after 1964 relied heavily on this mechanism of control to eliminate members of the opposition from unions. Between 1964 and 1979 the Ministry of Labour intervened to cancel the electoral mandates of the entire boards of directors of 1,202 unions. In all cases the interventions were in either professional or blue-collar organizations. In no case did the military intervene in employers' associations. For a detailed account of trade-union interventions after 1964, see Alves, *Estado e oposição no Brasil*, in particular table 8.3, p. 244.

8. I refer here to Law 4,330, of 1 June 1964, published in *Diario oficial da união*, 102, no. 104, of 3 June 1964.

9. See Mericle, 'Conflict Regulation', 130–1.

10. Erickson, *Brazilian Corporative State*, 159.

11. *Cadernos do CEAS* (São Paulo, 1977), 34–6.

12. I refer here to the Time-in-Service Guarantee Fund (Fundo de Garantia por Tempo de Serviço), a programme meant to provide readily available funds for compensation. For details on the workings of this programme, see Alves, *Estado e oposição no Brasil*, 97–9; and Vera Lucia B. Ferrante, *FGTS: Ideologia e repressão* (São Paulo: Editora Atica, 1978).

13. For an analytical history of all the different salary-control legislation since the policy was first introduced by the military government on 19 June 1964, see 'Dez anos de politica salarial', in *Departamento Intersindical de Estudos Estatisticos e Socio-Economicos—DIEESE*, Publication 3 (São Paulo, Aug. 1975). The indexing formula of one

raise per year was changed to raises every six months after the major wave of protest strikes in 1979, which involved over three million workers. Subsequent changes in salary legislation were imposed by the International Monetary Fund after the agreements with the Brazilian Government in 1983. An account of the changes in 1983 may be found in Alves, *Estado e oposição no Brasil*, 289–314.

14. In Alves, *Estado e oposição no Brasil*, I analyse in detail the Doctrine of National Security and Development and follow its implications for Brazilian society over the years. Here I shall merely summarize the specific components of the Doctrine that are relevant to the development of the 'culture of fear' in Brazil.

15. One of the most important theoreticians of the Brazilian Doctrine of National Security and Development was General Golbery do Couto e Silva. His best-known work on the subject is *Geopolítica do Brasil*, first published in the 1950s and released in a new edition that includes speeches and other writings on the subject of political control and military strategy. See do Couto e Silva, *Conjuntura política nacional, o poder executivo & geopolitica do Brasil* (Rio de Janeiro: Livraría Jose Olympio, 1981).

16. Estado Maior das Forças Armadas–Escola Superior de Guerra, Departamento de Estudos, *Manual básico da Escola Superior de Guerra* (Rio de Janeiro, 1976), 319–20.

17. For detailed information on the repressive apparatus and the mechanisms of political control and of censorship, see Maria Helena Moreira Alves, 'Mechanisms of Social Control of the Military Governments of Brazil (1964–1980)', in David H. Pollock and A. R. M. Ritter (eds.), *Latin American Prospects for the Eighties: What Kinds of Development?* (New York: Praeger, 1983), 240–303.

18. These programmes were modelled on the American military strategy of pacification developed during the Vietnam War. One such pacification programme was carried out in the conflicted region of Araguaia (in mineral-rich areas of the Amazon basin) and involved the use of an estimated 20,000 troops. For an account of this episode, see Alves, *State and Opposition in Military Brazil*, ch. 6, 'Armed Struggle and the National Security State'.

19. There is a great deal of documentation on torture and on the activities of underground organizations. For a detailed list of sources and a bibliography, see Maria Helena Moreira Alves, 'The Formation of the National Security State: The State and the Opposition in Military Brazil', Ph.D. diss. (Massachusetts Institute of Technology,

1982), 546–9. In addition, see my article on the activities of the death squadrons: Maria Helena Moreira Alves, 'Organizações paramilitares: Assassinos de aluguel', *Retrato do Brasil*, 22 (Jan. 1985), 258–60.

20. I refer here especially to Institutional Acts 1, 2, and 5, and their complementary acts.

21. For information on the history of trade unions in Brazil during the pre-1964 period, see Leoncio Martins Rodrigues, *Trabalhadores, sindicatos e industrialização* (São Paulo: Editora Brasiliense, 1974); Albertino, *Sindicato e desenvolvimento*; Francisco Weffort, 'Partidos, sindicatos e democracia: Algumas questões para a historia do periodo 1945–1964', Master's thesis (University of São Paulo, 1973); Boris Fausto, *Trabalho urbano e conflito social (1880–1920)* (Rio de Janeiro: Difel/Fifusao, 1977); Paulo Sergio Pinheiro, *Politica e trabalho no Brasil* (Rio de Janeiro: Editora Paz e Terra, 1975); Heloisa Martins, *O estado e a burocraticação do sindicato no Brasil* (São Paulo: Hucitec, 1979); and Luiz Werneck Vianna, *Liberalismo e sindicato no Brasil* (Rio de Janeiro: Editora Paz e Terra, 1978).

22. For an excellent historical analysis of the events in Contagem and Osasco in 1968, see Francisco C. Weffort, 'Participação e conflito industrial: Contagem e Osasco, 1968', *Caderno CEBRAP*, 5 (1972).

23. An acronym derived from the initials of the most important industrial towns of Brazil: Santo Andre, São Bernardo do Campo, São Caetano, and Diadema, all located in metropolitan São Paulo.

24. For an interesting analysis of the strikes of this period, see Bernardo Kucinski, *Brazil, State and Struggle*, published in 1982 by the Latin American Bureau, 1 Amwell Street, London EC1R 1UL. See also *Brazil: The New Militancy, Trade Unions and Transnational Corporations*, Transnationals Information Exchange Report, 17 (1978); these reports are published by Transnationals Information Exchange: TIE Europe, Paulus Potterstraat 20, 1071 DA, Amsterdam, The Netherlands.

25. See *Brazil*, 30.

26. Kucinski, *Brazil, State and Struggle*, 68.

27. For details on the strikes, including demands, results, and government reaction, see Alves, *Estado e oposição no Brasil*, 249–66.

28. Goffredo Carlos da Silva Telles, Jr., *Carta aos brasileiros: Em homenagem ao sesquicentenario dos cusos juridicos no Brasil* (São Paulo: Brazilian Bar Association, 11 Aug. 1977), in *Revista da Faculdade de Direito da Universidade de São Paulo*, 72 (1977), 411–23.

29. Metal workers, urban transport workers, civil construction workers, wheat and mill workers, textile workers, bakers, workers in the food industry, workers in clubs and service industries, workers in ceramics, gravediggers, workers in gasoline refining and distribution, natural gas workers, paper and pulp workers, garbage collectors, miners, electrical workers, commercial and shop workers, health workers, bank workers, teachers (all levels), public employees (all sectors), doctors, journalists, and rural agricultural workers.

30. Rio de Janeiro, São Paulo, Bahia, Pernambuco, Paraiba, Espirito Santo, Parana, Santa Catarina, Minas Gerais, Goias, Mato Grosso, Ceara, Rio Grande do Norte, Rio Grande do Sul, and Brasilia (federal district).

31. See details in Alves, *State and Opposition in Military Brazil*.

32. See *Brazil*, 48.

33. See Lucinski, *Brazil, State and Struggle*, 85–6.

34. See the table in *Brazil*, 30.

35. Although I refer mainly to São Bernardo do Campo, the analytical points are pertinent to all the towns of the ABCD area. The analysis that follows is based on participant observation and innumerable conversations with leaders, union members, and regular residents of the ABCD area. I frequently visited the ABCD region during 1978, 1979, and 1980, participated in most assemblies in the Vila Euclides and in the union, and was active in the organization of support for the strike. In 1982 and 1983 I lived in the area and could, therefore, closely participate in and observe the developing process of self-liberation that deeply transformed those who had felt oppressed and afraid.

36. At Volkswagen, management created a company committee and tried to legitimize it through holding elections in the plants. The workers of Volkswagen used the opportunity to organize a vast educational campaign on the importance of unity, of collectivity, and, more importantly, of autonomy. The elections turned into an impromptu plebiscite that rejected, by a margin of over 90%, the proposal for a factory commission tied to the company.

37. Note of the Comissão Provisoria de Trabalhadores da Ford, cited in Jose Carlos Aguiar Brito, *A tomada da Ford: O nascimento de un sindicato livre* (Rio de Janeiro: Vozes, 1982), 46–7. Aguiar Brito is one of the union organizers at Ford. The book is a detailed account of the process by which they gained the factory commission. It is based on taped interviews with workers.

38. See Aguiar Brito, *A tomada da Ford*, 21–3, 63. This description comes from a collective account by members of the Ford factory commission.

39. 'Estatutos da Comissão de Fabrica dos Operarios da Ford', in Aguiar Bito, *A tomada da Ford*, 120.

40. The regulations of the CLT established that members of a board of directors who had been removed from office by official intervention could not run for any union post for the rest of their lives.
41. Strikes in São Bernardo do Campo had become so frequent, and production so disrupted, that management, in despair, often drafted agreements signed by both Lula and Meneguelli (who, of course, had no legal authority) and then themselves convinced the union *interventores* to sign under the names of the recognized leaders. They found that workers refused to go back to work unless agreements were signed by people they considered to be their legitimate leaders.

25

Popular Mobilization under the Military Regime in Chile: From Invisible Transition to Political Democratization

Manuel Antonio Garretón Merino

IN 1973, the democratically elected Chilean president, Salvador Allende, was overthrown in a *coup d'état* by the commanders-in-chief of the armed forces. As a result General Augusto Pinochet ruled from 1973 to 1990. His repressive regime ended the left coalition's efforts to promote a 'Chilean road to socialism' and decades of democratic stability. For ten years, civilian opposition to the regime was expressed under the umbrella of the Catholic Church, or through the underground activities of political parties and social organizations. It was almost impossible to organize public and massive expressions of dissent and opposition, and when such outbreaks occurred they were severely repressed. However, beginning in May 1983, some mass mobilizations occurred, which came to be known as national protests. This essay addresses the significance of these challenges to the military's authoritarian rule and their role and limits in unleashing the transition to democracy in 1988 and ending the dictatorship in 1990.

What role could the protests play in restoring democracy? The opposition to the military was divided on this matter. Some opponents saw the protests as a means of wearing down the regime and forcing it to negotiate a democratic transition, while others believed that the mobilizations would in and of themselves destabilize the regime to the point where it would be forced to relinquish power. Social mobilizations became ever more frequent in the years after the first protest, and some regime opponents remained committed to them as the unique strategy of opposition. Others attributed Pinochet's success in retaining power to the failure of this strategy (see Garretón, 1987, 1991).

The political debate was connected to a more academic debate about the role of social mobilization in transitions from authoritarian to democratic regimes (see O'Donnell and Schmitter 1986; Garretón 1995, 1996a). Is mobilization indispensable for redemocratization, or does it result in greater repression and greater

This is a revised, enlarged, and updated version of an essay that appeared under the title 'Popular Mobilization and the Military Regime in Chile: The Complexities of the Invisible Transition', in Susan Eckstein (ed.), *Power and Popular Protest: latin American Social Movements* (Los Angeles and Berkeley: University of California Press, 1989), 259–77. The first version was translated by Philip Oxhorn and Susan Eckstein. By permission of the University of California Press.

consolidation of military power? How does it relate to other aspects of political transitions, such as regime decomposition, external influences, or internal mediations between regime and opposition? If mobilizations can play an instrumental role in the transition process, is the timing of the defiance consequential and are some types of mobilization better suited than others?

Popular mobilizations can be analysed through a third perspective, i.e. the debate over new social movements. It has been argued that structural changes in industrial or post-industrial societies, as well as in less developed and dependent countries, including Latin America, have generated new types of expressions of defiance and mobilizations for change (see Touraine, 1978, 1989; Slater, 1985; CLACSO-ILET, 1986; Garretón 1996b; Proposiciones, 1987; Escobar and Alvarez, 1992). Did the social mobilizations in Chile constitute something more than mass discontent with authoritarian rule? Did they represent the seeds of a new breed of social movements involving new political actors, premissed on a redefinition of the relationship between politics and society?

The 'invisible transition' to democracy, entailing the recomposition and reorganization of civil society (see Garretón 1989a), must be distinguished from the formal transition to political democracy. The latter involves specific measures that are designed ultimately to end military rule. Since civilian groups mobilized much more in Chile than under repressive governments in other Latin American countries in the 1970s, and since the military remained in power in Chile until 1990, whereas elsewhere in the region they returned to their barracks earlier, the Chilean case raises the question of the potential and limits of 'invisible transitions' to democracy.

In the first part of this article I discuss the general characteristics of social mobilization under military regimes. In the second and third parts, I review the evolution of mobilizations in Chile previous to military government and the main characteristics of the military dictatorship in this country. Subsequently, I analyse the mobilization under the Chilean military government, considering the period preceding social protest, the protest movement, the sectoral mobilizations, the limits and lessons of popular mobilizations, and the political reformulation of mobilization for the transition to democracy. Finally, I conclude

with some general remarks about the relations between mobilizations and the political process of democratization.

Social Mobilizations and Military Regimes

A review of the characteristics and evolution of the military regimes that emerged in the 1960s and 1970s in the Southern Cone of Latin America is not possible here (see Collier, 1979; Garretón, 1989a). For present purposes, several features need only be noted.

First, the regimes were hostile to 'popular' mobilizations. Having imposed themselves on highly politicized societies, they sought to demobilize Civil Society. They did not even attempt to build up political bases of their own. Their concern with depoliticization led them not only to dismantle established forms of mobilization but also to prevent new forms and new social actors from arising. Under such conditions, the presence of social mobilizations represents an authoritarian regime's inability to rule through repression and the existence of pockets of space in civil society for the reconstitution of collective action.

Secondly, the likelihood and nature of mobilizations are partially contingent on the phase of regime evolution. When military regimes first come to power, they are especially repressive: if there is any civilian mobilization, it is minimal and limited to testimonials or defensive expressions by groups directly affected by the repression. Even such restricted mobilization generally occurs under the shelter of such powerful institutions as the Catholic Church. However, once the regimes do not merely stake out their claims to power, but also try to transform the economic and social order and establish new bases of hegemony, sectors adversely affected by the transformations begin to mobilize in opposition. Should the regime's transformative project show itself not to work, mobilizations may become massive.

The first large-scale mobilizations reflect a loss of fear (Martínez, 1992). However, these mobilizations and the demands that participating groups make do not produce by themselves a sufficient crisis to undermine the regime. To bring down the regime and usher in a democratic transition, political leadership and co-ordination are also needed. The political leadership must address the multiple and varied expressions of discontent and the aspirations of protesters in a manner that unifies the groups opposing the regime.

Large-scale mobilizations can be found under different types of regimes. However, under dictatorship and highly repressive governments, the movements are aimed, explicitly, at the termination of the regimes, and they are shaped by the institutional context, which prohibits or constricts their activity. As a consequence, protests and mobilizations under dictatorships and authoritarian regimes have an emotional and 'heroic' aspect, which contributes to their politicization. Moreover, social movements under military governments represent efforts of groups in civil society that were eliminated, weakened, and denied political expression to regroup and reassert themselves, and the mobilizations may exacerbate regime crises and unleash or accelerate the process of redemocratization. The latter two processes do not always coincide, and they may involve contradictory dynamics.

Thirdly, social mobilizations may also assume a diversity of meanings and functions under military regimes. One type of mobilization is expressive and symbolic, with a strong ethical and emotional component. Above all else, this type of mobilization affirms or defends an identity and a community that have been threatened, and it involves rebellion for its own sake. Fasts and hunger strikes in defence of the right to live by families whose kin have 'disappeared' under the military exemplify this type of mobilization. The other types of mobilizations are more instrumental and oriented towards specific ends. One centres on mobilization as a means of strengthening organizational identity, autonomy, and 'legitimacy, as illustrated by organizational elections. Another is the classical mobilization in which protesters assert demands to improve their level of well-being. Land seizures and strikes for higher wages exemplify this type. Still another type of mobilization is explicitly political and aimed directly at the termination and replacement of authoritarian regimes. The Brazilian movement for 'direct elections' is illustrative of this type of mobilization.

The analytically distinguishable mobilizations in practice often exist in combination. The mixes, however, vary. One of the basic problems for an opposition movement is to combine the different types of mobilizations without having them identified with the particularistic concerns of any one participating group or class. Another problem is to avoid excessive politicization, which may make mobilizations unappealing to many.

Finally, the political significance of mobilizations depends on the effect they have on the state. This significance is not inherently determined either by the level or the type of mobilization. Social mobilizations do not in themselves bring about transitions from authoritarian to democratic rule. They can play a critical role in such a transition, but they are not 'the' source of change. For the transition to be completed, the governing bloc must decompose. Some negotiation between power holders and opposition must occur, and normally some external mediation must effectively press for such negotiation. For such reasons, the opposition strategy must take the 'state effect' of mobilizations into account. If it does not, the effect is likely to be determined by power holders. The relationship between civilian mobilizations and political negotiation is, therefore, of crucial significance (see Garretón, 1987, 1989*a*).

Social Mobilizations and Changing State Societal Relations in Chile before 1973

From the 1930s on, three processes occurred concomitantly in Chile. First, there was a process of political democratization involving progressive citizen participation, with a party system including the complete spectrum from the right to the left. Second, Chile also experienced social democratization: social-welfare benefits were progressively extended to the middle classes and, to a lesser extent, organized labour (workers in the so-called formal sector). The state played a crucial role in the extension of social benefits. Until the 1960s the peasants and urban *marginales* (impoverished workers in the 'informal sector' living in *poblaciones*, or shanty-towns) were excluded from social-welfare benefits. During the 1960s, however, peasants were mobilized for an agrarian reform and were unionized by the state and political parties, while the urban *marginales* mobilized through state-linked local organizations for goods and services. Finally, Chile also experienced advanced capitalist import-substituting industrialization, with a strong state presence in the economy. The country industrialized, while remaining dependent for trade revenue on the copper industry which was dominated by foreign capital until the 1960s (see Garretón, 1987, 1989a; Moulián, 1983).

The three processes implied a gradual and institutionalized, but conflictual, basis for the integration of socio-economic groups into the body politic and highly politicized struggles for benefits. While the political parties, including those on the left, accepted the political system, their involvement in social and economic struggles made for an exceptionally politicized society. This gave a distinctive mark to the Chilean integration process: a strong emphasis on social organization with political party ties, and a major emphasis placed on demanding benefits from the state. It also resulted in legal and semilegal mobilizations, with the state being the main focus of collective action.

In the 1960s, the dynamics of dependent capitalist development seemed to come into conflict with the process of democratization. The political parties radicalized, and the party system became highly polarized. The traditional right unified in the National Party, with increasingly authoritarian and nationalistic tendencies. The centre consolidated in the Christian Democratic Party, which had a strong messianic and transformative orientation. The Christian Democrats believed that they were the only party capable of ruling and promoting social change while also maintaining democratic institutions, and they accordingly resisted political alliances. Their vision of politics was exceedingly ideological and non-pragmatic. Meanwhile, the two major parties of the left—the Socialists and the Communists—formed an alliance, together with some groups that splintered off from the centre; they pressed for radical socialist changes within the rules of democracy. In general the ideological climate was radicalized leftward by the Cuban revolution.

In 1964 President Eduardo Frei, a Christian Democrat, initiated a process of capitalist modernization and democratization that incorporated previously excluded peasants and urban *marginales*. His government promulgated an agrarian reform and partially nationalized the copper industry. Midway through his term, though, the reform process bogged down. The state failed to respond adequately to the demands of the highly politicized civilian population. There were widespread illegal strikes and urban land seizures, sometimes severely repressed by the government. The increasing isolation of the Christian Democrats from both the right and the left, in a social and ideological context that legitimated radical social change, helped the left win the 1970 presidential election.[1]

Salvador Allende's Government of the left (Unidad Popular) tried to reverse the capitalist model of development by implementing reforms that benefited the 'popular' sectors at the expense both of foreign capital and large-scale Chilean capital. As a consequence, from the inception of his administration, the vast 'popular' sectors, the government, and the parties of the left were, in conflict with the right and the middle and upper classes. The confrontations drove the centre and, most importantly, the middle classes into militant opposition to the Allende Government which was unable to make broad social and political alliances because its strategy alienated potential allies. Little by little, the middle and upper classes abandoned institutional politics, culminating in their support of a *coup d'état*. Massive mobilizations ensued among all groups, and society became progressively polarized. Although Allende's Unidad Popular received 44 per cent of the vote in the March 1973 parliamentary elections, once the middle sectors and the Christian Democrats sided with the right, the government was left isolated. Allende's problems were further compounded by the US Government's refusal to continue to extend much-needed economic aid and by its strategy of political destabilization. In the context of economic crisis, polarization, and deinstitutionalization, the armed forces under the leadership of General Pinochet usurped power on 11 September 1973. They did so under the pretext of 'restoring the broken institutional system'.[2]

The Main Features of the Chilean Military Regime

The military regime had to address the crisis of Chilean capitalism as well as confront and repress the highly politicized Unidad Popular, and the socio-economic sectors it represented. It promoted a technocratic free-market economic restructuring, relying on the 'Chicago boys', close associates of Milton Friedman (see Vergara, 1984). It accordingly abolished programmes that favoured the 'popular' sectors. Politically, it relied heavily on repression and confronted the church, which denounced human rights violations and protected political victims. It eliminated all channels of collective political expression. The parties of the left were outlawed, and the activities of the other parties were highly restricted as well. Power, including control of the armed forces, was personalized in the hands of Pinochet. Eventually he began a gradual process of political institutionalization, resulting in a new Constitution in 1980. The Constitution allowed a 'transition' from the military dictatorship to an authoritarian regime, under the presidency of General Pinochet until 1989, during which time the governing junta had legislative power and no political activity was permitted.[3]

Beginning in 1981, the military's economic and political programmes ran into difficulty (Vergara, 1984). Many domestic businesses were hard hit once the *laissez-faire* economy made them exceptionally vulnerable to a global economic recession. The living standards of the middle classes plunged, while conditions in the 'popular' sectors deteriorated even further than during the first years of Pinochet's rule. The regime's civilian support, in turn, began to wither, to the point where the military was increasingly isolated politically. Meanwhile, 'popular' expressions of protest and pre-*coup* political parties—which had never been inactive but were necessarily limited in their actions—reasserted themselves. In 1983 massive political and social protests began. The discontented middle classes supported the demonstrations and other forms of mobilizations. The regime responded with repression, although it allowed for limited, informal channels of political articulation (wrongly called *aperturas*).

Although the opposition became increasingly outspoken, it remained fragmented. Various efforts to establish alliances among the reconstituted political parties were short-lived, and they failed to bring about a common strategy of transition. This gave room for the military regime to recompose the economic model and to enforce

the institutionalization designed in the 1980 Constitution, which was designed to assure authoritarian rule after 1989. In accordance with the Constitution, in 1988 the junta proposed Pinochet as the single candidate in the plebiscite for the next presidential period which would extend until 1997. He was defeated by the unified democratic opposition in the plebiscite in October 1988.[4]

The result of the plebiscite unleashed the transition process with some changes in the Constitution that allowed the first democratic elections in 1989 after seventeen years of dictatorship. Patricio Aylwin, the candidate of the centre-left coalition (Concertación de Partidos por la Democracia) was elected President. His inauguration in March 1990 culminated the transition process and started a new period that encompassed the consolidation of a democratic regime.[5]

This is the political and economic context in which Chilean social mobilizations must be understood.

Protest under the Chilean Military Regime 1973–83

Until 1983 anti-regime activity was rarely expressed through large mass mobilizations. When social mobilizations did occur, they addressed either government abuses or the specific concerns of individual socio-economic sectors. The mobilizations included defensive protests against assassinations, detentions, torture, and 'disappearances' (missing people). These protests took the form of fasts, hunger strikes, and quick, limited public rallies.

When the people mobilized around economic issues, concerns varied. In low-income neighbourhoods, people organized for subsistence needs: they set up soup kitchens and their own employment agencies, and they pressed local authorities for land and housing. Workers and university students, in turn, pursued their own sets of claims. Workers pressed for higher wages and changes in labour legislation through slowdowns and other work disruptions. Students began to mobilize through cultural activities, and they held short rallies to protest high fees for education, the presence of repressive agents in the universities, and, more generally, military intervention in academic life. There were also some mobilizations of a more explicitly political nature, for example rallies to celebrate the International Day of Labour (1 May) and to protest the 1980 plebiscite on the Constitution proposed by the military.

The diverse mobilizations between 1973 and 1983 had certain features in common. First, they were isolated incidences, erratic, and generally brief in duration. The size, irregularity, and brevity of the demonstrations reflected people's fears of government reprisals. Secondly, the mobilizations were rarely directed at anyone in particular in the expectation that demands would be satisfied. Instead, they reflected the efforts of groups to assert themselves. When groups attempted to press specific claims and their efforts were repressed, the mobilizations ended. Thirdly, many of these mobilizations occurred under the institutional protection of the Catholic Church. This protective environment helped social organizations to reconstitute themselves gradually as autonomous entities. Fourthly, militants and activists associated with the political parties, human rights groups, and church organizations gave a certain degree of continuity to the mobilizations during this period. They operated somewhat autonomously of their respective institutions, and they were always more radicalized than rank-and-file members. This emergent intermediate political class, linked with groups in civil society, laid the groundwork for the massive 1983 protest and was always present in the mobilizations that were to follow.

The Cycle of Protests and Strikes since 1983

The first so-called national protest occurred in Santiago in May 1983. The Copper Workers' Confederation (CTC) had initially called for a national strike. However, important union locals refused to support the strike. The CTC decided to call instead for a broad-based protest to capitalize on the growing discontent among the population at large. The CTC was well situated to lead such a mobilization because it is centred in a crucial sector of the Chilean economy. Since Chile is dependent on copper for foreign-exchange earnings, the CTC's political import-

ance has always been far greater than the size of its membership would suggest.

On the day that the CTC called for the national protest, there were strikes, high rates of absenteeism, work slowdowns, and demonstrations at work centres. At the universities, there were assemblies and demonstrations. Younger children stayed away from school. In the city centre and on main thoroughfares drivers honked their horns and people staged brief demonstrations. In middle- and lower-class neighbourhoods alike, residents boycotted stores; at night they turned out their lights and banged their pots and pans. The middle classes, who had used their kitchen utensils to express their opposition to Allende, now used them to symbolize their opposition to the very regime they had helped bring to power. Some shanty-towns erected barricades. Although the government had sought to ignore the protest, it responded with force once it became apparent that the mobilization had broad support and was politically threatening: two died, fifty were injured, and three hundred people were detained.[6]

Subsequent national protests were called almost monthly. There were also political rallies, marches, and campaigns for human rights and 'the right to live'. By July 1983 the mobilizations also began to involve cities other than Santiago. They varied in their success and the groups who joined in. More barricades were set up, and black-outs became increasingly frequent. The government responded, in turn, with greater use of force. For example, Pinochet announced the presence of 18,000 soldiers in the streets of Santiago during the fourth protest in August 1983. A large number of the troops were sent into low-income neighbourhoods, and hundreds of protesters were either detained or sent into internal exile.

The eleventh protest, in October 1984, turned into a kind of general strike. The government responded by imposing a state of siege which nearly ended the cycle of protests. While there were new calls for protests after the government lifted the state of siege half a year later, the mobilizations were smaller than in 1983 and generally involved only limited sectors of the opposition. However, at the end of 1985, the Democratic Alliance (Alianza Democrática), a political bloc including the Christian Democrats, some small parties of the right, a socialist party, and other socialist groups, called for a big mass rally. The Popular Democratic Movement, which included the other social party, the Communist Party, the Left Revolutionary Movement (MIR), and other groups of the left, also supported the rally.

In autumn 1986 steps were taken to strengthen the protest movement under the aegis of a newly formed group with ties to the political parties, the Civic Assembly (Asamblea de la Civilidad). The new group organized a politically effective two-day national strike. The regime responded, as in the past, with repression. This time, though, the burning of two youths gained international as well as national attention. None the less, the detention of the protest leaders, division within the opposition over the role such protest should play in political strategy, the discovery of arsenals among pro-insurrectional groups, and the imposition once again of a state of siege following an attempted assassination of Pinochet, undermined the protest movement.

What accounts for the emergence of the protest movement on the one hand and for its inability to sustain itself on the other? We have already mentioned the groundwork laid by the emergent intermediate political class of militants and activists linked to political parties. The surprising success of the initial protest can be traced to three additional factors: its multi-class base, the involvement of Chile's most powerful union, and the stress on broadly based defiance. Since the Pinochet regime was more reluctant to use force against the middle classes than the working and lower classes, middle-class participation reduced fears that protest would end in a massacre. The 1983 protest marked the first time in decades that the middle and 'popular' classes had allied themselves; under Allende, in particular, they militantly opposed one another. The involvement of the Copper Workers' Confederation, in turn, was important, not only because the union has been always very influential, but also because it included representatives of diverse opposition parties. For all these reasons it could mobilize large-scale support, which further minimized participants' fear of government reprisal (Martínez, 1992). The call for a mass protest, rather than for more limited and traditional expressions of defiance such as strikes, moreover, created a feeling of autonomy among participants.

The three elements were not equally present in the subsequent protests. The CTC, for example, later assumed a less activist role because the government severely repressed it for involvement in the first protest. Instead a broader-based organization, the National Workers' Command, sought to mobilize organized labour. While it incorporated unions in diverse economic sectors, including peasants, salaried white-collar employees, the owners of small businesses, and workers in both the state and private sectors, unions at the time lacked their former ability to mobilize the rank and file. Moreover, the National Workers' Command had to share leadership of the opposition movement with political parties, which became increasingly influential in subsequent mobilizations. As the political parties gained force, massive mobilizations were contingent on support from the entire spectrum of parties. The parties, meanwhile, came to assign different meanings to the social mobilizations. The more centrist parties increasingly viewed the protests as a means of forcing the armed forces to negotiate democratization. The parties on the left, by contrast, believed that mobilizations in and of themselves could destabilize the regime to the point where it would collapse. In addition, groups within the opposition movement were not equally committed to militant mobilizations, and their styles of activism differed. As the party blocs distanced themselves from one another, it became more difficult to mobilize support for the protests.

The formation of the Asamblea de la Civilidad was an effort to overcome differences within the opposition movement. Though dominated by middle-class groups, the Asamblea included representatives of a wide variety of socio-economic groups and political parties, among them the National Workers' Command, groups of *pobladores* and university students, professional organizations, truck-drivers, women's associations, human-rights organizations, and the Group for Constitutional Studies (Grupo de Estudios Constitucionales).

Most of the organizations in the Asamblea were committed to pluralism. Moreover, all the opposition parties participated in the Asamblea, although the Christian Democrats dominated it. With such a broad base, there was support for a two-day protest in July 1986. But the protest movement had undergone changes since 1983. The middle class, for one, became increasingly reluctant to support mobilizations. The combination of government repression, some concessions to middle-class *gremios* (unions), and expectations that the government would negotiate with the opposition made the middle class increasingly disinclined to oppose the regime openly. Labour remained more favourably predisposed towards the protests, but government repression undermined its capacity to mobilize. As a consequence, over time, students and, especially, young urban *pobladores* came to constitute the core of the protest movement. They tended to express themselves more aggressively than had the middle and organized working classes, and they distrusted political negotiation and coordination (E. Valenzuela, 1984). Shanty-town dwellers were also radicalized by groups such as the Frente Patriótico Manuel Rodríguez and Milicias Rodriguistas (both linked to the Communist Party) and MIR, which began to organize in their neighbourhoods. These groups pressed for violent confrontation and insurrection. They viewed the protests as heroic moments of confrontation and liberation, but their tactics served to isolate them from the rest of society.

The limits of the protests notwithstanding, their impact was substantial. They enlarged the field of collective action in a highly repressive environment. As a result of the mobilizations, people became less fearful of the military, and the relationship between civil society and the state changed. Moreover, the protests compelled the military to make some economic concessions, above all to the middle classes in order to co-opt them. Pressures from civil society forced the government to modify certain labour practices and aspects of its *laissez-faire* economic model. The protests resulted in some political changes as well. After the fourth protest, the regime began to combine a political logic with its military strategy. The government appointed an old politician of the right as minister of the interior. The new minister initiated a partial 'opening' (*apertura*) to mobilize civilian support for the regime and to institutionalize its rule, but he concomitantly limited the political options of the opposition. Although this political 'project' did not have the intended effect, the government granted certain political concessions, e.g. allowing some exiles to return and some opposition journals to be published.

From the viewpoint of the opposition, the protests allowed for the public appearance and revitalization of political parties and their grouping into larger political blocs (such as the Democratic Alliance and the Popular Democratic Movement). However, once the parties assumed leadership of the protests, differences in goals and strategies adversely affected the fate of the mobilizations. None of the parties provided a basis for consensus among the opposition. Although all the parties attached considerable importance to the mobilizations, none offered a coherent opposition strategy to put an end to the military regime. In the absence of any consensus among the party-dominated opposition, general goals and slogans such as 'Democracy Now' helped the civilian population overcome their fears and isolation. But such general goals did not provide a basis for transforming the mobilization of the civilian population into a more stable political force (Garretón, 1991).

The mass mobilizations since 1983 changed the face of the society. They allowed people to overcome fear. They revealed the military's failure to dissolve collective identities and inhibit collective action, and they reintroduced political 'space' for civil society. They also forced some concessions from the regime. Yet they failed to bring about the widely desired transition to democracy. This would be a process with a very different dynamic.

Sectoral Mobilizations

A full understanding of the dynamics of the protest movement requires a more detailed analysis of the role of specific socio-economic sectors (CLACSO-ILET, 1986). The leadership of organized labour (Campero and Valenzuela, 1984) played a central role in the convocation of the protest movement. However, rank-and-file workers did not play a particularly forceful role in the mobilizations. Their relatively weak presence was rooted in the military regime's impact on the labour movement. The economic crisis brought about by the *laissez-faire* economy of the 'Chicago boys' cost many workers their jobs, and the military drastically restricted labour's capacity to organize and defend its own interests. The unionized labour force declined by 54 per cent between 1972 and 1981, leaving only about 9 per cent of the labour force unionized after a decade of military rule. No doubt many of the workers who were fortunate enough to hold on to their jobs feared that they might be fired if they defied the government. Political divisions at the union leadership level also undercut labour's capacity to organize. Labour leaders were divided even on whether to organize into a single labour confederation. The Democratic Workers' Central associated with the Christian Democrats advocated independent, ideologically distinct *centrales* (national labour associations), while the National Union Co-ordinating Organization associated with the left (but including the more progressive sectors of the Christian Democrats) sought a unitary labour organization. These divisions among the labour leadership affected labour relations at the base level.

The critical role that the urban *pobladores* (Campero, 1987; Proposiciones, 1987; Oxhorn, 1995) came to assume had a major effect on the dynamics and impact of the protest movement. The core group to mobilize tended to be young *pobladores*, perhaps the sector most adversely affected by repression and government educational, employment, and housing policies. The mobilizations gave the socially and economically marginalized poor a sense of participation and belonging and affirmed their individual and social identities. The protests were of expressive and symbolic significance to the young *pobladores*, whose style was aggressive and, on occasion, violent. They set up barricades, burned tyres, and engaged in rock throwing. Their style, however, isolated them from other socio-economic groups.

As important as their participation came to be to the protest movement, most *pobladores* mobilized for specific demands, in particular land and housing. Their participation tended to be short-term, whether or not their demands were satisfied, and their demands did not provide a base on which other protesters could build. Moreover, the sector was so large that efforts to unify it failed.

The political parties added to the problem of mobilizing shanty-town dwellers. The Christian Democrats, the Christian left, the Communist Party, and MIR all tried to build up their own political bases in low-income neighbourhoods; in the process, they created divisions among the poor. The competition among the political parties contributed to the failure of such efforts as that of the Unitary Congress of Urban Pobladores to organize the shanty-town dwellers collectively in 1986. The *pobladores* therefore remained without a broad-based organization of their own through which their activities could be co-ordinated and their interests collectively articulated.

Within the middle class, at least three sectors must be distinguished: the petty bourgeoisie, professional groups, and university students. Small and medium-sized businesses, including independent truck drivers, and shop owners were hard-hit by the 'Chicago' economic model. Yet these groups tended to support mobilizations, through particular *gremios*, only when they felt that they could thereby negotiate concessions for themselves from the regime. They never supported sustained anti-regime activity, nor did they co-ordinate their efforts with other socio-economic sectors (Campero, 1984).

Professional *gremios* mobilized in defence of their own interests. They accordingly organized against regime legislation restricting professional association activities and against repression suffered by their membership. They pressed for association rights, including the right to select their own association leadership; they thus strengthened their organizations and their autonomy from the state, on the one hand, and politicized their organizations, on the other. Candidates linked to the opposition movement won *gremio* elections in nearly all associations affiliated with the Federation of Professional Associations. The action of the opposition was not, however, confined to the internal affairs of the *gremios*. Professional associations were, for example, active in the Civic Assembly, through which they issued declarations and called rallies in opposition to specific military legislation or abuses.

Student federations came closest to constituting a sustained social movement through which specific interests of the group's social base are linked with more general goals of democratization (Agurto, Canales, and de la Maza, 1985). In the first part of the 1980s, the student federations associated with anti-Pinochet political parties had successfully consolidated in all the universities. By 1985, twenty-two of the twenty-four student federations were led by democratically elected opposition leaders. Within the university system, the student groups pressed for reduced tuition fees, as well as for more radical changes, such as a revision of the system of rectorship appointments. Student activities were not, however, confined to the universities. Students also played an important role in the 'popular' protests. Sometimes they seized campuses, bringing university activity to a halt. They also organized street rallies. The linkages with national political parties, however, had the same adverse effect on students as on other mobilized sectors: they divided and weakened the student movement. For example, whereas the first democratic university elections under Pinochet, in 1984 and 1985, centred around alignments with the government or the opposition, in 1986 the opposition split. Electoral lists represented the diverse national political blocs. Tactics divided students as well. The non-politicized mass of students opposed disruptive activity. Many faculty and researchers, whose support was necessary for any substantial change in the universities, were also opposed to disruption of their work and became alienated by a sense that the university had become ungovernable.

Women emerged as a distinctive social force in the opposition movement (Meza, 1986). The failure of the 'Chicago' economic model had increased the presence of women in the informal sector: as men lost their jobs or experienced a decline in their earning power, many women were forced to take on low-wage jobs. Women became active participants in 'popular' organizations and the protests, although surveys showed many of them still to be conservative in their opinions and cultural visions.

The mobilization strategy of women was particularly effective. They mobilized more independently of political parties than other social sectors and stressed unity over partisan fragmentation. A good example was the Mujeres por la Vida (Women for Life) movement. In December 1983 this group sponsored the most unified massive rally against the regime. The group involved women of different socio-economic classes and diverse opposition parties. However, women maintained their autonomy vis-à-vis the political parties only during the pre-transition process, and there was little specificity and autonomy left to women's demands and participations at the beginning of democratic government.

In sum, different groups mobilized separately for their own sets of concerns and collectively in the protests. Participation in the protests was impressive, but how widespread was support for other anti-regime activities? One of the results of the protest and mobilization cycle, and the consequent opening of spaces in civil society, was the incipient emergence of public opinion as a significant factor. Public opinion surveys indicated that most Chileans supported peaceful defiance that would lead to negotiations to end the military regime while rejecting violent and disruptive activity (Huneeus, 1987). In general, there was widespread approval of strikes, petitioning of authorities, marches, and *caceroleos* (banging pots and pans at designated times). There was little support for bombings, black-outs, land seizures, and traffic blockages.

The Limits and Lessons of the Invisible Transition

The military had tried to eliminate collective identities, collective organization, and collective action, but had failed. The Pinochet regime had restricted the impact of protests through the use of force, but defiance won concessions for many groups. Civil society had reasserted itself to the point where it had room to organize and express itself. The recomposition of civil society constitutes what I have called the invisible transition to democracy.

This invisible transition largely involved groups that were active and politicized prior to the *coup*. The principal new social forces that emerged were women, youth, and social, cultural, and religious groups born in direct response to subsistence needs and government abuses, in particular human-rights violations.

But there were several limits to this invisible transition. First, the structural and institutional transformations introduced by the military reduced, weakened, and atomized the organizational 'space' of economic and social groups. Under the Brazilian military regime, by contrast, there was much more room for organization (albeit of a corporatist nature). The contraction of the formal sector and rising unemployment, in particular, undermined the capacity of previously organized groups to mobilize against the regime in Chile. The proportion of wage-earners in the economically active population declined from 53 per cent in 1971 to 45 per cent in 1980 and 38 per cent in 1982 (Martínez and Tironi, 1985), and the proportion of the economically active population who were either unemployed or employed in jobs paying less than the minimum wage, and offering no stable employment or social security, rose from 14 per cent in 1971 to 25 per cent in 1980 and 36 per cent in 1982. The young and women were especially hard hit by the economic contraction.

Secondly, the economic dislocations shifted the bases of mobilization somewhat from the 'classes' to the 'masses' and 'ascriptive' groups, i.e. from the more organized and formal sectors of society to the more amorphous or marginalized ones. Thus women and youth, who had been so adversely affected by the military regime's economic policies, became active in the opposition movement. They mobilized through neighbourhood and women's groups—not through work-based groups, the historical loci of economic struggles in Chile.

Thirdly, because society had become so fragmented, each sector assigned its own meaning to and promoted its own form of mobilization, which at times was in conflict with other sectors. The call for a goal with broad appeal such as 'Democracy Now' was an effort to overcome fragmentation. However, when the goal was not attained, the opposition movement was debilitated. Moreover, the repressive environment encouraged highly expressive and emotional mobilizations, not instrumental mobilizations designed to attain specific and negotiable demands.

Military rule had taken its toll on society. It had modified relations between the state, the political party system, and social movements (Garretón, 1987, 1989a, 1989b). While the military regime had been unable to create the political system it intended, it had disarticulated the previous system. The political forces that surfaced under the military maintained some continuity with the past, but they had great difficulty in political negotiations. The newly politicized 'popular' sectors and emerging social forces were poorly represented. While some activists and militants had gained prominence, they remained isolated from both the more established leadership of parties and the social bases of the 'popular'

organizations. Moreover, ideological and organizational differences among the 'political class' impeded the formation of a strong, unified opposition movement that could transcend the particularistic concerns of diverse social and economic groups.

At the end of the cycle of protest and of this re-emergence of civil society (O'Donnell and Schmitter, 1986) some lessons could be drawn from both the strengths and the weaknesses of mobilization as a strategy for bringing about a transition to democracy. Social mobilizations are undoubtedly indispensable for such a transition. They help to reconstitute civil society and to transform military regimes. They can 'deepen' the 'invisible transition' and obtain some political concessions. But by themselves they will not bring about the array of institutional changes necessary for the restoration of full democracy. Political direction and co-ordination are essential. In the absence of a consensual and coherent political strategy for change, ideological and expressive differences fragment groups similarly committed to democracy and limit the impact of the mobilization. Thus, mobilization must be combined with other political processes, such as negotiation and regime decomposition, before democratization is likely.

Political Mobilization for the Transition

The Chilean opposition to the military regime learned the lessons of the invisible transition only slowly and unevenly (Garretón, 1991). The delays allowed the regime to reassert itself and to enforce the process of institutionalization that would culminate in the plebiscite to legitimize Pinochet for another presidential period.

In 1986 the discovery of the weapons arsenal of the Frente Patriótico Manuel Rodríguez (which had its origins in the Communist Party, yet remained autonomous), their attempt to assassinate Pinochet, and the state of siege that ensued, demobilized the opposition. The protest movement had already eroded, and the notion spread that the regime could not be overthrown. The opposition leadership understood also that because of the time lost and its incapacity to present an institutional formula for a transition, they would be forced to adjust to the formula proposed by the regime. Henceforth a strictly political vision to end the regime oriented all social movements, subordinating the more symbolic and heterogeneous mobilizations.

The discussions in the social and political opposition whether or not to participate in the institutions created by the dictatorship culminated in February 1988 when all the main political parties, except the Communist, signed an agreement to participate in the plebiscite in October 1988 to vote against Pinochet (Concertación de Partidos por el NO). The main aim then became to obtain optimal conditions to defeat Pinochet in the plebiscite (voter registration, freedom to campaign, access to television, end of the state of siege, return of the exiles). Public opinion overwhelmingly wanted to end the regime by political and negotiated means. It had to be reoriented so as to oppose the dictatorship in the plebiscite. So, opposition focused exclusively on the campaign.

In this more political than social scenario, the issue was to channel social organizations and to convert social demands into political ones. In 1988 the workers' movement strengthened its autonomy through the creation of the Central Unitaria de Trabajadores, integrating a new actor into the process of transition. At the same time, the Acuerdo Social por el NO was established as the heir of the Asamblea de la Civilidad to participate in the plebiscite, and various organizations representing different social sectors were organized but remained subordinate to the political leadership: the Concertación for the No Vote. Parties and social organizations overcame their past conflicts.

The triumph of the opposition in the plebiscite of October 1988 unleashed a process of transition characterized, first of all, by the attempt of the military to conserve their main prerogatives in the future democratic regime and the struggle of the opposition to moderate the authoritarian element of the Constitution of 1980.[7] Secondly, the Concertación for the No Vote was transformed into a political coalition for government (Concertación de Partidos por la Democracia) with a common political programme and candidates for the presidential and parliamentary elections of December 1989. During this period, the elaboration of the governmental programme and the electoral campaign were the main links between social organizations and political par-

ties, processing social demands often mediated by the non-governmental organizations that flourished during the dictatorship (Garretón, 1989b, 1993; Concertación, 1989). This transition process culminated in the first democratic elections after seventeen years of dictatorship in December 1989, won by Concertación, and the inauguration of the democratic government headed by President Patricio Aylwin in March 1990.

Conclusions

Popular mobilizations under military regimes have two main meanings (Garretón, 1996a). On one hand, they are manifestations of the reconstruction of the social fabric, the loss of fear, the reconstitution of social subjects and actors, the expression of the sense of belonging and also of concrete material and non-material social demands. In that sense, they constitute what I have called the invisible transition to democracy, and they cannot be reduced to partisan or narrowly political ideas. The diversity and the heterogeneity of their meanings, along with their emotional and symbolic weight, account for these aspects.

On the other hand, popular and social mobilizations express the rejection of a dictatorship and the aspiration to terminate it. This demands a coherent political strategy that, due to the characteristics of contemporary military and authoritarian regimes, has to be oriented towards negotiations and institutional mediations, thus subordinating and orienting the various expressions of popular protest.

Popular mobilizations are not able by themselves to end military regimes and achieve a transition to democracy. Effective political mobilization for a democratic transition cannot emerge without roots in popular mobilizations. The connection between these two processes is problematic, and their principal actors differ. Indeed, these processes often appear as contradictory. Their co-ordination in a single strategy to end the dictatorship requires political co-ordination and leadership. The transformation of popular mobilization into a political movement for democracy, even if it is a necessary and indispensable condition to end the dictatorship, entails the loss of the heroic, symbolic, and autonomous dimensions of collective action. Here we find one source for the disenchantment all new democracies have to face sooner or later.

Notes

1. On the Frei period, see Moulián (1983).
2. On the Allende period and the military coup, see A. Valenzuela (1978); and Garretón and Moulián (1994).
3. For general accounts of the Chilean military regime until the mid-1980s, see Garretón (1989*a*); and A. Valenzuela and S. Valenzuela (1985).
4. For an analysis of the Chilean military regime through the 1980s, see Drake and Jáksic (1991); and Garretón (1987).
5. For an analysis of the Chilean transition to democracy and the first period of democratization, see Concertación (1989); CIEPLAN (1990); Drake and Jáksic (1991); and Garretón (1993, 1995).
6. For general descriptions and analyses of the protest cycles, see de la Maza and Garcés (1985); Martínez (1992); Agurto, Canales, and de la Maza (1985); and Proposiciones (1987).
7. For an analysis of the plebiscite and its political consequences, see CIS (1989); Drake and Valenzuela (1989); and Garretón (1991, 1993, 1995).

References

Agurto, Irene, Canales, Manuel, and de la Maza, Gonzalo, (1985) (eds.), *Juventud chilena: Razones y subversiones* (Santiago: ECO (Educación y Comunicación)–FOLICO (Formación de Líderes Cristianos Obreros)–SEPADE (Servicio Evangélico para el Desarrollo)).

Campero, Guillermo (1984), *Los gremios empresariales: Comportamiento socio-político y orientaciones ideológicas* (Santiago: ILET (Instituto Latinoamericano de Estudios Transnacionales)).

—— (1987), *Entre la sobrevivencia y la acción política* (Santiago: ILET).

—— and Valenzuela, José Antonio (1984), El movimiento sindical en el régimen militar chileno (Santiago: ILET).

CIEPLAN (1990), *Transición a la Democracia: Marco político y Económico* (Santiago: CIEPLAN).

CIS (1989), *La campaña del NO vista por sus creadores* (Santiago: Editorial Melquíades).

CLACSO-ILET (1986), *Los movimientos sociales en Chile y la lucha democrática* (Santiago: ILET–CLACSO (Consejo Latinoamericano de Ciencias Sociales)).

Collier, David (1979) (ed.), *The New Authoritarianism in Latin America* (Princeton, Princeton University Press).

Concertación de Partidos por la Democracia (1989), *Programa de Gobierno* (Santiago: Documentos Diario La Epoca).

De la Maza, Gonzalo, and Garcés, Mario (1985), *La explosión de las mayorías: Protesta nacional, 1983–1984* (Santiago: ECO).

Drake, Paul W., and Jáksic, Iván (1991) (eds.), *The Struggle for Democracy in Chile 1982–1990* (Lincoln and London: University of Nebraska Press).

—— and Valenzuela, Arturo (1989), *The Chilean Plebiscite: A First Step toward Democratization*, Lasa Forum 19 (Winter).

Escobar, Arturo, and Alvarez, Sonia (1992) (eds.), *The Making of Social Movements in Latin America: Identity, Strategy and Democracy* (Boulder, Colo.: Westview Press).

Garretón, M., Manuel Antonio (1987), *Reconstruir la Política: Transición y consolidación democrática en Chile* (Santiago: Editorial Andante).

—— (1989*a*), *The Chilean Political Process* (Boston: Unwin and Hyman).

—— (1989*b*) (ed.), *Propuestas políticas y demandas sociales*, 3 vols. (Santiago: FLACSO).

—— (1991), 'The Political Opposition and the Party System under the Military Regime', in Paul W. Drake and Iván Jaksic (eds.), *The Struggle for Democracy in Chile 1982–1990* (Lincoln and London: University of Nebraska Press), 211–50.

—— (1993), 'La redemocratización política en Chile: Transición, inauguración y evolución', *Estudios interdisciplinarios de América Latina y el Caribe*, (University of Tel-Aviv), 4 (1): 5–25.

—— (1995), *Hacia una nueva era política: Estudio sobre las democratizaciones* (Santiago: Fondo de Cultura Económica).

—— (1996*a*), 'Social Movements and the Process of Democratization: A General Framework', *International Review of Sociology*, 6 (1): 41–9.

—— (1996*b*) (ed.), Monographic section, Social Movements in Latin America in the Context of Economic and Socio-Political Transformation', *International Review of Sociology*, 6 (1): 39–156.

Antonio, Manuel, and Moulián, Tomas (1994), *La unidad popular y el conflicto político en Chile* (Santiago: Ediciones CESOC).

Huneeus, Carlos (1987), *Cambios en la opinión pública: Una aproximación al estudio de la cultura política en Chile* (Santiago: CERC (Centro de Estudios de la Realidad Contemporánea), Academia de Humanismo Cristiano).

Martínez, Javier (1992), 'Fear of the State, Fear of Society: On the Opposition Protests in Chile', in Juan E. Corradi, Patricia Weiss, and Manuel

Antonio Garretón (eds.), *Fear at the Edge: State Terror and Resistance in Latin America* (Berkeley: University of California Press), 142–60.

Martínez, Javier and Tironi, Eugenio (1985), *Las clases sociales en Chile: Cambio y estratificación, 1970–1980* (Santiago: SUR).

Meza, M. A. (1986) (ed.), *La otra mitad de Chile* (Santiago: Ediciones CESOC, Instituto para el Nuevo Chile).

Moulián, Tomás (1983), *Democracia y Socialismo en Chile* (Santiago: FLACSO).

O'Donnell, Guillermo, and Schmitter, Philippe (1986), *Transitions from Authoritarian Rule: Tentative Conclusions about Uncertain Democracies* (Baltimore: Johns Hopkins University Press).

Oxhorn, Philip (1995), *Organizing Civil Society. The Popular Sectors and the Struggle for Democracy in Chile* (University Park, Penn.: Pennsylvania State University Press).

Proposiciones (1987), *Marginalidad, movimientos sociales y democracia*, No. 14 (Santiago: Ediciones SUR).

Slater, David (1985) (ed.), New Social Movements and the State in Latin America (Amsterdam, Paris).

Touraine, Alain (1978), *La Voix et le regard* (Paris: Seuil).

——(1989), *América Latina: Política y Sociedad* (Madrid: Espasa Calpe).

Valenzuela, Arturo (1978), *The Breakdown of Democratic Regimes: Chile* (Baltimore: Johns Hopkins University Press).

—— and Valenzuela, Samuel (1985) (eds.), *Military Rule in Chile: Dictatorship and Opposition* (Baltimore: Johns Hopkins University Press).

Valenzuela, Eduardo (1984), *La rebelión de los jóvenes* (Santiago: Ediciones SUR).

Vergara, Pilar (1984), *Auge y caída del neo-liberalismo en Chile* (Santiago: FLACSO).